2nd Baltic Conference for Students and Young Researchers (BalCon 2018)

MATEC Web of Conferences Volume 219 (2018)

Gdynia, Poland
20 - 23 April 2018

Editor:

Elzbieta Urbanska-Galewska

ISBN: 978-1-5108-7405-3

Printed from e-media with permission by:

Curran Associates, Inc.
57 Morehouse Lane
Red Hook, NY 12571

Some format issues inherent in the e-media version may also appear in this print version.

Printed by Curran Associates, Inc. (2018)

For additional information, please contact EDP Sciences – Web of Conferences
at the address below.

EDP Sciences – Web of Conferences
17, Avenue du Hoggar
Parc d'Activité de Courtabœuf
BP 112
F-91944 Les Ulis Cedex A
France

Phone: +33 (0) 1 69 18 75 75
Fax: +33 (0) 1 69 28 84 91

contact-edps@webofconferences.org

Additional copies of this publication are available from:

Curran Associates, Inc.
57 Morehouse Lane
Red Hook, NY 12571 USA
Phone: 845-758-0400
Fax: 845-758-2633
Email: curran@proceedings.com
Web: www.proceedings.com

TABLE OF CONTENTS

edp sciences

Web of Conferences

Statement of Peer review

In submitting conference proceedings to *Web of Conferences*, the editors of the proceedings certify to the Publisher that

1. They adhere to its **Policy on Publishing Integrity** in order to safeguard good scientific practice in publishing.
2. All articles have been subjected to peer review administered by the proceedings editors.
3. Reviews have been conducted by expert referees, who have been requested to provide unbiased and constructive comments aimed, whenever possible, at improving the work.
4. Proceedings editors have taken all reasonable steps to ensure the quality of the materials they publish and their decision to accept or reject a paper for publication has been based only on the merits of the work and the relevance to the journal.

Title, date and place of the conference 2ND BALTIC CONFERENCE FOR STUDENTS AND YOUNG RESEARCHERS BALCON2018, 20-23.04.2018, STENA LINE FERRY TERMINAL, GDYNIA, POLAND

Proceedings editor(s): ELŻBIETA URBAŃSKA-GALEWSKA

Date and editor's signature 27.09.2018

17, avenue du Hoggar - PA de Courtabœuf – BP 112 - 91944 Les Ulis Cedex A (France)
Tél. : 33 (0)1 69 18 75 75 - Fax : 33(0)1 69 07 45 17 – www.edpsciences.org

Timber frame houses resistant to dynamic loads - seismic analysis

Marcin Szczepański[1,*], *Wojciech* Migda[2] ,

[1] Gdansk University Of Technology, Narutowicza 11/12, 80-233 Gdańsk Department of Civil and Environmental Engineering; marszcze@pg.gda.pl

[2] Gdansk University Of Technology, Narutowicza 11/12, 80-233 Gdańsk Department of Civil and Environmental Engineering; wmigda@pg.gda.pl

Abstract. The aim of the article is to present results of seismic analysis results of two real-sized timber frame buildings subjected to seismic excitations. The first model was insulated with mineral wool, the second one with polyurethane foam. Technology and specifications involved in both models construction is based on the previously conducted experimental research on timber frame houses, including wall panels tests, wall numerical models and study on material properties and precisely reflect results of the those research. During the seismic analysis reference node located in buildings were selected. In selected node displacement values were measured and compared between two analyzed models. The results of the numerical analysis presented in the article indicate that the application of polyurethane foam for a skeleton filling of the timber-frame building leads to the increase in stiffness as well as damping of the whole structure, which results in a considerable increase in the seismic resistance of the structure.

1 Introduction

While analysing the technology of single-family houses resistant to dynamic loads, including seismic excitations and traffic impact[1-4] it becomes evident that timber is one of the best materials for that kind structures. Development of different solutions is further escalated with the spread of ecology and energy-efficient related topics, as well as functionality of such structures as building passive houses. Timber frame buildings constructed with accordance to the technological viewpoint are usually quite resistant to dynamic effects, including earthquakes [5]. Timber-based boards, like OSB (multifunctional or OSB/3 oriented strand board) are the most frequently elements for covering of ceilings, roofs as well as most of walls of the timber frame houses. These boards have very good durability properties, which allow us to stiffen the whole structure and reduce the forces occurring during dynamic excitations [6 - 9]. Data gathered in Japan and South America shows that timber frame houses can often withstand mining shakes and even catastrophic earthquakes with just minor damage [5].

* Corresponding author: marszcze@pg.edu.pl

Thermal insulation is extremely important in relation to the timber frame house building technology [10, 11]. Materials used for thermal insulation of such structures may be incompatible with the timber frame or may significantly stiffen it due to their parameters [12]. Testing mechanical parameters of different insulation materials proves their usability for insulation and stiffening purposes. Mineral wool is the most popular insulation material, not co-operating mechanically with the timber frame. Poliurethan foam is another, different kind of insulation material. Due to the closed cell structure and mechanical properties polyurethane foam increases the rigidity of the timber frame elements [12].
The aim of the article is to present the results of dynamic numerical analysis focused on earthquake-induced response of two timber frame buildings insulated with mineral wool and with polyurethane foam. Both numerical models show identical geometrical properties and differ only in the used thermal insulation material.

2 Experimental and numerical studies of exterior walls models

Two models of exterior walls were constructed using two different insulation materials to verify the behaviour through the experimental study (see [12] for details) Both experimental models were made in the traditional technology of timber houses. One of them was filled with mineral wool and the other with polyurethan foam.

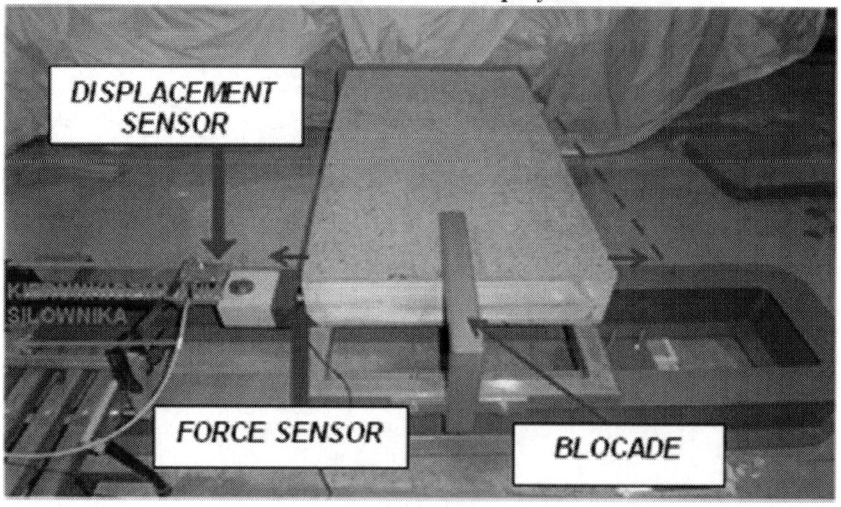

Fig. 1. View of the experiment setup with mounted specimen [13]

A specially designed test setup was used in the experimental study (Fig. 1). The tests were performed for the following excitation frequencies: 0.5 Hz, 1.0 Hz, 2.0 Hz and 5.0 Hz, for various values of displacement. The hysteresis loops were obtained for different displacements for a given frequency excitation. On the basis of the experimental results, numerical models of two wall elements were prepared. The results of numerical analysis were consistent with the results from the experimental study confirming the accuracy of the models created [12].

3 Numercial models of timber frame houses - dynamic analysis

A single-storey timber frame building of the following parameters:

- storey height: 2.80 m
- length of the building: 12.0 m
- width of the building: 12.0 m

was taken into consideration in the numerical analysis (see numerical models at Fig. 2 and Fig. 3). A building timber frame was made of a conifer timber of C24 class surrounded with OSB3 boards on both sides. Numerical model 1 was created based on the assumption that the structure is insulated with mineral wool, while numerical model 2 was dedicated to building insulated with polyurethan foam. The model of the real structural model was created based on numerical models of exterior walls which, in turn, were verified based on experimental study [12].

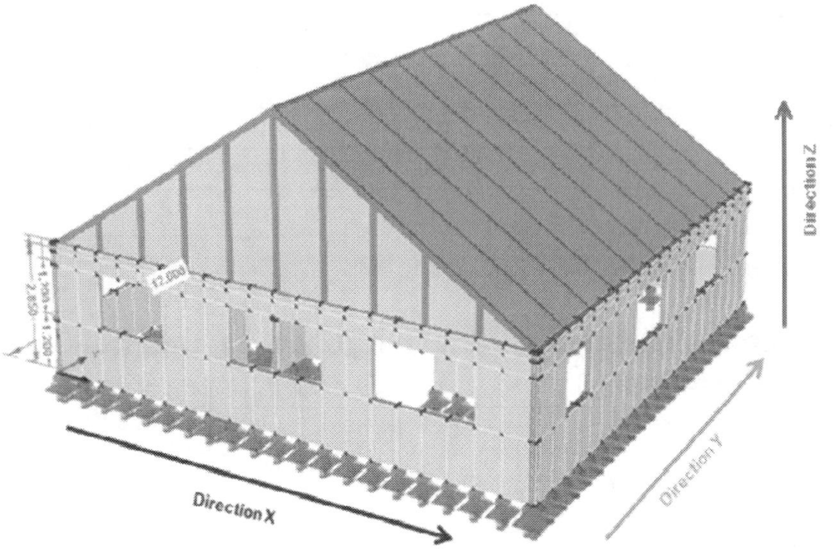

Fig. 2. Numerical model of the building showing a typical wall element

The natural vibrations modes, together with the corresponding frequencies, were obtained as the results of the modal analysis of the timber frame buildings filled with mineral wool and polyurethane foam. The examples of the results in the form of the second natural vibration modes are shown in Fig. 3 and Fig. 4. The values of the first 10 natural frequencies are also summarized in Table 1 (see [13] for simmilar procedure).

Fig. 3. Second natural vibration mode for both models (mineral wool first from the left and polyurethane foam second), second natural frequency:
f_1= 1.975 Hz, f_2 =3,358Hz

Table 1. Natural Frequencies for first 3 Natural Vibration Modes

Model filled with mineral wool (Hz)	Model filled with polyurethane foam (Hz)
1,89	3,136
1,975	3,358
2,543	4,337

The results of modal analysis of both models show that the timber frame model insulated with polyurethane foam displays larger stiffness.

4 Numercial analysis of the response of the building under seismic excitation

The next stage of the study was devoted to the numerical analysis focused on structural behaviour under seismic excitation. During the analysis, both models were subjected to the Loma Prieta earthquake of October 17th 1989 (see Table 2.) acting in all three directions simultaneously. Visualisation of reference node no. 2177 selected to analysis shown in Fig. 5.

The displacement time histories in all three directions for selected node no. 2177 of model insulated with mineral wool are shown in Fig. 6. The corresponding results for the model with polyurethane foam are presented in Fig. 7. The detailed comparison between the peak displacements at node no. 2177 for both models in all three directions is shown in Tables 3 - 5.

Tabel 2. Earthquake parameters

Earthquake	Date	Magnitude	Measuring station	Component	Exstreme acceleration value [m/s^2]
Loma Prieta	17.10.1989	6,9	Capitola – Fire Station	NS	3,907
				EW	4,629
				UD	5,000

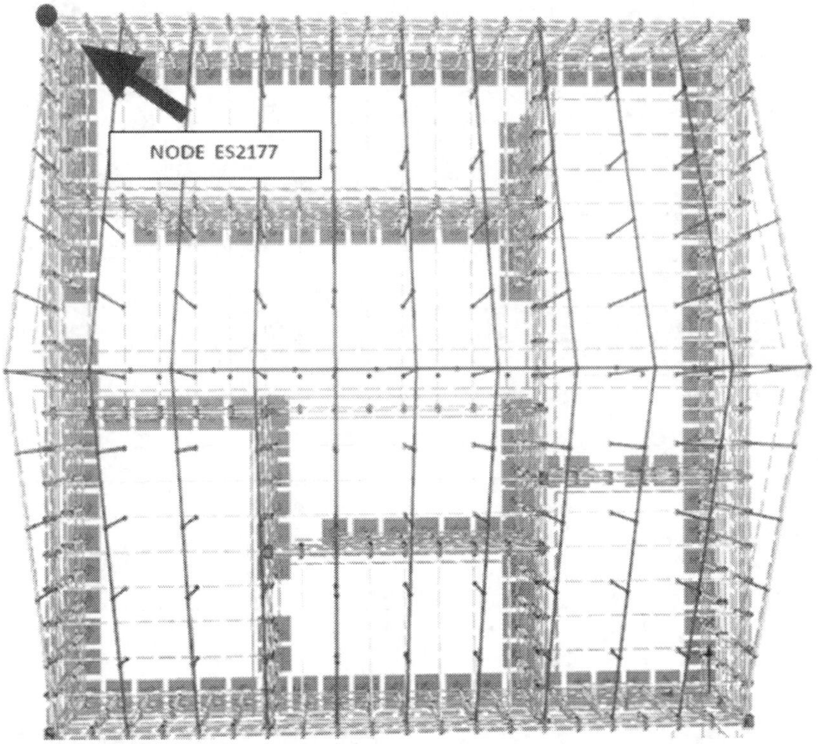

Fig. 4. Visualisation of refence node selected to analysis

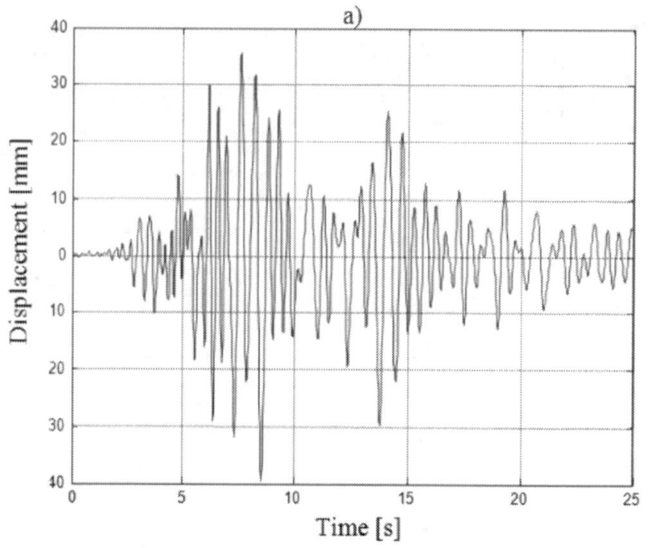

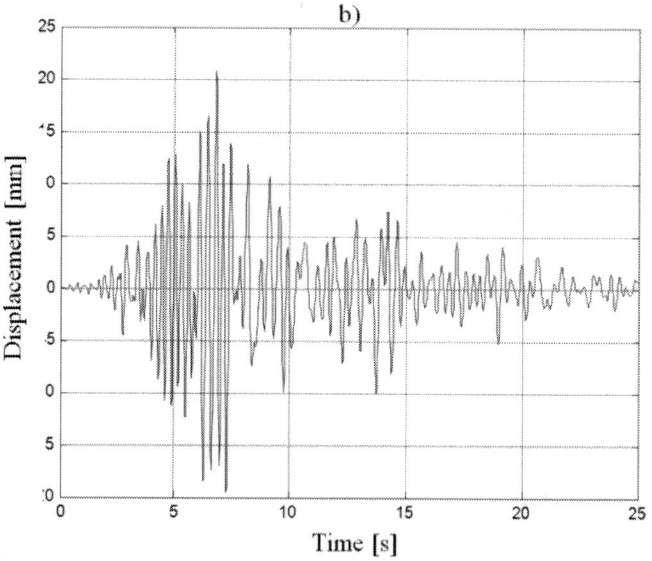

Fig. 5. Displacement time histories for both models - node 2177 direction: x (a- mineral wool; b- polyurethane foam)

Tabela 3 Peak displacements at node no. 1 - x direction

Model with mineral wool [mm]	Model with polyurethane foam [mm]	Reduction (%)
40,10 mm	20,50 mm	48,8

Tabela 4 Peak displacements at node no. 1 - y direction

Model with mineral wool [mm]	Model with polyurethane foam [mm]	Reduction (%)
34,50 mm	12,50 mm	63,7

Tabela 5 Peak displacements at node no. 1 - z direction

Model with mineral wool [mm]	Model with polyurethane foam [mm]	Reduction (%)
2,80 mm	1,45 mm	48,2

5 Conclusions

The dynamic numerical analysis focused on the response of timber frame house under seismic excitation has been presented in this paper. The results of the study confirm that using a polyurethan foam leads to much lower displacement values comparing to the case when the mineral wood is applied as an insulation material. The reductions of peak values for node no.2177 is as large as 48,8%, 63,7 % and 48,2 % for x, y and z direction, respectively. It is especially important considering the fact that the reduction in displacements is directly related to the reduction in damage of structures exposed to dynamic loads (see also [13]). Therefore, based on the results, a positive outcome of using polyurethane foam insulation instead of mineral wool in timber frame structures exposed to seismic excitations is well observed.

References

[1] T. Toratti, Seismic Design of Timber Structures, FEMA, 1994.

[2] A.K. Chopra, Dynamic of Structures: Theory of Applications to Earthquake Engineering, Prentice-Hall, USA, 1995.

[3] ECS, Eurocode 8: Design Provisions for Earthquake Resistance of Structures. Brussels, European Committee for Standardization, Belgium, 1998.

[4] A. Jakubczyk–Gałczyńska, A. Kristowski, R. Jankowski, Comparing the effectiveness of ANNs and SVMs in forecasting the impact of traffic–induced vibrations on building. Geodetic Congress (BGC Geomatics), Gdańsk, Poland, 121–125 (2017)

[5] J-M. Seo, I-K. Choi, and J-R. Lee, "Experimental study on the aseismic capacity of a wooden house using shaking table," Earthquake Engineering and Structural Dynamics, vol. 28, pp. 1143-1162, 1999.

[6] J. Kiyono and A. Furukawa, "Casualty occurrence mechanism in the collapse of timber-frame house during an earthquake," Earthquake Engineering and Structural Dynamics, vol. 33, pp. 1233-1248, 2004.

[7] J. Vessby, Analysis of shear walls for multi-storey timber buildings, Linnaeus University Dissertations, No 45/2011, 2011.

[8] S. Pei and J.W Van de Lindt, "Coupled shear-bending formulation for seismic analysis of stacked wood shear wall systems," Earthquake Engineering and Structural Dynamics, vol. 38, pp. 1631-1647, 2009.

[9] R. Zarnic and B. Dujic, "Study of rateral resistance of massive X-lam wooden wall system subjected to horizontal loads," Earthquake Engineering on Timber Structures, Portugal 2006.

[10] W. Nitka, Mój dom z drewna, Centrum Informacyjne Lasów Państwowych, Warszawa, 2010.

[11] Radziszewska-Zielina E.: Analiza porównawcza parametrów materiałów termoizolacyjnych mających zastosowanie jako izolacja ścian zewnętrznych. Konferencja Naukowo-Techniczna Inżynieria Przedsięwzięć Budowlanych, nr 4, s. 32-37, Kraków 2014.

[12] M. Szczepański, W. Migda, and R. Jankowski, "Construction technology of timber-frame houses resistant to dynamic loads - study on models of exterior walls", Advances in Science and Technology - Research Journal, vol. 9, no. 28, pp. 75-80, 2015.

[13] M. Szczepanski, W. Migda, R. Jankowski, Modal analysis of real timber frame houses with different insulation materials. *Advances in Science and Technology Research Journal*, vol. **10**, 215-221 (2016)

System-reliability analysis of steel truss towers

Wojciech Mochocki [1,*], *Paulina* Obara [1] , and *Urszula* Radoń [1]

[1] Kielce University of Technology, Faculty of Civil Engineering and Architecture, al. Tysiąclecia PP 7, 25-314 Kielce, Poland

Abstract. Steel truss towers are important structures in civil engineering, used as, for example, antenna towers, architectural towers, transmission towers, chimney towers, light towers, viewing towers, or water towers. In the paper, the latter structural type is analysed. Three types of truss towers with different topology are discussed. The towers are designed according to the standards. Next, reliability analysis is performed and the influence of type of truss topology on reliability is analysed. Reliability analysis concerns formulation of the limit state, calculation of failure probability and determination of reliability index. In the paper, the system approach is proposed, definitely more complete than that obtained using approximation-simulation methods. In the case of structures analysed in the paper, the types of reliability models are determined. Identification of reliability models relies on studies on the transformation of the safe structural system into the geometrically variable system (mechanism). The investigations made it possible to determine the kinematically admissible failure mechanisms which contain minimal critical sets of elements. In order to identify the mechanism, spectral analysis of the linear stiffness matrix is used. In the case of the towers of concern, the formulas specifying the number of mechanisms for any number of repeating sections are determined.

1 Introduction

Towers are stand-alone structures rigidly fixed to the base. A number of tower types can be distinguished due to their use [1]. Antenna towers are tall structures designed to support antennas for telecommunications and broadcasting, architectural towers commemorate historic events, transmission towers (electricity pylons) support power lines, chimney towers provide support for flue systems, light towers offer the possibility of installing powerful lighting, viewing towers have observation decks providing extensive views, and finally water towers are used to pressurize a water supply system for water distribution.

In many cases, steel truss tower is considered the most effective structural option. Such a solution makes it possible to build to greater heights than the case of reinforced concrete or masonry structures. Truss towers use lower amounts of material than solid structures, and additionally, they are easy to construct and assemble. However, as in the case of pin-jointed multi-element steel structures, reliability assessment is an issue of key

* Corresponding author: wojtek1391@gmail.com

importance [2]. The structure is reliable if it performs functions specified in the design over the entire service life. A usual measure of reliability is the reliability index β [3].

The methods commonly used for the structural reliability assessment can be classified as follows [4]: Level I, Level II, Level III methods and the system approach. Level I, i.e. semi-probabilistic methods involve the verification of structural reliability, but they do not determine the exact failure probability. Those methods use partial safety factors. Level II methods include approximation (FORM, SORM) and simulation (Monte Carlo, Importance Sampling) methods. In Level II methods, boundary probability distributions of individual random variables are used [3, 4]. Level III methods are fully probabilistic and they provide an extension of Level II methods. In this case, the cumulative probability distribution of random variables is known. Another group contains system approach (serial, parallel and mixed systems). In this case structural characteristics of the building must be taken into account, i.e. interaction of the structural elements in the load-carrying system [3, 5-8]. In the paper, the system approach is used to assess the reliability.

2 Truss tower design

In the paper, reliability analysis is conducted for towers exhibiting different truss types. Water tower with a water tank located at the top of the structure is designed in accordance with current standards [2, 9-11]. The structural model is shown in Fig.1, the truss types can be seen in Fig. 2.

According to the design, the tower is situated in the wind speed zone I, in the roughness class 2. The structure foundation is at A=170 m a.s.l. The wind load is modelled in accordance with the loading scheme appropriate for this structure [11, 12]. The main permanent load is a cylindrical steel sheet water tank that is filled with water. Additionally, the dead load, the weight of the 3x3 m steel deck, on which the tank is installed, and the deck service load of 2 kN/m^2 are taken into account.

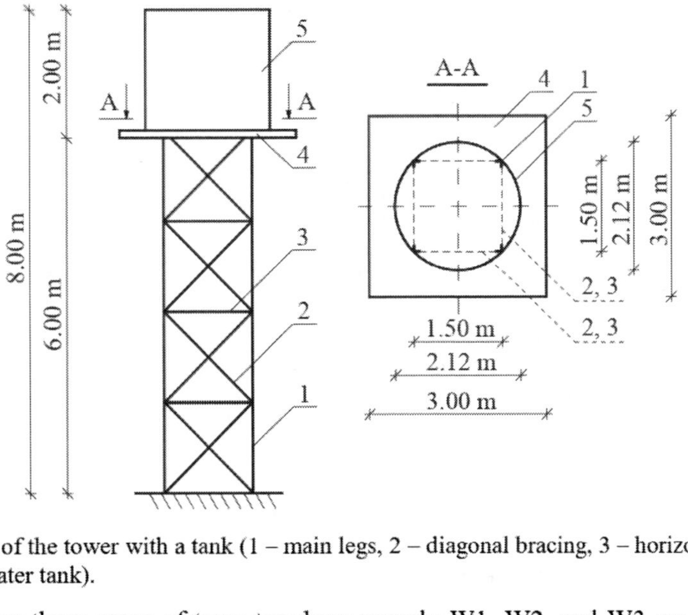

Fig. 1. Scheme of the tower with a tank (1 – main legs, 2 – diagonal bracing, 3 – horizontal bracing, 4 – deck, 5 – water tank).

In the paper three cases of truss topology namely W1, W2, and W3 are discussed. In Fig. 2 truss topologies for n=4 repeatable sections are presented. Table 1 shows steel sections, numbers of the most stressed elements and their stress intensity level depending on the assumed truss type. The towers are made of S235 steel of the yield strength f_y=235 MPa.

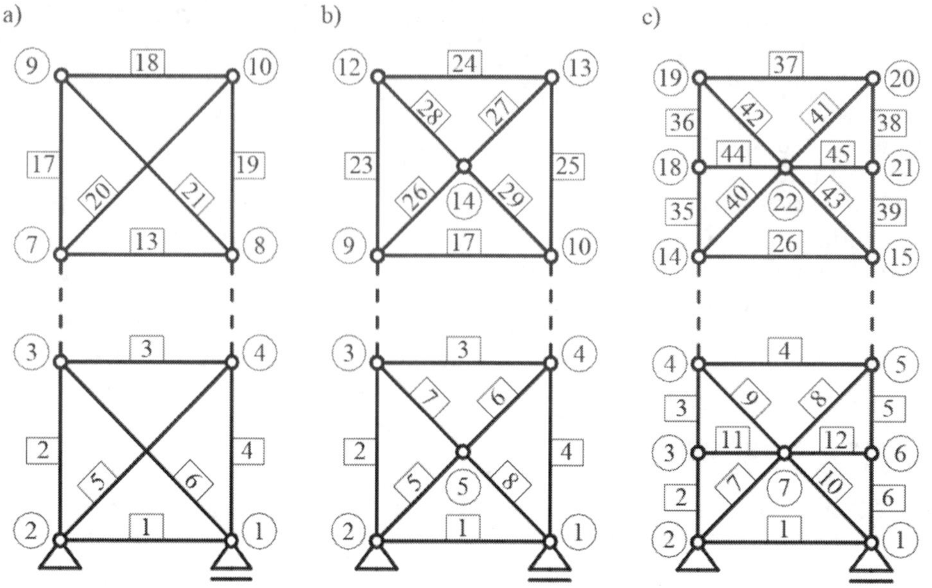

Fig. 2. Truss topology: a) W1, b) W2, c) W3.

Table 1. Structural elements of truss towers.

Truss topology	Element type	Element profile	Most stressed element	
			Element number	Stress intensity level [%]
W1	main leg	L50x6	2	78
	diagonal bracings	L40x5	5	73
	horizontal bracings	L40x4	3	11
W2	main leg	L50x6	2	78
	diagonal bracings	L40x5	5	73
	horizontal bracings	L40x4	3	11
W3	main leg	L40x4	2	65
	diagonal bracings	L45x6	7	81
	horizontal bracings	L40x4	4	20

3 Reliability analysis of the truss

3.1 System-reliability

When the system approach is applied to assess the reliability of a rod structure, it is necessary to consider its structural arrangement and the interaction of structural elements. This way, the reliability model, which is the structure eigen value, can be identified. In order to define the reliability model, the kinematically admissible failure mechanisms (KAFMs) should be specified. They contain the minimal critical sets of elements (MCSE), which have a specified number of causative elements (l). In reliability theory, three system types are distinguished, serial, parallel and mixed (parallel-serial and serial-parallel) systems (Fig. 3).

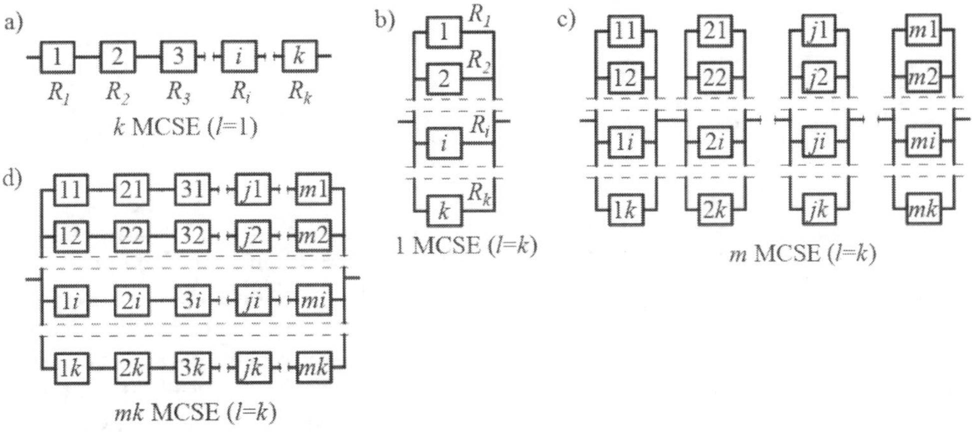

Fig. 3. Types of reliability models: a) serial system, b) parallel system, c) parallel-serial system, d) serial-parallel system.

The serial system (Fig. 3a) is the one which undergoes failure if even only one element fails (becomes inoperable), making the system becomes geometrically variable. The serial system corresponds to statically determinate structures, in which the capacity of the weakest element is decisive for the capacity of the entire structure. The characteristic feature of the serial system is that with the growth of the number of its elements, the capacity and reliability decrease. Reliability of the system R is calculated from the formula:

$$R = \prod_{i=1}^{k} R_i \tag{1}$$

where:
R_i reliability of a single element, k – number of the system elements.

The serial system consists of k minimal critical sets of elements (k MCSE), each of which has one causative element (l=1).

The parallel system (Fig. 3b) is operable if at least one element is operable. The parallel system corresponds to certain statically indeterminate structures. The capacity and reliability of this system increase with the number of the elements. In this case, one minimal critical set of elements (1 MCSE) is found, which consists of k causative elements (l=k), and the reliability of the system is calculated as follows:

$$R = 1 - \prod_{i=1}^{k} \left(1 - R_i\right) \tag{2}$$

In the case of mixed (parallel-serial and serial-parallel) systems, in order to identify a proper reliability format of the structure it is necessary to analyse how the elements are coupled, i.e. what relations hold between them. The parallel-serial system (Fig. 3c) consists of m MCSE, each of which includes k causative elements (l=k). The reliability of such a system is computed applying a formula:

$$R = 1 - \prod_{j=1}^{m} \left[1 - \prod_{i=1}^{k} \left(1 - R_{ji}\right) \right] \tag{3}$$

In contrast, in the case of the serial-parallel system (Fig. 3d), the number of MCSE is $m \cdot k$, and each MCSE consists of k causative elements ($l=k$). The reliability of such a system is computed acc. formula:

$$R = 1 - \prod_{i=1}^{k} \left[1 - \prod_{j=1}^{m} R_{ji} \right] \tag{4}$$

Complex structures may contain systems with different types of connections presented above. That also refers to truss towers, as shown in the paper.

In order to determine reliability of the system, it is also necessary to compute the reliability of a single element. The procedure that leads to the determination of the reliability of element R_i was presented, among others, in the paper [7]. In order to determine the element reliability, it is necessary to determine random variables of the effect of action E_i, and of capacity N_i. Random variables are described by means of normal distribution with the parameters μ (expected value) and σ (standard deviation). With respect to standard deviation for effect of actions σ_{Ei}, variation coefficient is assumed at the level of 6%, whereas due to standard deviation for capacity σ_{Ni}, its level is 10%. A subsequent step is determining the safety margin Z_i, which is a difference between the random variable of the effect of action E_i and of capacity N_i. As a result, the safety margin Z_i is also a random variable described by means of normal distribution. The parameters of this system are the expected value of the safety margin $\mu_{Zi} = \mu_{Ni} - \mu_{Ei}$ and standard deviation of the safety margin $\sigma_{Zi} = \sqrt{\sigma_{Ei}^2 + \sigma_{Ni}^2}$. Next, reliability index for a single element $\beta_i = \mu_{Zi} / \sigma_{Zi}$ is computed. On its basis, probability of the element failure can be estimated $P_{fi} = \Phi(-t_i)$, where Φ is the Laplace function. The final step involves the computation of the element reliability $R_i = 1 - P_{fi}$ and the reliability of the system R. In order to compute the reliability of the system, it is necessary to identify the reliability model.

3.2 Kinematically admissible failure mechanisms for truss towers

Defining kinematically admissible failure mechanisms (KAFMs) that contain minimal critical sets of elements (MCSE) consists in specifying the possible mechanisms. To that end, spectral analysis of the linear stiffness matrix is performed:

$$(\mathbf{K_L} - \lambda \mathbf{I})\mathbf{q} = \mathbf{0} \tag{5}$$

where:

$\mathbf{K_L}$ linear stiffness matrix, λ – eigen values, $\mathbf{q}$ – displacement vector.

Eigen values λ describe the system energy state, whereas the eigen vectors corresponding to eigen values describe the form of deformation. The system is not a mechanism if all eigen values λ have a positive value. The mechanism will be found if any of eigen values λ_i equals zero, and the mechanism form is defined by the eigenvector corresponding to the value λ_i [6]. In order to carry out spectral analysis of the linear stiffness matrix the program developed by the authors in the Mathematica environment is applied.

In the paper, truss towers with three truss topology are analysed (Fig. 2). On the basis of the analyses, formulas are derived for the number of mechanisms lm for each truss (W1, W2, and W3), depending on the number of elements le and the number of repeating sections n. With respect to W1 and W2 topologies, the number of mechanisms is given by the formula:

$$lm = \frac{(le-(n-1))!}{2!((le-(n-1))-2)!} \tag{6}$$

whereas for W3 topology:

$$lm = \frac{(le-(n-1)-2n)!}{2!((le-(n-1)-2n)-2)!} + 2n \tag{7}$$

For instance, Table 2 shows the number of mechanisms lm depending on truss topology, for the number of repeating sections from 1 to 4.

Table 2. Number of mechanisms.

Truss topology	Formula for the number of tower elements le	Number of repeating sections n			
		1	2	3	4
		Number of mechanisms lm			
W1	$6+5(n-1)$	15	45	91	153
W2	$8+7(n-1)$	28	91	190	325
W3	$12+11(n-1)$	47	157	331	569

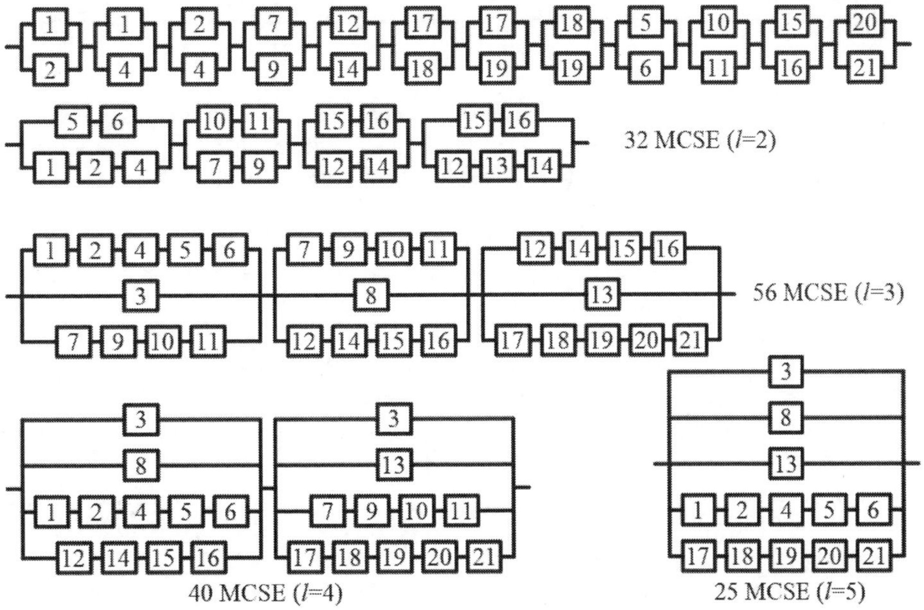

Fig. 4. Kinematically admissible failure mechanisms for the truss W1 with n=4 repeating sections.

When the analysis is reduced to simple formulas, it is possible to quickly evaluate the number of mechanisms in the truss structure. That is important because machine computations for the number of sections greater than four (n>4) are highly time-consuming. That results from a quick increment in the number of mechanisms with an increasing number of rods, which is decisive for the duration of computations. However, formulas (6) and (7) do not reveal the elements contained in a specific mechanism. Additionally, they do not make it possible to establish the type of relation between elements, i.e. whether they are serial, parallel or mixed systems. Thus it is necessary to perform a full analysis. Such an analysis for the tower W1 with 4 repeating sections is presented graphically in Fig. 4.

3.3 Reliability analysis of the truss

In order to compare the truss towers with respect to reliability, the reliability index β is computed for each tower of concern taking into account the most disadvantageous arrangement of loads. Reliability indexes are determined both for the assumption that all elements of the tower are connected serially, β_1, and considering actual connections and interactions between the rods (mixed system) taken into account. The reliability analysis considering mixed connections is performed by means of the program developed by the authors in the Mathematica environment. The reliability index β_2 is computed for the assumption that only parallel-serial systems are found. It is a certain simplification, allowing for a full automation of the computational process. To account for serial-parallel systems that occur in the structures, equations for individual KAFMs are worked out, and the reliability index β_3 is computed. It should be noted that it is a highly time-consuming process, to possibly create a serious difficulty for a larger number of elements. The results of reliability indexes for computational approaches quoted above are shown in Table 3.

Table 3. Reliability indexes of the structure.

Truss topology	*le*	*lm*	β_1	β_2	β_3
W1	21	153	1.500	3.454	3.454
W2	29	325	1.446	3.293	3.294
W3	45	569	1.320	2.938	2.938

4 Conclusions

Truss towers, classified as multi-element pin-jointed structures, require special attention with respect to the assessment of their reliability. The analyses carried out for the study lead to some valid conclusions.

The spectral analysis of the linear stiffness matrix of truss towers makes it possible to specify the number of mechanisms and to identify parallel-serial mechanisms. On this basis, formulas for computing mechanisms for a greater number of repeating sections can be derived.

In the towers of concern for each topology, namely W1, W2, and W3, an increase, by over 100%, in the reliability index is observed when the mixed system is accounted for (β_2 and β_3), compared with the situation when the reliability index is computed like for the serial system (β_1).

In each case the reliability index is computed, tower W1 showed the highest, whereas - tower W3 – the lowest reliability value. That is related to the fact that with an increase in the number of members, the reliability of the structure decreases.

The simplification of computations proposed in the study, which consists in taking into account only the parallel-serial mechanisms in the mixed system, does not noticeably affect the reliability index of well-designed truss towers. In the case of W2 tower, the difference between β_2 and β_3 is 0.001. Instead, this approach allows full automation of the computational process. Also, it is possible to avoid the necessity of working out all the equations for individual KAFMs found in the system.

References

1. K. Rykaluk, *Steel structures. Chimneys, towers, masts* (in Polish), Oficyna Wydawnicza Politechniki Wrocławskiej, (Wrocław 2005)
2. PN-EN 1990:2004 Eurocode: Basis of structural design.
3. S. Woliński, K. Wróbel, *Reliability of building structures* (in Polish), Oficyna Politechniki Rzeszowskiej, (Rzeszów 2001)
4. O. Ditlevsen, H. O. Madsen, *Structural Reliability Methods*, Department of Mechanical Engineering Technical University of Denmark, June-September (2007)
5. A. Biegus, *Probabilistic analysis of steel structures* (in Polish), Wydawnictwo Naukowe PWN, (Warszawa-Wrocław 1999)
6. J. Kłosowska, P. Obara, J. Turant, IOP Conference Series: Materials Science and Engineering, Kinematically Admissible Failure Mechanisms for Plane Trusses, **245** (2017), doi:10.1088/1757-899X/245/2/022022
7. K. Kubicka, U. Radoń, Archives of Civil Engineering, Proposal for the assessment of steel truss reliability under fire conditions, **61**, 4, 141-154 (2015)
8. P. Thoft-Christensen, M. Baker, *Structural reliability theory and its applications*, Springer-Verlag (1982)
9. PN-EN 1991-1-4 Eurocode:2008 1. Actions on structures. General actions. Wind actions.
10. PN-EN 1993-1-1 Eurocode:2006 3. Design of steel structures. Part 1-1: General rules and rules for buildings.
11. PN-EN 1993-3-1. Eurocode:2006 3. Design of steel structures. Part 3-1: Towers, masts and chimneys - Towers and masts.
12. K. Rykaluk, Inżynieria i Budownictwo, Wind load on steel towers according to Polish and European standards (in Polish), **65**, 7, 367-374 (2009)

Load carrying capacity of the eccentric joint in the truss made of open cross-sections

Małgorzata Gordziej-Zagórowska[1,*], and *Elżbieta* Urbańska-Galewska[1]

[1]Gdańsk University of Technology, Faculty of Civil and Environmental Engineering, Narutowicza 11/12, 80-233 Gdańsk, Poland

Abstract. The influence of eccentricity at intersections of truss members on the load carrying capacity of the truss joint is presented in the paper. The research truss elements were designed as cold-formed open cross section. Analytical calculations, numerical analysis and experimental research were conducted to reveal how the eccentricity affects the effort of material in the joint area. The results of analysis and investigations are compared and discussed. The main achievement of the tests carried out is statement that slender plane members of the compression chords are safe compared with the results of analytical calculations.

1 Introduction

Cold-formed steel members constitute a significant part of steel building structures. A widespread application of these elements in various types of steel constructions causes that new research problems are still defined. Both individual structural elements such as beams or columns [1-2] and parts of the structure, including lattice girders, are tested [3-6].

In practice, in truss girders joints eccentricities often occur for technological reasons, causing local bending and shear forces in the truss chords [7]. In the case of structures made of closed cross-sections, the influence of the eccentricities on the load capacity of truss joints has already been thoroughly tested [8], [9]. However, in the case of trusses made of open cross-sections, such knowledge has not been acquired yet. In order to check the load-carrying capacity of a truss chord cross-section of a bending truss joint, the designer formerly had to use simplified dimensioning methods [10-12]. This led to uneconomical solutions.

The lack of knowledge about the work of eccentric truss joints made of open cross-sections was a direct reason for the presented research. The purpose of these studies and analysis was to check the extent the brace members stiffening the walls of the hat-section chord, as well as how it affects the load capacity of the joint.

* Corresponding author: malgor@pg.edu.pl

2 Scope of the work

In order to study the above mentioned problem, analytical and numerical calculations as well as experimental research of the truss segment made of cold-formed open cross-section profiles were carried out. The chords were designed from the hat-section H 39/117/106/117/39x2, and the brace members of channel-section C 17/75/100/75/17x2 (Fig. 1b, c). The assumed sections were made of galvanized steel S350GD with zinc surface Z275MA. The geometry of the research model used for testing is shown in Figure 1. The joint with the eccentricity of e = 103.92 mm located in the compressed chord was taken for the analysis, because only the walls of the compression elements may lose local stability.

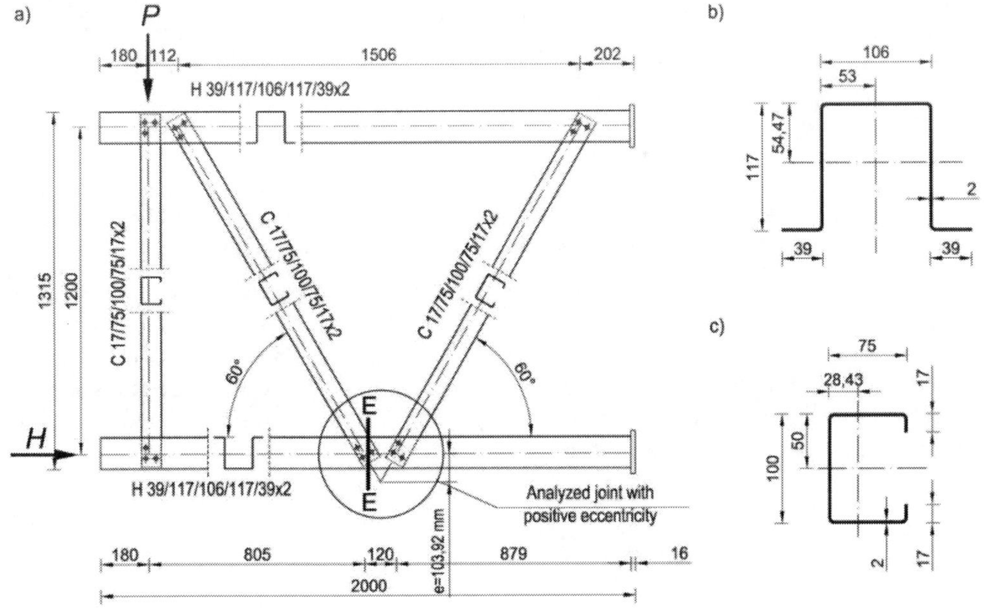

Fig. 1. a) research model; dimensions of profiles used in experimental research for: b) chords, c) brace members.

The research model was loaded with two concentrated forces *H* and *P*. The force *H* caused the compression of the bottom chord. The force *P* caused local bending of the bottom chord in the joint area through the natural forces distribution in the eccentric joint.

In addition, the scope of analytical and numerical calculations has been extended by wall thickness values equal to: 1 mm, 1.5 mm and 4 mm. A summary of performed types of analyses is presented in Table 1.

Due to the use of the same model geometry as well as the load history in analytical and numerical analyses, it was possible to compare the results of calculations.

Table 1. Types of carried out analyses for the eccentricity e=103.92 mm.

		analytical	experimental	numerical
Wall thickness (C-section and hat-section)	1.0 mm	x		x
	1.5 mm	x		x
	2.0 mm	x	x	x
	4.0 mm	x		x

3 Analytical calculation

In engineering practice, dimensioning of truss chords made of cold-formed open cross-sections profiles, with positive eccentricities at intersections is carried out using the formulas specified in the standards [10]–[13]. In the considered case, the truss chord is divided into several members connected at the intersection points with the brace members, as shown in Figure 2.

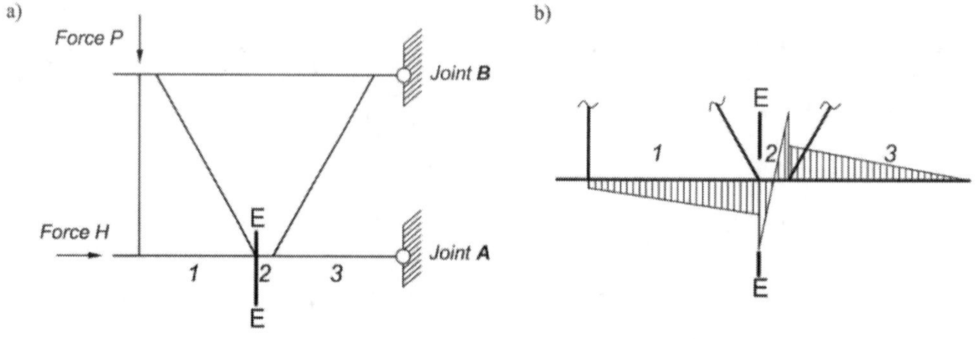

Fig. 2. a) Static scheme of the analysed truss along with the numbering of the analysed members of the compression chord, b) the graph of bending moments on the compression chord.

The values of internal forces can be determined in each chord member, for example by means of commercial computer programs. In the analysed case due to the existing positive eccentricity, apart from normal forces, shear forces and bending moments in the member No. 2 occur. Considering the forces occurring in all the members of the compression chord (No. 1÷3), the effort of material at the particular cross-sections was calculated based on the elastic verification in accordance with equation (1) [11].

$$\frac{N_{Ed}}{N_{c,Rd}} + \frac{M_{y,Ed}+\Delta M_{y,Ed}}{M_{cy,Rd,com}} \leq 1{,}0 \qquad (1)$$

where:

- N_{Ed} compression force,
- $M_{y,Ed}$ bending moment (resulting from the eccentricity at the joint),
- $\Delta M_{y,Ed}$ additional moment due to the movement of the centre of gravity of the section (for class 4),
- $N_{c,Rd}$ design resistance of the compressed cross-section,
- $M_{cy,Rd,com}$ effective design resistance for bending.

The effort of material at the cross-section of the bent and compressed member No. 2 (in the middle of the joint) is always greater than the material effort at members No. 1 and No. 3. This is related to the occurrence of shear forces and bending moments (consequence of the positive eccentricity). Therefore, it is logical that in the design situation exceeding of the joint resistance is a signal for the constructor to increase the cross-section of an entire chord or make strengthening covers in the joint area, as presented in [14].

Figure 3 shows a plot of the effort of material at the cross-section E-E (Fig. 1a) developed on the basis of calculation results (equation (1)) of joint with eccentricity e=103.92 mm for all analysed wall thicknesses: 1 mm, 1.5 mm, 2 mm and 4 mm. According to the classification included in the standard [8], sections with wall thickness of 1 mm, 1.5 mm and 2 mm are class 4 sections and a 4 mm - class 2 section. Figure 3 shows a significant decrease in the joint resistance of class 4 sections in terms of class 2 section.

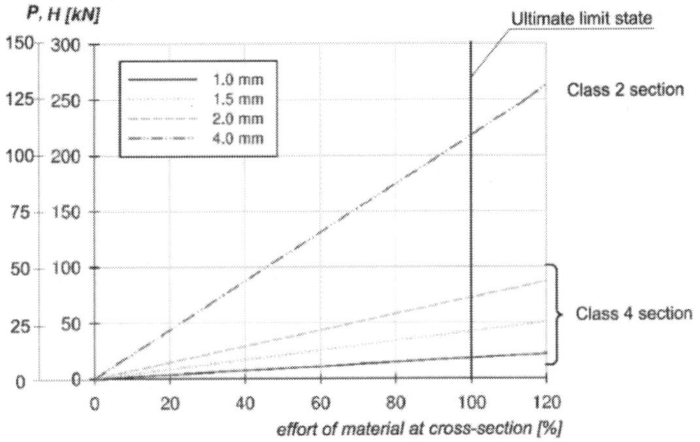

Fig. 3. Diagram of the relationship of the effort of material at cross-section E-E in dependence on the applied load, determined on the basis of analytical, elastic calculations.

4 Experimental research

Experimental research of the load capacity of a truss joint with the eccentricity e=103.92 mm was conducted on truss girder models in 1:1 scale (Fig. 1). The destructive tests of 5 similar truss models were performed.

A detailed description of the course of the experimental tests is presented in [7].

The purpose of the test was to determine the deformation forms and strains of the hat-section in the joint area, thus the investigations were carried out to register the strains in several measurement cross-sections, including the place of maximum bending moment (E-E section - Fig. 1a).

During the destructive tests local instability of hat-section in the E-E cross-section was observed in the all research models.

The origin of local instability of the walls of the hat-section in the E-E cross-section was defined as the critical state of the cross-section. When the critical load had been reached, the structure worked at its postcritical state, until it reached the limit state. In the analysed joint area, the onset of yield stress of the hat-section corners was identified as the ultimate limit state (see Figure 4).

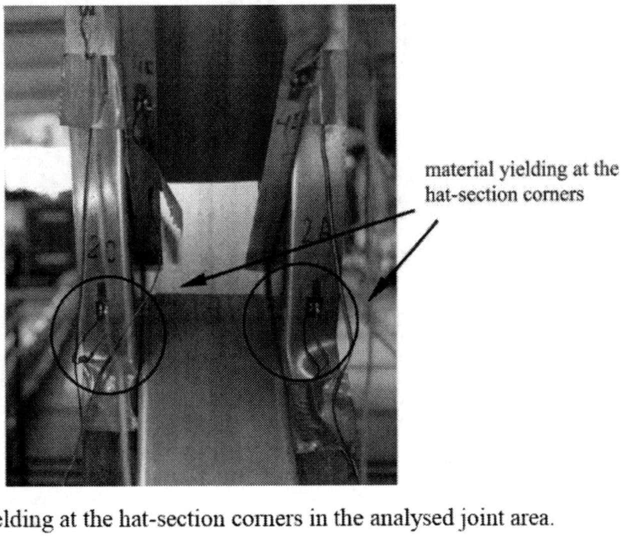

Fig. 4. Material yielding at the hat-section corners in the analysed joint area.

Based on the destructive tests of the five research models (M2 - M6), the critical load values were obtained as a combination of forces: P and the corresponding force H, summarized in table 2. The standard deviation of the critical load was calculated for models M2 to M5. The results obtained from the M6 model were neglected due to significant differences in the recorded strains at the measurement cross-sections. The M1 model was used to test the research station.

Table 2. Combination of forces P i H_{cor} corresponding to the critical load of the hat-section.

Research model number	Force P [kN]	Average P_{av} [kN]	Standard deviation s_P [-]	Force H_{odp} [kN]	Average $H_{cor,av}$[kN]	Standard deviation s_H [-]
M2	8.5	6.98	1.5	96	91.75	4.4
M3	7.7			89		
M4	5.0			95		
M5	6.7			87		
M6	5.0	-	-	106	-	-

The aim of the experimental research, apart from obtaining the information about the work of the truss segment with the positive eccentricity in the joint, was also the validation of the computational model. Therefore, the obtained results in the form of a register of strains and displacements of the model as well as deformations of the joint constituted the basis for the validation of the numerical model.

5 Numerical analysis

The numerical model of the truss segment was made in the MSC Marc program using the finite element method - FEM. Figure 5 shows the geometry of the numerical model. Numerical research has been divided into two stages:

- Stage I - validation of the calculation model,
- Stage II - numerical analysis.

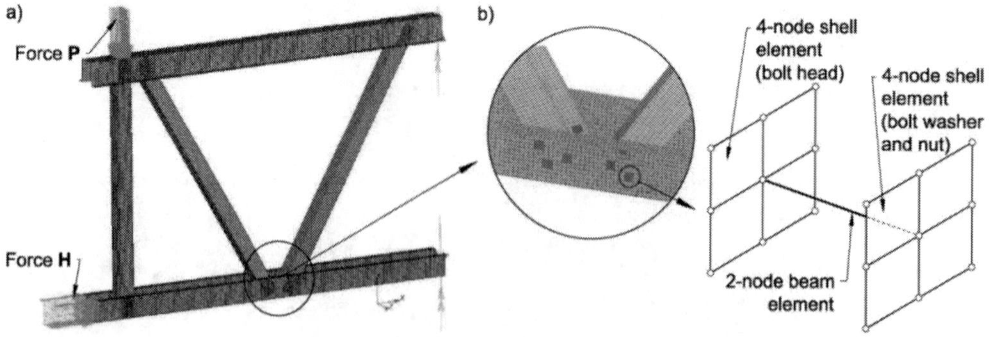

Fig. 5. Numerical model: a) truss structure, b) detail of the bolted connection [15].

The validation of the numerical model was carried out to check out if whether the numerical model correctly describes the real behaviour of the structure [15]. It was possible because of the data obtained from experimental research. When the validation of the computational model was completed, appropriate numerical analyses were started. In these analyses, the wall thickness values of the research models members were taken separately, similarly to the case of analytical calculations (point 3).

In order to numerically model the cold-formed members the 4-node shell elements were applied, the 2-node beam elements were applied for the bolt model. The developed bolted connection model consisted of the 2-node beam element imitating the bolt shank and two stiff areas consisted of four 4-node elements, imitating the bolt head on one side (on the web of the hat-section) and a washer with a nut on the other (as a part of the channel section flange). The rigid areas were modelled by thickened four 4-node shell elements (Figure 5b). The bolted connection models were taken at the locations of real bolts in order to obtain the point connection of the channel brace members and hat-section chord.

The research model was loaded with two forces: the force H and the force P in a ratio of 2:1, similarly as in the experimental research.

Figure 6 a-d, presents the deformations of the research model corresponding to the state of failure for all considered variants of models. Two cross-sections were analysed:

- E-E located at the place of maximum bending moment of the analysed joint,
- M-M defined as the cross-section in which the greatest deformations occurred (place of destruction – Figure 6 a-d).

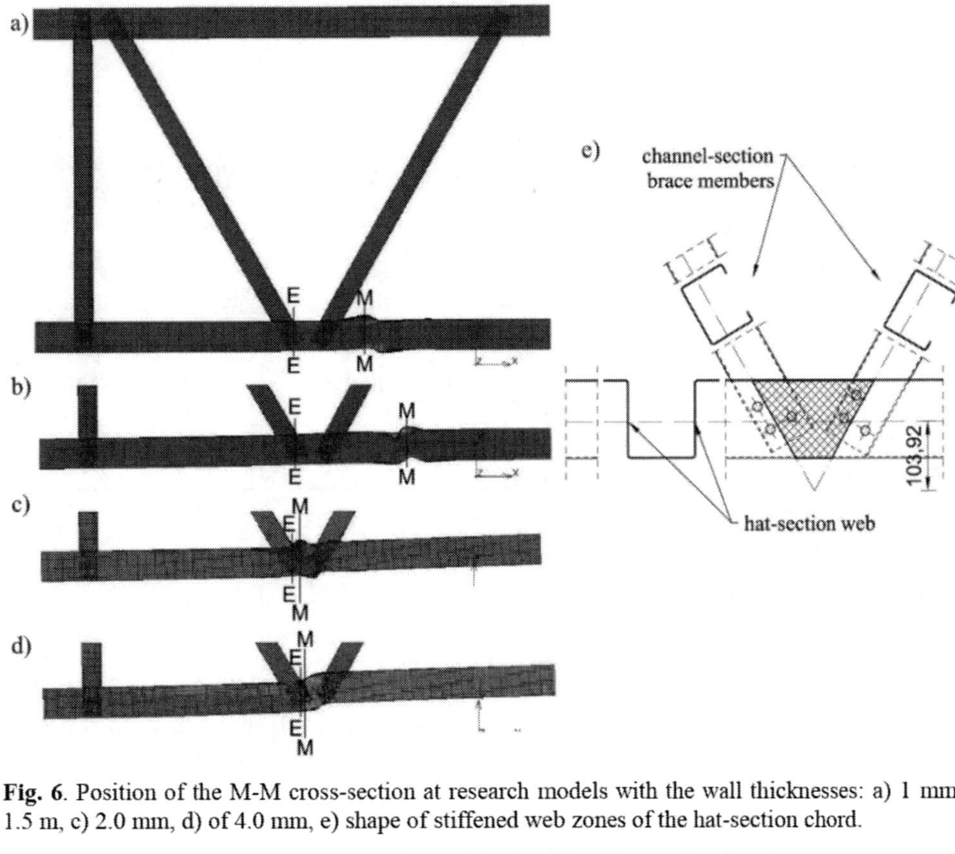

Fig. 6. Position of the M-M cross-section at research models with the wall thicknesses: a) 1 mm, b) 1.5 m, c) 2.0 mm, d) of 4.0 mm, e) shape of stiffened web zones of the hat-section chord.

Based on numerical analyses, the critical and ultimate load values summarized in Table 3 were obtained. Critical load was obtained on the basis of linear elastic bifurcation analysis (LBA), the limit load on the basis of geometric and material nonlinear analysis (GMNA).

In the case of profiles with a wall thickness of 1 mm and 1.5 mm, an increase in the critical load value due to the ultimate load value is observable. In the case of profiles with a wall thickness of 4 mm, the value of critical load corresponds to the global buckling form and is larger than the ultimate load, what is consistent with the limit load theory of compressed members. However, in the case of a 2 mm thick section, although it meets the standard [10] class 4 criteria and according to this classification should be subjected to the theory of postcritical plate resistance, however the critical load is greater than the ultimate load. Therefore, at this stage of the research it was concluded that the results for the 2 mm thick section are not unambiguous. In the future, a more detailed numerical model of the node is planned.

Table 3. Comparison of critical and ultimate load values for all analysed variants.

	Critical load (based on the LBA analysis)		**Ultimate Load** (based on the GMNA analysis)		**Load increase**	
Wall thicknesses	P_{cr} [kN]	H_{cr} [kN]	P_u [kN]	H_u [kN]	*P* [%]	*H* [%]
1.0 mm	7.62	15.22	16.2	32.33	112.6	112.4
1.5 mm	24.87	49.69	32.69	65.23	31.5	31.3
2.0 mm	58.14	116.17	55.62	111.23	-4.3	-4.3
4.0 mm	306.38	612.60	133.13	265.71	-56.5	-56.6

6 Comparison of results

Comparing the analytical calculations with the results of numerical analysis, it can be observed that the value of the ultimate load determined numerically (X_{num}) is higher than the value of the truss design load (X_{an}) (Table 4) in all variants. It should be noted that at this stage of research, imperfections have not been yet included in numerical analysis, however, affecting on the increase of the result values.

Table 4. Comparison of limit load values determined on the basis of analytical calculations and numerical analysis.

	analytical calculations X_{an}		**numerical analysis** X_{num}		**Relative ratio** $(X_{num}-X_{an})/X_{an}$	
Wall thicknesses	**P_u [kN]**	**H_u [kN]**	**P_u [kN]**	**H_u [kN]**	**P [%]**	**H [%]**
1.0 mm	9.1	18.3	16.2	32.33	78	78
1.5 mm	21.4	42.7	32.69	65.23	53	53
2.0 mm	37.0	74.0	55.62	111.23	50	50
4.0 mm	112.2	224.5	133.13	265.71	19	19

The increase of ultimate resistance determined numerically in relation to the values obtained by analytical methods is particularly large at class 4 sections. It confirms their high imperfection sensitivity.

The most important result of the numerical analysis was that in the cases of 1 mm and 1,5 mm walls thickness, considering local stiffening of the hat-section webs through the channel-section walls (Fig. 6 e), the failure (M-M section - Figure 6 a-d) of the compression chord occurs beyond the analysed joint, while the approximate analytical calculations show that the cross-section at the joint is a critical one (regardless of the thickness of the profile walls).

7 Conclusions

The analysis of the results of the tested truss segment with the eccentricity e=103.92 mm in the joint, obtained on the basis of analytical, experimental and numerical tests of various wall thicknesses allowed to state that following cases occur due to a chord web (plate element) slenderness:

- $\lambda > 70$: the channel-sections brace members stiffen the web walls of the hat-section chord in the joint area.
- $\lambda < 70$: no influence of local stiffeners on the stiffness of the web walls of the hat-section was observed.

In order to confirm the conclusions, further analyses are expected for other eccentricity values in the joint. It will also be necessary to carry out analyses including imperfections.

References

1. D. Dubina, Foreword, Thin-Walled Struct., 42, 2, 149–152, (2004)
2. P. Deniziak, K. Winkelmann, Shell Struct. Theory Apl., 4, 533–536, (2018)
3. J. Jankowska-Sandberg, J. Kołodziej, Eng. Struct., **46**, 165–172, (2013)
4. D. Visy, S. Adany, A. L. Joo, *Eighth Int. Conf. Adv. Steel Struct.*,1–18, (2015)
5. J. L. Dawe, Y. Liu, J. Y. Li, J. Constr. Steel Res., **66**, 4, 556–565, (2010)
6. M. Krajewski, P. Iwicki, *Recent Prog. Steel Compos. Struct. Proc. 13th Int. Conf. Met. Struct. Zielona Góra Pol.*, 417–424, (2016)
7. M. Gordziej-Zagórowska, E. Urbańska-Galewska, P. Deniziak, Ł. Pyrzowski, Civ. Environ. Eng. Rep., **25**, 2, 107–123, (2017)
8. J. Bródka, M. Broniewicz, *Konstrukcje stalowe z rur.* (Arkady, Warszawa, 2001).
9. J. Bródka, M. Broniewicz, *Konstrukcje stalowe z kształtowników zamkniętych*, t. I. (Polskie Wydawnictwo Techniczne, Rzeszów, 2016)
10. EN 1993-1-1, *Eurocode 3: Design of steel structures - Part 1-1: General rules and rules for buildings*, Eur. Comm. Stand., (2005)
11. EN 1993-1-3, Eurocode 3: *Design of steel structures - Part 1-3: General rules - Supplementary rules for cold-formed members and sheeting*, Eur. Comm. Stand., (2006)
12. EN 1993-1-8, Eurocode 3: *Design of steel structures - Part 1-8: Design of joints*, Eur. Comm. Stand., (2005)
13. EN 1993-1-5, Eurocode 3: *Design of steel structures - Part 1-5: General rules - Plated structural elements*, Eur. Comm. Stand., (2006)
14. M. Gordziej-Zagórowska, E. Urbańska-Galewska, Ł. Pyrzowski, P. Deniziak, A. Łukowicz, *Recent Prog. Steel Compos. Struct. Proc. 13th Int. Conf. Met. Struct. Zielona Góra Pol.*, 425–432, (2016)
15. M. Gordziej-Zagórowska, E. Urbańska-Galewska, Ł. Pyrzowski, *Shell Struct. Theory Apl.*, **4**, 417–420, (2018)

On modelling of the buckling resistance of welded I-section columns

Radosław Szczerba[1,*], *Marcin* Gajewski[1], and *Marian* Giżejowski[1]

[1]Warsaw University of Technology, Faculty of Civil Engineering, Pl. Politechniki 1, 00-661 Warsaw, Poland

Abstract. This paper analyzes the influence of geometrical and material imperfections on the buckling resistance of welded I-section columns subjected to axial compression through numerical and analytical models. The paper is divided into two parts. The first part recalls a FEM parametric study of members under compression taking into account different slenderness ratios, as well as different amplitudes of initial crookedness and different values of postwelding residual stresses. The formulation of analytical approach is the main issue of the second part of the paper. Analytical formulation of the buckling resistance is based on a statistical hypothesis of the minima value approach, called the Marchant-Rankine's-Murzewski approach (M-R-M). Calibration of imperfection factors included in the analytical formulation is made using the results of FEM simulations performed in the first stage of research investigations.

1 Introduction

Recent design strategies are based on advanced analysis applying the stiffness degradation concept to account for the combined effects of plasticity and imperfections on the nonlinear behaviour of structures. Many standards, including Eurocode 3 [1, 2], permit the use of non-linear finite element method (FEM) simulations. Geometrically nonlinear refined plastic hinge analysis of GMNIA type (**g**eometrically and **m**aterially **n**onlinear **a**nalysis with **i**mperfections) is the method accurate enough in the resistance assessment for engineering practice.

The main aim of the paper is to assess the influence of geometrical and material imperfections on the buckling resistance of welded I-section columns subjected to axial compression. To achieve this goal the authors divided the paper into two parts. In the first part the results of a FEM parametric study (with the use of GMNIA) of compressed members were recalled [3]. The presented outcomes form the basis for the analytical formulation of the buckling resistance of welded I-section columns which is the subject of the second part of the paper. The analytical approach is based on a statistical hypothesis of the minima value approach [4, 5], called the Marchant-Rankine's-Murzewski approach (M-R-M). Calibration of the imperfection factors contained in the analytical formulation is

* Corresponding author: r.szczerba@il.pw.edu.pl

performed with the use of the results of FEM simulations recalled in the first stage of research investigations.

2 Effect of I-section postwelding residual stresses on the equivalent stress-strain model

Let us consider a perfectly straight column made of steel grade S355 with an elastic -plastic model of parent steel material, and with the same stress strain characteristics for tension and compression (Fig. 1a). A low value of the hardening modulus of $E_{har}=E/1000$ is adopted. The initial self-equilibrated postwelding stresses distribution is approximated by adopting a standard residual stress block postulated in [3] (Fig. 1b). The buckling direction is about z-z axis.

Table 1 presents the parameters $\alpha_f t_f$ and $\alpha_w t_w$ (the dimensions of respectively flange and web parts under tension) for the discrete values of ψ_{ten} equal to 1.0 for the common steel grade S355 and ψ_{com} from the range of (0.1, 0.5) with the interval of 0.1.

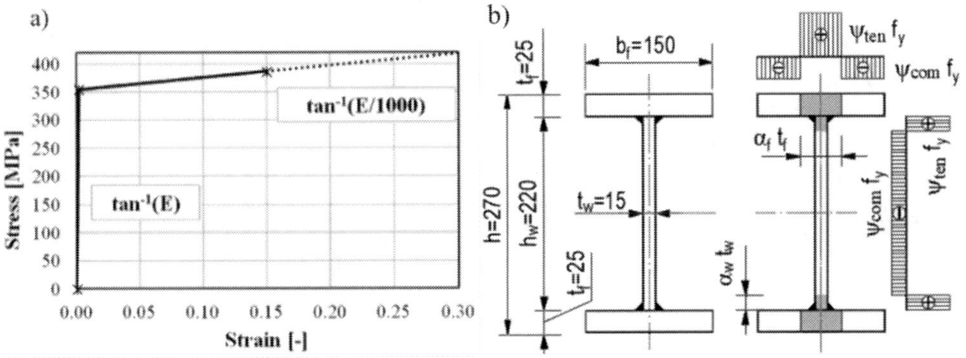

Fig. 1. Material characteristic and postwelding residual stresses distribution.

Table 1. The parameters describing the distribution of postwelding residual stresses [3].

ψ_{ten}	ψ_{com}	$\alpha_f\ t_f$ [mm]	$\alpha_w\ t_w$ [mm]	$\psi_{ten}\cdot f_y$ [MPa]	$\psi_{com}\cdot f_y$ [MPa]
1.0	**0.1**	13.6	10.0	355.0	**35.5**
1.0	**0.2**	25.0	18.3	355.0	**71.0**
1.0	**0.3**	34.6	25.4	355.0	**106.5**
1.0	**0.4**	42.9	31.4	355.0	**142.0**
1.0	**0.5**	50.0	36.7	355.0	**177.5**

When an I-section column is subjected to axial compression, firstly the column behaves elastically so that the first piece of multi-linear equivalent σ_{eff} - ε_{eff} relationship is $\sigma_{eff}=E\varepsilon_{eff}$. The first yielding of compression zones starts when $N_{Ed}=(1-\psi_{com})N_{pl,Rk}$ that corresponds to the reduced yield stress $\sigma_{y,red}=(1-\psi_{com})f_y$. The column tangent stiffness is then suddenly degraded from its elastic value of EI_z to its tangent value of $E_T I_z$.

The stress-strain relationship can be therefore formulated deterministically as a minimum of three variables being the functions of ε:

$$\sigma = \min\left(E\,\varepsilon,\quad \sigma_{H,eff} + E_T\,\varepsilon,\quad \sigma_{y,red} + E_{har}\,\varepsilon\right) \tag{1}$$

in which $\sigma_{H,eff}$ is the value of initial postwelding stress (see Fig. 2a) and the initial stress value of rigid-plastic-strain-hardening part of the parent steel stress-strain relationship is expressed as:

$$\sigma_{y,red} = \left(1 - \frac{E_{har}}{E}\right) f_y \tag{2}$$

The results of finite element simulations of the stub test representing the equivalent stress-strain model ($\sigma_{eff} - \varepsilon_{eff}$ model) for a considered steel grade of S355 are given in Fig. 2a, together with the estimates of E_T and $\sigma_{H,eff}$ values ($\sigma_{H,eff,0.1}$ is for $\psi_{com} = 0.1$, $\sigma_{H,eff,0.2}$ is for $\psi_{com} = 0.2$, etc.).

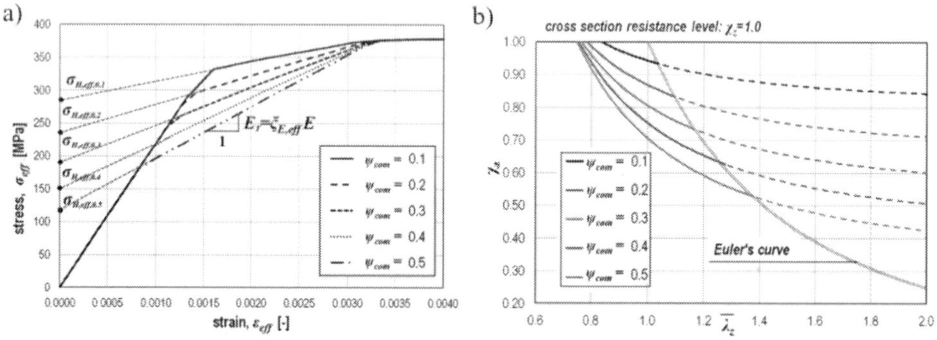

Fig. 2. Construction of equivalent stress-strain diagram for the post-welding steel (a), graphical interpretation of the buckling reduction factor (b).

The parameters of the equivalent stress-strain characteristic are given in Table 2.

Table 2. Parameters for steel grade S355.

ψ	$\sigma_{H,eff}$ [MPa]	$E_{har,eff}$ [GPa]	$\xi_{E,eff}=E_T/E$	$\xi_{H,eff}=\sigma_{H,eff}/f_y$
			FEM	FEM
0.1	287.4	27.93	0.133	0.810
0.2	233.9	43.97	0.209	0.659
0.3	189.3	57.32	0.273	0.533
0.4	151.0	68.75	0.327	0.425
0.5	117.6	78.74	0.375	0.331

3 Buckling curve analytical formulation

Making use of the equivalent stress-strain model for postwelding steel, derived in the previous section, the upper bound buckling curve formulation of perfectly straight column with material imperfections considered through the standard pattern of postwelding residual stresses from Fig. 1b takes the following nominal format (the relationship is graphically presented in Fig. 2b using dimensionless coordinates):

$$N_{b,nom} = \min\left(N_{cr}, N_{cr,eff}, N_{cr,har}\right) \tag{3}$$

in which N_{cr} is the Euler elastic critical force, and the inelastic critical forces corresponding to a piece-wise equivalent stress-strain relationship given by Eqn. (1) are calculated as

follows: $N_{cr,eff} = N_{H,eff} + (E_T/E)\, N_{cr}$ is the critical load with residual stresses taken into account, $N_{cr,har} = N_{H,har} + (E_{har}/E)\, N_{cr}$ is the critical load referred to the hardening stiffness, $N_{H,eff} = \sigma_{H,eff} A$ and $N_{H,har} = \sigma_{y,red} A$.
Using the conventional dimensionless coordinates used in Eurocode 3 [1], Eqn. (3) may be presented in the rearranged format:

$$\chi_z = \min\left[\frac{1}{\overline{\lambda}_z^{2}}, \xi_{H,eff} + \frac{\xi_{E,eff}}{\overline{\lambda}_z^{2}}, (1-0.001) f_y + \frac{0.001}{\overline{\lambda}_z^{2}}\right] \tag{4}$$

It can be seen that the intersection of the Euler's curve for the residual stress free perfectly straight column and that of the perfectly straight column affected by welding residual stresses travels towards larger values of the slenderness ratio when ψ_{com} increases. The slenderness ratio corresponding to the intersection point is greater than unity and calculated as follows:

$$\overline{\lambda}_{z,ref} = \sqrt{\frac{1-\xi_{E,eff}}{\xi_{H,eff}}} \tag{5}$$

The slenderness given by Eqn. (5) is tabulated in Table 3, where $\xi_{H,eff}$, $\xi_{E,eff}$ from Table 2.

While the members are affected by residual stresses in a lower extent, $\xi_{H,eff} \rightarrow 1$ and $\xi_{E,eff} \rightarrow 0$, so the modified slenderness $\overline{\lambda}_{z,ref} \rightarrow \overline{\lambda}_z$.

Table 3. Reference slenderness ratio.

ψ	0.1	0.2	0.3	0.4	0.5
$\overline{\lambda}_{z,ref}$	1.035	1.095	1.168	1.258	1.374

The real members are affected by material and geometric imperfections. The global effect of imperfections and their interaction during the progressive yielding under applied axial loads on the buckling resistance may be modeled using the semi-probabilistic minima value approach in which the variables N_{cr}, $N_{cr,eff}$ and $N_{cr,har}$ of Eqn. (3) are considered independent random variables following a Weibull distribution. Such a statistical hypothesis of the minima value approach, resulting from the randomization of variables used in Eqn. (3) is called the Merchant-Rankine's-Murzewski approach (M-R-M approach [5]). In the case study presented herein, such approach leads to a three dimensional Weibull minima distribution of the buckling resistance. As a result, the following equation is obtained for the reduction factor χ_z of an imperfect column subjected to residual stresses and initial bow deformations:

$$(\chi_z)^{-n} = \overline{\lambda}_z^{2n} + \left(\xi_{H,eff} + \xi_{E,eff}\overline{\lambda}_z^{-2}\right)^{-n} + \left(0.999 + 0.001\overline{\lambda}_z^{-2}\right)^{-n} \tag{6}$$

in which the n is the imperfection factor being an inversion of the equivalent Weibull coefficient of variation u, i.e. $n=1/u$. The u parameter plays the same role as the imperfection factor α in the Eurocode 3 [1]. While the values of u and α are higher, the resistance reduction factor χ_z is lower and vice versa.

4 Imperfection factor calibration based on FEM simulations

The results of finite element simulations are presented in [3] for imperfect columns subjected to the residual stress block presented in the previous sections (for five different values of the compressive residual stress ordinate defined as 0.1, 0.2, 0.3, 0.4 and 0.5) and to geometric imperfections according to the lowest buckling mode (for seven different values of the dimensionless geometric amplitude $\overline{e_0} = e_0/L$, namely 1/10000, 1/5000, 1/4000, 1/3000, 1/2000, 1/1000 and 1/750, where L is the column length). The smallest geometric imperfections amplitude corresponds to a quasi-straight column while the largest one to the fabrication tolerance level. Nine values of the column slenderness ratio ranging from the slenderness domain <0.6,1.4> (an interval of 0.1 applied) were used, giving all together over 300 FEM results of the buckling resistance simulations for the steel grade S355 as well as for the selected values of column imperfection parameters and slenderness ratio values. It was observed that for each residual stress arrangement, the buckling reduction factors decrease with the increase of $\overline{e_0}$ values. Comparing the results for the same $\overline{e_0}$ value but with different ψ_{com} values one can conclude that the buckling reduction factors also decrease for the increase of the compressive residual stress ordinate. Both imperfections, material and geometric have the degrading effect on the column buckling resistance. Their influence on the reduction factor is magnified for the slenderness ratio around $\overline{\lambda}_{z,ref}$. Similar simulations of residual stress influence on the flexural buckling of welded I-girders were presented also in [6, 7]. The FEM calculations of compressed elements with the use of concept of equivalent geometric imperfections were described in [8-11].

4.1 Imperfection parameter for different material and geometric imperfections

The results of $\chi_{z,FEM}$ presented in [3] are used hereafter to compute the imperfection factor of Eqn. (6), denoted as n_{FEM}, from the following equation:

$$\chi_{z,FEM} = \left[\overline{\lambda}_z^{2n_{FEM}} + \left(\xi_{H,eff} + \xi_{E,eff} \overline{\lambda}_z^{-2} \right)^{n_{FEM}} + \left(0.999 + 0.001 \overline{\lambda}_z^{-2} \right)^{n_{FEM}} \right]^{-1/n_{FEM}} \tag{7}$$

Eqn. (7) cannot be solved explicitly for n_{FEM}, therefore an iterative procedure of a Newton type need to be applied in order to do so. Results indicate that there are noticeable differences between n_{FEM} for different values of the slenderness ratio.
The following conclusions can be made:

1. Since the practical value of n_{FEM} is not less than 1 and not greater than those around 10, the large values of n_{FEM} indicate that any value of n in the neighbourhood of 10 may give an acceptable approximation of FEM results from an engineering point of view.
2. Regarding the practical range of n_{FEM} values, the imperfection factor is generally of a greater value for the slenderness ratio around $\overline{\lambda} = \overline{\lambda}_{ref}$ then for the tail values of the slenderness ratio.

4.2 Approximation with use of a slenderness independent imperfection parameter approach

The imperfection parameter n is calibrated as a slenderness ratio independent factor n_{opt} with the calibration made using Mathematica function Non-Linear Model Fit. In all cases

the convergence of the NMinimize method was very good allowing for finding the global minimum of the goal function inside the domain. The results are presented in Table 4.
The results presented in Table 4 prove that the imperfection parameter n_{opt} decreases with the increase of both imperfection factors, $\bar{e}_0$ and ψ_{com} (or increase with the equivalent coefficient of variation $u_{opt} = 1/n_{opt}$). Applying the results given in Table 4, the interpolation function n_{opt} ($\bar{e}_0$, ψ_{com}) is calibrated with the use of the Mathematica software. The following linear function can be postulated:

$$n\left(\psi_{com}, e_0\right) = a_0 + a_1\psi_{com} + a_2 e_0 \tag{8}$$

Parameters a_0, a_1 and a_2 are determined as equal to: 4.94, -5.82, and -1039.62, respectively. Such approximation is not accurate enough, so a non-linear approximation is assumed in the form:

$$n\left(\psi_{com}, e_0\right) = b_0 + b_1\psi_{com} + b_2 e_0 + b_3 \frac{1}{e_0\psi_{com}} \tag{9}$$

with parameters b_0 to b_3 determined as: 3.50, -3.86, -351.69, 0.00002889. The graphical interpretation of both approximations are presented in Fig. 3.

Table 4. Imperfection parameter n_{opt} for S355 (ψ_{ten} = 1.0).

Dimensionless compressive residual stress factor ψ_{com}	Dimensionless amplitude of initial deflection profile $\bar{e}_0$						
	1/10000	1/5000	1/4000	1/3000	1/2000	1/1000	1/750
0.1	5.86	4.88	4.54	4.17	3.63	2.81	2.52
0.2	3.93	3.51	3.38	3.16	2.86	2.35	2.15
0.3	2.92	2.69	2.61	2.50	2.33	2.00	1.87
0.4	2.24	2.14	2.09	2.04	1.94	1.74	1.65
0.5	1.84	1.77	1.76	1.72	1.67	1.54	1.49

a) b)

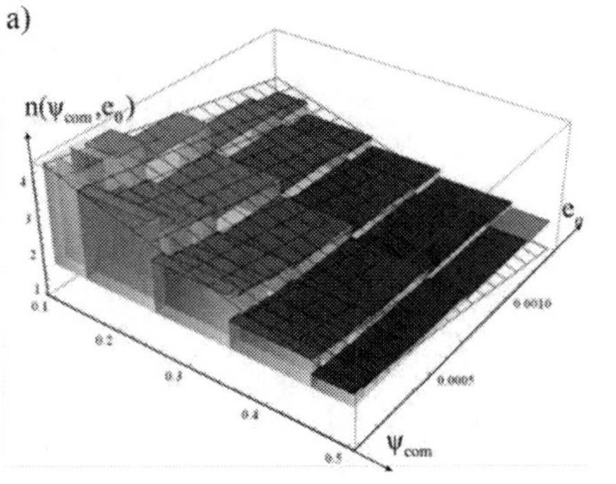

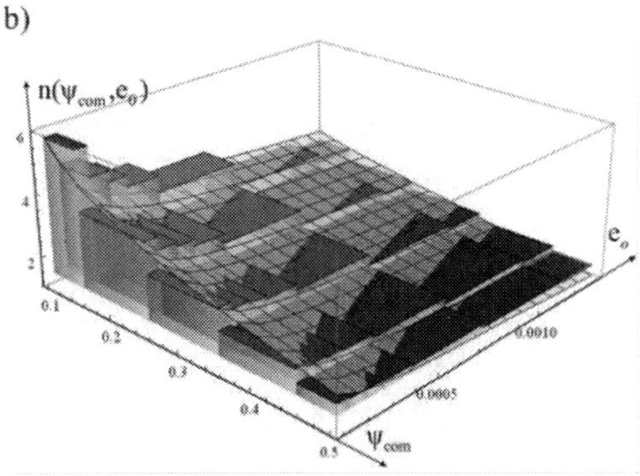

Fig. 3. Linear approximation (a), non-linear approximation of n_{opt} as a parameter for model described by Eqn. (7) (b).

Fig. 4 presents the results of χ_z from Eqn. (6) based on the analytical formulation and the imperfection parameter n from Eqn. (9), and comparing them with discrete FEM results. Only the extreme cases of $\bar{e}_0$ for each ψ_{com} values are considered, namely the white circles represent the results corresponding to the lowest geometric imperfection amplitude while the black ones to the highest amplitude. The doted curves given in Fig. 4 in colour correspond to function variables in Eqn. (3), N_{cr} in blue, $N_{cr,eff}$ in red and $N_{cr,har}$ in green. From the comparison, the following conclusions can be drawn:

1. While applying a slenderness independent n-value, the accuracy level obtained is unevenly distributed over different slenderness ratio ranges.

2. When the slenderness independent imperfection factor n is used, FEM results corresponding to higher geometric imperfection amplitudes may be reproduced more closely by the proposed analytical model than those for the lowest geometric imperfection amplitude.

3. The buckling reduction factor formula given by Eqn. (6) would approximate more closely the FEM results for small geometric imperfections if the imperfection parameter n is a function of the slenderness ratio like in the Eurocode's imperfection parameter η is, namely in [1] $\eta = \alpha\left(\bar{\lambda} - 0.2\right)$.

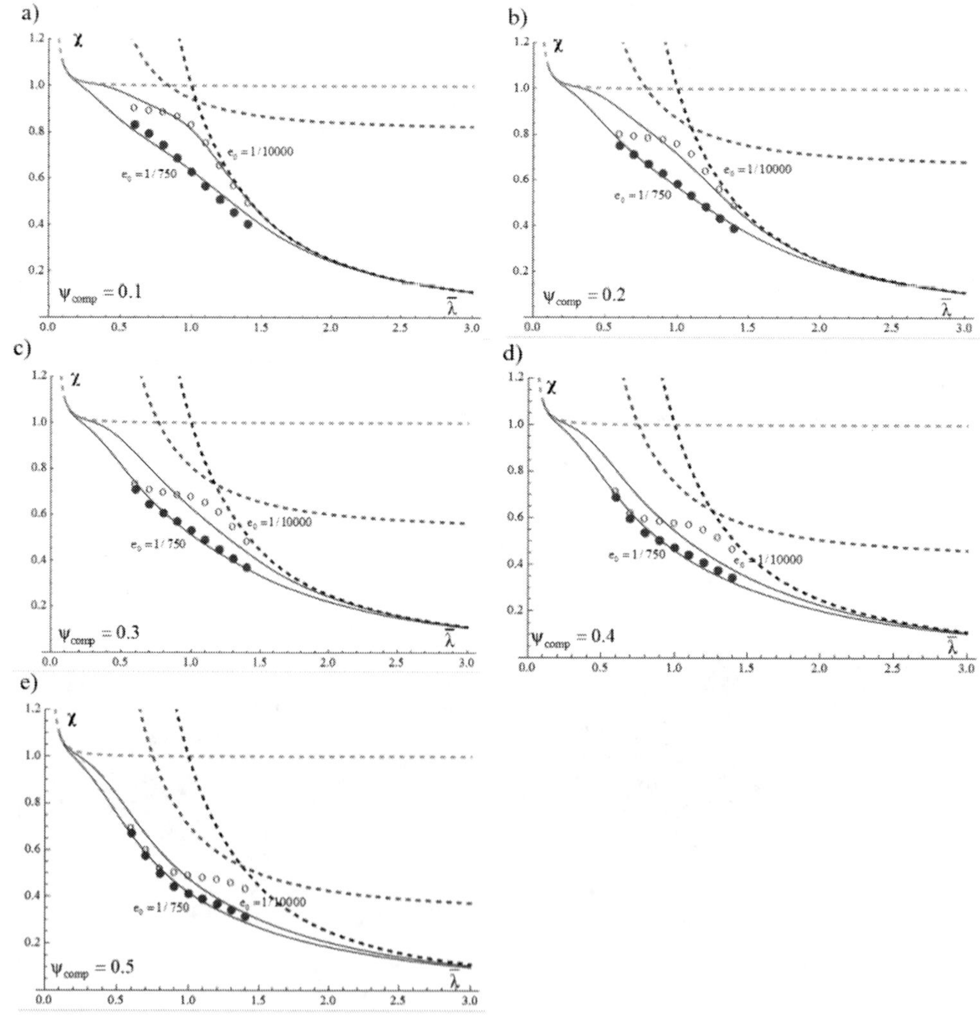

Fig. 4. Reduction factors χ_z calculated analytically and $\chi_{z,FEM}$ from FEM simulations for steel grade S355.

5 Summary and conclusions

The authors presented the issue of buckling resistance modelling of welded I-section columns in a complementary way. The results of FEM calculations were recalled herein and were used to develop the buckling curve analytical formulation. The application of statistical hypothesis of the minima value approach (M-R-M approach) gave a good fit of the analytical solutions and the FEM calculations outcomes. The results of linear and non-linear approximations were described in detail and summarized in particular subsections. One could observe a higher imperfection factor $\bar{e}_0$ triggers a better fit of the analytical solutions and FEM results. The adoption of the very small value of the dimensionless geometric amplitude $\bar{e}_0$ equal to 1/10000 leads to the analysis of a practically perfect element. In this case the compliance of results was not satisfactory. The perfect match was observed in case of the value $\bar{e}_0$ equal to 1/750.

The results presented in the paper lead to the conclusion that the calculations should be also executed for other types of steel, especially high strength steels. Presently, the authors perform simulations of elements made of steel grade S 690 to check the applicability of M-R-M approach not only in case of common used steels.

References

1. *EN-1993-1-1. Eurocode 3: Design of steel structures Part 1-1: General rules and rules for buildings* (Brussels, European Committee for Standardization, 2005)
2. L. Simoes da Silva, R. Simoes, H. Gervasio, *Design of Steel Structures, Eurocode 3: Design of Steel Structures, Part 1-1: General Rules and Rules for Buildings* (ECCS Eurocode Design Manual, Ernst & Sohn, Berlin, 2010)
3. B. Launert, R. Szczerba, M. Gajewski, M. Rhode, H. Pasternak, M. Giżejowski, Mat. test., **59**, No. 1, 47-56 (2017)
4. J. Murzewski, *Reliability of engineering structures* (Arkady, Warszawa, 1989) [in Polish]
5. A.M. Barszcz, M.A. Giżejowski, Arch. of Civ. Eng., **LII**, 1, 59-86 (2006)
6. B. Launert, M. Rhode, A. Kromm, H. Pasternak, T. Kannengiesser, Mat. R. Proc. **2**, 109-114 (2016)
7. Z. Kala, J. Valeš, Eng. Str., **134**, 37-47 (2017)
8. T. Tankova, L. Marques, L. Simões da Silva, A. Andrade, J. of Constr. St. R., **128**, 839-852 (2017)
9. Z. Kala, Arch. of Civ. and Mech. Eng., **15**, 1098-1107 (2015)
10. L. Simoes da Silva, L. Marques, C. Rebelo, J. of Constr. St. R., **66**, 575-590 (2010)
11. F. Papp, Eng. Str., **106**, 124-136 (2016)

Experimental study of mechanical properties of steel cables

Renata Orzłowska[1,*]

[1]Koszalin University of Technology, Faculty of Civil Engineering, Environmental and Geodetic Sciences, Śniadeckich 2 St. 75-453 Koszalin, Poland

Abstract. Guyed masts are structures particularly sensitive to various types of nonlinarities. Static calculations of masts should be performed in accordance with EN 1993-3-1 using global elastic analysis. This is justified in the case of pre-stretched cables on mast guys, characterized by a constant value of the modulus of elasticity. In the case of non-stretched cables this approach is inappropriate, because permanent rope elongations caused by the applied load can significantly reduce the stiffness of the mast. The paper presents the preparation process of rope samples for experimental tests and preliminary results of tests aimed at determining the actual stress – strain relation of the rope and estimation a constant elastic modulus E of the cable after a few cycles of loading and unloading. The results of the obtained tests are introductory to further numerical analysis of guyed masts, including physical nonlinearity of guys.

1 Introduction

In engineering practice a constant value of elastic modulus E for cables is taken into account in order to calculate guyed masts. This approach is correct in the case of pre-stretched cables on mast guys. It happens, however, that despite the constant modulus of elasticity assumed in calculations, new, non-stretched cables are used to build the guyed masts. This solution may constitute a serious factual error. Brand new cables are characterized by a non-linear stress – strain characteristic $\sigma - \varepsilon$, therefore the variable elastic modulus depends on current stresses level and may be less (even by 30%) [1,2] than the constant value of E [3]. This issue acts directly on the stiffness of mast elastic supports and may produce permanent strains corresponding to the mast non-vertical position. Therefore the computations should consider actual mechanical properties of non-stretched cables, determined on the basis of experimental static tensile test results [4]. The paper presents a detailed process of preparing samples of a certain steel rope for tests. The research methods in this field and the standard requirements are also discussed. The results of the research will be used directly to analyze guyed masts in the light of physical nonlinearity.

* Corresponding author: renata.orzlowska@tu.koszalin.pl

2 Purpose and scope of the experimental research

The experimental research was conducted on a spiral strand rope – type 1x37 of diameter equal 20.1 mm and a minimal breaking force equal 367 kN, stated by the manufacturer [5,6] (Fig.1).

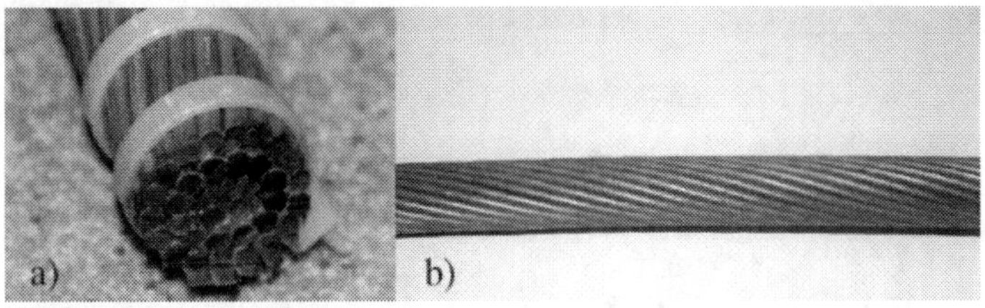

Fig. 1. Rope accepted for the research: a) cross section, b) view of the tested rope.

The scope of the research included:
- Determination of the stress – strain characteristic during static tensile test based on measured values: force – elongation of cable,
- Finding the constant elastic modulus *E* of cable after a few cycles of loading and unloading,
- Determination of the real value of a cable breaking force.

3 Preparation of the cable samples to research

The current standard regulations [7] present recommendations for the cable samples to conform in a research, concerning the determination of the real value of a cable breaking force. In the case of spiral strand ropes, the minimal length of a sample depends on its nominal diameter, it should be equal 1000 mm, disregarding the rope termination at both ends. They should be anchored to make the anchorage resistance at least equal to the resistance of an anchored cable (Table 1).

Table 1. The minimal length of a cable sample without anchorages.

Nominal diameter of cable (mm)	d≤6	6<d≤20	20<d≤60	d>60
Minimal length of spiral strand rope (mm)	500	1000	50xd	3000

Due to the possibility of placing a cable sample in a tensile test machine, an anchorage in the form of a conical bush using socketing was selected. The minimal dimensions of the conical bush based on guidelines [1,8] are shown in Fig. 2.

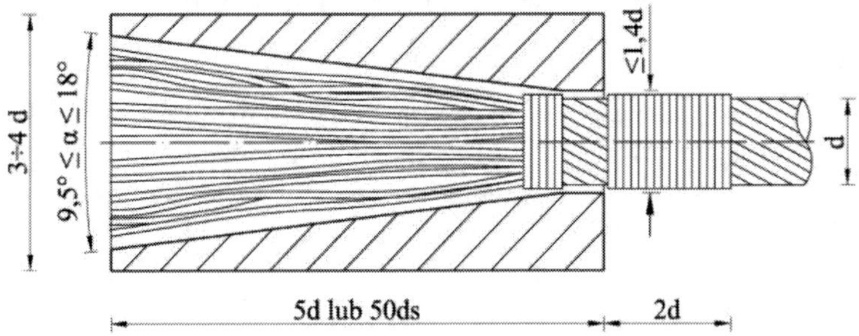

Fig. 2. Minimal dimensions of the conical bush.

According to [9], the width of the hole should be at least 1.2d + 3 mm. Fig. 3, shows geometrical dimensions of the samples prepared to the tests

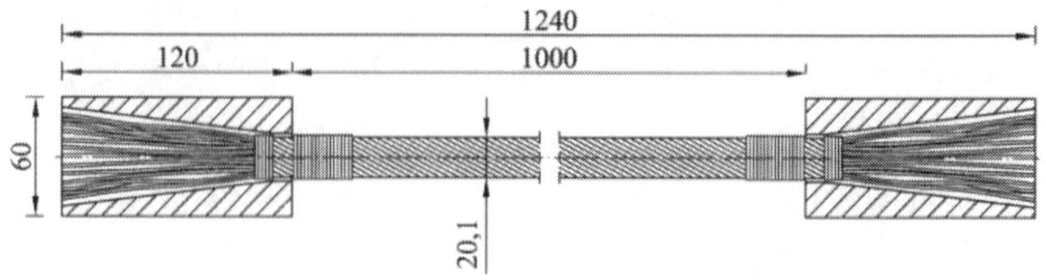

Fig. 3. Dimensions of cable sample to tests.

The conical bushes, anchoring the cable section prepared to tests were made of 355 steel grade. The original cold socketing compound named WIRELOCK (resin for conical sockets) was used as a socketing media. The preparation process of cable sample anchorage was performed according to the standard guidelines [8] and recommendations of the resin manufacturer's instructions [10]. After measuring and marking the length of a cable section to cut, separate sections on both ends should be secured to prevent their unlaying. For this purpose the serving bands were used. Before placing the cable in the bush it is necessary to perform the primary serving – best from a soft annealed iron wire of small diameter. In accordance with [10], primary serving should be applied at distance equal to the length of the bush basket, measured from the cable end and must be at least twice as wide as the minimal diameter of the rope. In order to unlay the end of the cable, one should pull the cable through the conical bush, remove the secondary servings at the end of the cable and unlay individual wires in such a manner that the ending of the cable would resemble a broom. In accordance with [8], the spreading angle should not be wider than 45° from the vertical axis of the cable, while according to the resin manufacturer's instructions [10], this angle should be limited to 30 degrees. Once the wires are bent they should be cleaned from grease and dirt. Before placing the broom into the bush, the wires should be completely dry. The conical bush should be put on the broom, in such a way that the base of the broom sticks out of the bush no more than half of the minimal diameter of the cable. It is very important to evenly distribute the wires in the bush basket.

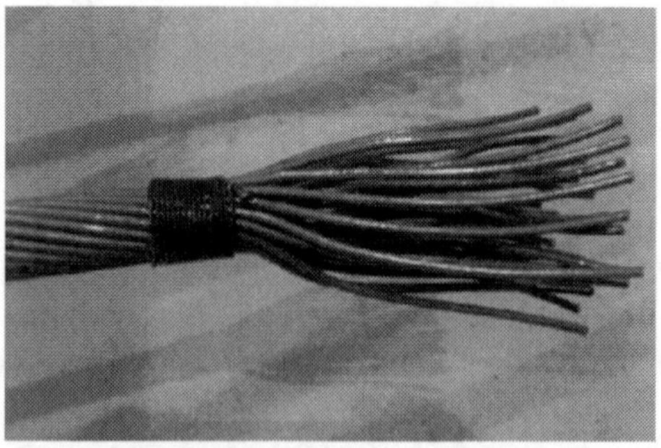

Fig. 4. Unlaying of the spiral strand rope into individual wires.

According to [8], collinearity of the clamp rope should be ensured in relation to the bush on length equal three coils of cable. This length can be reduced by means of appropriate clamps. The bush should be secured before changing the location. The rope end in the form of a broom and the bush should be placed vertically in one axis. It is necessary to properly seal the hole of the bush prior to socketing (the resin should not be allowed to leak out). In accordance to [10], putty or plasticine should be used for sealing. The rope termination prepared this way is ready for socketing.

Fig. 5. The socketing of rope termination using WIRELOCK resin.

The procedure of conical bush socketing using WIRELOCK resin was done according to [8,10]. In order to do so, the two-containers socketing compound, consisting of polyester resin and granular compound in WIRELOCK technology were used in the tests. After mixing both components a chemical process, which begins a curing process of resin takes place. According to [10], mixing of all contents from both containers of the individual kit is required. After pouring all of the resin to the container with granular compound, one should thoroughly mix them for at least 120 second using a flat wooden paddle. After obtaining a homogeneous consistency of the mixture, one should slowly pour it into the conical bush until it is completely filled. It is especially important to maintain a sufficiently low pouring speed due to the possibility of the occurrence of air bubbles, which reduce the strength properties of resin. The socketing material should be appropriately fluid, in order to efficiently penetrate into any narrow crevices. During socketing, any leakage of resin from the conical bush should be carefully observed. In the case of leaks, it is required to stop them as well as to refill the resin loss in the conical bush. The bonding speed depends on the air temperature during socketing. The WIRELOCK resin congeals within 15 minutes at the temperature of 18°C to 24°C. An increase in temperature by 10°C shortens the gelling time by approximately 50%, in turn a decrease in temperature by 10°C lengthens the gelling time by approximately 100%. Rope termination should be motionless until it is possible to perform a scratch test, which involves scratching with a sharp object over the surface of the resin. The result of this test is accepted only if not more than a thin, bright-colored crack is visible on the surface. Before testing, the sealing material securing the resin from leaks must be removed from the bush. The resin component in the original phase, after mixing the ingredients occurs in the form of a dark blue – green liquid. After the curing process it is visible in the form of a bright green compact solid.

Fig. 6. WIRELOCK resin after the curing process.

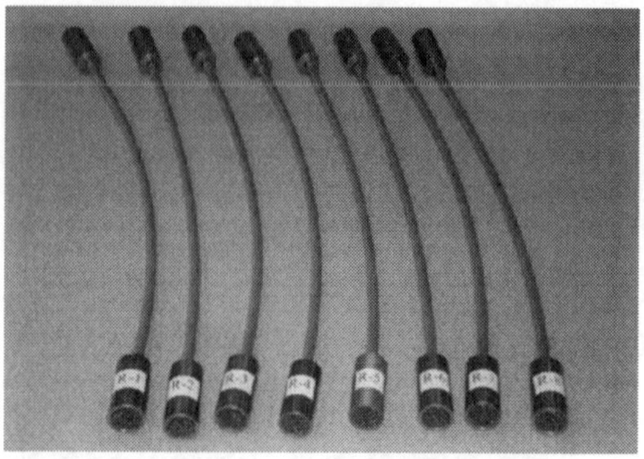

Fig. 7. Preparing the cables sample to tests.

3 Experimental tests on ropes

Experimental tests were carried out on the Inspekt 600 testing machine shown in Fig. 8.

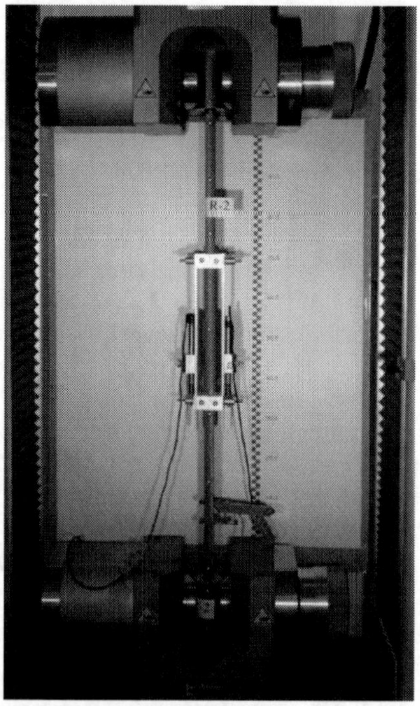

Fig. 8. View of the rope sample in tensile test machine.

First of all, experimental tests of longitudinal strains and elastic modulus of the rope were carried out. According to [11], the pre-stretching process of steel mast guys should be carried out by means of cyclical rope loading and unloading in the range from 10% to 50% of the breaking force. The test performance requires at least 10 cycles. The tested rope samples were loaded and unloaded 12 times at a rate of 1 kN/s. In order to eliminate the initial non-straightness, the tested rope sections anchored in the handles of the testing machine should be loaded with a preliminary force. In experimental tests, this value was taken at the level of 10 kN. The measurement of the force value from the tensile test machine in the conducted tests was read in a time interval equal to 0.5 s. Regardless of the exact graph of characteristic $\sigma - \varepsilon$ from the testing machine, the elongations of the rope samples were additionally measured using two extensometers with a nominal accuracy equal 0.001 mm. A measurement base on length 300 mm was assumed. Moreover, in order to determine the cable elongation, an optical image correlation system [12] was used. This system allows to measure element displacements produced by loading with the similar accuracy equal to accuracy of extensometers.

4 Results of experimental tests

In order to determine the constant value of elasticity modulus E of the cable after its pre-stretching, the stress-strain characteristic $\sigma - \varepsilon$, shown in Fig. 9 was obtained in experimental tests.

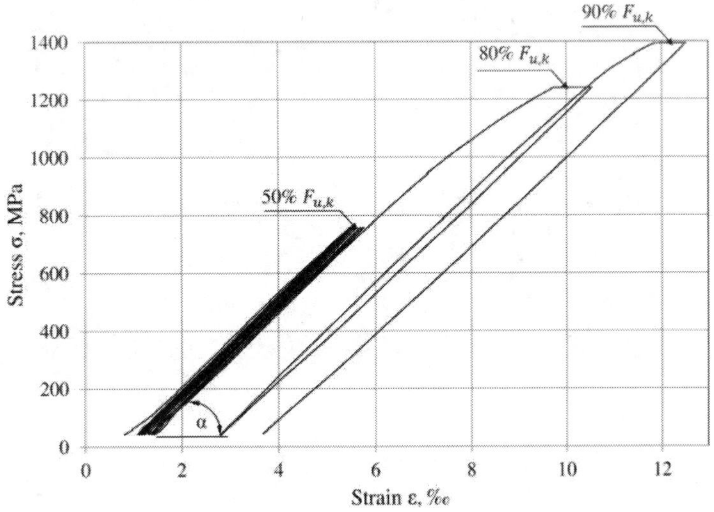

Fig. 9. Graph of non-linear dependence stress-strain during cyclic loading and unloading.

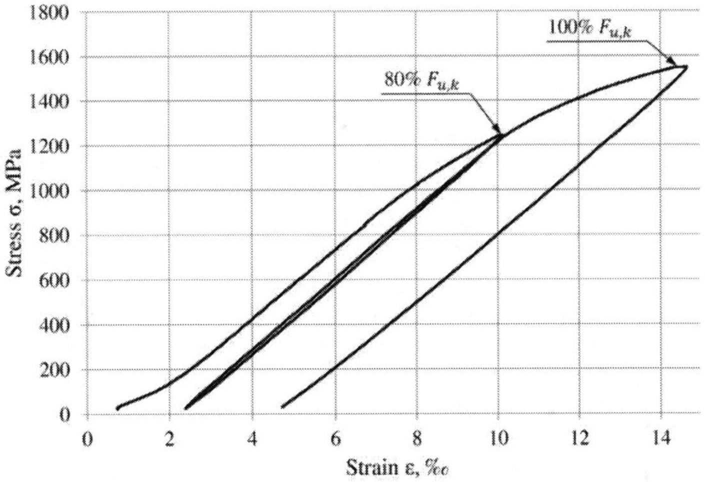

Fig. 10. Graph of stretching the factory rope in two consecutive loading and unloading cycles.

The standard (constant) elasticity modulus E of the cable is referred to as $E=tg(\alpha)$. The real value of this modulus, determined by the tests after the pre-stretching (twelve times), is equal to 158,28 GPa for the R6 sample. Thus, the above dependence confirms a good compliance of the obtained results (the standard [3], the E material value of the considered type of ropes is equal 150±10 GPa). It can be observed on Fig. 9 that during the first rope loading, the stress – strain characteristic $\sigma - \varepsilon$ is curvilinear, the curve graphs in the loading and unloading curves differ significantly. A rope, especially a new one, is permanently lengthen in the stretching process as a result of elimination of spaces between twisted wires in the cross section. Repeating loading and unloading of the cable causes its permanent elongations, which has an influence on a significant increase of elasticity modulus E. Fig. 10 shows the stress – strain characteristic $\sigma - \varepsilon$, obtained during the static tensile test based on measured values of force – elongation of the rope.

The rope stretching graph (Fig. 10) shows the non-linear dependence of $\sigma - \varepsilon$, of the tested sample under the influence of the applied load equal 80% of the nominal breaking force (in the first cycle) and 100% of this force (in the second cycle).

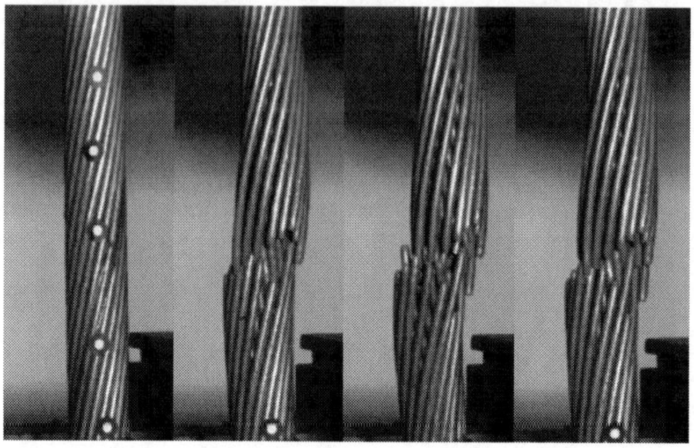

Fig. 11. Stages of destruction of a rope sample in an experimental test of the actual rope breaking force, obtained with the use an optical image correlation system [12].

Fig. 11 presents the process of breaking the sample in the test machine in order to determine the real cable breaking force. In the tensile process, a rope broke while reaching a force equal to 427 kN.

5 Summary

The paper describes the process of preparing samples of a particular specific steel rope for the experimental tests. The process of socketing the rope terminations with modern WIRELOCK resin guide is precisely described. The use of this technology guarantees high strength and a good adhesion to steel. The socketing process can be conducted in field conditions in the temperature range from +9°C to 43°C, without a necessity of using any additional improvement. A relatively short time to prepare resin for socketing and achieve full strength in the terrain is a great advantage. For this reason, the above-mentioned technology is increasingly used in industry and civil engineering around the world. The results obtained from the research concerning determination of the standard modulus value of rope deformation after initial pre-stretching, are similar to those provided by the rope manufacturer [5]. Comparing the obtained value of the modulus equal 158,28 GPa, it can be seen that it is range of the value assumed in the design computations (160±10 GPa – manufacturer's data, 150±10 GPa – standard [3]).

The actual breaking force of a cable with value 427 kN is greater than the nominal breaking force specified by the manufacturer (367 kN). The non-linear dependence of $\sigma - \varepsilon$ of the tested rope sample, obtained during the static tensile test, is a characteristic of non-stretched cable. In this case, correct determination of standard elastic modulus of the cable must refer to a specific cable load level and it is therefore a variable value. It should be noted that experimental studies have not been completed. The results of the obtained tests are preliminary for a further numerical analysis of guyed mast structures, including physical nonlinearity of the guys.

Acknowledgements
I would like to express my sincere thanks to Mr. John Dodd (Millfield Wirelock) and Mr. Daniel Jacheć (PFEIFER SEIL – UND HEBETECHNIK GMBH) for the support and financial assistance that contributed to carrying out the presented research. I want to express my appreciation and thanks to SANGER METAL Sp. z o. o. for valuable instructions and technical support.

References

1. M. Matuszkiewicz, *Static analysis of cable structures in elasto-plastic range,* Doctoral dissertation, Koszalin University of Technology (2001) [in Polish]
2. Sz. Pałkowski, *Cable structures,* WNT (1994) [in Polish]
3. Eurocode 3: Design of steel structures – Part 1-11: Design of structures with tension components (2008) [in Polish]
4. J. Hankus, GIG, *Structure and mechanical properties of steel ropes* (1990), [in Polish]
5. Pfeifer Seil –und Hebetechnik GmbH, *Pfeifer Tension Members* (2015)
6. PN-EN 12385-10+A1: Steel wire ropes – Safety – Part 10: Spiral ropes for general structural applications (2008) [in Polish]
7. PN-EN 12385-1+A1: Steel wire ropes – Safety – Part 1: General requirements (2009) [in Polish]
8. PN-EN 13411-4: Terminations for steel wire ropes – Safety – Part 4: Metal and resin socketing (2011)
9. K. Feyrer, Springer-Verlag Berlin Heidelberg New York, *Drahtseile. Bemessung, Betrieb, Sicherheit* (1994)
10. Millfield Enterprises (Manufacturing) Limited, *WIRELOCK. Technical data manual* (2017)
11. Eurocode 3: *Design of steel structures – Part 3-1: Towers, masts and chimneys – Towers and masts* (2008) [in Polish]
12. GOM GmbH Brauschweig, *PONTOS User Manual* (2011)

Investigation of laser heat treated Monel 400

Mateusz Kukliński[1,*], *Aneta* Bartkowska[1] , and *Damian* Przestacki[1]

[1] Poznań University of Technology, Faculty of Mechanical Engineering and Management, Pl. Marii Skłodowskiej-Curie 5, 60-965 Poznań, Poland

Abstract. In this research, Monel metal was laser heat-treated for microstructural, microhardness and roughness investigation. The treatment is an initial stage for welding Monel without additional material for structural elements. The treatment was carried out with diode laser TruDiode 3006 which allows to reach a power of 3 kW. The material was treated with a constant laser beam power, equal to 1400 W, and four different laser beam velocities: 5, 25, 50 and 75 m/min. The distance between single laser tracks was 0,5 mm in every experimental series. It was found that laser heat treatment of Monel does not influence its hardness. The depth of melted areas is decreasing with an increasing laser beam velocity. The melted area manufactured with laser beam velocity equal to 5 m/min is about 350 μm. Increasing the laser beam velocity to 75 m/min causes depth reduction to about 100 μm. The melted areas are built with column crystals oriented in the direction of heat dissipation perpendicular to the heating direction.

1 Introduction

Monel 400 alloy, due to its high corrosion resistance, is mainly applied for constructions and functional parts where aggressive environment occur. These could be acid, alkali and salt but also seawater and polluted urban air. Monel 400 shows excellent resistance among others against sulfuric, hydrofluoric, phosphoric and organic acids. Moreover, this material is easy to shape by forming and machining processes, contrary to other nickel superalloys [1,2-4].

It was air contamination that forced architects and engineers to use corrosion-resistant material for buildings covering. Monel 400 was a perfect solution for this requirement. The first major roofing installation made of Monel was Pennsylvania Railroad Station in New York City in 1909. The potential life of this roof was estimated as 300 years and lasted the life of the building. Some other buildings with Monel covering are Museum of Art in Philadelphia, PA, Pentagon Building, National Cathedral and National Art Gallery – all three of these are placed in Washington, D.C. [5,6].

Laser technologies which are applied within a broad spectrum of applications, are also used for processing construction elements. The ability to propagate in a straight line and the possibility to easy adjust beam's power make laser the most precise heat source. Moreover, due to its concentration in small area, laser provides only minor changes in material

* Corresponding author: mateusz.kuklinski@doctorate.put.poznan.pl

microstructure. Four major categories of laser material processing are forming, joining, machining and surface engineering. In manufacturing building roofs and facades laser is mainly used for welding, cutting and surface finishing [7-9,10].

This research focuses on analyzing the influence of laser beam scanning velocity on Monel 400 properties after laser heat treatment. The survey is aimed at prospective welding construction elements made of Monel without additional material.

2 Research methodology

2.1 Material

The material investigated in this experiment was Monel 400. Its chemical composition is given in Table 1. Microstructure of this alloy is built of nickel-copper solid solution. Because of great solubility of nickel and copper in all proportions, it is a single-phase alloy. Before the laser heat treatment, material was cut to obtain specimens of dimensions 30 mm x 20 mm x 12 mm.

Table 1. Chemical composition of Monel 400 alloy.

Si [%]	Cu [%]	Fe [%]	Mn [%]	C [%]	S [%]	Ni [%]
max 0.5	28-34	max 2.5	max 2.0	max 0.3	max 0.024	bal.

2.2 Laser heat treatment

Laser heat treatment was carried out with a diode laser TRUMPF TruDiode 3006. This laser is based on single diode modules which allow to reach power of 3 kW. To manipulate the location of laser beam, it was integrated with KUKA KR16-2 robot. During laser track manufacturing, laser beam was started above and turned off behind the specimen area to make sure that laser beam scanning velocity is equal on an entire laser track. The scheme of described system is presented in Figure 1.

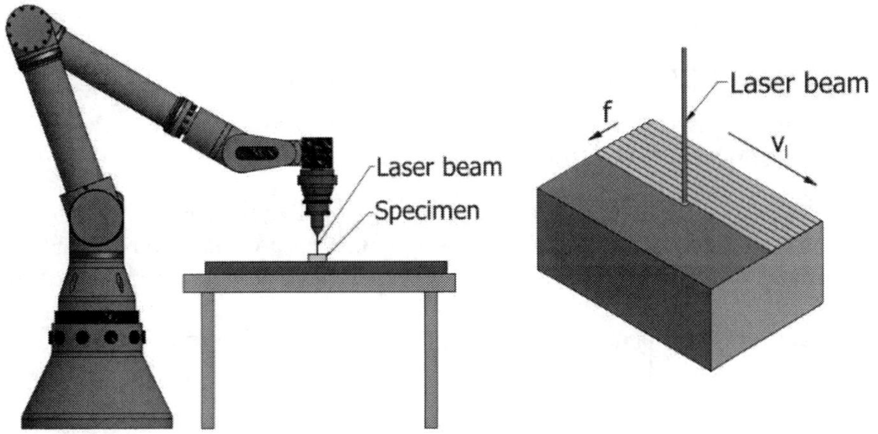

Fig. 1. Scheme of Monel 400 laser heat treatment process.

Each sample was laser heat treated on side of dimensions 30 mm x 20 mm. Constant parameters were laser beam power $P = 1400$ W, laser beam diameter $d_l = 1$ mm and distance between laser tracks $f = 0.5$ mm. The samples were treated with different laser beam scanning velocity v_l ranged from 5 m/min to 75 m/min. A detailed summary of laser heat treatment parameters is given in Table 2.

Table 2. Laser heat treatment parameters.

P [W]	q [kW/cm^2]	v_l [m/min]	d_l [mm]	f [mm]
1400	178.3	5	1	0.5
		25		
		50		
		75		

2.3 Samples preparation

In order to obtain metallographic microsections for microstructural investigation, samples were cut across laser tracks and cutting surfaces were grinded with abrasive papers of grit ranged from 120 to 2000. Then, specimens were polished for 20 minutes and etched with Marble's reagent.

2.4. Visual investigation

First step of visual investigation was observation of laser treated surface. For this purpose stereomicroscope Stereo Discovery V.20 was used. Pictures for microstructural investigation were taken using light microscope Opta-Tech of series LAB40. Besides from analysis of microstructure, the obtained pictures were used for measuring dimensions of laser tracks and microhardness imprints. These measurements were carried out using Axio Vision software.

2.5 Microhardness and roughness testing

Indentations for microhardness measurement were fabricated with Zwick 3212B tester with load $F = 0.9807$ N. Roughness of laser tracks was measured using profilographometer Hommel Tester H500.

3 Results and discussion

3.1 Laser heat treated surface investigation

The first stage of investigation specimens after laser heat treatment was taking pictures of processed surface. The chosen images are shown in Figures 2a-d. Based on these images, it was initially concluded that increasing laser beam scanning velocity from 5 m/min to 25 m/min results in decreasing the width of laser tracks. However, further decreasing the size of laser tracks was not found with enhancing laser beam scanning velocity to 50 m/min and 75 m/min. Furthermore, laser tracks shown in Figure 2a seem to overlap each other, unlike the others.

Besides from laser tracks size, it can be also seen that surface treated with $v_l = 5$ m/min is probably more rough than these shown in Figures b, c and d. Therefore, surface quality was separately investigated, the results are described in paragraph 3.2.

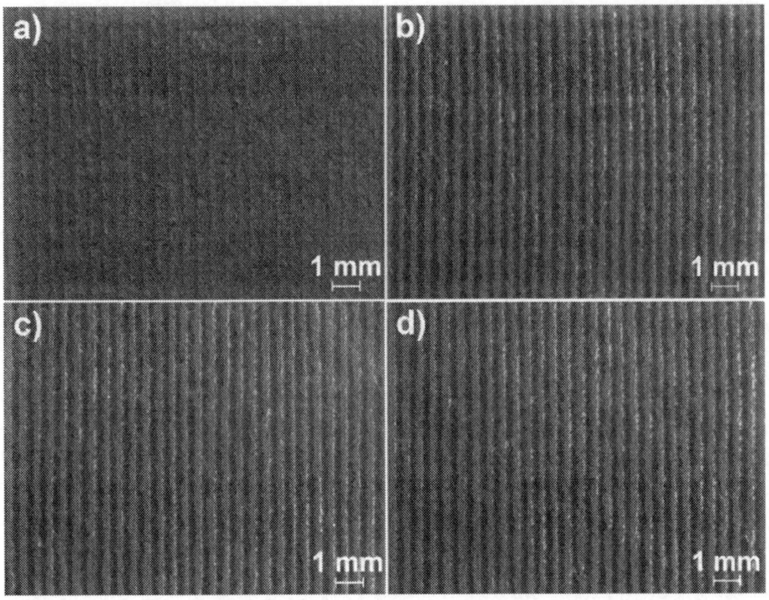

Fig. 2. Surface of Monel 400 alloy after laser heat treatment with laser beam scanning velocity: a) 5 m/min, b) 25 m/min, c) 50 m/min, d) 75 m/min.

3.2 Roughness testing

Roughness was measured to determine Monels' surface quality after laser heat treatment. The obtained results in the form of roughness parameters S_a and S_z are shown in Figure 3 with their graphical representation for easier comparison. S_a parameter is an average roughness deviation and S_z is a ten-point surface irregularities height. Additionally, exemplary 3-dimensional profile of surface processed with scanning laser beam velocity $v_l = 25$ m/min is shown in Figure 4.

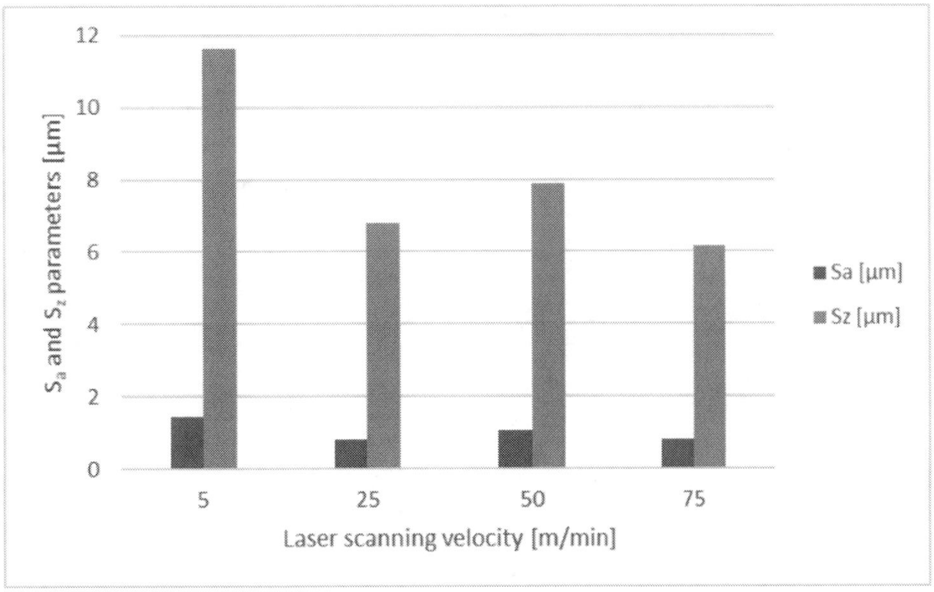

Fig. 3. The influence of laser beam scanning velocity of roughness paramteres S_a and S_z.

Considering the obtained values of roughness parameters it is clear that, as aforementioned, surface is more smooth on surfaces treated with higher scanning laser beam velocities. However, it seems that increasing laser beam scanning velocity from 25 m/min to 75 m/min does not still lower the roughness significantly. Average roughness parameter S_a of surfaces produced with scanning speed greater than or equal to 25 m/min is about 36% lower than for 5 m/min. Average S_z parameter of these is approximately 40% lower than S_z of surface produced with the lowest laser beam scanning velocity.

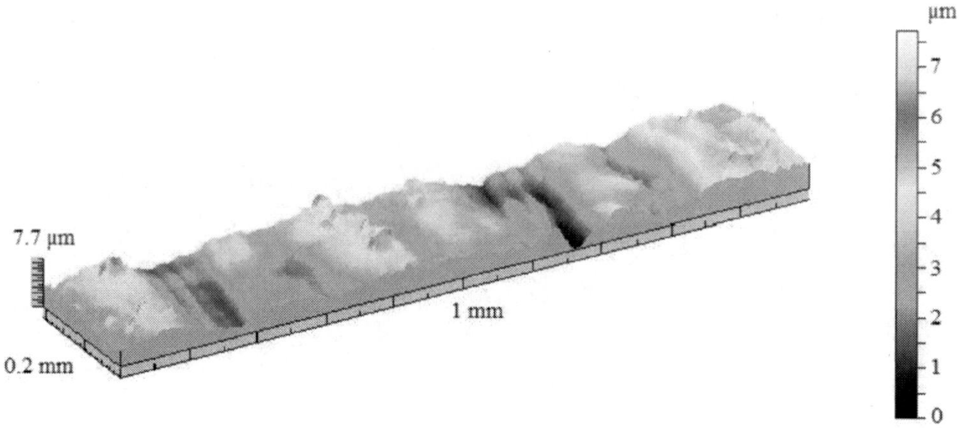

Fig. 4. 3-dimensional roughness profile of surface processed with laser beam scanning velocity v_l = 25 m/min.

3.3 Microstructure investigation

Cross-section microstructure of laser heat treated surfaces are shown in Figure 5. They are sorted in ascending laser beam scanning speed order. Figure 6 presents more detailed views of laser tracks produced with scanning velocities equal to 5 m/min and 75 m/min. The areas shown in Figure 6 are marked with white rectangles in Figure 5.

It can be clearly seen in Figure 5 that increasing laser beam scanning velocity results in decreasing the depth of laser tracks. Their depths were measured and obtained results and average values are given in Table 4. It is worth noting that laser track depth lower approximately 74% with enhancing laser beam scanning velocity from 5 m/min but remain almost constant with increasing the scanning speed up to 75 m/min.

Table 3. Laser tracks' depths for different laser beam scanning velocities.

v_l [m/min]	Number of measurement, results [µm]					Average [µm]
	1	2	3	4	5	
5	355	366	366	363	360	**362**
25	93	94	90	99	94	**94**
50	93	96	90	94	92	**93**
75	92	94	94	35	93	**94**

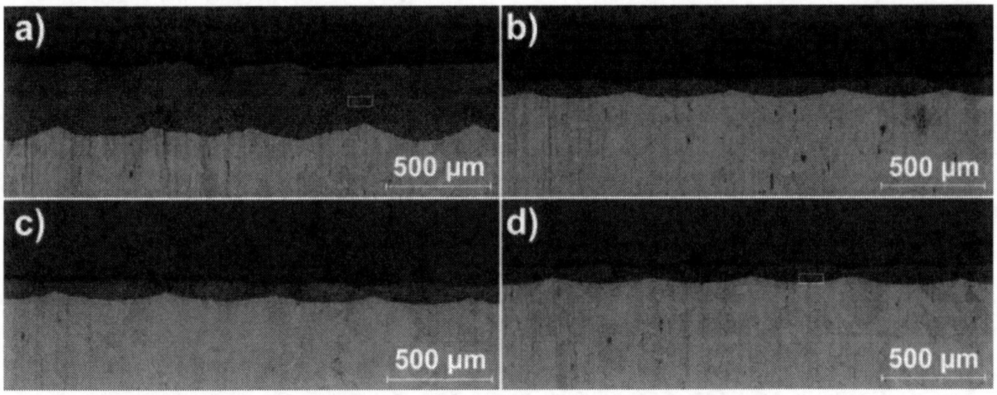

Fig. 5. Microstructures of multiple laser tracks manufactured with: a) $v_l = 5$ m/min, b) $v_l = 25$ m/min, c) $v_l = 50$ m/min, d) $v_l = 75$ m/min.

As observed in Figure 6, the morphology of laser tracks includes column crystals of different lengths. These crystals are mostly oriented in the direction of heat dissipation. Column crystals in Figure 6a are larger than these in Figure 6b. Heat dissipation is directed mainly to the surface but also to the boundary between laser track and substrate.

It was found that all laser tracks are free of cracks and voids. This provides evidence that if the welding process is carried out with parameters used in this research, there are no critical stresses in material during quick solidification. For the sake of confirmation that after laser heat treatment properties of Monel 400 remain constant microhardness measurement was carried out, its results are described in paragraph 3.4.

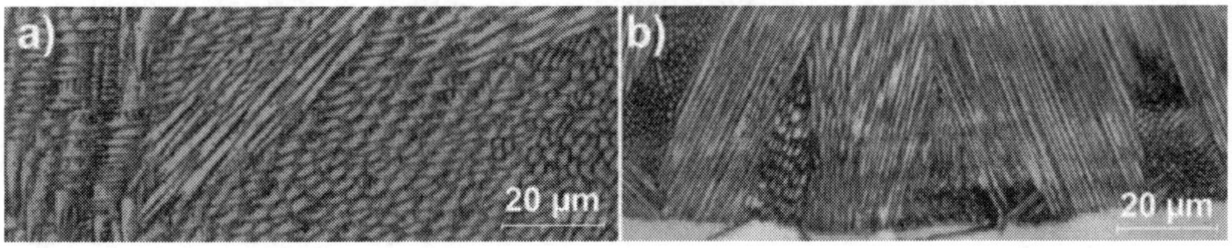

Fig. 6. Microstructures of laser tracks manufactured with: a) $v_l = 5$ m/min, b) $v_l = 75$ m/min.

3.4 Microhardness measurement

Results of microhardness testing carried out on cross-sections of Monel 400 after laser heat treatment are shown in Table 4. Every specimen shows three indentations fabricated in laser track area and three in the substrate. In Figure 7 a dark-field picture of specimen treated with $v_l = 50$ m/min is shown for comparison of microhardness indentation sizes in laser-heated area and the substrate.

Average microhardness of laser tracks is 165.3 HV0.1 and of substrate – 163.2 HV0.1. Average microhardness of laser tracks was calculated on the basis of values given in Table 4. An insignificant difference between these two values leads to conclusion that laser heat treatment of Monel 400 does not affect its microhardness. However, microhardness of laser tracks produced with $v_l = 5$ m/min is slightly lower than the others and it was interpreted as a result of larger column crystals occurring.

Table 4. Average values of microhardness measurement on specimens laser heat treated with different laser beam scanning velocity.

v_l [m/min]	Average laser track's HV0.1	Average substrate's HV0.1
5	159.5	163.2
25	168.9	
50	162.8	
75	169.9	

Fig. 7. Microhardness indentations fabricated in laser track area and the substrate of specimen processed with v_l = 50 m/min.

4 Conclusions

The paper presents an investigation of influence of laser heat treatment carried out with different laser beam scanning velocities on Monel 400 alloy. The research results led to the following conclusions:

1. Increasing the laser beam scanning velocity from 5 m/min to 25 m/min causes the laser track dimensions reduction by approximately 74% but further enhancing of the scanning speed does not affect the depth of heated zone.
2. Laser heat treated surfaces are more smooth if laser beam scanning velocity equal to 25 m/min is applied for the process. However, increasing laser beam scanning velocity to 50 m/min and 75 m/min have no influence on the roughness.
3. Laser tracks manufactured in this research are free of cracks. It proves that there are no critical stresses in material during quick solidification if the process is carried out with aforementioned parameters.
4. Microhardness of laser heat treated Monel 400 does not change as a result of material remelting. It is crucial for producing weldments whose mechanical properties are identical to base metal parameters.

5 References

1. http://www.specialmetals.com/assets/smc/documents/alloys/monel/monel-alloy-400.pdf: Monel alloy 400, Publication Number SMC-053, Special Metals Corporation (2005)
2. K.D. Ramkumar, N. Arivazhagan, S. Narayanan, D. Mishra, Hot Corrosion Behavior of Monel 400 and AISI 304 Dissimilar Weldments Exposed in the Molten Salt Environment Containing Na_2SO_4 + 60% V_2O_5 at 600°C, Materials Research **17(5)**, 1274-1284 (2014)
3. J.H. Rosenfeld, N.J. Gernert, Life Test Results for Water Heat Pipes Operating at 200°C to 300°C, AIP Conf Proc **969**, 123-130 (2008)
4. D.B. Sarraf, W.G. Anderson, Heat pipes for high temperaturę thermal management, Proceedings of the ASME InterPack Conference **1**, 707-714 (2007)
5. L.E. Shoemaker, G.D. Smith, A century of Monel metal: 1906-2006, JOM-J Min Met Mat S **58(9)**, 22-26 (2006)
6. http://www.upenn.edu/pennpress/gallery/prr/prrpix.html: The Pennsylvania Railroad, Volume 1: Building an Empire, 1846-1917, Albert J. Churella (2012)
7. B. Adak, P. Nash, D. Chen, A. Swinglo, Microstructural characterization of laser cladding of Cu-30Ni, J Mater Sci **40**, 2015-2054 (2005)
8. A. Bartkowska, M. Kukliński, P. Kieruj, The influence of laser heat treatment on the geometric structure of the surface and condition on the surface layer and selected properties of Waspaloy, MATEC Web Conf **121**, 03006 (2017)
9. D.B. Sarraf, W.G. Anderson, Heat pipes for high temperaturę thermal management. Proceedings of the ASME InterPack Conference **1**, 707-714 (2007)
10. D. Przestacki, A. Bartkowska, M. Kukliński, Influence of laser heat treatment on microstructure and properties of surface layer of Waspaloy aimed for laser assisted machining. Int J Adv Manuf Tech **93(9-12)**, 3111-3123 (2017)

Stability of roof trusses stiffened by trapezoidal sheeting and purlins

Natalia Korcz-Konkol[1,*], and *Piotr* Iwicki[1]

[1]Gdańsk University of Technology, Faculty of Civil and Environmental Engineering, Narutowicza 11/12, 80-233 Gdańsk, Poland

Abstract. In the article linear buckling analysis of a set of steel trusses braced by purlins and trapezoidal sheeting are conducted. The buckling load factor due to the height of a corrugated sheeting profile is investigated in parametric studies. The minimal height of trapezoidal sheeting required for preventing the sheeting and chords of the trusses against the buckling is obtained. Two groups of models are considered: "axial" model as a simple one and "eccentric" model as a more complex one. In the second group of models, eccentricity between the top chord of the trusses and purlins is considered, by means of equivalent beam elements. The differences between models are indicated and the results are discussed.

1 Introduction

It is well-known that sheeting can stabilize purlins against distortional buckling [1]. Additionally, according to [2], if fastening of the sheeting to the structure meets specific requirements, it can be considered as the diaphragm, which affects the stiffness and spatial character of work of the structure and, by extension, deflections and cross-sectional forces of particular structural members. As a result, the corrugated sheets of a roof apart from bearing vertical loads are part of bracing which stabilizes not only purlins but also an entire structure, e.g. roof trusses. Large research has been conducted on bracing requirements or effect of corrugated sheeting stiffness on buckling load of various elements of structures (e.g. purlin, top chord of the truss) connected directly to sheeting, as e.g. [3, 4, 5, 6]. The problem of bracing requirements of trusses was also investigated in experimental research [7, 8]. In [9] the influence of the diaphragm effect on the behaviour of portal frames with purlins and cladding was analysed.

The present research focused on the buckling study of a set of the trusses stiffened by purlins and trapezoidal sheets. The aim is to find parameters of the roof diaphragm (the height of the trapezoidal sheet profile) which stabilize the cladding and the chords of the trusses against buckling. The eigenvalue problem is solved in order to obtain the roof buckling load factor. The influence of the corrugated sheeting and the purlins on buckling load factor of the roof is observed. Additionally, the differences in two groups of models of the roof (simple and complex one) are indicated.

* Corresponding author: natkorc1@pg.edu.pl

Two-side fastening of the sheeting (instead of four-side fastening) is taken into account, in the case of roofs with purlins it occurs more frequently in engineering practice. The sheeting is fastened to the purlins, which rest on the top chords of the trusses. There is no direct connection between sheeting and trusses except the connection with gable trusses, which is performed using special elements, the so-called shear connectors.

2 Structure description

The roof shown in Fig. 1 and Fig. 3 is considered in the parametric analysis. The model was built following the structure studied in [5]. The main structural elements of the roof are five 24 m long trusses with the depth in the middle equal to 1,61 m. Truss chords and two compressed diagonals near the supports are made of 2L profiles (top chord - 2L 90×90×9, bottom chord - 2L 80×80×8 and two diagonals - 2L 65×65×7). Other diagonals consist of U65 profile. The truss spacing is 6 m. Roof decking made of trapezoidal sheets rests on Z-purlins located every second truss joint (2,4 m spacing, Z 250×60×2 profile). The geometry of the trapezoidal cladding profile is presented in Fig. 2. It was assumed that the height h of the corrugated sheeting profile may vary between 20 mm and 70 mm, other geometric parameters are constant.

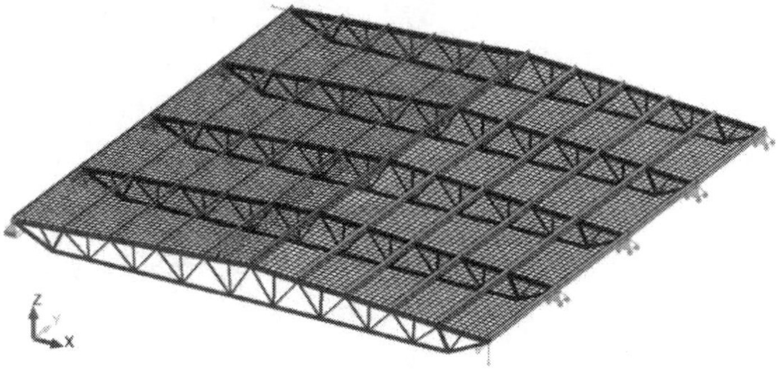

Fig. 1. Roof structure - axonometric view.

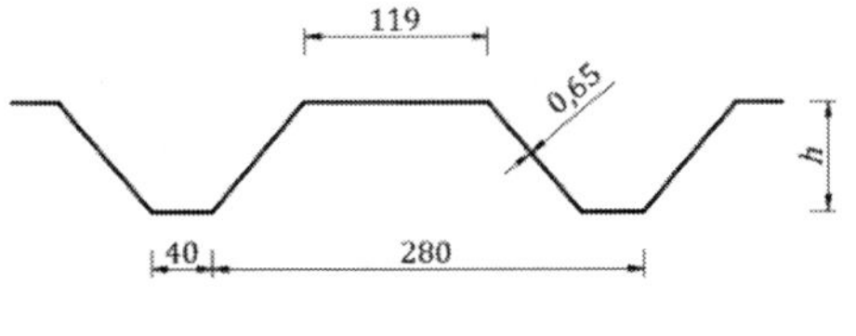

Fig. 2. Trapezoidal sheeting geometry [mm].

3 Numerical model

Linear buckling analyses of the roof are carried out by means of program Robot Structural Analysis Professional [10]. Truss chords and purlins are applied in numerical models as spatial beam elements with six degrees of freedom in node. The connections between truss elements are rigid except the diagonals that are connected to the truss chords by in-plane hinges. Corrugated sheets are substituted by equivalent orthotropic shell elements.

Trusses, as parts of a single-storey steel building, are supported according to the scheme presented in Fig. 3. The load, corresponding to self weight and snow, is applied on the sheeting surface in vertical direction with the value 1.74 kN/m^2.

Two groups of models are built: the first group with sheeting, purlins and top chord of the trusses modelled at the same plane ("axial" models) and the second group – distinguishing the levels of the roof elements ("eccentric" models).

In the "eccentric" models the approach based on [11, 12] is adopted in order to include the fact of two sides fastening of the sheeting. Equivalent spatial beam elements are implemented with the length corresponding to the eccentricity between the top chord of the truss axis and purlin axis (see Fig. 4). Element with the length of 160 mm is applied. The sheeting is connected directly to purlins (in the axis plane, without any eccentricity included). The spacing of the nodes connecting purlins and sheets corresponds to the spacing of the purlin-sheeting fasteners (case of fastening in every corrugation is taken into consideration). As a result, two layers of the roof structure are built. Cantilevered elements are applied, with rigid connection to the truss top chords and with hinged connection to the purlins (axial and truss-in-plane rotation is released). The scheme of the releases is the result of the approximation of the connection detail of the top chord of the truss (2L profile), the purlin (Z profile) and the sheeting. In the present study flexibility of the fasteners and connections are not included. The cross-section of the equivalent element is not calculated, the rigid profile is chosen (HEA 100 profile).

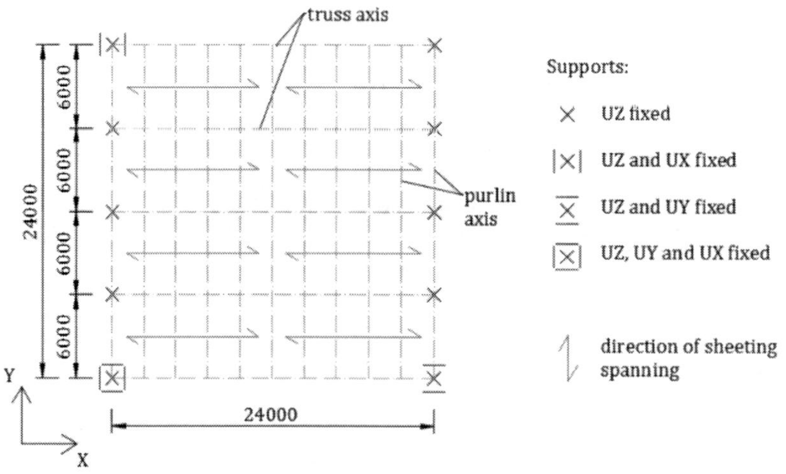

Fig. 3. Roof structure - static scheme [mm].

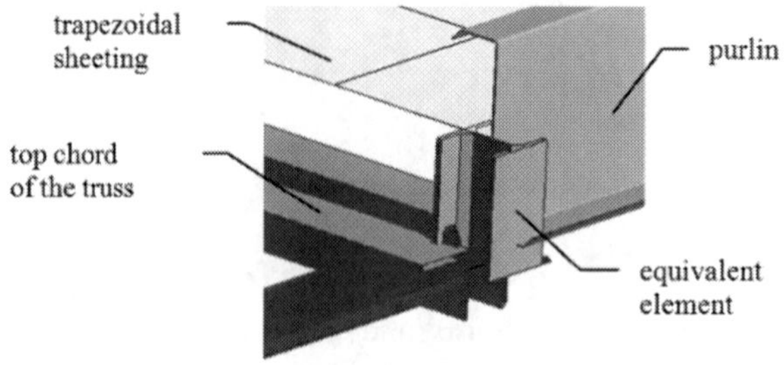

Fig. 4. Detail of the eccentricity connection (top chord – purlin - sheeting).

Substitution trapezoidal sheeting by the equivalent orthotropic shell model is performed using a tool implemented in ARSA software. This tool, available also in many other computational programs for numerical 3D analyses of the structures, computes the stiffness matrix terms based on the trapezoidal plate geometry selected by the user (the user is not obliged to calculate the matrix terms himself [10]. Obviously, this approach is only an approximation. This issue is widely discussed in [13, 14, 15].

4 Results of numerical simulations

The values of the roof buckling load factor due to the height of the corrugated sheeting, obtained for both "axial" and "eccentric" models, are presented in Table 1 and Fig. 5. Samples of the buckling modes corresponding to the lowest values of buckling load acting downward are presented in Fig. 6a-c.

Table 1. Buckling load factor λ of the roof due to the height of the corrugated sheeting.

The height h of the sheeting [mm]	"Axial" model		"Eccentric" model	
	Buckling load factor λ	Buckled element	Buckling load factor λ	Buckled element
20	0,48	sheet	0,42	sheet
25	0,75	sheet	0,65	sheet
30	1,08	sheet	0,94	sheet
35	1,25	diagonal	1,28	sheet
40	1,25	diagonal	1,49	diagonal
70	1,25	diagonal	1,49	diagonal

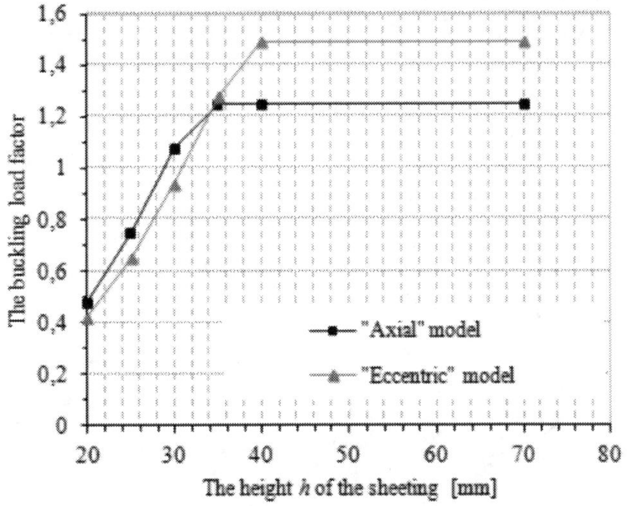

Fig. 5. The buckling load factor λ of the roof due to the height of the corrugated sheeting.

The increase of the buckling load of the roof with the increase of corrugated sheet height was observed in both groups of models. For "eccentric" models it was also found that for sheets with the height 40 mm and higher the critical load does not increase (in the case of "axial" model it occurred for sheets 35 mm high). Minimal depth of the roof decking required for stabilization of the roof is therefore 40 mm according to "eccentric" models and 35 mm according to "axial" models. The buckling mode for a 40 mm high cladding corresponding to downward loading is presented in Fig. 6c. Buckling of the compressed U-profile diagonal occurred. Further increase of buckling load can be achieved increasing the cross-section of this diagonal.

The results of present studies can be compared to the results obtained in [5], where the same trusses were analysed, but in a slightly different support scheme and with the cladding system without purlins (instead of a purlin roof structure). In this case the minimal height of the roof cladding required for stabilization of the truss was 50 mm.

The lowest values of buckling load for corrugated decking with the profile higher than 30 mm in case of "eccentric" model (and 25 mm in case of "axis" model) correspond to upward loading of the roof caused e.g. by wind suction.

Fig. 5 shows the difference between the results obtained for two groups of models. As long as the sheeting is the crucial element in the light of buckling analyses, the "axial" model gives a higher buckling load factor than the "eccentric" model. It is induced by the fact that "axial" model does not consider two sides fastening of the sheeting (fastening only in two edges - to the purlins), which is accounted in "eccentric" models. Instead, when the diagonal begins to be the crucial element, the relation is opposite: the "axial" model gives lower buckling load factor than the "eccentric" model. It can be explained by the fact that the eccentricity of the connection between rafter and purlin included in a numerical model using equivalent element leads to the stiffness increase in the truss nodes. The diagonals buckle mainly in the truss plane, but also slightly out of plane. That can be the reason why the models with equivalent elements applied show the diagonals more rigidly supported compared to diagonals in "axial" models.

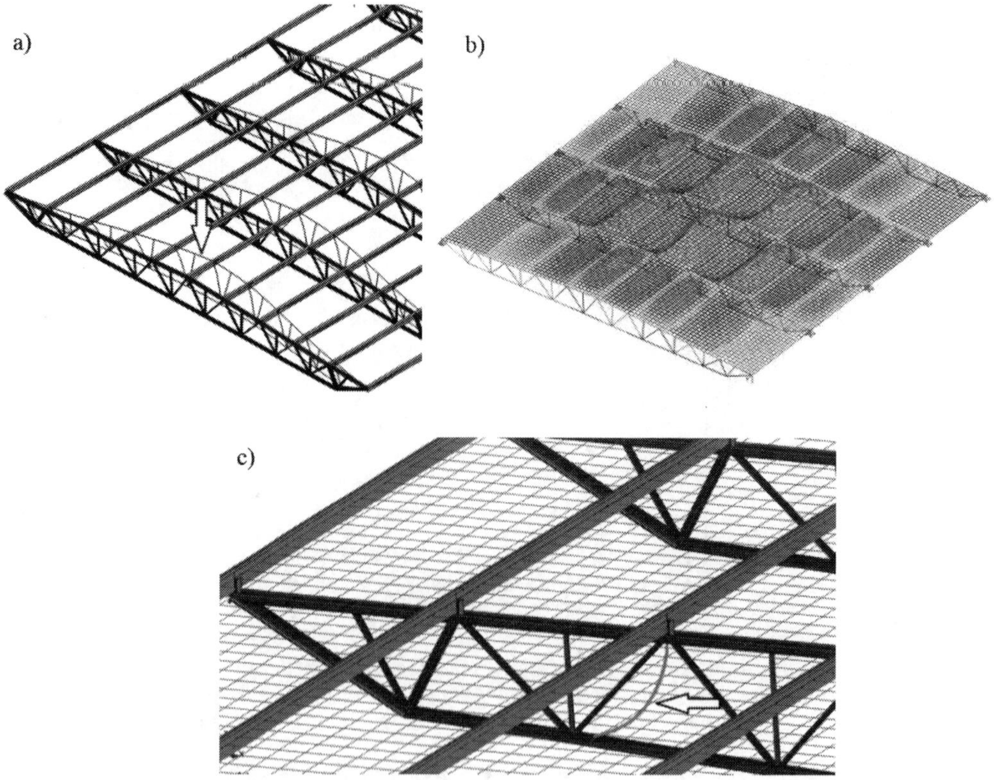

Fig. 6. The first buckling mode corresponding to: a) roof without trapezoidal sheeting ("eccentric" model, λ=1,17), b) roof with trapezoidal sheeting (h = 30 mm, "axial" model), c) roof with trapezoidal sheeting (h = 40 mm, "eccentric" model).

5 Conclusions

The paper presents the results of stability studies on the set of trusses stiffened by purlins and corrugated cladding. In the parametric analysis of the roof a minimal height of the trapezoidal sheeting required for preventing the sheeting and the chords of the trusses against the buckling was obtained and the buckling load factor due to the height of the corrugated sheeting was investigated. Two groups of roof models ("axial" model as a simple one and "eccentric" model as a more complex one) were considered. Their differences were discussed consecutively.

As it was mentioned, applying trapezoidal sheeting by the equivalent orthotropic shell model and including the eccentricity of the members using equivalent spatial beam elements lead to approximation. In the second group of models, the eccentricity between the top chord of the trusses and the purlins is considered and the spacing of the nodes connecting sheeting and purlins corresponds to the spacing of the sheet - purlin fasteners. However, the flexibility of fasteners and connections, e.g. truss - purlin connection, purlin - sheets connection, seam fasteners (sheet - sheet connection) is omitted. Obtaining real flexibility of the connection and evaluation of the influence of them on the stability of the roof requires extra studies. Non-linear static analysis of an imperfect structure and shell model analysis of an entire structure are planned to verify the LBA results.

References

1. J. Bródka, R. Garncarek, K. Miłaczewski, *Corrugated sheets in steel building* (Arkady, Warszawa, 2009)
2. European Recommendations for the Application of Metal Sheeting Acting as a Diaphragm. Stressed Skin Design. ECCS - TC7, TWG 7.5 (1995)
3. B. Gosowski, *Spatial stability of braced thin-walled members of steel structures*, Journal of Constructional Steel Research, **59**, 839-865 (2003)
4. A. Biegus, D. Wojczyszyn, *Buckling length of chords out of the truss plane.* Inżynieria i Budownictwo, **11/2004** (2004)
5. P. Iwicki, *Stability of roof trusses stiffened by corrugated sheets,* In W. Pietraszkiewicz, I. Kreja (Eds), *Proceedings of the 9th SSTA Conference "Shell Structures. Theory and Applications, Vol. 2"*, 113-116 (CRC Press/Balkema, 2010).
6. A. Biegus, *Trapezoidal sheet as a bracing preventing flat trusses from out-of-plane buckling*, Archives of Civil and Mechanical Engineering, **15**, 735-741 (2015)
7. J. Jankowska-Sandberg, J. Kołodziej, *Experimental study of steel truss lateral-torsional buckling,* Engineering Structures, **46**, 165-172 (2013)
8. M. Krajewski, P. Iwicki, *Analysis of brace stiffness influence on stability of the truss,* International Journal of Applied Mechanics and Engineering, **20**, No.1, 97-108 (2015)
9. Zs. Nagy, A. Pop, I. Mois, R. Ballok, *Stressed Skin Effect on the Elastic Buckling of Pitched Roof Portal Frames,* Structures, **8**, 227–244 (2016)
10. Robot Structural Analysis Professional, *User Manual* (Autodesk Inc 2015), http://help.autodesk.com/view/RSAPRO/2015/ENU/
11. J. Goczek, J. Juchniewicz, Ł. Supeł, *Wpływ pionowego stężenia rygla pełnościennego na zachowanie się ramy płaskiej.* In M. A. Giżejowski et al. (Eds): *Proceedings of the XIth ICMS "Progress in Steel, Composite and Aluminium Structures"* (T&F Group, 2006)

12. M. Gryniewicz, J. K. Szlendak, *FEM model of the steel building roof includes stressed skin diaphragm action effects,* In M. A. Giżejowski et al. (Eds): *Proceedings of The XIII ICMS "Recent Progress in Steel and Composite Structures"* (CRC Press, 2016)
13. D. Wennberg, P. Wennhage, S. Stichel, *Orthotropic Models of Corrugated Sheets in Finite Element Analysis*, ISRN Mechanical Engineering, **2011**, Art. ID 979532 (2011)
14. Y. Xia, M.I. Friswell, E.I. Saavedra Flores, *Equivalent models of corrugated panels*, International Journal of Solids and Structures, **49**, 1453–1462 (2012)
15. N. Korcz, E. Urbańska-Galewska, *Influence of fasteners and connections flexibility on deflections of steel building including the stressed skin effect,* Technical Science, **21(2)** (2018)

Influence of sheet/purlin fasteners spacing on shear flexibility of the diaphragm

Natalia Korcz-Konkol[1,*], and *Elżbieta* Urbańska-Galewska[1]

[1]Gdańsk University of Technology, Faculty of Civil and Environmental Engineering, Narutowicza 11/12, 80-233 Gdańsk, Poland

Abstract. The paper presents the influence of sheet/purlin fasteners location (in reference to trapezoidal profile cross section) on shear flexibility of the cladding acting as a diaphragm. Analytical procedures were presented and their limitations were discussed. Next, selected schemes of fasteners location, known from engineering practice, but not included in European codes and recommendations, were analysed numerically in order to observe the differences in shear stiffness of the panel. The cantilevered diaphragm as a part of the roof of a single storey building was chosen for the analysis. The analysis was carried out for the valley-fixed trapezoidal sheeting with wide pans. Corrugated sheets were built by two types of models: full 3D shell model and equivalent orthotropic 2D shell model. Flexibility of fasteners and connections was included in the calculations using equivalent beam finite elements. The results were discussed from the stressed-skin design point of view.

1 Introduction

Trapezoidal cladding of the steel building as a part of the diaphragm affect the stiffness and spatial character of work of the structure and, by extension, deflections and cross-sectional forces of particular structural members. Although the idea of the stressed skin design (the diaphragm design) has been widely known for many years [1], it is still rarely included in engineering practise. Simultaneously the development of numerical tools which support engineers in structural analysis brings new possibilities to take into consideration the stressed-skin effect. More and more effective, verified and simple to apply methods of including the diaphragm effect in structural calculations are still being searched and developed. Moreover, new technological and constructional solutions are the topics of current research studies. [2-7]

The aim of the analysis presented in the paper is to evaluate the influence of the sheet/purlin fasteners spacing (in reference to trapezoidal profile cross section) on shear flexibility of the roof cladding acting as a diaphragm. Selected schemes of fasteners location, known from engineering practice, but not included in European codes and recommendations [1,8], are analysed numerically using computer program called Autodesk Robot Structural Analysis Professional 2015 (ARSA).

* Corresponding author: natalia.korcz@pg.edu.pl

2 Model of the structure – general description

A cantilevered diaphragm built of cold-formed members was taken into consideration, as in [6]. Main axial dimensions were 6,0 x 6,0 m, purlin spacing - 1,5 m (Fig. 1). The analysed structure included two top chords of the trusses (hat cross section profile with dimensions 114 x 121 x 6 mm), five Z-purlins (dimensions 250 x 60 x 2 mm) and trapezoidal cladding of the height h = 45 mm and two thickness variants t (0,6 mm or 0,7 mm). Trapezoidal sheeting with wide pans, fixed to the purlins in wider part of profile was analysed. The valley-fixed cladding was considered as it is recommended in the stressed-skin design. Geometry of the trapezoidal sheeting is presented in Fig. 2.

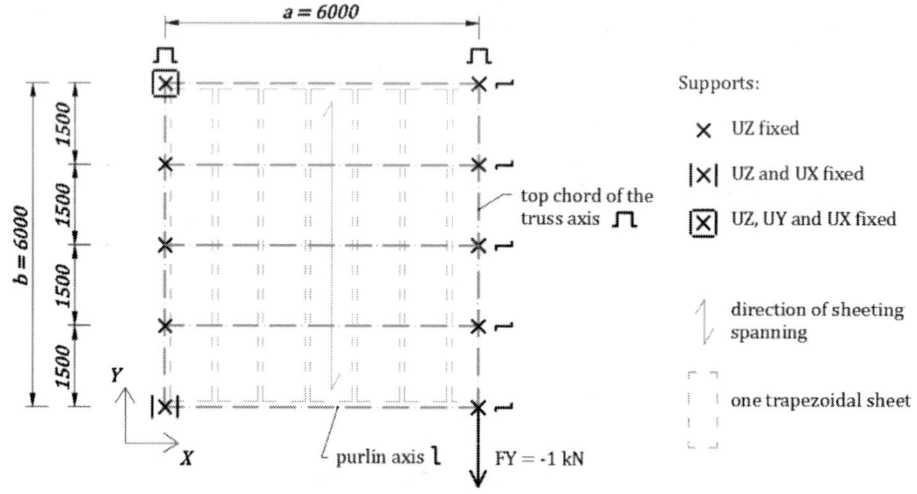

Fig. 1. Static scheme of the diaphragm used in the analyses [mm].

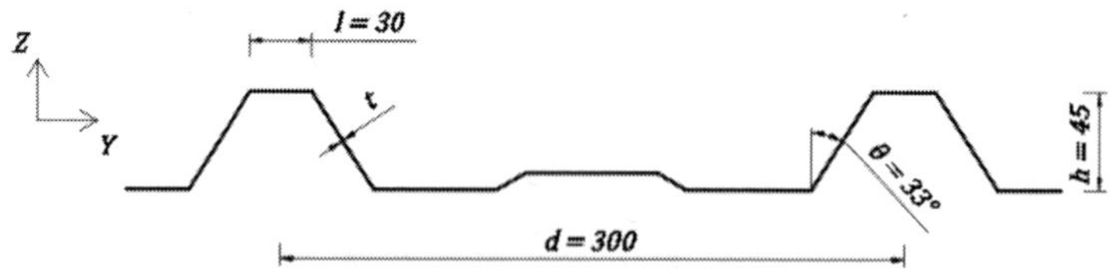

Fig. 2. Trapezoidal sheeting geometry [mm].

Fasteners are important components of the diaphragm. Sheet/purlin fasteners and purlin/rafter connections were taken into account in the analysis. Seam fasteners were neglected. The case of two sides fastening of the sheeting (instead of four sides fastening) was analysed, as the one of a frequent practical occurrence.

Shear load (force with the value of 1 kN) was axial-pointed to the top chord of the truss, as shown in Fig. 1. The supports ensured the cantilevered work of the panel. Moreover, in the points where diagonals join the top chord in the truss girder, vertical displacements were fixed (it was detected that spring supports were not necessary in this particular case).

3 Analytical procedures

The stressed skin design procedures accompanied by practical instructions were provided as recommendations by European Convention for Constructional Steelwork in 1995 [1]. The current version of Eurocode 3 also refers to [1] and discusses diaphragm effect very briefly [8].

An analytical procedure of calculating shear flexibility c of the diaphragm presented in [1] is a component method: flexibility c is the sum of components due to sheet deformation, fasteners and connections deformation and flange forces in the edge members (Table 1). The procedures distinguish two schemes of sheet/purlin fasteners spacing, depending on the location of fasteners in cross-section of trapezoidal sheeting: fastening in every trough (1F) and fastening in alternate troughs (2F), as shown in Fig. 5e-f. The values of particular components of shear flexibility of the diaphragm, calculated for two variants of the thickness t (0,6 mm and 0,7 mm) and for two cases of fastening (1F) and (2F) are presented in Table 1.

Table 1. Components of shear flexibility of the diaphragm calculated analytically according to [1].

Components of shear flexibility - due to:		**Values of shear flexibility [mm/kN]**			
		t = **0,6 mm**		t = **0,7 mm**	
		1F	**2F**	**1F**	**2F**
$c_{1.1}$	profile distortion	0,1491	0,7516	0,1014	0,5112
$c_{1.2}$	shear strain in the sheet	0,0268		0,0230	
$c_{2.1}$	slip in the sheet/purlin fasteners	0,0350	0,0700	0,0350	0,0700
$c_{2.2}$	slip in the seam fasteners	0,0000	0,0000	0,0000	0,0000
$c_{2.3}$	purlin/rafter connections	1,1800	1,1800	1,1800	1,1800
$c' = c_{1.1} + c_{1.2} + c_{2.1} + c_{2.2} + c_{2.3}$		1,391	2,028	1,339	1,784
c_3	axial strain in purlins	0,023			
$c = c' + c_3$		**1,414**	**2,052**	**1,363**	**1,808**

The results of shear flexibility of the analysed diaphragm (Table 1) showed that the difference between fastening in every corrugation (1F) and in alternate corrugations (2F) results most from $c_{1,1}$ component (formula 1), so the component due to profile distortion. This component depends, among other things, on sheeting constant K: K_1 in the case (1F) and K_2 in the case (2F). Moreover, constant K depends on geometry of the sheeting profile – on the dimension ratios l/d, h/d and on web angle θ (Fig. 2).

What is more, the choice between fastening scheme (1F/2F) acts upon the flexibility components due to fasteners and connection deflection. Firstly, fastening every alternate troughs (2F) doubles – in comparison to (1F) fastening - $c_{2,1}$ component (formula 2), so the component due to sheet/purlin fasteners. In addition, the choice between (1F/2F) fastening influences – to much lesser degree – the value of $c_{2,2}$ component (formula 3), so the one due to seam fasteners (omitted in these analyses) and the value of $c_{2,3}$ component (formula 4) due to purlin/rafter connection (in the analysed example this influence is not revealed).

$$c_{1,1} = \frac{a \cdot d^{2,5} \cdot \alpha_1 \cdot \alpha_4 \cdot K}{E \cdot t^{2,5} \cdot b^2} \tag{1}$$

$$c_{2,1} = \frac{2 \cdot a \cdot s_p \cdot p}{b^2} \tag{2}$$

$$c_{2,2} = \frac{2 \cdot s_s \cdot s_p \cdot (n_{sh} - 1)}{2 \cdot n_s \cdot s_p + \beta_1 \cdot n_p \cdot s_s} \tag{3}$$

$$c_{2,3} = \frac{2}{n_p} \cdot (s_{pr} + \frac{s_p}{\beta_2}) \tag{4}$$

where:

a, b	diaphragm dimensions (Fig. 1),
d, t	trapezoidal sheeting geometry (Fig. 2),
K	sheeting constant (explanation in the text),
p	pitch of sheet/purlin fasteners (d in case 1F and $2d$ in case 2F),
E	Modulus of elasticity,
$\alpha_1 - \alpha_3$	factors to allow for the effect of intermediate purlins,
α_4	factors to allow for the number of sheets and fastener position (e.g. 1F/2F),
β_1, β_2	factors to allow for the number of sheet/purlin fasteners per sheet width;
n_{sh}	number of sheets per the dimension a,
n_s	number of seam fasteners per side lap (excluding those which pass through both sheets and the supporting purlin),
n_p	number of purlins,
s_p	slip per sheet/purlin fastener per unit load [mm/kN],
s_s	slip per seam fastener per unit load [mm/kN],
s_{pr}	deflection of top of purlin at purlin/rafter connection per unit load [mm/kN].

Note: parameters to allow for the number and location of purlin/rafter fasteners are underlined.

Fastening in the centre line of valley corrugation is assumed in both cases of sheet/purlin fastening (1F/2F) according to [1]. In the case of trapezoidal sheeting with wide pans, fasteners are usually located close to webs of sheeting due to low plate stiffness of the valley corrugations (Fig. 5.1a-d). As a result the rotational spring stiffness of the purlin C_D and resistance for bending of the purlin stabilised by the cladding increase [9]. In the article the effect of lateral stabilisation of the purlin by sheeting was not considered. Instead, the influence of purlin/rafter fasteners location (close to the web of the sheeting) on shear flexibility of the panel was analysed, which can be important in the stressed-skin design.

4 Numerical model

Axonometric view of the numerical model of the analysed diaphragm is presented in Fig. 4. Top chords of the truss girders and purlins were applied as 3D beam finite elements with real cross sections, corrugated sheets - as shell elements (section 4.1). Fasteners and connections were substituted by equivalent beam finite elements (section 4.2).

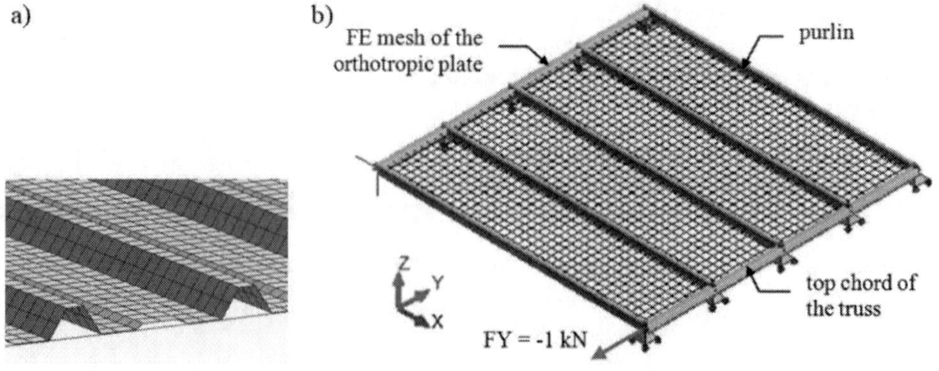

Fig. 4. Diaphragm used in the analyses – axonometric view in ARSA: a) part of fully 3D-modelled trapezoidal sheeting, b) whole structure with orthotropic 2D shell model of trapezoidal sheeting.

4.1 Models of trapezoidal sheeting

Corrugated sheeting was applied in the numerical model using shell elements. Two types of trapezoidal sheeting models were built:

- fully 3D-modelled trapezoidal sheeting - „3D" (Fig. 4a),
- equivalent orthotropic 2D shell model - „Orto".

Substitution trapezoidal sheeting by the equivalent orthotropic shell model is schematically presented in Fig. 5. The idea of this process is to find the equivalent stiffness matrix of the orthotropic 2D shell which reflects different stiffness of steel trapezoidal sheeting in perpendicular directions.

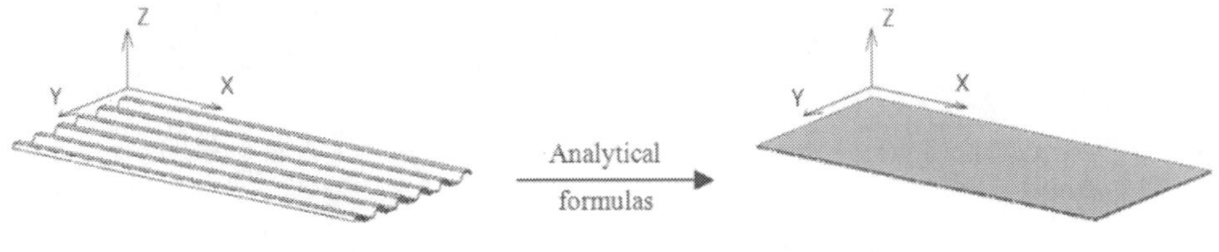

Fig. 5. The idea of substitution the fully 3D-modelled trapezoidal sheeting by the orthotropic 2D shell model [7], based on [10].

The computational programs for numerical 3D structural analysis, e.g. ARSA and RFEM, implement the tools (predefined orthotropic shells/plates) allowing the user to select the trapezoidal plate geometry instead of define stiffness matrix values of equivalent orthotropic shell. The matrix values are calculated by the program and the user is not obliged to calculate them himself. This tool was used in "Ortho" models of the shear panel. However it is worth noting that the analytical procedures for the stiffness matrix of equivalent orthotropic plate for trapezoidal sheeting, both implemented in numerical programs and known from the literature, diverge considerably. This issue is widely discussed in [6, 7].

In comparison to fully 3D-modelled trapezoidal sheeting "3D", applying equivalent orthotropic 2D shell model decreases computing power requirement, time of the calculations and the size of the results file, which is particularly important in case of complex numerical models. Moreover, it reduces the work consumption of the modelling. Obviously this approach is only an approximation. One of the assumptions is the symmetrical cross-section of the cladding (while in practise asymmetrical cross-sections of sheeting are more common). The local stiffness changes are not recognised. It was suspected that equivalent orthotropic 2D shell model "Ortho" is too simplified to observe the influence of the sheet/purlin fasteners spacing (in reference to trapezoidal profile cross section) on shear flexibility of the diaphragm. That is why the analysis was carried out for two types of cladding models: "Ortho" and "3D".

4.2 Connections and fasteners in numerical model

Fasteners and connections flexibilities are the components of shear flexibility of the diaphragm (Table 1). In order to take into consideration these flexibilities with simultaneously slight complication of numerical model of the panel and increase of the file size, the approach introduced by [2] was adopted. In this approach theoretical slip s of fasteners and connections and their placement is modelled using equivalent cantilever finite beam elements. Calculations of the cross sections of equivalent finite element were performed in [6,7].

5 Scheme of the analyses

Trapezoidal sheeting (in two variants of the thickness *t*: 0,6 mm and 0,7 mm) was built using two types of models: full 3D model and equivalent orthotropic model (respectively: „3D" and „Ortho" in Table 2), what is explained in section 4.1 in detail. In the case of both models, 6 schemes of sheet/purlin fasteners spacing were analysed and presented in Fig. 5. Schemes a) – d) are the examples of fastener location not included in [1], schemes e) – f) are the examples related to the procedures based on [1] (explained in section 3). While computing shear flexibility of the panel according to [1], local stiffeners of the sheet are not taken into account. That is why stiffeners were neglected also in the numerical models (1F) and (2F).

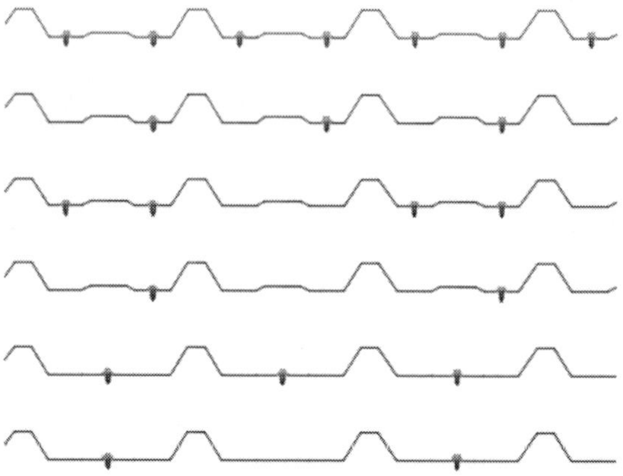

Fig. 5. Schemes of sheet/purlin fasteners spacing: a) scheme 1F/2, b) scheme 1F/1, c) scheme 2F/2, d) scheme 2F/1, e) scheme 1F, f) scheme 2F.

6 Results and conclusions

Values of shear flexibility of the diaphragm obtained for two types of trapezoidal sheeting model: 3D model ("3D") and orthotropic shell model ("Ortho") are presented in Table 2. The differences between "3D" and "Ortho" values (column "Δ") and percentage values of these differences related to the "3D" values (column "%") were calculated. The results are also shown in Fig. 6.

Table 2. Shear flexibility of the diaphragm.

Fasteners Spacing	Shear flexibility of the diaphragm [mm/kN]							
	t = 0,6 mm				*t* = 0,7 mm			
	3D	Ortho	*Δ*	%	3D	Ortho	*Δ*	%
1F/2	1,321	1,207	0,114	8,6	1,217	1,197	0,020	1,6
1F/1	**1,454**	**1,454**[a)]	0,000	0,0	**1,374**	**1,392**[a)]	-0,018	-1,3
2F/2	**2,047**	**1,477**	0,570	27,8	**1,765**	**1,396**	0,369	20,9
2F/1	2,432	2,225[b)]	0,207	8,5	2,023	1,920[b)]	0,103	5,1
1F	1,470	1,454[a)]	0,016	1,1	1,396	1,392[a)]	0,004	0,3
2F	2,230	2,225[b)]	0,005	0,2	1,941	1,920[b)]	0,021	1,1

[a)] the spacing (1F/1) and (1F) are identical in case of „Ortho" models
[b)] the spacing (2F/1) and (2F) are identical in case of „Ortho" models

The values of shear flexibility of the panel in the case of schemes (1F) and (2F) (Table 2) refer to analytical values calculated according to [1] (Table 1). In the case of every corrugation fastening (1F), the differences between analytical and numerical values of the flexibility (related to analytical values) did not exceed 4% and in the case of fastening in alternate corrugation (2F) – did not exceed 9%. A proper agreement between analytical and numerical results was the starting point to analyse the schemes of fastener spacing, not included in [1].

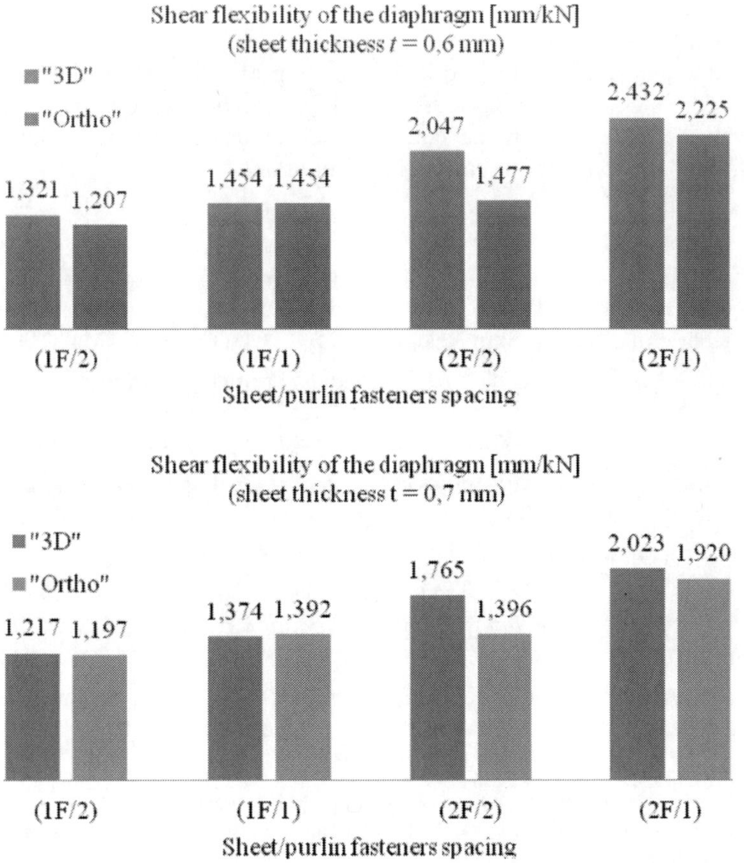

Fig. 6. Shear flexibility of the diaphragm depending on the scheme of sheet/purlin fasteners spacing for two numerical models of the sheeting: „3D" and „Ortho" (description in the text).

Small differences between values of the flexibility achieved for schemes (1F/2) and (1F/1) signalize that - in case of analysed structure - increasing the number of purlin/rafter fasteners (from one per corrugation to two per corrugation) from the stressed skin design point of view is not cost-effective.

The analysis showed that in the case of the analysed structure, the diaphragm in scheme (2F/2) is more flexible than in scheme (1F/1), while the number of sheet/purlin fasteners per the length of the purlin is the same („3D" model: 41% increase of flexibility value was noted for the sheet thickness $t = 0{,}6$ mm and 28% increase for $t = 0{,}7$ mm). The scheme of sheet/purlin fasteners, beneficial for the rotational spring stiffness C_D of the purlin stabilised by the cladding is disadvantageous from the stressed-skin design point of view. The increase of flexibility values were observed only for „3D" models. For "Ortho" models the influence weren't revealed (Fig. 6). Discrepancy between values obtained for both models is interpreted as the confirmation that „Ortho" model includes number of fasteners (component $c_{2.1}$) but is not sensitive to their location (particularly do not consider fasteners location in regard to webs of sheeting and their effect on the sheet deformation - $c_{1.1}$ component).

The "Ortho" models brought good results for schemes of fasteners close to assumptions based on [1], e.g. for the scheme (1F/1). For schemes, which depart from the assumptions of [1], for example for the scheme (2F/2), increase of discrepancies between models "Ortho" and "3D" were observed.

The effect of stabilisation of the purlin by sheeting was not considered in the article. However, it can be derived from the results that it is worth to analyse the influence of the irregularity of sheet/purlin fasteners spacing (e.g. in the scheme (2F/2) as in [9]) on the linear stiffness of the purlin restrained by cladding.

Taking into consideration the work consumption of fully 3D-modelling of trapezoidal sheeting "3D" it is obvious that this model cannot be used in engineering practise. That is why the awareness of the consequences of using equivalent numerical models (as "Ortho" model) among the engineers is so important.

References

1. European Recommendations for the Application of Metal Sheeting Acting as a Diaphragm. Stressed Skin Design. ECCS - TC7, TWG 7.5 (May 1995)
2. M. Gryniewicz, J. K. Szlendak, *FEM model of the steel building roof includes stressed skin diaphragm action effects*, in: M. A. Giżejowski et al. (Eds): *Proceedings of The XIII ICMS "Recent Progress in Steel and Composite Structures"* (CRC Press, 2016)
3. A. L. Joó, L. Dunai, *Full-scale experimental tests on steel frames with various claddings*, in: *Proceedings of The Eighth International Conference on Advances in Steel Structures* (2015)
4. A. Lendvai, A. L. Joó, *Test based finite element development for diaphragm action.* in: *Proceedings of The International Colloquium on Stability and Ductility of Steel Structures* (2016)
5. Zs. Nagy, A. Pop, I. Mois, R. Ballok, Stressed Skin Effect on the Elastic Buckling of Pitched Roof Portal Frames, Structures, **8**, 227–244 (2016), http://dx.doi.org/10.1016/j.istruc.2016.05.001
6. N. Korcz, Modele numeryczne uwzględniające tarczową pracę pokrycia dachowego z blach trapezowych, JCEEA, **XXXIV**, **64 (4/I/17)** (2017). DOI: 10.7862/rb.2017.207
7. N. Korcz, E. Urbańska-Galewska, Influence of fasteners and connections flexibility on deflections of steel building including the stressed skin effect, Technical Science, No. **21(2)2018** (2018)
8. EN 1993-1-3:2006. Eurocode 3. Part 1-3
9. M. Gajdzicki, J. Goczek, Ł. Kwiatkowski, Wpływ rozmieszczenia łączników mocujących poszycie z blachy trapezowej na sztywność obrotową podparcia sprężystego. Inżynieria i Budownictwo, **6** (2017)
10. D. Wennberg, P. Wennhage, S. Stichel, Orthotropic Models of Corrugated Sheets in Finite Element Analysis, ISRN Mechanical Engineering, **2011** (2011) DOI:10.5402/2011/979532

Comparative analysis of different numerical models of a steel radial gate

Krzysztof Brusewicz[1] , and *Robert* Jankowski[1,*]

[1]Gdańsk University of Technology, Faculty of Civil and Environmental Engineering, Narutowicza 11/12, 80-233 Gdańsk, Poland

Abstract. Hydrotechnical structures are important components in water management system and general flooding safety. Their reliability should be ensured since potential damage might lead to catastrophic consequences. Weir gates are considered to be highly vulnerable elements of each hydro power plant, with regard to its dynamic resistance. The aim of the paper is to compare different numerical models and their influence on the results of a computational modal analysis of a steel radial gate. The investigation has been conducted for models using beam and shell elements, while assuring the same geometry, material properties and boundary conditions. The results of the comparative computational analysis indicate that the eigenmode shapes are similar for both models, while the corresponding eigenfrequencies are considerably different. These differences are important from the point of view of dynamic analysis, especially that the first eigenfrequency falls within the energy range typical for earthquakes and mining tremors.

1 Introduction

Hydrotechnical structures are planned for long term operation and require a large investment of money and time to build and manage, which is not the same as for other structures [1, 2]. Their placement and reliability are a vital component in a country water management system and general flooding safety. In this regard, hydro power plants are considered among the most important structures. They provide the country with cheap and accessible electricity; they are also a part of general water management system. Their reliability is of great importance, therefore they should be able to withstand extreme conditions. In order to secure the infallibility of the structure, several analyses should be carried out, including the response to dynamic loads caused by earthquakes or mining tremors.

One of the most vulnerable elements of each hydro power plant, with regard to dynamic resistance, is the weir gate. Weir gates provide the necessary water accumulation level needed for energy production [3]. The failure of a weir gate may cause loss in production or even lead to major damage, with even catastrophic consequences. As mentioned by Peilert in [4], steel radial gates are superior to normal vertical plates with regard to both their weight and construction time. When comparing both constructional solutions for the same

* Corresponding author: jankowr@pg.edu.pl

weir, the radial gates tend to have mass up to 50% lower than vertical plate gates. Similar statements on this subject can also be found in other works (see [3, 5], for example).

In constant development of new tools and software, the struggles of engineers as well as designers take a new shape. The previous restrictions connected with limited hardware strength limiting the calculations have almost completely vanished. The modern researcher can apply various software to suit the needs well, however it is still required not to trust them blindly and to verify the obtained results. The aim of the article is to compare two typical element types used in a numerical model of a steel radial gate in order to find the differences between them in light of modal analysis. The Lanczos method within the ABAQUS software has been used in the study [6, 7].

2 Numerical models

Two geometrically identical (see Figure 1) numerical models have been created. Both of them have the same boundary conditions and affiliated loads. The only difference lies in the chosen element types. The first computational model has been constructed with the use of beam profiles defined by the user. The second model consists of shell elements only, meaning that each beam is made of several shells combined together into a single profile. The connections in both models have been modelled as ties, both with the same nodal values.

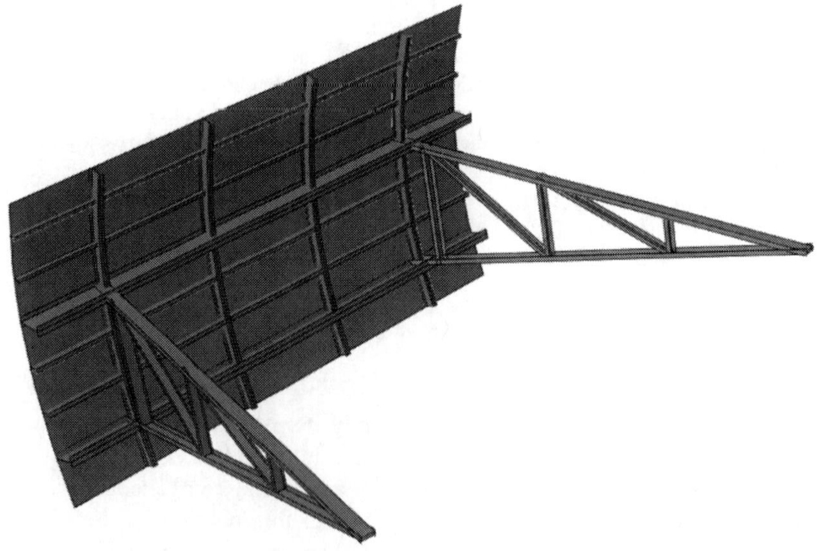

Fig. 1. The model geometry used for modal analysis.

2.1 Geometry and materials

The total height of the steel radial gate, considered in the study, is 4.3 m. Its width is equal to 8.0 m. The radius from the bearing to the radial plate is 6.5 m. The radial plate, which is 7 mm thick, is supported by vertical I180 beams and horizontal C80 and C140 beams. The structure main bearing beams are two I360 girders connected with the arms of the gate. The arms are C180 and C140 profiles. The details concerning the geometry and profiles can be found in [1]. The dimensions of all the profiles are presented in Table 1 (see also description of symbols in Figure 2).

Table 1. Profiles used in the gate.

Profile	h	s	z	t
	[mm]	[mm]	[mm]	[mm]
I180	180.00	82.00	10.00	6.90
I360	360.00	143.00	20.00	13.00
C80	80.00	45.00	8.00	6.00
C140	140.00	60.00	10.00	7.00
C180	180.00	70.00	11.00	8.00

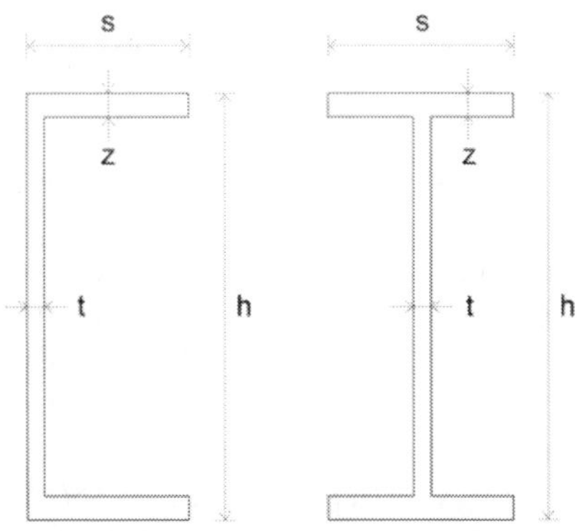

Fig. 2. Used element profiles.

2.2 Element types and mesh

The discretization of the computational mesh for both models is similar. The mesh has been created using global size control. The analysis has been conducted for elements of approximate area of 50 mm^2 for beams and 100 mm^2 for the radial plate. The first model has two types of elements. It employs shell elements (4-node doubly curved shell with reduced integration, hourglass control, finite membrane strains - ABAQUS codename S4R) so as to model the radial plate. On the other hand, beam elements (2-node linear beam in space with the codename B31) were applied for modelling the supporting beam profiles. The second model applies only shell elements for modelling all members of the structure.

2.3 Boundary conditions

An important part of the numerical model of steel radial gates is related to boundary conditions. Due to the possible movement between the radial gate and pillars (the gate is connected to them via bearings), the horizontal displacement is allowed. It is also important to mention that the gate cannot move from its bearings (if such a case occurs, a major disaster could take place). Therefore, two types of boundary conditions have been applied in the model. The first one includes fixed displacements in the bearings (see Figure 3). The second boundary condition allows the vertical displacements of the radial bottom edge to be blocked.

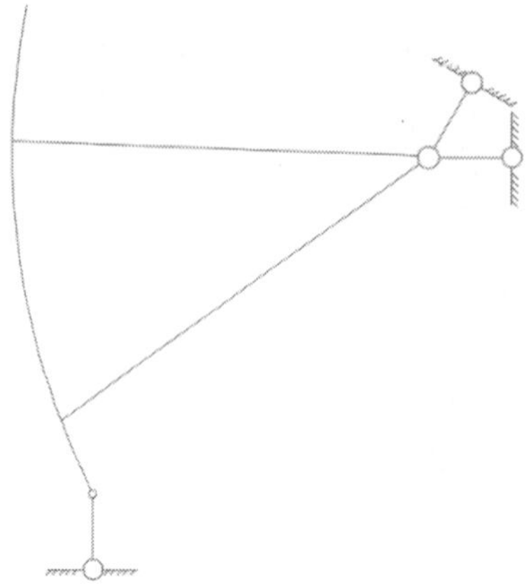

Fig. 3. Boundary conditions in the model.

3 Results and discussion

The conducted modal analysis has resulted in 10 eigenmodes and eigenfrequencies for each of the models. Selected eigenmode shapes are presented in Figures 4-7. The comparison between the eigenfrequencies for both models is shown in Table 2.

The results shown in Figures 4-7 indicate that the eigenmode shapes are similar for both models. On the other hand, as clearly seen in Table 2 that the corresponding eigenfrequencies are considerably different. In the case of the first eigenfrequency, the difference is equal to 0.84 Hz, what stands for the increase by as much as nearly 62% when comparing the shell model with the beam model. This result is very interesting and important from the point of view of the dynamic properties of a structure [8, 9]. The low frequency suggests that the analysed structure is susceptible to dynamic loads, such as earthquakes or mining tremors, which carry most of their energy in the 1-5 Hz range [10-13]. It indicates, that in the case of such dynamic loads, it is possible for the radial plate to pound against pillars (compare [14-16]) causing unsealing, deformation, other difficult to repair damage [17] or even the destruction of the structure. From this point of view, relatively large difference in the eigenfrequency value between the beam and shell model is really important for obtaining the accurate response from the dynamic analysis.

a) b)

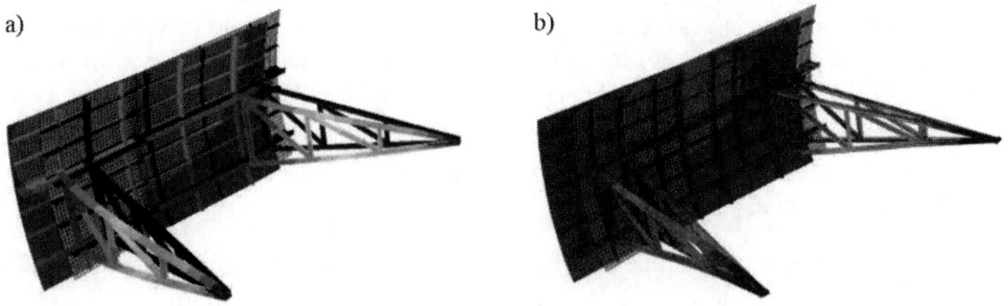

Fig. 4. First eigenmode shape: a) beam model b) shell model.

a) b)

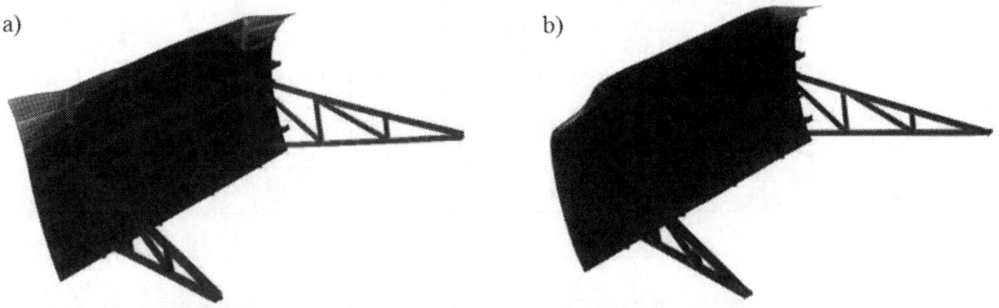

Fig. 5. Second eigenmode shape: a) beam model b) shell model.

a) b)

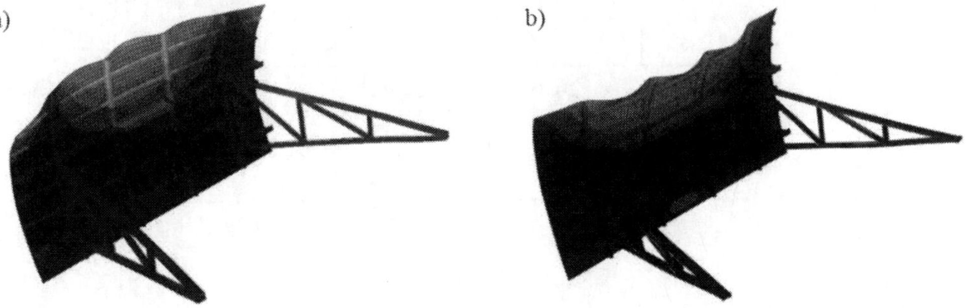

Fig. 6. Fourth eigenmode shape: a) beam model b) shell model.

a) b)

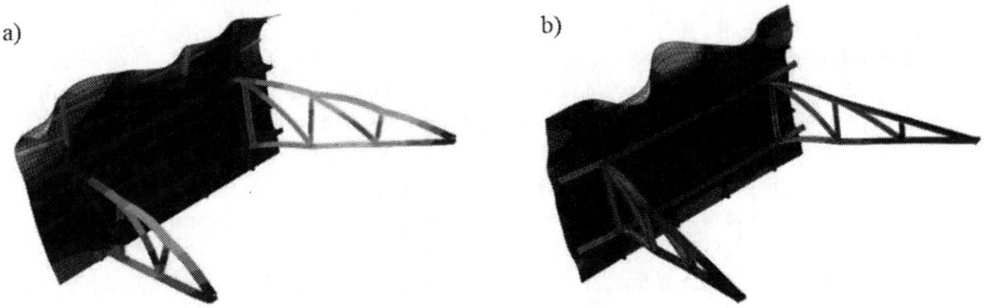

Fig. 7. Tenth eigenmode shape: a) beam model b) shell model.

Apart from the first natural frequency, the rest of the eigenfrequencies have values larger than 10 Hz. Such high frequencies occur rather seldom as components of external excitations [8], so the probability that they will be induced in the structure is very low. Also, it can be seen from Figures 4-7 that the shapes of the higher eigenmodes represent mainly local forms of vibrations (see also [10]).

While comparing the frequencies in Table 2 we can see the cases when one model shows greater stiffness than the other one. The first eigenfrequency value is higher for the shell model than for the beam model, suggesting that the arms modelled with shell elements have a higher horizontal resistance in comparison to beam elements. The frequency values for the next two eigenmodes are lower for the shell model than for the beam model. Since these modes represent vibrations of the radial plate upper corners, the results indicate that the application of beam elements results in the increase in stiffness of the supporting vertical members. However, in the case of eigenmodes no. 4-10, the frequency values are higher for the shell model than for the beam model, as it can be observed in Table 2.

Table 2. Eigenfrequencies for both models.

No.	Beam model	Shell model	Difference
	[Hz]	[Hz]	[%]
1	1.34	2.17	61.94
2	15.23	13.64	10.44
3	15.40	13.67	11.23
4	15.98	16.59	3.82
5	17.44	20.49	17.49
6	18.40	21.86	18.80
7	20.86	22.46	7.67
8	22.92	23.45	2.31
9	23.18	23.68	2.16
10	24.49	25.54	4.29

4 Conclusions

The results of a modal analysis of a steel radial gate, using two different numerical models, have been presented in this paper. The computational investigation has been conducted for models with beam and shell elements, while assuring the same geometry, material properties and boundary conditions.

The results of comparative analysis indicate that the eigenmode shapes are similar for both models, while the corresponding eigenfrequencies are considerably different. These differences are important from the point of view of analysis of structure exposed to dynamic loads. The above statement is especially important in the case of the first eigenfrequency, since it falls within the energy range typical for earthquakes and mining tremors. On the other hand, the rest of the determined eigenfrequencies are somehow large and the probability of being induced is relatively low, since such frequencies occur rather seldom as components of external excitations.

Generally speaking, the application of shell elements allows us to model different structures more precisely comparing to the case when beam elements are used. The resulting differences in eigenfrequencies between two models are mainly attributed to the way of modelling using different element types in ABAQUS software. Moreover, boundary conditions and connections between elements have been idealized in both models. It should be underlined, that accuracy verification of the numerical model should always be conducted comparing the numerical results with the experimental ones. Therefore, further study, focused on experimental tests, is planned to be conducted so as to confirm the numerical results obtained. After experimental verification, a full non-linear dynamic analysis are also scheduled (including some stochastic issues [18-20]) so as to verify the seismic resistance of different types of steel radial gates.

The authors would like to thank TASK CI for the opportunity to use ABAQUS software. Calculations were carried out at the Academic Computer Centre in Gdańsk

References

1. Z. Boretti, *Konstrukcje stalowe w budownictwie wodnym* (Arkady, Warszawa, Poland, 1968)
2. D. Kowalski, B. Grzyl, A. Kristowski, The cost analysis of corrosion protection solutions for steel components in terms of the object life cycle cost, Civil and Environmental Engineering Reports, **26**, 5-13 (2017)
3. K. Fanti, *Budowle piętrzące* (Arkady, Warszawa, Poland, 1972)
4. F.W. Peilert, *Zamknięcia jazowe o napędzie mechanicznym* (Arkady, Warszawa, Poland, 1957)
5. W. Depczyński, A. Szamowski, *Budowle i zbiorniki wodne* (Oficyna Wydawnicza Politechniki Warszawskiej, Warszawa, Poland, 1999)
6. O.C. Zienkiewicz, R.L. Taylor, *The Finite Element Method* (Butterworth-Heinemann, Oxford, UK, 2002)
7. K. Brusewicz, R. Jankowski, Modal analysis of a fish-belly flap type of steel water gate, 11th Conference "Shell Structures: Theory and Applications", Gdańsk, Poland, 11-13.10.2017, **4**, 351-354 (2018)
8. A. Chopra, *Dynamics of Structures: Theory and Applications to Earthquake Engineering* (Prentice Hall, Englewood Cliffs, USA, 1995)
9. M. Ebrahimian, M.I. Todorovska, T. Falborski, Wave method for structural health monitoring: testing using full-scale shake table experiment data. Journal of Structural Engineering, 143, article no. 04016217 (2016)
10. W.F. Chen, C. Scawthorn, *Earthquake Engineering Handbook* (CRC Press, Boca Raton, USA, 2003)
11. J. Przewłócki, W. Knabe, Settlement of a soil stratum subjected to an earthquake, International Journal for Numerical and Analytical Methods in Geomechanics, **19**, 813-821 (1995)
12. Z. Zembaty, Rockburst induced ground motion - a comparative study, Soil Dynamics and Earthquake Engineering **24**, 11-23 (2004)
13. E. Maciag, K. Kuźniar, T. Tatara, Response spectra of the ground motion and building foundation vibrations excited by rockbursts in the LGC region, Earthquake Spectra, **32**, 1769-1791 (2016)
14. C.G. Karayannis, M.J. Favvata, Inter-story pounding between multistory reinforced concrete structures, Structural Engineering and Mechanics **20**, 505-526 (2005)
15. H. Naderpour, R. C. Barros, S.M. Khatami, R. Jankowski, Numerical study on pounding between two adjacent buildings under earthquake excitation, Shock and Vibration, **2016**, article ID 1504783 (2016)
16. P.C. Polycarpou, L. Papaloizou, P. Komodromos, An efficient methodology for simulating earthquake-induced 3D pounding of buildings, Earthquake Engineering and Structural Dynamics, **43**, 985-1003 (2014)
17. T. Falborski, R. Jankowski, A. Kwiecień, Experimental study on polymer mass used to repair damaged structures, Key Engineering Materials, **488-489**, 347-350 (2012)
18. R. Jankowski, H. Walukiewicz, Modeling of two-dimensional random fields, Probabilistic Engineering Mechanics, **12**, 115-121 (1997)
19. J. Przewłócki, *Problemy stochastycznej mechaniki gruntów. Ocena niezawodności* (Dolnośląskie Wydawnictwo Edukacyjne, Wrocław, Poland, 2006)

20. P. Sorn, J. Górski, J. Przewłócki, Probabilistic analysis of a space truss by means of a multidimensional variable description, Archives of Civil Engineering, **61**, 99-123 (2015)

The buckling analysis of a cold-formed steel C-sectional column resting on an elastic foundation

Patryk Deniziak[1,*] and *Piotr* Iwicki[1]

[1] Gdańsk University of Technology, Faculty of Civil and Environmental Engineering, Narutowicza 11/12, 80-233 Gdańsk, Poland

Abstract. The paper is focused on stability analysis of different models of steel cold-formed C-section column. The analysed element may be applied as a primary load-bearing member in steel trusses or silo columns. The column can be used as a support for a wall of roof plates made of corrugated sheets. The wall sheeting that restrains the column against buckling may be considered as an elastic foundation. Due to the design solutions the foundation may be modelled as a series of lateral translational or rotational elastic supports. Such model is used for example in silo column analysis. In the paper linear bifurcation analysis (LBA) of axially compressed columns with different type of elastic foundation is carried by means of ABAQUS software. The analysed thin-walled column (C 250×100×33×4) is regarded as a load-bearing element in a steel hall. The column was modelled by shell elements, the foundation modelled by means of elastic supports. The column length was 5000 mm. The authors present a 3D image of an influence of lateral translational and rotational elastic supports on the buckling resistance of the column.

1 Introduction

In modern steel structures cold-formed elements are no longer used only as secondary sections like roof purlins or wall girts. Many companies patented their own systems that allow to erect complete steel halls including major beams, lattices and columns using only thin-walled sections. The cold-formed element stiffness is smaller than its hot-rolled equivalent what may have an effect on global stability of these slender members, the effect is described in more detail in articles [6, 10]. Consequently, the presence of adjacent elements is taken into account in order to increase the stiffness. The wall or roof sheeting can be analysed as an involved structural part as presented in articles [2, 8]. This approach to the issue is theoretically easy to apply but practically complicated to verify.

The authors have chosen a thin-wall cross-section that is implemented by an international steel structure manufacturer. They decided to analyse the elastic foundation influence on a member resistance. The paper is focused on stability analysis of various models of steel cold-formed C-section column. The analysed element may be also applied

* Corresponding author: patdeniz@pg.edu.pl

as a primary load-bearing member in steel trusses or silo columns. This column can be also used as a support for a wall or roof plates made of corrugated or trapezoidal sheets.

2 Issue description

The wall sheeting that restrains the column against buckling may be considered an elastic foundation located along the element web. Due to the design solutions the foundation may be modelled in the form of lateral translational or rotational elastic supports. This approach is well-known in the process of silo column design [5, 7, 9], included in the EC 1993-4-1 [4].

The analysed thin-walled column cross-section (C 250×100×33) is shown in - Figure 1a. The authors assumed the thickness of the element wall equal to 4.0 mm. The element is 5000 mm high and the location and types of the elastic supports are presented in Figure 1b.

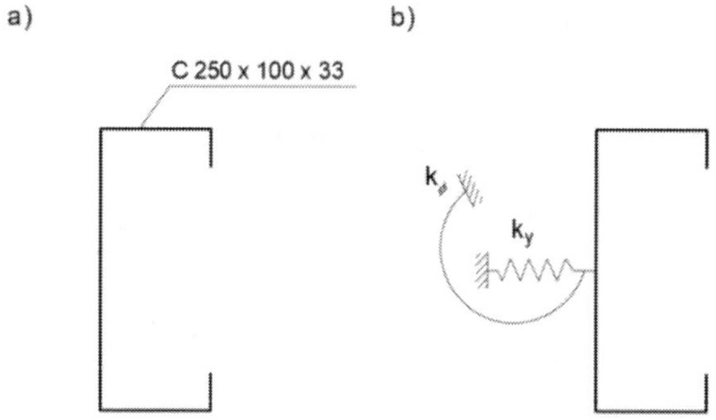

Fig. 1. The a) analysed built-up column cross-section, b) location of elastic supports.

The procedure of determining the elastic foundation stiffness provided by wall sheeting is shown in Figure 2.

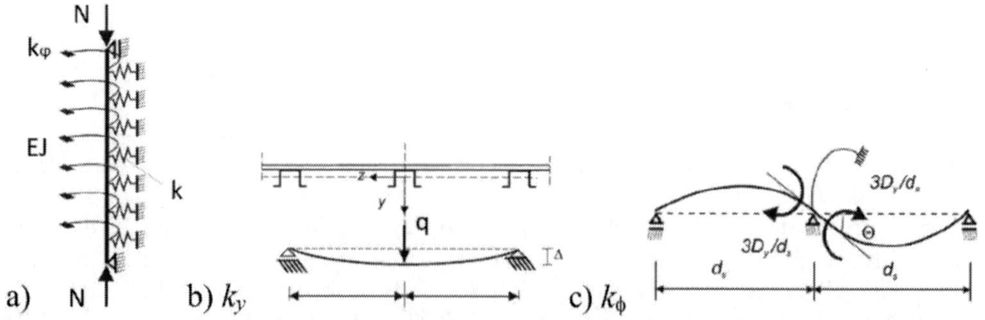

Fig. 2. Thin walled column resting on elastic foundation provided by wall sheets, (a) axially compressed column supported by elastic foundation in lateral direction, stiffness determination of elastic foundation in (b) lateral and (c) rotational direction.

In the paper linear bifurcation analysis (LBA) of axially compressed columns with different types of elastic foundation is carried out using ABAQUS [1] software. Elastic S350 GD+Z steel material was used, according to the EC 1993-1-3 [3]. A column double hinged static scheme is assumed. Top edge shell load is applied of a 1000N resultant force.

The column was modelled by 11.322 shell S4R elements, the foundation modelled by means of point elastic supports. The web arrangement of 17 point lateral translational and rotational elastic supports (spaced every 312,5 mm due to the location of connectors) is presented in Figure 3. The top and bottom edge boundary conditions are marked in a Figure 3 too (the restrained degrees of freedom are highlighted). While the shape variation of the lower edge cross-section is restrained the situation resembles the case of a welded bottom plate. Blocking the possibility of changing the shape of the lower edge cross-section is corresponding to the actual support in the case of welded bottom plate.

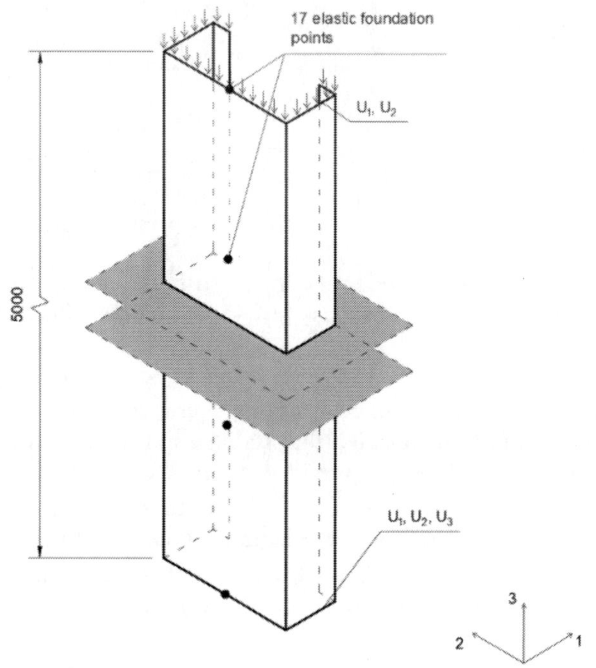

Fig. 3. The web arrangement of point elastic support and restrained degrees of freedom for element top and bottom edges.

In the case of column restraining by steel sheeting between two other columns (Figure 2b and 2c) the lateral translational and rotational stiffness can be compared to simple computational situations commonly known in structural mechanics as a slope and deflection method (Figure 4). A doubled reaction from two adjacent columns (for displacement or rotation equal to one) can be compared to the steel sheeting stiffness assuming rigid connection to the column.

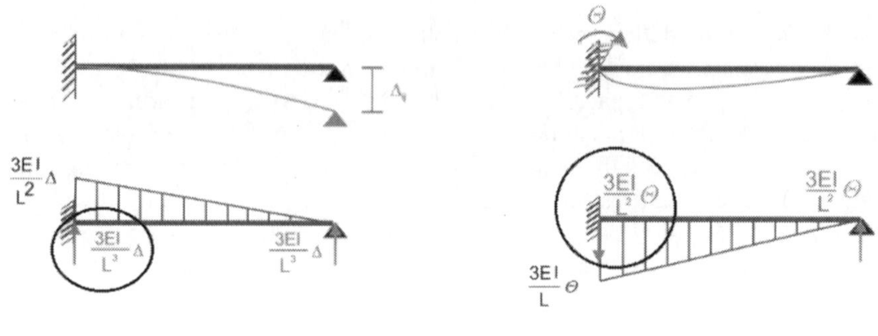

Fig. 4. The slope and deflection method computational situations of the assumed schematic scheme.

3 Results

3.1 Relation between foundation stiffness and the critical load

Several combinations of a different k_y and k_ϕ stiffness values were taken into account. The Abaqus linear bifurcation analysis provided suitable critical load factor for each combination.

Figures 5 and 6 present the evolution of buckling critical load of the column related to the point foundation stiffness. Increasing the lateral stiffness (k_y) of the foundation column enhances the column load capacity, that stabilizes at the level of 350 kN. Reaching lateral stiffness (k_y) equal to 100 N/mm provides the most effective stiffening. Furthermore, while changing the rotational (k_ϕ) stiffness alone does not affect the column resistance. Subsequently, the authors conducted a combined analysis to check the interaction between the lateral (k_y) and rotational (k_ϕ) stiffness related to the total critical load value, what is presented on 3D graph in Figure 7.

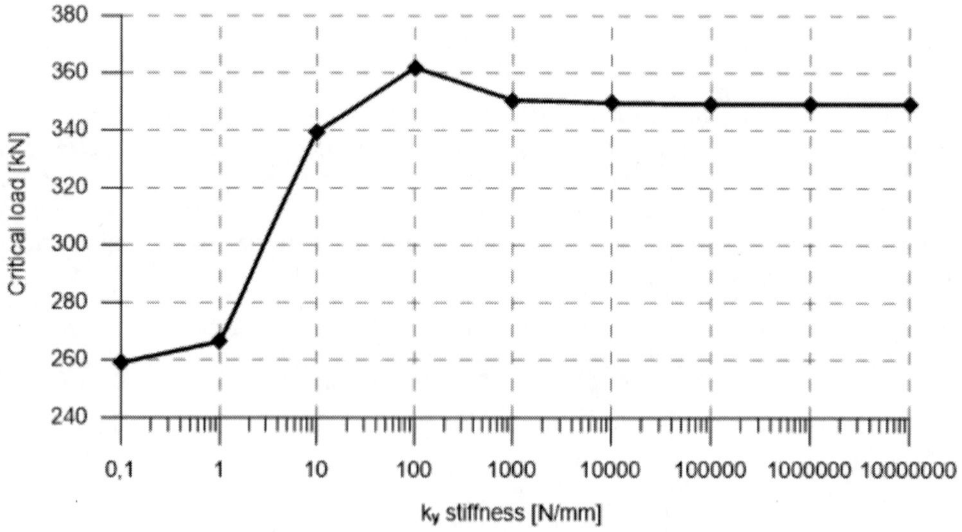

Fig. 5. Critical load dependence on k_y lateral stiffness change ($k_\phi = 0$).

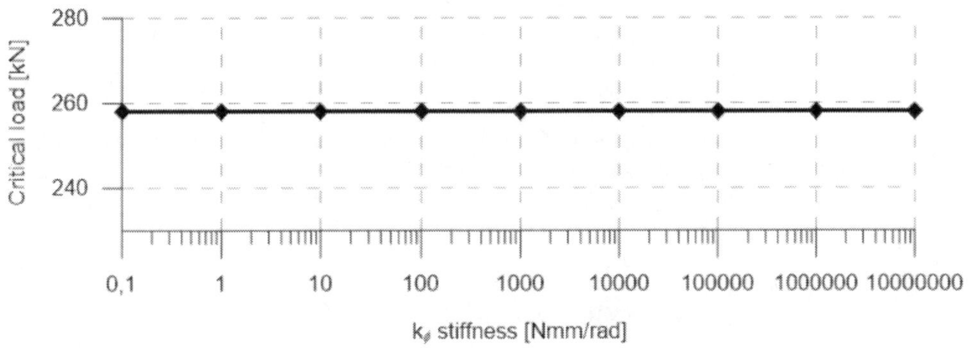

Fig. 6. Critical load dependence on k_ϕ rotational stiffness change ($k_y = 0$).

An existing steel wall or roof sheeting is characterized by both lateral and rotational stiffness. The combination of these two stiffnesses together gives higher critical load than single factor analysis especially after reaching 1.000 N/mm and 10.000 Nmm/rad. Exceeding the values of 10.000 N/mm and 1.000.000 Nmm/rad does not improve the column resistance. Local buckling occurs the column walls, thus further stiffening is not effective anymore.

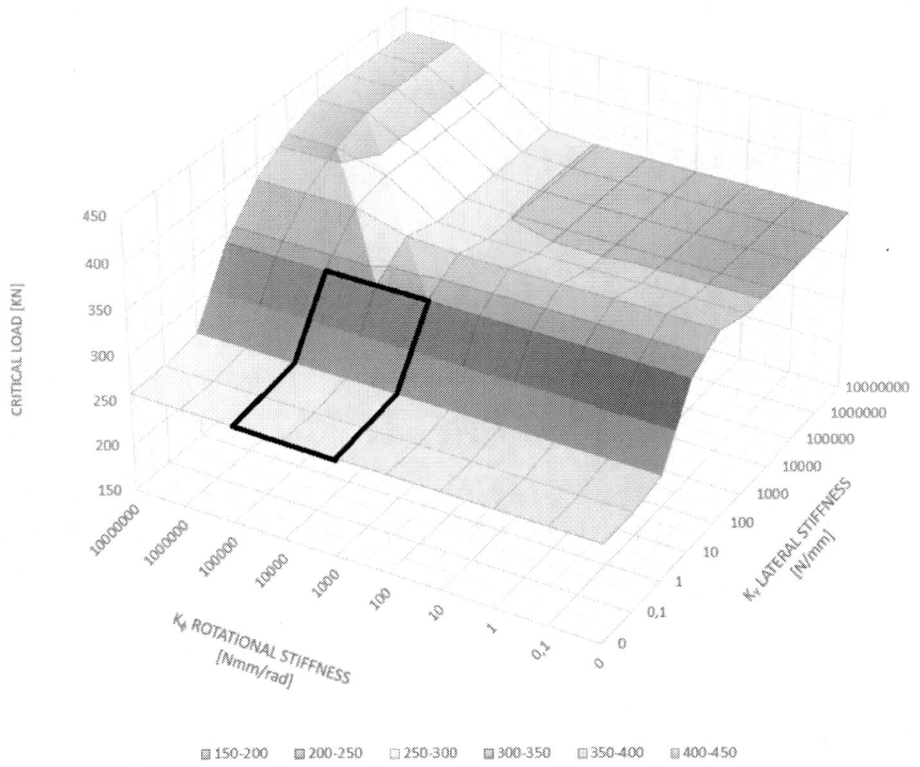

Fig. 7. 3D representation of combined critical load dependent on the lateral and the rotational stiffness.

3.2 Relation between foundation stiffness and the buckling forms

While changing lateral and rotational stiffness different forms of stability loss were observed: flexural, torsional and local buckling appeared.

In the case of–low lateral stiffness (k_y) of the column foundation a global flexural buckling mode was obtained. For a higher foundation stiffness rotational buckling modes were observed. The change in rotational (k_ϕ) stiffness alone does not influence the flexural buckling form until both factors are combined together. Increasing both stiffnesses leads ultimately to over-stiffening and the element buckles locally. The combined relation between the foundation stiffnesses and the buckling forms is shown in a Table 1.

3.3 Selected wall sheeting stiffness checking

The authors selected two trapezoidal sheeting cross-sections (TR 18.124.868 t=0.50 mm and TR 55.235.940 t=1.0 mm) to check a practical range of the stiffness. A moment of inertia for steel sheeting was calculated accordingly to the distance between points where an elastic foundations was applied making the results comparable with the previous assumptions of elastic point foundation. The obtained practical stiffness range is marked in Figure 7 with a black thick line and bold in Table 1.

The ordinary sheeting application increases the analysed column resistance by 32%. Also the buckling form is changing. Thicker and bigger trapezoidal sheeting can prevent against global deformation. This information can be profitable for manufacturers dealing with steel hall erection.

Table 1. Combined relation between the foundation stiffnesses and the element buckling forms.

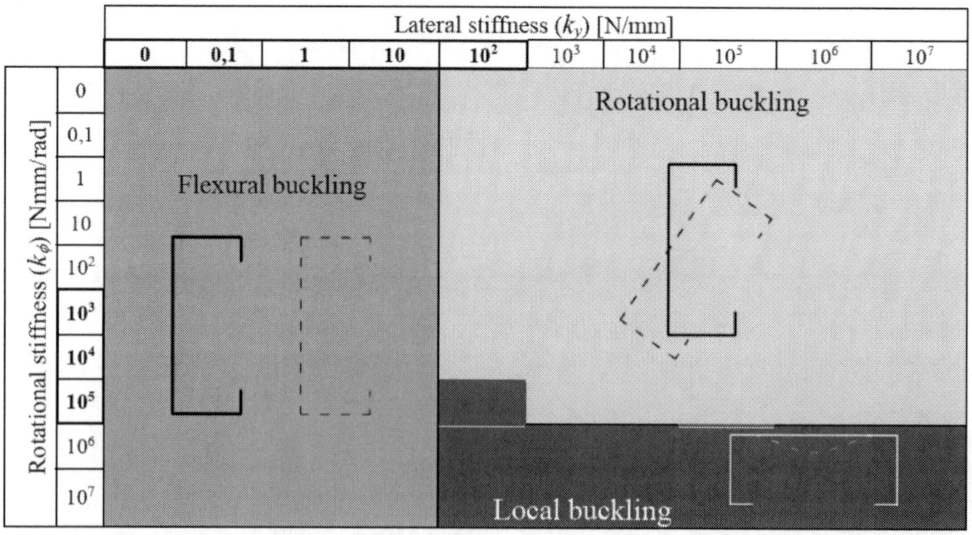

In the upcoming future the authors plan to check other thin-walled columns making the cross-section thickness and the height variable, thus creating a set of clues for designers. The model will be developed and the sheeting will be implemented as a full-scale shield shell Abaqus-based model [1]. It is also planned to analyse a build-up cross-section of C-profile with battens connecting the chords.

In the opinion of the authors the topic seems to be serviceable and perspective as evidenced by numerous state-of-the-art publications of this kind.

References

1. *ABAQUS, Theory Manual, Version 6.8*, Hibbit, Karlsson & Sorensen Inc (2008)
2. C. Basaglia, D. Camotim, R. Gonçalves, A. Graçab, GBT-based assessment of the buckling behaviour of cold-formed steel purlins restrained by sheeting, Thin-Walled Structures, **72** (2013)
3. EN 1993-1-3, Eurocode 3: Design of steel structures - Part 1-3: General rules - Supplementary rules for cold-formed members and sheeting
4. EN 1993-4-1, Eurocode 3: Design of steel structures. Part 4–1: Silos.
5. P. Hajko, J. Tejchman, M. Wójcik, Investigations of local/global buckling of cylindrical metal silos with corrugated sheets and open-sectional column profiles, Thin-Walled Structures **123** (2018)
6. Y. Lu, W. Li, T. Zhou, H. Wu, Novel local buckling formulae for cold-formed C-section columns considering end condition effect, Thin-Walled Structures **116** (2017)
7. K. Rejowski, P. Iwicki, Buckling analysis of cold formed silo column, Mechanics and Mechanical Engineering **20** (2016)
8. J. Santos Franco, E. Miranda Batista, Buckling behavior and strength of thin-walled stiffened trapezoidal CFS under flexural bending, Thin-Walled Structures **117** (2017)
9. M. Wójcik, P. Iwicki, J. Tejchman, 3D buckling analysis of a cylindrical metal bin composed of corrugated sheets strengthened by vertical stiffeners, Thin-Walled Structures **49** (2011)
10. X. Zhou, Z. Liu, Z. He, General distortional buckling formulae for both fixed-ended and pinned-ended C-section columns, Thin-Walled Structures **94** (2015)

Influence of nonlinearities on the efficiency and accuracy of FEM calculations on the example of a steel build-up thin-walled column

Patryk Deniziak[1,*] and *Karol* Winkelmann[1]

[1] Gdańsk University of Technology, Faculty of Civil and Environmental Engineering, Narutowicza 11/12, 80-233 Gdańsk, Poland

Abstract. Due to the increase of computing capabilities of standard processing units, it is possible to perform complex analyses, considering a number of nonlinearities, such as geometric, material and boundary (contact) even on personal computers. In the paper, the authors have analysed the efficiency and accuracy of standard PC's FEM calculations performed in Abaqus CAE 2017 software on the example of a critical load assessment of a thin-walled steel column element with selected nonlinearities. A cross-section shape of a built-up column used by an international steel structures manufacturer was adopted. The analysis serves to check the behavior of their product. Several types of FEM analysis, strictly based on the EN standard regulations were performed. In turn, the relation of computational time to the adopted analysis type was obtained. Moreover, the produced load values in different types of independent calculation were compared and analysed. A possible future development in the field, based mainly on full-scale experimental tests, was also highlighted.

1 Introduction

The computation power of a standard processing unit (a single computer station) is still increasing and has now reached a level that allows solving quite complicated computational tasks even on a personal computer. This introduced a broader adaptation and popularization of numerical modelling in civil engineering. Modern commercial CAE software uses mainly the Finite Element Method (FEM) to solve the structural mechanics problem and efficient graphical post-processing environments to visualize to stress-strain dependencies. The FE methods are strongly affected by computer processing parameters, FE model detailing and undertaken analysis mode, the last factor being the subject of the paper.

The authors have analysed the efficiency and accuracy of FEM calculations performed in Abaqus CAE software (v. 2017) on an example of a structural response of a thin-walled steel column element with an addition of selected nonlinearities. The analysis is step-wise, observing said parameters after each load increment. The Abaqus CAE software allows to consider material, geometric and boundary condition (contact) nonlinearities, which are

* Corresponding author: patdeniz@pg.edu.pl

proven to have significant impact on the model behaviour – their appropriate adoption in the model helps obtaining an exact image of a real structural response.

Authors would like to thank the Academic Computer Centre in Gdansk CI TASK (Polish: *Centrum Informatyczne Trójmiejskiej Akademickiej Sieci Komputerowej)* for sharing the Abaqus CAE software licence.

2 Element description

Since a cold-formed technology has become popular in steel constructions, many companies patented proprietary element shapes. Experimental laboratory tests of some cold-formed elements are presented in [1-5]. The authors have chosen a cross-section shape of a built-up thin-walled column designed and implemented by an international steel structures manufacturer. The analysis serves to check the behaviour of the product. Other manufacturer solutions were also tested and are presented in [6].

The analysed column consists of two chords – the thinnest elements has been chosen out of the company product catalogue. Cross-section wall thickness is equal to t = 3 mm. The inner chord (C 190x80x36x3) is located inside the main chord (C250x100x33x3) and is stiffening the whole element. C-sections are mutually connected by rows of M12 bolts (class of 8.8). The cross-section of the column is shown in Figure 1.

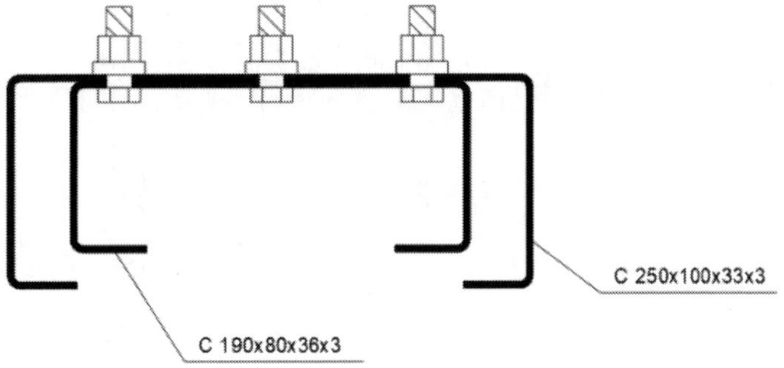

Fig. 1. The analysed built-up column cross-section.

Total lengths of the main elements are equal to 3350 mm (the outer one) and 2700 mm (the inner one). A length difference indicates that the outer chord is not stiffened along entire length. The overview of the column, presenting the spacing between rows of bolts on both elements, is given in Figure 2. A bolt rows arrangement influence on the column resistance was analysed and is presented in [7].

The element is made of S350 GD+Z steel. It should be noted, that the connection between the column and the bottom plate is designed in the same way they are assembled in the real-life structures of this kind.

3 FEM calculations

3.1 Model description

The authors have decided to check how the computational model complexity influences the efficiency and accuracy of the FEM calculation process. Nonlinearities are complement every forthcoming step of the numerical calculations, which makes the model more similar to the real-life construction, yet more complex at the same time.

The column has been represented with square S4R shell elements located on a grid of nodes, their total number is equal to 59.900. As bolts are much stiffer than thick plates, connectors have been modelled as super-stiffened wire elements connected to two circular shells. These circular shells were tied to both inner and outer chord. Stresses and deformations in bolts are not taken into account in this analysis.

A gap between closely-spaced shell planes is modelled as a distance between middle surfaces of shells. In case of chords, the gap is equal to 3 mm, representing the true distance between middle surfaces of steel elements. In fact, chords are closely-spaced and touch each other all along. This normal and frictional interaction was modelled by a proper contact law.

A bottom horizontal plate surface is fully fixed and the main chord top edge has its horizontal displacements blocked. The rest of degrees of freedom in the column head are free (including all rotations). The element is axially compressed by a nodal force subjected in the centroid of the main chord.

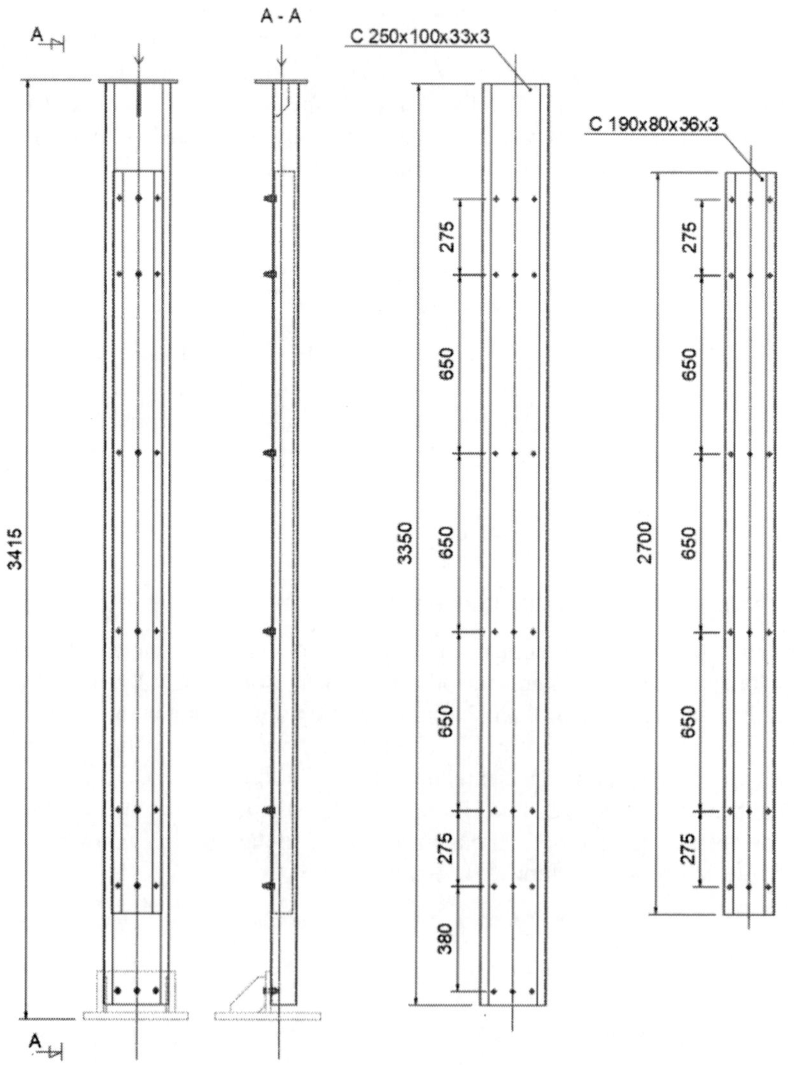

Fig. 2. The analysed built-up column overview, with both chords' dimensions given.

3.2 Types of analysis

Several types of FEM analysis are strictly based on the EN standard regulations. These standards recommend considering imperfections and nonlinearities in the computational model to make it more complex and exact. As previously mentioned, three different kinds of nonlinearities are taken into account: the material, geometric and boundary condition (contact) nonlinearities. The abbreviations used by the authors in the following chapters are taken accordingly to EN 1993-1-6 [8] guidelines. This set is then further expanded by the Authors' analysis considering a set of contact nonlinearities. The list of used types of analysis with their abbreviations are presented in Table 1.

Table 1. Considered types of linear and nonlinear analyses and their abbreviations according to EN 1993-1-6 [8].

Analysis type	**Abbr.**
Linear elastic Analysis	LA
Linear elastic Bifurcation (buckling) Analysis	LBA
Geometrically Nonlinear elastic Analysis	GNA
Materially Nonlinear Analysis	MNA
Contact Nonlinear Analysis	CNA
Geometrically and Materially Nonlinear Analysis	GMNA
Geometrically and Materially Nonlinear Analysis including Contact	CGMNA

The full linear analysis (LA) is the simplest one and should be performed at the beginning. This analysis is fast and does not require much computer resources but is relevant only when the load is relatively small (stresses do not exceed yield strength). This analysis gives an opportunity to check the correctness of all model assumptions and assembly.

Linear bifurcation analysis (LBA) has been carried out as well to determine eigenvalue buckling load. The biggest advantage of this analysis is its fast stability load value convergence, however it is still full linear. The issues of true material and structural behaviour have not been considered yet.

Nonlinear analyses are characterized by the fact that load is divided into increments and the stiffness matrix has to be aggregated again every time after inputting these increments. Furthermore, stiffness is changing with the load increments. This is a feature shown by all real-life engineering structures. Also, the stability path is no longer linear. Nonlinear analysis represents a more realistic response of the considered structure than its linear equivalent.

Table 2 presents availability of the FE most common model used nonlinearities with regard to the type of analysis.

Table 2. The availability of nonlinearities in dependence on the type of particular analyses in FEM.

Abbr.	**Full linear analisis**	**Large deformation theory**	**Nonlinear elasto-plastic material law**	**Boundry conditions nonlinearity**
LA	**YES**	NO	NO	NO
LBA	**YES**	NO	NO	NO
GNA	NO	**YES**	NO	NO
MNA	NO	NO	**YES**	NO
CNA	NO	NO	NO	**YES**
GMNA	NO	**YES**	**YES**	NO
CGMNA	NO	**YES**	**YES**	**YES**

3.2.1 Material nonlinearity

Materially nonlinear analysis employs (among others) an elasto-plastic material law. Steel stress-strain relation is linear only when yield stresses are not reached. Reaching after this point, the curve becomes more unstable and unpredictable. Elasto-plastic material model allows to consider permanent plastic deformation influence on global stability of the element.

Figure 3 represents the stress-strain curve taken into account in the mentioned materially nonlinear analysis. Yield stress is equal 350 MPa and an ultimate strength is assumed as 420 MPa according to [8]. The stress-strain curve is a simplification of a S350 GD+Z steel curve found in [9].

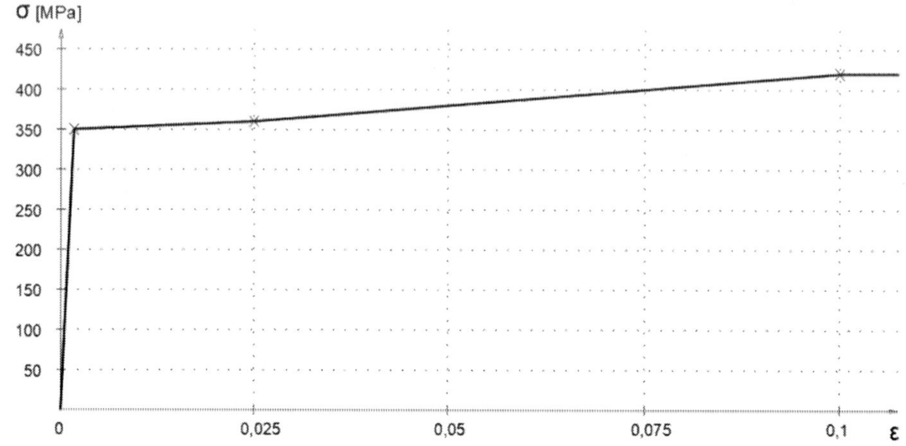

Fig. 3. Stress-strain curve taken into account in the undertaken materially nonlinear analyses.

3.2.2 Geometric nonlinearity

Geometrically nonlinear analysis incorporates a large deformation theory. The element shape is subjected to large-scale deformations, which is considered during the analysis in the form of a relevant variation in the element stiffness. The position of the load is also changing, following the element deformation. The change in the shape of the structure is sometimes very rapid. For example, some convex shells may snap through during the loading processes, changing the stiffness significantly. This phenomenon is presented graphically in [10].

3.2.3 Boundary condition nonlinearity

Boundary condition nonlinearity includes the contact between model parts what is shown in [11]. Without the contact added, both chords are able to penetrate through each other, an effect which is not possible in real-life structure. The contact enables the occurrence of frictional forces, which allows for a shared response of both chords – since the parts of the model work accordingly, the stiffness rises rapidly. Also, for this analysis scenario, the structural deformation is no longer linear.

3.2. Mixed analyses

Combined mixed analyses have also been carried out. Geometrically and materially analysis (GMNA) and geometrically and materially analysis with contact (CGMNA) steps were created. Since the most accurate mapping of the structural response is obtained in these modes of the analysis, the solution of these methods may act as an reference solution.

4 Results and conclusions

All computational steps were solved on a traditional PC, equipped with i7-6700 CPU unit, with a processor working at 3.40GHz. All 4 cores and 8 threads of the processing unit were involved. Before reliable results were received, the CPU unit was being warmed-up. Individual details of each type of the analysis are presented in Table 3.

Table 3. Analysis load, time and maximal deformation obtained for each considered analysis type.

	Load	Time	Max deformation	Max vertical deformation
LA	160 kN 400 kN	50 s. 51 s.	5 mm 12 mm	2 mm 4 mm
LBA	161 kN (minimum value)	1 min 51 s.	1 mm	-
GNA	160 kN 250 kN 319 kN*	1 min 23 s. 1 min 21 s. 2 min 25 s.	7 mm 19 mm 48 mm	2 mm 4 mm 9 mm
MNA	160 kN 250 kN 319 kN 350 kN 400 kN	1 min 20 s. 1 min 19 s. 2 min 30 s. 2 min 01 s. 3 min 45 s.	5 mm 7 mm 24 mm 65 mm 337 mm	2 mm 4 mm 17 mm 46 mm 237 mm
CNA	160 kN 250 kN 319 kN 350 kN 400 kN	4 min 29 s. 4 min 46 s. 5 min 11 s. 5 min 27 s. 5 min 12 s.	5 mm 8 mm 10 mm 11 mm 13 mm	2 mm 3 mm 3 mm 4 mm 4 mm
GMNA	160 kN*	1 min 33 s.	7 mm	1 mm
CGMNA	162 kN*	9 min 19 s.	8 mm	2 mm

* the cases where further loading was not possible due to model's global instability

In all cases of static general analysis in Abaqus, the load applied to the element was gradually increased up to 400 kN. This is a value equal to the maximum load value of a planned experimental compressive tests due to the limitations of both the test stand and the equipment. Although, it should be pointed out, that reaching a load value of 400 kN was not possible in some analysed scenarios due to the global instability of the model. The first load increment was set as 10% of the maximum load. Further increment values were automatically selected by the software.

Geometrical analysis, either free of nonlinearities or combined with nonlinearities, identifies a significant role of second–order effects on the total load–carrying capacity of the column. The bottom connection eccentricities initiates an unintended bending moment that is not neglected in successive load increments. Taking the above mentioned phenomenon into account, a second–order effect – based decrement of the maximum load that the element can carry should be considered. Considering the most complex analysis (GMNA and CGMNA) indicates the total load-carrying capacity of the column equals approximately to 160 kN. Therefore, the level of 160 kN has been set as reference load level for each step.

The first critical load value, obtained for the analysed column is equal to 161 kN. What is worth mentioning, a local deformation of the main chord in its bottom area is observable instead of a global stability loss – it indicates the importance of the second–order effects in the supports both in the model and in planned experimental tests.

The total calculation time increases together with model complexity, as expected. Amongst many factors, the contact consideration was proved to put the strongest impact on the extension of the calculation time – the computational process was exceeding the linear elastic analysis computational time by over 500%. Furthermore, the CGMNA scenario increases the total calculation time by more than 10 times.

In this specific case, the GMNA analysis turned out to be the most favourable one from the time-consuming point of view.

Disregarding the calculation time, the correctness of all mentioned model assumptions should be verified, for example by planned natural scale experimental compressive tests.

5 Future development of the topic

In the presented calculations the load values are obtained solely on the basis of numerical calculations. However, the FEM calculations should only act as a representation of the real-life structural response. The values of numerical calculations should not be relating to each other, but they should be referred to an experimentally obtained value. This said, a full scale experimental test of the column is being prepared, and the results of computations will be then compared to the obtained results.

Also, the calculations should be performed parallelly on two or more computers of similar parameters, this way two interesting conclusions may be specified – the relation of computational time to the computing power of the PC and the mean value of the load-time relation produced on the basis of independent calculation. Moreover, for the power-time relation, such multiple tests on different processing units may indicate if this relation is linear or not.

In the next step, it is also worth checking the calculation scenario of constant load increment for every analysis type blocking the possibility of automatic load increment selecting because small differences may depend on this factor.

An addition of geometrical imperfections in the form of first eigenmode of the column elements to the model is also planned, in order to make the model even more similar to real-life structures of this kind in terms of its response. Implementing this imperfection should verify the obtained load-carrying capacity of the column model and should result in interesting observations. Example of geometric imperfection measurements on cold-formed steel channels is presented in [12].

Due to the fact, that the column is designed and implemented by an international steel structures manufacturer and the analysis serves to verify the structural response of their product, the considered topic may be further extended in many directions, according to the needs of the manufacturer. Thus, the column may be a potential subject of further works.

References

1. M. Dabaon, E. Ellobody, K. Ramzy, Nonlinear behaviour of built-up cold-formed steel section battened columns, Journal of Constructional Steel Research, **110** (2015)
2. Y. H. Lee, C. S. Tan, S. Mohammad, In-plane sway behaviour of slender cold-formed steel bolted frames, Thin-walled Structures, **114** (2017)
3. K. Słowiński, W. Wuwer, *Investigations of closely spaced build-up bars with flexible joints,* The 6th International Conference on Thin Walled Structures (2011)
4. R. Zaharia, D. Dubina, Stiffness of joints in bolted connected cold-formed steel trusses, Journal of Constructional Steel Research, **62** (2006)
5. J. Zhang, B. Young, Numerical investigation and design of cold-formed steel built-up open section columns with longitudinal stiffeners, Thin-walled Structures, **89** (2015)
6. M. Gordziej-Zagórowska, E. Urbańska-Galewska, Ł. Pyrzowski, The validation process of truss model with joint eccentricities, Shell Structures: Theory and Applications, **4** (to be published)
7. P. Deniziak, K. Winkelmann, TS-based RSM-aided design of cold-formed steel stiffened C-sectional columns susceptible to buckling, Shell Structures: Theory and Applications, **4** (to be published)
8. EN 1993-1-6, Eurocode 3 – Design of steel structures – Part 1-6: Strength and Stability of Shell Structures.
9. M. Gordziej-Zagórowska, PhD thesis, *Wpływ mimośrodów na nośność węzłów dźwigarów kratownicowych wykonanych z kształtowników giętych na zimno o przekrojach otwartych* (2017) [in Polish]
10. G. Watts, M.K. Singha, S. Pradyumna, Nonlinear bending and snap-through instability analyses of conical shell panels using element free Galerkin method, Thin-walled Structures, **122** (2018)
11. P. Mika, *Analiza MES zagadnień sprężysto-plastycznych – program Abaqus*, **13** (2011) [in Polish]
12. S. Selvaraj, M. Madhavan, Geometric imperfection measurements on cold-formed steel channels, Wiley Online Library (2017)

Study on 19th-century cast iron columns from the former financial office building in Kwidzyn

Michał Kwasek[1], *Aleksander* Piwek[1], and *Robert* Jankowski[2,*]

[1]Gdańsk University of Technology, Faculty of Architecture, ul. Narutowicza 11/12, 80-233 Gdańsk, Poland
[2]Gdańsk University of Technology, Faculty of Civil and Environmental Engineering, ul. Narutowicza 11/12, 80-233 Gdańsk, Poland

Abstract. The article presents the results for cast iron columns used as floor supports in the former financial office building in Kwidzyn (Poland). The stocktaking drawings provided the general and detailed data about the 19th century designer's decisions. During the research, architectural and structural analysis was carried out. The graphics were used as a starting material for preparing 3D calculation models. The FEM-based numerical simulations were conducted using ABAQUS software. It was shown that the selection of such a structural solution in the building allowed arranging the interior space in an innovative way. The results of the study prove the thesis put forward of a low application extent of the load-bearing capacity of cast iron column used.

1 Introduction

The methods to obtain metal from ores were already known many centuries BC [1], but only the 18th and the 19th century brought technological opportunities that made it possible to cast extremely resistant structural elements of iron alloy, often with beautiful decorative forms [2-5]. The popularity of this material influenced the 19th-century art, utility objects and architecture [6]. The new material was used in construction of new buildings and the old ones partially or completely rebuilt during that period [7].

In the 19th century, cast iron structural elements were frequently selected on the basis of design tables. That method made it possible to determine the required cross-sectional values based on the element span. The simplified tactic increased the efficiency of the design process significantly. It was also related to the manufacturing method of cast iron elements. Due to the technological improvements at the time, identical columns and beams could be quickly obtained. They were efficiently selected and installed in various structures. The drawback of this design method was related to the fact that the load-bearing capacity of elements was not optimized. It is difficult to determine whether the designers made a full application of cast iron resistance. A few publications deal with this issue in general (e.g. [8-13]), even less in relation to the area of Poland (e.g. [14]).

[*]Corresponding author: jankowr@pg.edu.pl

Cast iron was also used to build floors of buildings in the 19th century. Metal beams (often made of cast iron) were supported on brick walls. The filling between the beams was also made of ceramic elements. Quite often, floor beams were also supported on intermediate supports in the form of cast-iron columns. All such solutions made it possible to use longer beams and obtain a spacious interior. The impression of a large space was not obliterated by intermediate supports, which, due to high compressive strength of cast iron, resulted in relatively small cross sections.

The article is focused on cast iron columns used as floor supports in the former financial office building in Kwidzyn (Poland), used now as the municipal library. This property is located at Piłsudski 21 Street, erected in the eclectic style in the decade of 1890s.

2 Architectural analysis

The analyzed building was originally intended for the headquarters of the financial office. It is a brick structure, three-storey high (the top storey is a applicable attic), with basement. In the symmetrical plan, it is possible to distinguish the central part with the entrance and staircase as well as two side wings (Fig. 1, Fig. 2).

2.1 Construction solutions

The floors were probably made of bricks with metal infills using Klein technology (no exposures were made during the site inspection). In some parts of the floor, the barrel vault is visible. In the lateral parts of the building, spacious rooms are located both on the ground floor and on the first floor. These rooms show a nearly rectangular shape, its dimensions are 7.1 x 8 m. Cast iron columns have been installed in pairs in their central part (see Figure 3). They underpin the binding joists providing support to other metal beams. In each room, both columns are positioned in the main beam axis, with an axial distance from each other of 84 cm. The columns on the ground floor and first floor are above each other. Cast iron columns were partially or completely walled up in one of the rooms on the ground floor and in one of the rooms on the first floor, as a results of contemporary rebuilding. The main girder beam, supported by the columns, shows an overall length of approximately 7.4 m and a maximum span of 3 m. The height of the beam is around 270 mm. Its width of about 46 cm was obtained by spreading apart the metal elements and most likely filling the in-between space between them with bricks. Such an approach brought about increasing the width of the support of the perpendicular beams.

2.2 Columns

The columns are 3.5 m high, showing a round shaft of outer diameter of 20 cm. The decorations are modest. The columns have a clearly marked base and a capital, but they do not present any decorative band on the shaft section (Figure 3). The base, of a 25 cm diameter, contains a simple profiling in the upper part, which forms the boundary between it and the shaft. The capital also does not either show any rich form. Visually, it can be divided into two parts, in both parts a cylinder can be indicated. The lower part is separated within the shaft using a half-round moulding of the same diameter. At the top, it ends with a simple profile creating a socket, with a flat disc above. The second part of the capital is set on it, its construction clearly results from structural reasons.

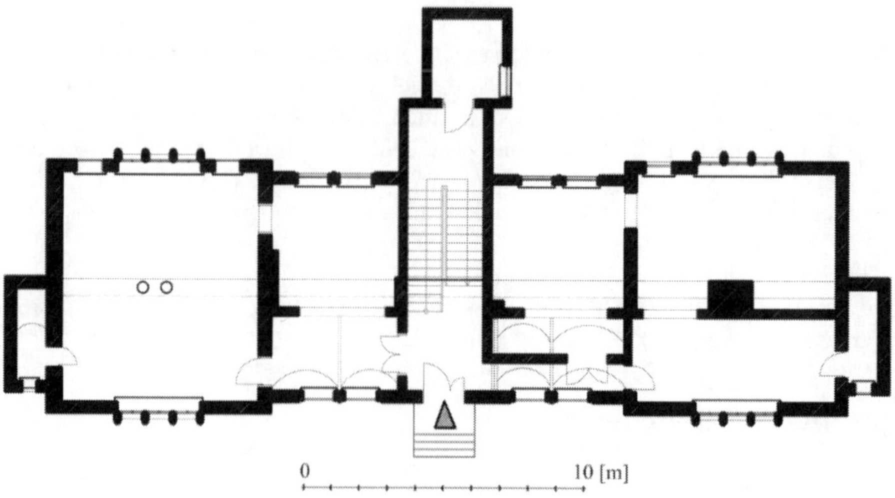

Fig. 1. Plan of the ground floor of the former financial office in Kwidzyn (stocktaking 2017).

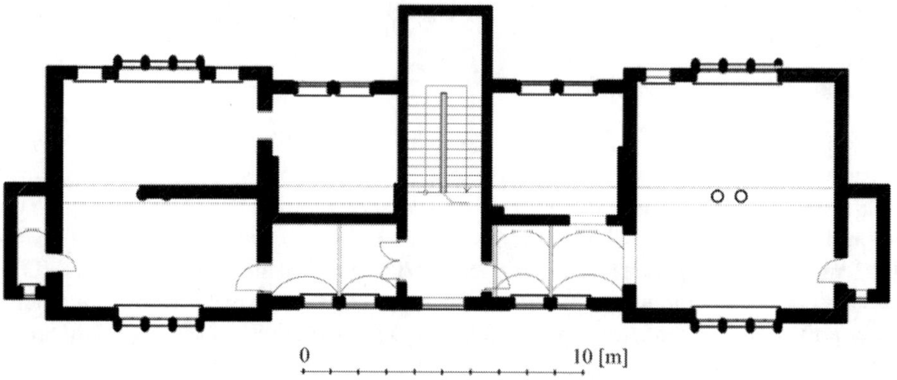

Fig. 2. Plan of the first floor of the former financial office in Kwidzyn (stocktaking 2017).

Each column is topped with a separate cast iron slab (Figure 3A) with dimensions of 26 x 48 cm and a thickness of 4.5 cm. Two stiffening ribs are symmetrically added to the cylindrical element, whose diameter in this part has been increased to 24 cm. They protrude from the cylinder on its whole height. Ribs take the form of 3 cm thick plates with outer curved edges which are mostly "S" shaped (Figure 3B). There is a protrusion at the connection to the plate which forms additional stiffening. The ribs are positioned along the axis of the plate parallel to its longer side. The same axis is set perpendicular to the axis of the binding joist. Such a solution makes the beam's support area bigger. All profiles, the upper plate and stiffening ribs are integral parts of the column. They were created at the stage of casting the pillar.

Increasing the diameter of the pipe (shaft) and adding stiffening elements increased additionally the load-bearing capacity of the support by securing the top plate against bending in the case of non-axial loading forces. The columns are connected to the main beam in the corners of the plate using four bolts passing through the plate and the bottom flanges of the section of the main floor beam.

Fig. 3. Cast iron columns in the building of the former financial office in Kwidzyn: A - view of the beam's double support, B – detail of the column capital, C - base detail, D - general view.

3 Structural analysis

Strength calculations were carried out of the cast iron column, complemented by a simplified numerical analysis using the FEM-based ABAQUS software.

The analyzed column on the ground floor carries loads from the first floor and the usable attic. The calculation of loads was made mostly on the basis of the applicable standard PN-EN 1991-1-1 2004 [15]. The weight of the heavy-type ceramic-metal Klein floor was taken into account. The imposed load was assessed for reading and office rooms. Safety factors were applied according to the standard PN-EN 1990:2004/AC[16]. The total resultant load acting on a single column was assumed 235 kN, the actual load was uniformly distributed on the surface of the head plate. Stress check was conducted in the most critical column cross section. In the numerical analysis, the column was fully fixed at the bottom, allowing a moderate rotation of the capital.

3.1 Numerical simulation

The computational model has been created (see Figure 4). It can be considered the representative one for all columns used in the building. The model consists of a vertical tube (3,46 m length), a horizontal slab and stiffening ribs. The elements are fixed to each other by means of contact faces. All dimensions and profiles, defined in the numerical model, are based on the stocktaking drawings. The 8-node standard solid elements available in ABAQUS software have been used. The mesh has been created proceeding global size control. The analysis has been conducted for elements of an approximate planar area of 100 mm^2 for most vertical elements, excluding the base, which showed the planar area around

1000 mm^2. In the case of horizontal parts (plate and profiled parts of the shaft), the approximate planar area of 400 mm^2 was set to the analysis. Due to the lack of exact material characteristics (no material tests were conducted due to technical limitations), the cast iron material properties were assumed based the average data typical for 19th century elements [17,18].Modulus of longitudinal elasticity was set E = 100 000 MPa while modulus of transverse elasticity G = 38 000 MPa [18] and Poisson's ratio ν = 0,27 [19].

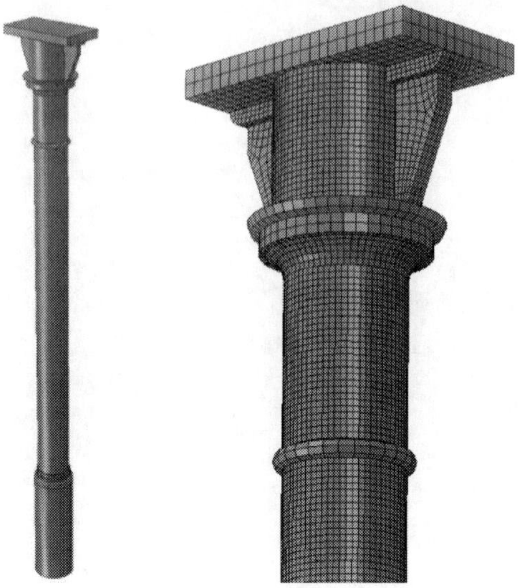

Fig. 4. The model geometry used for numerical analysis.

3.2 Structural calculations

In the first step of the study the column was loaded axially. The maximum compressive stresses determined in this way were used to compare them with the allowable values. In order to calculate the load-carrying capacity σ_c of the cast iron column, the method developed in the second half of the 19th century by Ludwig von Tetmajer was used [17]. The method is based on the value of slendernessλ calculated for the analyzed column as equal to:

$$\lambda = \frac{h}{i} = \frac{351}{6{,}4} = 54{,}8 \qquad (1)$$

While the elements detect a low slenderness up to 80, the Euler's formula of critical stresses is not valid, instead the following polynomial formula should be used[18]:

$$\sigma_c = 776 - 12 \cdot \lambda + 0{,}053 \cdot \lambda^2 = 776 - 12 \cdot (54{,}8) + 0{,}053 \cdot (54{,}8)^2 = 277{,}6 \text{ MPa} \qquad (2)$$

Additionally, the safety factor of n=1/3 was recommended [17], therefore the allowable stress σ_a can be calculated as:

$$\sigma_a = n\sigma_c = \frac{277{,}6}{3} = 92{,}5 \text{ MPa} \qquad (3)$$

The calculated value is similar to the one described by Tetmajer in his monograph [17]. Established magnitudes have changed over the years, and throughout the countries. Anyway, in most of the cases they were around 80-90 MPa.

Comparable values of allowable compressive stress were also described in German standard DIN 1051 issued in 1937 [20]. It allowed evaluating the load-bearing capacity of cast-iron elements, permissible compressive stresses for columns amounted to 90 MPa. According to this standard, stresses in a cast iron column should be determined from the formula [20]:

$$\sigma = \frac{P \cdot \omega}{A} \leq \sigma_a \tag{4}$$

$$\omega = \frac{\sigma_a}{\sigma_b} \tag{5}$$

where σ_b is the allowable stress for buckling, while the column slenderness is not higher than 80, it holds [20]:

$$\sigma_b = 90 - 0{,}01005 \cdot \lambda^2 \tag{6}$$

Considering the analysed column the value is:

$$\sigma_b = 90 - 0{,}01005 \cdot (54{,}8)^2 = 59{,}8 \text{ MPa} \tag{7}$$

$$\omega = \frac{\sigma_a}{\sigma_b} = \frac{90}{59{,}8} = 1{,}51 \tag{8}$$

The value of the maximum compressive stress in the column was also determined on the basis of a simplified numerical analysis applying the model shown in Figure 4. In the case of axially loaded column, the maximum compressive stress was equal to σ = 23,91 MPa (see Figure 5A). It should be noticed that the bearing capacity of the element determined on the basis of the Tetmajer method [17] significantly exceeds the actual compressive stresses, when the axial compression is assumed.

In the second stage of numerical simulations, load eccentricity was introduced. The results of the analysis indicated that the allowable compressive stresses of 90 MPa was reached starting from the eccentricity of 16 cm. Due to this eccentricity value, the compressive stress was as large as 92.61 MPa (see Figure 5B).

4 Conclusions

The architectural and structural analysis of cast iron columns of the former financial office building in Kwidzyn (Poland) has been described in this paper. The computational investigation has been conducted using the 19th and 20th century methods, the results have been compared with the numerical analysis based on FEM.

The results of the study indicate that the low application extent of the load-bearing capacity of the cast iron column takes place while applying both historical methods. While the Tetmajer method shows around 25% the total value, the German points it at around 39%. The outcome of the investigation, showing a distinctly small use of the cast iron column load-bearing capacity, is extremely important. It is because of the fact that cast iron, widely used in architecture and construction in the past, is still a little-known material. It is not in a common use for such purposes as in previous centuries, but still encountered in the historical structures. Thus the outcome of the study can be useful for the professional evaluation calculations during the adaptation of buildings for new purposes.

It should be added that the use of cast iron supports made it possible to create a more spacious interior in the analyzed building. The designers have replaced more massive brick columns that would visually divide the room. Such a solution was not desirable due to the original purpose of the structure. It was proved that cast iron columns with tubular cross section have high load-bearing capacity. It can be therefore confirmed that structural considerations have decided to use cast iron to provide a room with better functionality in the case of analyzed building.

The investigation described in this article was based on a number of simplifications. Therefore, further study, focused on experimental tests and numerical simulations, is planned to be conducted, so as to fully confirm the results obtained. Detailed tests on samples taken from the analyzed columns are important from the point of view of determination of material properties. After the experimental verification, the advanced non-linear numerical analyses are also scheduled, including some stochastic issues involved (see [21-23], for example).

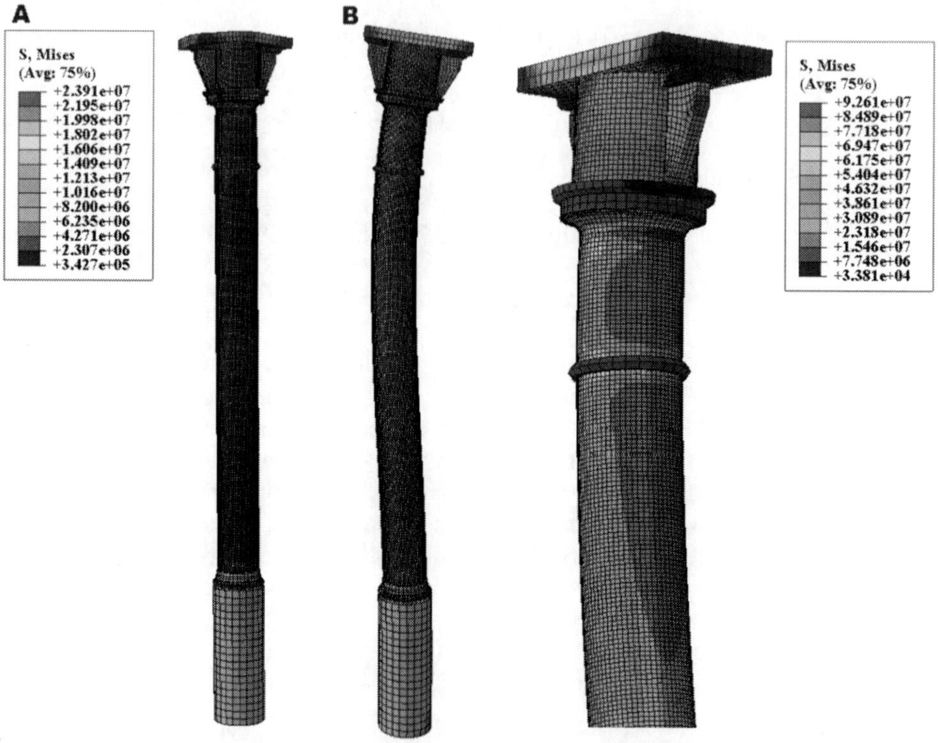

Fig. 5. Results of a simplified numerical analysis performed in ABAQUS software: A - stress map for an axially loaded column for actual loads values, B – stress map for deformed column loaded eccentrically (eccentricity of 16 cm);Stress maps show equivalent stresses due to Huber-Mises-Hencky criterion.

The authors would like to thank TASK CI for the opportunity to use ABAQUS software.

References

1. K. Gierdziejewski, *Zarys dziejów odlewnictwa polskiego (Outline of Polish foundry history)*, (National Technical Publishing House, Stalinogród, Poland, 1954)
2. M. Kwasek, A. Piwek, Cast iron staircase in Aleksandrów Kujawski (Poland) - history, construction, architectural form, WMCAUS, Procedia Engineering, **161**, 2147-2154 (2016)

3. M. Kwasek, A. Piwek, Lightweight glass and metal roof structures in the 19th century. Architecture and construction (XXII Lightweight Structures in Civil Engineering, Olsztyn, Poland, 31-36, 2016)
4. Ł. Lewandowski, Żeliwne schody kręcone w zespole fabrycznym Izraela Poznańskiego w Łodzi z drugiej połowy XIX wieku (Cast iron spiral stairs in the complex of Izrael Poznanski Factory in Lodz, the 2nd half of 19th century), Materiały Budowlane, **5**,104-105 (2017)
5. Ł. Lewandowski, *Konstrukcja i dekoracja schodów żeliwnych XIX wieku na wybranych przykładach (Construction and decoration of cast iron stairs of the 19th century on selected examples)*, Integracja sztuki i techniki w architekturze i urbanistyce (Wydawnictwo UTP, Bydgoszcz, Poland, 253-268, 2015)
6. M. Kwasek, A. Piwek, Cast iron in the 19th-century building equipment, WMCAUS, IOP Conf. Ser.: Mater. Sci. Eng., **245**, 042086 (2017)
7. J. Gloag, *A history of cast iron in architecture* (George Allen & Unwin, London, 1948)
8. I. Brych, M. Sýkora, Assessment of cast-iron columns using analytical models, Applied Mechanics and Materials, **821**, 782-788 (2016)
9. D. Friedman, Cast-iron-column strength in renovation design, Journal of Constructed Facilities, **9(3)**, 220-230 (1995)
10. R. Käpplein, Untersuchung und Beurteilung alter Gußkonstruktionen, Stahlbau **66, Heft 6**, 318-332 (1997)
11. B. König, W. Klingsch, *Untersuchungen zur Erarbeitung eines Sicherheitskonzeptes für historische Gußeisenstützen* (IRB-Verlag, 1994)
12. B. König, *Historische gusseiserne Stützen – Eine zerstörungsfreies Beurteilungsverfahren für die Belastbarkeit bei Normaltemperatur und im Brandfall* (Institut für Konstruktiven Ingenieurbau, Bergische Universität GH Wuppertal, 1995)
13. J. Rondal, K.J.R. Rasmusen, On the strength of cast iron columns, Journal of Constructional Steel Research, **60(9)**, 1257-1270 (2014)
14. J. Marcinowski, Z. Różycki, Reinforcement of existing cast-iron structural elements by means of fiber reinforced composites, CEER, **20(1)**, 37-45 (2016)
15. PN-EN 1991-1-1:2004, *Eurocode 1: Actions on structures - Part 1-1: General actions - Densities, self-weight, imposed loads for buildings* (2004)
16. PN-EN 1990:2004/AC *Basis of structural design* (2010)
17. L. Tetmajer, *Die Angewandte Elasttizitäts und Festigkeitslehre* (Franz Deuticke, Leipzig und Wien, 1905)
18. K. Czapliński, Dawne wyroby ze stopów żelaza (Former iron alloy products),(Lower Silesian Educational Publishing House, Wrocław, Poland, 21-28, 2009)
19. C. Podrzucki, *Żeliwo. Struktura, właściwości, zastosowanie (Cast Iron. Structure, properties, application)*, t. 2 (Wydawnictwo ZG STOP, Kraków, Poland, 11-32, 1991)
20. DIN 1051. *Berechnungs grundlagen für Guseisenim Hochbau*, (1937)
21. R. Jankowski, H. Walukiewicz, Modeling of two-dimensional random fields, Probabilistic Engineering Mechanics, **12**, 115-121 (1997)
22. J. Przewłócki, *Problemy stochastycznej mechaniki gruntów. Ocena niezawodności* (Dolnośląskie Wydawnictwo Edukacyjne, Wrocław, Poland, 2006)
23. P. Sorn, J. Górski, J. Przewłócki, Probabilistic analysis of a space truss by means of a multidimensional variable description, Archives of Civil Engineering, **61**, 99-123 (2015)

The influence of position of the post or its absence on the performance of the cable barrier system

Dawid Bruski[1,*], *Stanisław* Burzyński[1], *Jacek* Chróścielewski[1], *Łukasz* Pachocki[1], *Krzysztof* Wilde[1], and *Wojciech* Witkowski[1]

[1]Gdańsk University of Technology, Faculty of Civil and Environmental Engineering, ul. Narutowicza 11/12, 80-233 Gdańsk, Poland

Abstract. Road safety barriers are used to increase safety in potentially dangerous places on the roads. They are designed and installed on the roads to prevent any vehicle from getting outside the travelled way or from entering the opposite lane of the road. Barriers, which are used on European roads, have to undergo full scale crash tests according to the EN 1317 standards. Nowadays as a supplement to real crash tests, numerical simulations are commonly used. The work concerns the influence of position of the post or its absence on the crashworthiness of the cable barrier based on numerical study results.

1 Introduction

Nowadays, when a number of new roads and motorways are being built, the issue of proper location of road restraint systems becomes more and more important. These devices are not neutral i.e. vehicle impact with barrier can also be dangerous. Therefore they should be thoughtfully and reasonably installed along the road or in dangerous places. The basic road safety equipment are road barriers. Their main task is to successfully contain the vehicle without breakage of any of its principal longitudinal element and then re-direct the vehicle back on the road at the same time ensuring the lowest possible load on vehicle occupants.

Prior to mass production, safety barriers undergo appropriate crash tests and meet the requirements according to EN 1317 standards [1, 2]. However, these standards do not indicate the type of a safety barrier system to be used on a given road category, leaving these decisions to national or local road authorities.

In Poland, only safety barriers which fulfil the EN 1317 requirements can be used on national roads. The barriers used on national roads should be identical in every aspect to those that have successfully passed the full scale crash tests [2].

The EN 1317 standards consider a limited number of crash test configurations, in addition, they only apply to straight section of the safety barrier, with impact point in its middle section. Whereas in reality there are many different cases, these mentioned standards do not include, like barriers installed on horizontal and vertical road curves

*Corresponding author: dawid.bruski@pg.edu.pl

(concave and convex) [4], variation of vehicle type, velocity and angle of impact, weather or soil conditions (they may not be the same as in the real test) [5], the influence of distance between the barrier and the road or curb [6] or different barrier lengths, in a particular case the barrier may be short, thus not as secure as their full-length parent. In addition, simulations allow to analyse the influence of various obstacles such as truss supporting structure located directly behind the barrier. For these reasons nowadays, in parallel with real crash tests, numerical simulations are commonly used to evaluate and improve the properties of safety barriers. Their great advantage is that their costs are much lower compared to the costs of in-situ tests. They allow for a detailed analysis of the test and to consider many factors, e.g. the change of material properties. Some numerical simulations of crash tests were presented in the papers [7, 8]. It is worth noting that the current standard [9] allows numerical simulations to certify safety barriers modified under certain conditions. In recent years, numerical simulations of crash tests have been significantly developed. In 2012, the standards [10-13] have been published concerning guidelines for numerical simulation of crash test and their verification and validation.

A common problem during the installation of the barrier is a situation when some posts of the barrier have to be mounted in the location of others elements of the road (e.g. drainage wells). In these cases, the posts of the barrier are usually shifted and in this area the post spacing of the barrier is inadequate or simply the posts are not installed. The question remains open about safety of such a structure. Some examples of solutions of this issue are presented in Figs. 1-a, 1-b and 1-c.

In the work the influence of position of the post or its absence on the performance of the cable barrier system was studied. Cable barriers are generally perceived a beneficial construction, usually providing smaller decelerations for people inside the vehicle during the collision than other types of safety barriers. The properties of cable barriers shift the collision population into low-severe type, where occupants are not seriously injured and the vehicle after leaving the barrier can often continue to drive, opposed to the case of steel or concrete barrier [14]. Often road administrators prefer cable barriers instead of others barrier due to their lower cost, ease of maintenance, ease of snow ploughing operations or better see-through appearance and lower crash severity [15]. However, it should be highlighted that cable barriers are not universal devices, in some situations there is a need to use other types of barriers, e.g. where the bridge abutment is adjacent to the road.

In order to assess the influence of the local changes in post spacing, numerical FEM-based simulations of cable barriers were performed.

Fig. 1. An example of a) the local change in post spacing, b) the missing post in cable barrier c) the missing post in steel barrier.

2 EN 1317 Standards

The EN 1317 standards [1, 2] define impact test acceptance criteria for crash tests and test methods that should be subjected to safety barriers prior to their use on public roads. In order to pass the crash test, a safety barrier needs to satisfy a series of requirements e.g. the barrier is bound to successfully contain and redirect the vehicle, the vehicle is restrained from rolling over, any elements of the barrier cannot penetrate to vehicle cabin, and after collision, the vehicle must not cross a line parallel to the initial traffic face of the system before impact (exit box). It is emphasized that the standards do not specify conditions for the barrier geometry, dimensions or materials. Instead they describe the performance classes, giving ranges for the three major criteria: containment level, working width and impact severity level. Containment level defines the barrier capacity to contain vehicle impacting the barrier. For example, if a denotes a N2 Containment Level (table 1), two tests, TB11 and TB32, should be conducted (table 2). On the basis of the results from the tests above, the working width and the impact severity level are determined. The working width marks the ability of the barrier to deform during collision, defined as the maximum lateral distance between any part of the safety barrier on the undeformed traffic side and the maximum dynamic position of any part of the barrier or vehicle if its body deforms around the barrier. The impact severity level (table 3) reflects severity of the impact for people inside a vehicle. Two indicators, the acceleration severity index (ASI) and the theoretical head impact velocity (THIV) are required to determine this level. According to [2], severity index is reported for tests with cars and there is no obligation to evaluate this index for tests with heavier vehicles.

Table 1. Containment levels [2].

Containment levels		**Acceptance test**
Low angle containment	T1	TB21
	T2	TB22
	T3	TB41, TB21
Normal containment	N1	TB31
	N2	TB32, TB11
Higher containment	H1	TB42, TB11
	L1	TB42, TB32, TB11
	H2	TB51, TB11
	L2	TB51, TB32, TB11
	H3	TB61, TB11
	L3	TB61, TB32, TB11
Very high containment	H4a	TB71, TB11
	L4a	TB71, TB32, TB11
	H4b	TB81, TB11
	L4b	TB81, TB32, TB11

Table 2. Vehicle impact test descriptions [2].

Test	**Impact speed, km/h**	**Impact angle, °**	**Total mass, kg**	**Type of vehicle**
TB11	100	20	900	Car
TB21	80	8	1 300	Car
TB22	80	15	1 300	Car
TB31	80	20	1 500	Car
TB32	110	20	1 500	Car
TB41	70	8	10 000	Rigid HGV
TB42	70	15	10 000	Rigid HGV
TB51	70	20	13 000	Bus
TB61	80	20	16 000	Rigid HGV
TB71	65	20	30 000	Rigid HGV
TB81	65	20	38 000	Articulated HGV

Table 3. Impact severity [2].

Impact severity levels	ASI, -	THIV, km/h
A	ASI ≤ 1.0	≤ 33
B	1.0 < ASI ≤ 1.4	
C	1.4 < ASI ≤ 1.9	

3 Numerical model

The calculations were conducted using a finite element code LS-DYNA (MPP double precision R8.1.0) on supercomputer Tryton managed by Academic Computer Centre (CI TASK) in Gdańsk (Poland). The LS-DYNA system is a basic, widespread tool to perform the simulations of crash tests. This system employs a special form of explicit central difference method to integrate the equations of motion [16, 17].

3.1 Numerical model of vehicle

In numerical simulations, a Suzuki Swift (also called Geo Metro) numerical model was used (fig. 2). This model was taken from ROBUST project repository [18], next subjected to minor modifications. The mass of the numerical model is 928,7 kG and consists of 20 089 nodes and 16 291 finite elements. Near the centre of gravity of the vehicle, a special-purpose finite element to measure accelerations and angular velocities has been defined, on this basis severity indices are determined.

Fig. 2. Car Suzuki Swift (left) and its numerical model (right).

3.2 Numerical model of cable barrier

The model of the cable barrier system of a total length of 64.1 m was create for the study. The system overview is shown on figure 3.

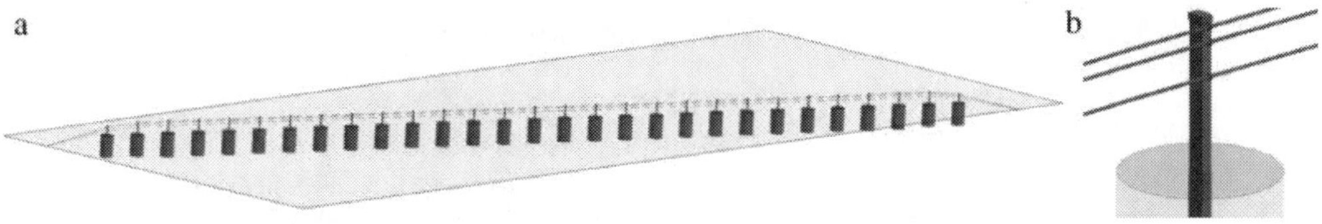

Fig. 3. Cable barrier model: a) general view, b) post and wire ropes view.

The system consists of three wire ropes mounted to the post with steel hooks. The height of the post is 1.7 m, it is embedded 0.95 m in the soil. The spacing between each two posts is 2 m. Two utmost posts at the both ends of the barrier are equipped with additional steel elements to stabilize the post in the soil. A numerical model of the barrier consists of

182 333 nodes and 206 064 finite elements. Since LS-DYNA is used as computational tool, all specific terms used below are described in detail in code theory documentation [16].

Steel posts are discretized by shell elements of Belytschko-Tsay type (characteristic dimensions ~13-15 mm), defined with Mat Piecewise Linear Plasticity material model. Hooks that keep the wire ropes at the correct height are discretized as beam elements. In order to discretize the wire ropes the beam elements with Belytschko-Schwer resultant beam formulation were used. The Mat Moment Curvature Beam material model was assigned to wire ropes. This method was successfully applied in the paper [19], which compares the simulation results to the experimental ones. The length of a single finite element of the wire rope is 25 mm. In order to perform pretension of the wire ropes, the beam elements are applied with discrete cable formulation and with Mat Cable Discrete beam material model. The soil was modelled in the form of cylinders, forming a background for post embedding. Similar solution was used in files available on NCAC public library [20]. The soil is represented by means of Mat Soil And Foam material model.

4 Numerical simulations description

In order to evaluate the influence of the position of the post or its absence on the performance of the cable barrier system, the following research protocol was effected:

1. determination of the most unfavourable impact location of the vehicle on the barrier,
2. change of the location of posts followed by the study of its influence on the results,
3. local removal of posts and its influence on the results.

In order to conduct the above plan, a series of simulations of TB11 crash test was carried out to determine critical impact location. In the TB11 crash test, the 900-kg vehicle impacts the barrier with the velocity of 100 km/h and the impact angle is 20° (table 2). The impact points were set between the posts 10 and 11 every 0.2 m. The test with the highest value of parameter ASI was then assumed as the most unfavourable configuration. After determining the final impact point further simulations were carried out, here the posts no. 10 and 11 were firstly shifted, next removed. The influence of the location change of the post or its absence was determined on the basis of these simulations.

5 Results

5.1 Determination of the impact location

The ASI results obtained for variable impact position are shown in fig. 4. The impact location has a slight influence on the ASI value. In all cases the indicators make it possible to qualify barrier to class A of impact severity level. Based on this results, the impact location was assumed 0.2 m measured from the post no. 10. and for the purpose of further analysis this simulation will be named tb11_00.

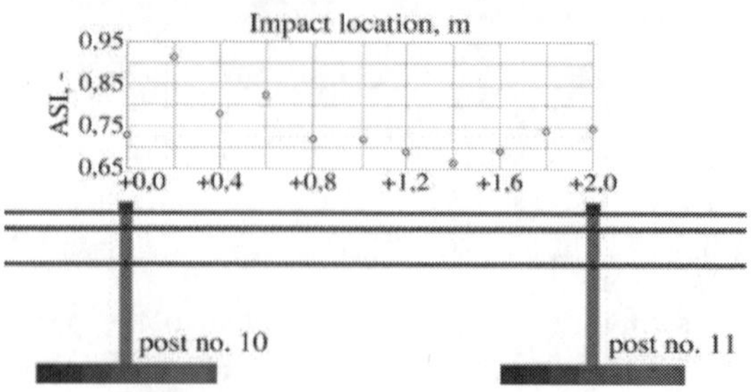

Fig. 4. ASI depending of the impact location.

5.2 The influence of position of the post

In order to determine the influence of the local shift of the post no. 10 (the first post before the impact point) the following values of displacement along the barrier of the post were analysed: -1.0 m, -0.5 m, +0.5 m, +1.0 m. The post after the impact point (no. 11) was analysed in the same way as the post no 10, thus it was also moved: -1.0 m, -0.5 m, +0.5 m and +1.0 m. The results are presented in table 4.

Based on the obtained results, it was found that the change in the location of the post shows a small influence on impact severity indices. In all considered cases, the values of ASI and THIV parameters are lower than for the base test tb11_00 (exception: ASI for the tb11_10_-10 test is higher). The obtained values are also comparable to parameters from the analysis of the determination of the impact location (section 5.1). The local post shift also affects the working width, as is shown in table 4. In the cases where the post before the impact point was moved -0.5 m and -1.0 m, and where the post after the impact point was moved +0.5 m and +1.0 m, the working width was determined by the element of the barrier. The same effect was observed in basic test tb11_00. In these tests deformation of the wire ropes was not significant. The post, which was bent by the car at the level of the ground, determines the working width. In other tests (tb11_10_+05, tb11_10_+10, tb11_11_-10, tb11_11_-05) the working width was determined by the front wheel of the vehicle. Figure 5 shows the comparison of the decisive moments for working width between the tests tb11_00 and tb11_10_+05. For a clearer appearance, the figure shows only the barrier elements above the ground and the vehicle tire, the visibility of other elements of the numerical model has been turned off. It can be noticed that the difference between working widths in these two test is not large, yet in tb11_00 test (dark colour) the vehicle was much closer to the traffic side during the impact (smaller displacement of the wire ropes) and in tb11_10_+05 (bright colour) the vehicle significantly penetrated the barrier. This is presented in fig. 5 which shows the deformation of the wire ropes and the position of the wheel. This fact may be very important if road accident occurs to cause local change in post spacing, e.g. when at a distance slightly greater than the working width determined from the base test tb11_00 (0.84 m), the road lamp will be installed.

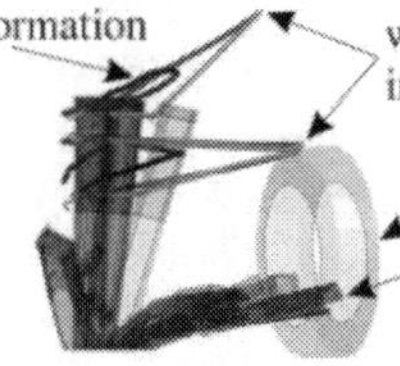

Fig. 5. Comparison of working width between tests tb11_00 and tb11_10_+05.

Table 4. Results of the influence of position of the post.

Post	Shift, m	ASI, -	THIV, km/h	W_N, m	nomenclature
Base test		0.92	28.3	0.84	tb11_00
10	- 1.0	0.93	26.8	0.92	tb11_10_-10
	- 0.5	0.83	27.7	0.89	tb11_10_-05
	+ 0.5	0.79	26.6	0.95	tb11_10_+05
	+ 1.0	0.83	27.0	0.89	tb11_10_+10
11	- 1.0	0.80	27.3	0.81	tb11_11_-10
	- 0.5	0.79	25.9	0.77	tb11_11_-05
	+ 0.5	0.86	26.8	0.86	tb11_11_+05
	+ 1.0	0.89	25.5	0.75	tb11_11_+10

5.3 The influence of absence of the post

In order to assess the influence of the local absence of the barrier post, two numerical simulations were carried out, the first in which the post no. 10 was removed, the second in which the post no. 11 was removed. Table 5 shows the obtained results.

Table 5. Results of the influence of absence of the post.

Post absence	ASI, -	THIV, km/h	W_N, m	nomenclature
Base test	0.92	28.3	0.84	tb11_00
10	0.84	25.9	0.89	tb11_10
11	0.79	23.5	0.77	tb11_11

The values of ASI and THIV in the considered cases are slightly lower than for the base test tb11_00. The working width for the tb11_10 was greater than the value obtained from the base test tb11_00, while for the test tb11_11 the working width is smaller. In these both considered cases the car element (front left tire) is decisive upon the working width. Additionally, the obtained values of the analysed parameters are similar to those obtained in the section 5.2. Figure 6 shows the selected result of test tb11_11.

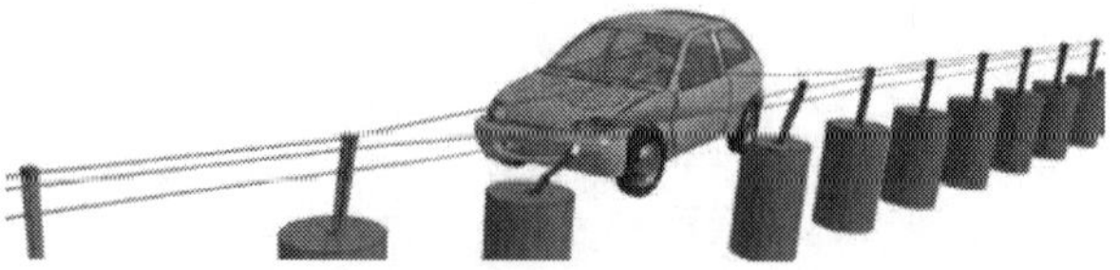

Fig. 6. View of the vehicle during collision in the test tb11_11 (0.14 s after impact).

6 Conclusions

The aim of the study was to investigate how the position change of the post or its absence affects the performance of the cable barrier. This type of problem often appears during the installation of barriers on roads or highways, where the road elements which has been already built impede the correct installation of the barrier.

Numerical simulations show that in the case of the analysed cable barrier the influence of local change of location of the post or its removal is insignificant for the values of the ASI, THIV and W_N parameters. However, it was noticed that variation of the post position (or the post missing) makes the working width increase, so this value might not be already determined by the element of the barrier but the vehicle element may be the most distant part that determines the working width value. This situation can be especially dangerous if there is an existing obstacle e.g. road lamp or some kind of supporting structure for road signs at the distance slightly exceeding the barrier's working width. The obtained results suggest that, when it is not possible to keep the basic spacing of the posts, the elements of road equipment such as street lamps should not be placed, even if their installation is designed at a distance greater than the working width of the barrier.

It should be emphasized that the posts in the cable barrier play a different role than in the steel barrier. In the steel barrier, posts cooperate with the longitudinal rail, whereas in the cable barrier the main role of the posts is to keep the wire ropes at an appropriate height. When the road accident occurs, the wire ropes should be easily detached from the posts. Additionally, as in [17], it is suspected (though it remains unknown) that longer spacing between the posts in cable barrier may cause a vehicle to underride or to pass between the wire ropes. However, these phenomena did not take place in the considered cases. When the vehicle impacted the barrier, the two wire ropes remained at a height above the wheels and the third highest wire rope slipped over the car's hood and then was kept approximately at the level of half of the height of the door's window (see fig. 6).

Numerical simulations were made using the Finite Element Method. This work presents a practical method of analysing the issue that can often be found on roads and highways. The performed simulations confirm the usefulness of numerical simulations to solve this kind of problems. It is noted that these analysis are only predictions and no real full scale crash test has been performed to confirm the numerical calculations.

This work was supported by the National Centre for Research and Development (NCBiR) and General Director for National Roads and Motorways (GDDKiA) under the research project "Life Cost Analysis of Road Safety Elements" (contract number DZP/RID-I-64/12/NCBR/2016). Calculations have been carried out at the Academic Computer Center in Gdańsk, Gdańsk University of Technology.

References

1. PN-EN 1317-1:2010. Road restraint systems – part 1
2. PN-EN 1317-2:2010. Road restraint systems – part 2
3. GDDKiA, *Wytyczne stos. drogowych barier ochronnych na drogach krajowych* (2010)
4. K. Wilde, K. Jamroz, M. Budzyński, D. Bruski, S. Burzyński, J. Chróścielewski, Ł. Pachocki, W. Witkowski, JCEEA, Vol. XXXIV, **64** (3/I/17) (2017)
5. D. Bruski, S. Burzyński, J. Chróścielewski, K. Skwira, K. Wilde., W. Witkowski, MI, **63** (2017)
6. K. Wilde, K. Jamroz, D. Bruski, M. Budzyński, S. Burzyński, J. Chróścielewski, W. Witkowski, AoCE, **63** / 2 (2017)
7. Z. Ren, M. Vesenjak, Eng. Fail. Analysis, **12** (2005)
8. K. Wilde, D. Bruski, S. Burzyński, J. Chróścielewski, W. Witkowski, Math. a. Num. Aspects of Dyn. Sys. Analysis (2017)
9. PN-EN 1317-5+A2:2012. Road restraint systems – part 5
10. PD CEN/TR 16303-1:2012. Road Restraint Systems. Part 1
11. PD CEN/TR 16303-2:2012. Road Restraint Systems. Part 2
12. PD CEN/TR 16303-3:2012. Road Restraint Systems. Part 3
13. PD CEN/TR 16303-4:2012. Road Restraint Systems. Part 4
14. L. Mikołajków, M. Autostr., **11** (2006)
15. S. Cooner, Y. Rathod, D. Alberson, R. Bligh, S. Ranft, D. Sun, *Development of guidelines for cable median barrier systems in Texas* (2009)
16. J. Hallquist, *Ls-Dyna Theory Manual* (2006).
17. T.J.R. Hughes, *The Finite Element Method: linear static and dynamics finite element analysis*. Mineola, New York, Dover Publications, Inc. (2000)
18. http://www.vegvesen.no/s/robust/, date of access 01.10.2016.
19. C. Stolle, J. Reid, Modeling Wire Rope Used in Cable Barrier Systems, 11th International LS-DYNA Users Conference (2010)
20. http://www.ncac.gwu.edu/vml/models.html, date of access 10.03.2016.

Comparison of theoretical deformation with real life model of a steel bridge span

Karol Federowicz[1]

[1]West Pomeranian University of Technology, Faculty of Civil Engineering and Architecture, al. Piastów 50, Szczecin, Poland

Abstract. The article presents the selection process of a static system and optimization for steel bridge span. The issue was addressed as an entry to a BRICO 2017 contests in designing steel bridges. The proposed solutions were checked in terms of their bearing capacity, technological assembly and economics. General guidelines for geometry selection were described. The designed construction was produced and tested under load to determine the deflection in both vertical and horizontal plane. Furthermore, the article details the joints and technological issues met during the design process. Theoretical calculations and actual measurements on a real life model were compared. The process of structural design of civil engineering structures is a complicated, time consuming multilayered issue. The key to a good design lies in a compromise between esthetics, functionality, capacity exhaustion and technical possibilities.

1 Introduction

Metal bridges were firstly used in the second half of 18th century in England. The first significant and documented structure was the Iron Bridge on the Severn river in Shropshire, England, built between 1779-1781 [1]. Due to technical difficulties caused by the lack of welding and screw joints, the bridge incorporated feather connection typical for wooden structures. Since then the methods of designing steel structures have been vastly updated. Due to development of material engineering applicable materials underwent fundamental changes [2].

Mechanical properties of steel such as isotropy and high compressive and tensile strength made it one of the most popular construction materials applied in bridge engineering [3]. Development of metallurgy significantly increased the properties of steel. New materials allowed to design structures with longer spans and to use thinner profiles.

Popularity of steel constructions was also boosted by the participation of academic environment. In the past 30 years various bridge design competitions for students gained popularity. The contests try to advert future designers about different aspects of structural design. Future engineers learn to calculate bearing capacity of bridges with maximum reduction in steel usage. Aesthetics and technical feasibility are also taken into consideration.

[1] Corresponding author: federowicz-karol@zut.edu.pl

This article presents the design and optimization process of a bridge submitted to BRICO 2017 Nordic Steel Bridge Competition.

2 Boundary conditions in accordance with BRICO 2017 rules

The contest requires design and preparing a steel bridge model with a span of 1000x5000 mm and clearance of 700 mm. The structure was limited to 2000 mm height.

Single elements were intended not to exceed the dimensions of 250x250x1250 mm and the weight of 7 kg. The structure should be joined only with welds and screws. The rules allowed the application of M6-M16 screws, up to 8.8 class. Pre-stressed joints and riffled elements or welded screws were forbidden [4].

Except of geometric limitations, the regulations strictly determined the assembly process. The contest site was divided into several zones for storing, assembling and building as seen in Figure 1 below.

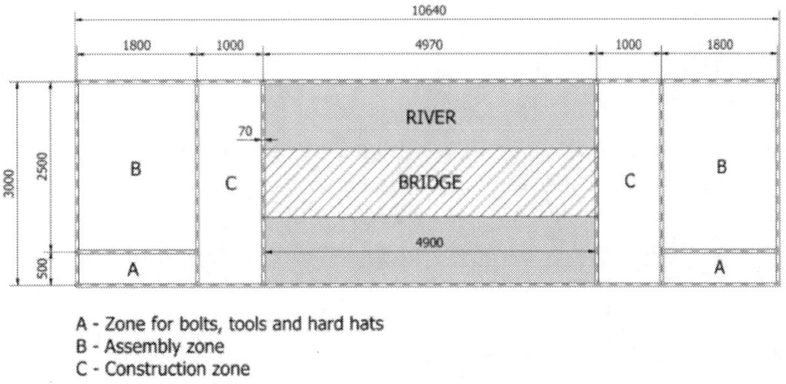

Fig. 1. Contest site division [4].

The B zone (Fig. 1) was intended for pre-assembly of maximum 3 parts at a time. The C zones and River zone were limited to a single team member only. Additionally, in order to reflect real-life conditions construction during assembling could not "touch" the River zone. Temporary supports were allowed, but had to ensure stability and could not exceed 1300x250x250 mm.

The assembly regulations forced the designers to divide the bridge into possibly smallest number of parts. This also required to maintain assembly simplicity and minimal number of joints.

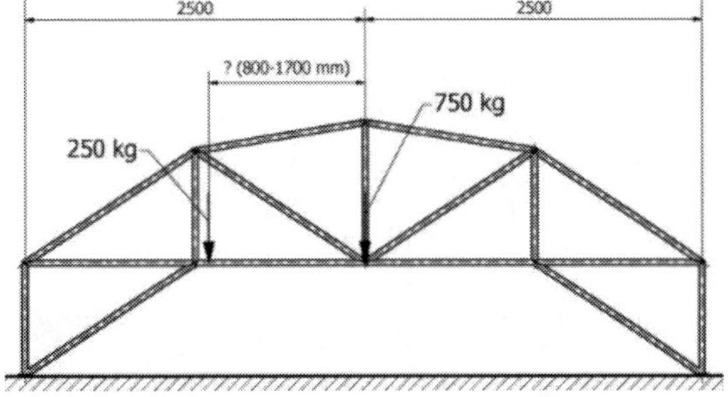

Fig. 2. Vertical loads applied to the bridge [4].

In order to evaluate the stability of the bridge, tests were carried out to measure the deformation under load in two-dimensions. The vertical deformation was measured in the center of the span as seen in Figure 2. Two-forces were applied to the bridge. The first one (750 kG) in the middle, the second (250 kG) randomly applied between 800-1700 mm from the center of the span.

Deformability in a horizontal direction was measured by applying 30 kG vertical load in the middle of the span and 25 kG of force horizontally. The deflections were measured with electronic dial gauge at the lowest point of the structure in the middle of the transversal span.

3 Selection of model geometry

The designing process was preceded by several concept calculated in Autodesk Robot Structural Analysis. Each solution was evaluated in terms of structural optimization. The assessment included division of the structure on shipping elements. Total weight of the structure and vertical and horizontal deflections were also taken into account. Advantages and disadvantages of each designed model are described in the further parts of this study.

3.1 Spatial truss bridge

The first analysis regarded a beam bridge with deck and supports made of spatial truss (Fig. 3). The solution was supposed to exhibit high rigidity in both planes [5]. However, due to geometrical restriction the design was ineffective, utilizing almost 200 kg of steel. The vertical deflection amounted to 8.2 mm.

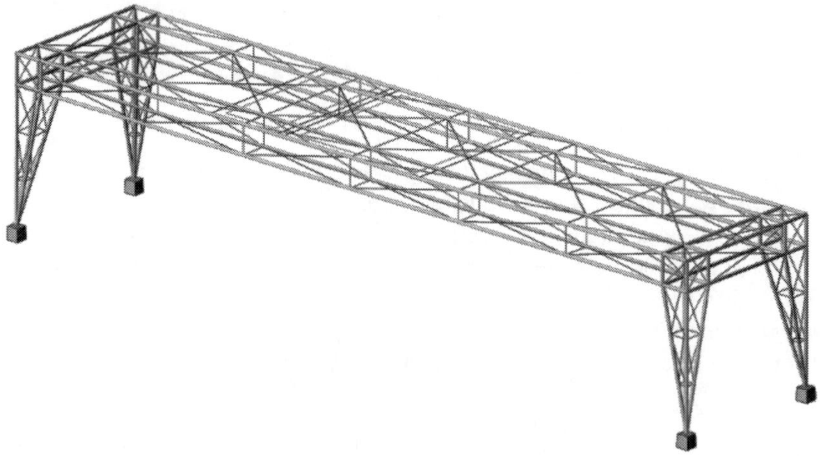

Fig. 3. Spatial truss bridge.

3.2 Single girded bridge

Single girder bridge design has reduced the number of elements compared to the previous solution (Fig. 4). This design is often used for reinforced concrete and pre-stressed bridges. The proposed design reduced the overall weight to 96 kg. Additionally, assembly elements were all repetitive, facilitating the assembly. However, reduction to a single central girder increased its susceptibility to torsion. Small dimensions of the spatial truss that build the main girder resulted in increased difficulty of assembly process. The joints between the girder and deck bars also posed additional difficulty for installation.

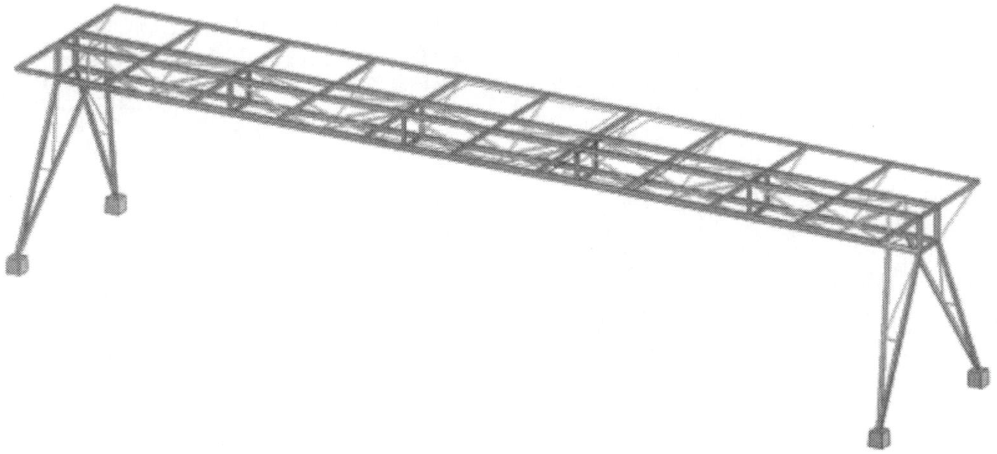

Fig. 4. Single girder bridge.

3.3 Arched bridges

Another proposed model consisted of arched bridge. Except of strength and stability, the model showed advantageous visual aesthetics. The arched design allowed the application of thinner steel profiles due to reduction of flexural element. Due to technological difficulty the girder was divided into separate segments (Fig. 5). In order to increase the rigidity in the vertical plane the authors decided to use truss girders. The obtained structure exhibited high rigidity (vertical deflection of 3.2 mm). Replacing the spatial trusses with flat girders made the structure more susceptible to horizontal deformation which in turn increased the number of bracings. The solution caused several assembly issues, particularly in the vicinity of joints. The feasibility evaluation proved this issue critical.

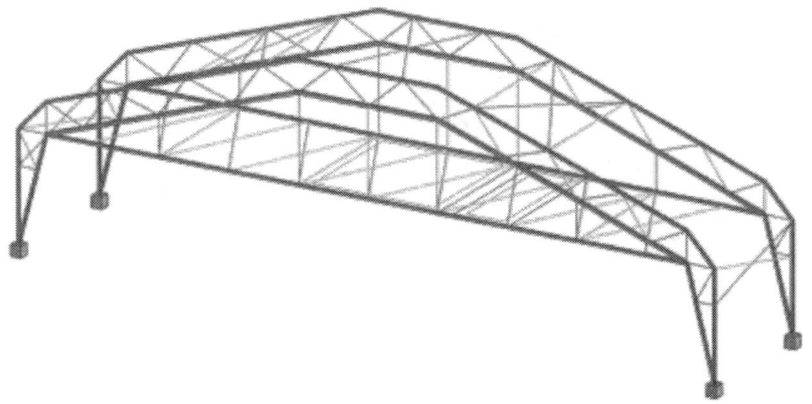

Fig 5. Arched bridge.

3.4 Proposed model

While performing simulations of the proposed models the final one has been chosen for production. The design once again covered of two symmetrical girders providing high rigidity in both planes (Fig. 6). The girder geometry was optimized by switching to triangular build-up. Showing two bars in the top flange and on in the bottom, the geometry had simplified the assembly of the structure. Additionally, the design facilitates the execution of butt joints, which shorten the assembly time [6,7]. Installation of the deck at the level of top flange allowed to reduce the number of bracing, without vertical rigidity decrement.

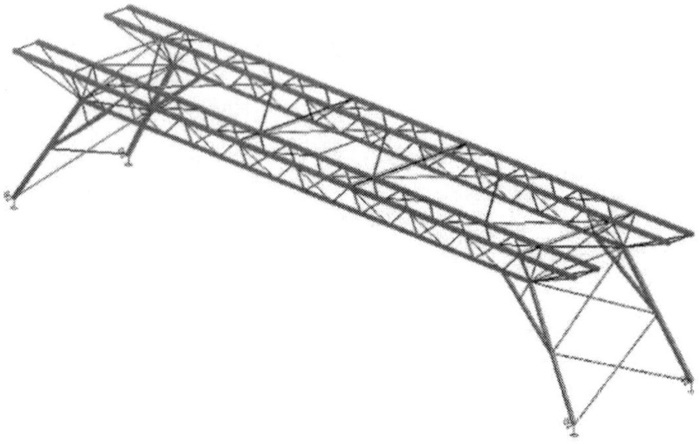

Fig. 6. Final design.

In order to further optimize the structure, the supports were inclined in both planes. This allowed to reduce the final span, thus resulted in decreasing the size of steel profiles. Unfortunately, this method increased the complexity of element production. It was necessary to prepare a 3D model and required chamfering with special equipment.

3.4.1 Profile selection

Designing an optimal construction requires balance between number of profiles and size. Proper optimization results in simplification of assembly and construction while the profiles showed their maximum effort. One of the most important aspects of design is to retain optimal dimension proportions. For freely supported beams [5] the height to span ratio should be between 1/20 to 1/25.

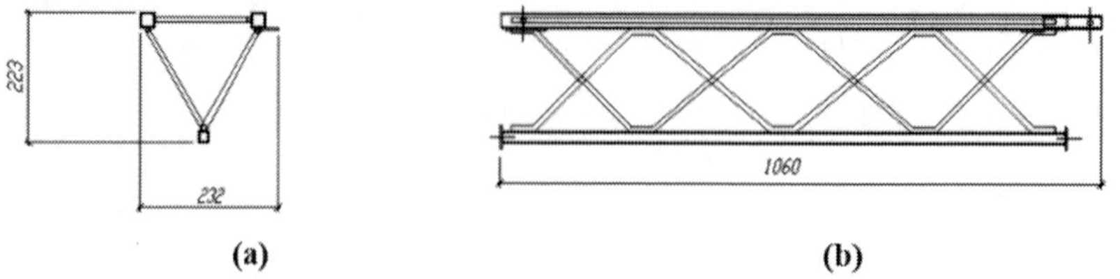

Fig 7. Assembly element – girder a) cross-section b) side view.

In the case of a proposed structure the assumed range was 20-25 cm. The structure included 5 profile types only: 3 round bars Φ6, Φ8 Φ10 for bracing and lattice, and square channels 20x2 for tension flange and 25x2 for compressed flange (Fig. 7a and 7b). Due to this setup the theoretical total weight amounted 75 kg.

It is necessary to maintain geometrical invariance in girders. It can be done by proper bracing layout during the design of spatial systems [8]. The girder design requires bracing in the compressed flanges. This allows to shorten the buckling length. Due to contest regulations and known horizontal force, the designers decided to use unsymmetrical system of bracing that works only under tension. In real-life structures where load can be applied in different directions, it is recommended to use symmetrical bracing [8]. There are no design guidelines regarding limit slenderness for tension bars. The authors assumed the limit value of λ=320.

4 Division of structure into shipped elements

At the design stage of prefabricated elements it is required to take into account their shipment and assembly. Regarding the contest regulations the bridge was divided into 16 main shipped elements and singular bracing bars. In order to highly shorten the assembly the bracing was screwed with a single bolt to gussets and supports. In order to connect to the girders, steel plugs were used in the top flange and gusset in the bottom. Chosen joints allowed for swift assembly and reduction of screws (Fig. 8a, b).

(a) (b)

Fig. 8 (a) Steel plug joint, (b) View of prepared structure.

5 Validation of proposed model in actual structure

In order to design complex spatial structures it is necessary to develop detail project. The structure, was modeled and calculated in two CAD software packages: Aveva Bocad and Tekla Structure. After modeling required joints, the total weight of structure amounted to 86 and 83 kG respectively. Theoretical mass calculated in Robot package was 75 kg. The difference occurred due to not incorporating the weight of gussets, screw and welds. It is necessary to consider all existing dead loads while considering calculation model.

The bridge was assembled in a steel prefabrication workshop, which highly increased the precision of the structure (Fig. 9). Structural elements were welded together with semiautomatic MIG welding. This allowed to produce high quality, aesthetic weld for 2 mm thick elements.

According to the contest regulation the bridge imposed load in two planes. After the test loading with a horizontal force, the measured horizontal deflection amounted to 0.5 mm. Given vertical loading in the center of the span and 1350 mm from the center (randomly chosen value in accordance to regulations) the deflection amounted to 7.45 mm. The calculated theoretical value for assumed system was 7.6 mm.

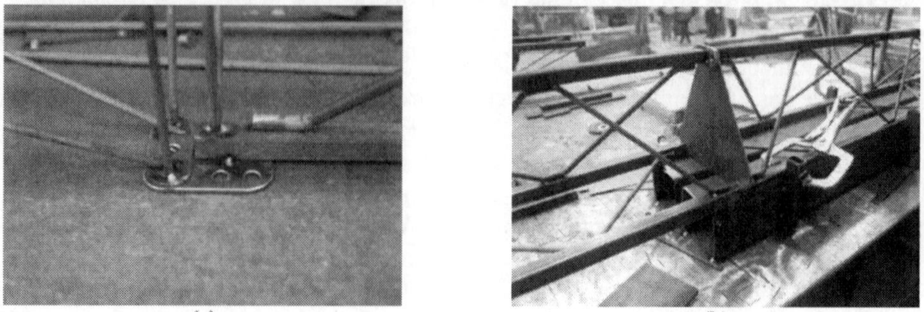

(a) (b)

Fig. 9. (a) Gusset, (b) Structure prefabrication.

6 Summary

Structural design is a multiphase analysis of different solutions in order to search for the optimal one. Regarding the BRICO 2017 where not only the esthetics, but also mass and rigidity were taken into consideration it was necessary to find a compromise. Table 1 presents the most important characteristics of the analyzed models. None of the models exceled in every manner. The proposed solution was located in-between rigid massive beam bridge and slender arched bridge.

Table 1. Characteristics of proposed models.

Bridge	Number of assembly parts	Calculated mass [kg]	Theoretical horizontal deflection [mm]	Theoretical vertical deflection [mm]
3.1	277	169	0.3	8.2
3.2	166	96	1.4	15.0
3.3	130	74	8.9	3.2
BRICO	189	75	0.3	7.4

The comparison of mass between the produced structure (84.5 kg) and theoretical values from Robot (75 kg) and Tekla (83 kg) software clearly shows that despite the calculation precision there are always slight shortcomings. The differences in the model mass calculated with software and actual mass were produced by thin galvanic coating.

Fig. 10. Assembled construction.

There are some differences in calculated values and actual measurement of deformations: 0.3 mm to 0.5 mm in terms of horizontal deformation and 7.6 mm to 7.45 mm in vertical deformation. There are several reasons for this combination. The most important is the simplification of node rigidity in calculations. The software always assumes ideal hinge or fixed support which constitutes to slight differences caused by backlash in actual structure (Fig. 11b). The same problem occurs with screws where there is always a bit of space between the actual screw and the hole. Additionally, unsymmetrical loading of the construction can create eccentricity. While subjected to unsymmetrical load or structural elements outside the plane, the structure starts to exhibit torsion. This issue is usually excluded from the calculations (Fig. 11a).

(a) (b)

Fig. 11. (a) Test loading, (b) Structure fixing.

The studied case of bridge design showed that it is necessary to include production and assembly technology. The use of continuous girder lattice significantly simplified the assembly of construction. Steel plug joints in girders allowed to shorten the assembly time to 14 minutes and 20 seconds.

The article clarifies that structural design is a complex, time consuming and multilayered the problem of structural designing problem indeed. A properly constructed structure combines esthetics, functionality, strength and feasibility. In case of civil engineering structures there are no perfect solutions that combine all of the above issues.

References

1. N. Cossons, B. Trinder, *The Iron Bridge: Symbol of the Industrial Revolution*, Bradforn-on-Avon: Moonraker Press (1979)
2. S. Smiles, *Industrial Biography: Iron Workers and Tool Makers,* Bremen: Europaeischer Hochschulverlag GmbH & Co. KG (2010)
3. K. Rykaluk, *Konstrukcje stalowe, podstawy i elementy*, Wrocław: Dolnośląskie Wydawnictwo Edukacyjne (2001)
4. BRICO 2017, 2016, Rules and regulations, Tallinn (2016)
5. J. Biliszczuk, et al. *Projektowanie stalowych kładek dla pieszych*, Wrocław: Dolnośląskie Wydawnictwo Edukacyjne (2004)
6. PN-EN 1993-1-1: 2006, Reguły ogólne i reguły dla budynków
7. PN-EN 1993-1-8: 2006, Projektowanie węzłów
8. A. Biegus, *Stalowe budynki halowe*, Warszawa: Arkada (2005)

The influence of high tensile bolts tightening torque on stress distribution in the bolted connection of rockfall barrier under impact load

Krzysztof Kuliński[1,*], and *Maciej* Major[1]

[1] Częstochowa University of Technology , Faculty of Civil Engineering, ul. Akademicka 3, 42-200 Częstochowa, Poland

Abstract. The paper presents the finite element method approach to determine stress distribution in rockfall barrier post with end to end-plate connection, assembled with high tensile bolts under two different types of impact load. For the analysis purposes typical fixed to the foundation, rockfall barrier post is adopted with 2,50 m height and sixty degrees slope to the horizontal ground line. One post end is fixed (welded) to the base plate – anchored to foundation, the second end is supported via retaining rope. The impact load applied exactly in the post mid-height and boulder caught via net in the mid-distance between two consecutive post are discussed. Four different locations of end to end-plate connection are analysed measuring from the fixed support i.e. dimension varies in the range of 0,30 m up to 0,60 m with 0,10 m step. Six high tensile bolts, zinc coated with 12.9 grade strength have been adopted in the connection. The stated boundary problem was solved by means of SolidWorks software.

1 Introduction

Rockfalls in mountains, on engineered slopes, mines etc. due to an unknown time of occurrence are hazardous phenomena causing serious threat for human lives, settlements, transportation and other infrastructure. The most significant influence on rockfall is triggered by the surrounding environment i.e. rainfalls [1], cycle freeze-and-thaw cycles [2] and progressive weathering [3]. The areas located at the junction of tectonic plates or volcanic grounds detect earthquakes as additional main factors [4]. It is worth noting that transportation infrastructure or facilities located nearby the rock slopes may also trigger a rock avalanche via emitted vibrations.

In the rockfall – exposed areas the major needs are an accurate prediction of rockfall runout, definition of possible threat to human lives or potential damage caused to infrastructure and then mitigation or possibly prevention of the hazardous phenomena. The crucial importance on planning the strength, supporting scheme and dimensions of barriers preventing individual or massive rock falling is observed with respect to in situ investigation of terrain and computer simulations [5-8]. An exemplary scheme of typical rockfall barrier with different supporting conditions is shown in Fig. 1.

* Corresponding author: kkulinski@bud.pcz.pl

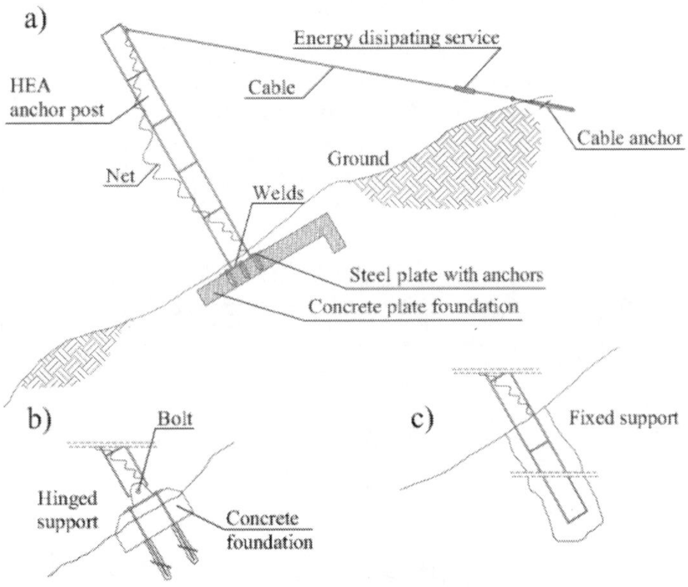

Fig. 1. Schemes of typically utilized rockfall barriers with different types of supports, a) barrier post fixed to the retaining concrete plate, b) hinged type of support, c) hammered post (fixed type).

In practice two different types of supports are applied i.e. the hinged support where barrier post may freely rotate in the slope direction (see Fig. 1b) and fixed support (see Fig. 1c). The barrier supporting scheme depends mainly on soil strength and the energy transferred via the falling rocks. Fixed supports due to transfer of momentum forces are utilized especially in bedrocks or in ground with relatively high bearing capacity to allow the posts to be grouted or even hammered into (see Fig. 1c). In the cases showing soil of a significantly lower strength than bedrock, gravels, sands special concrete retaining plate may be utilized to prevent barrier sliding (see Fig. 1a). It is worth noting that hinged barrier posts are easier to replace after stated exceeded deformations or any other stated damage than the fixed ones. In the hinged barriers energy is transferred mainly via the retaining ropes, while in self-supporting fixed posts energy is dissipated via profile deformation and net rigidity – the rest of the energy is transferred onto the concrete pedestal. The main advantage of self-supporting fixed posts is the possibility of omitting the retaining ropes, which eliminates the necessity of providing additional anchors upslope, however if the required ropes may significantly strengthen the system bearing capacity.

In the paper stress distribution is investigated in a fixed rockfall barrier post with end to end-plate connection under applied two different impact loads. The first type of loading concerns case in which boulder hits the post exactly, whereas the second type corresponds to the case in which boulder is caught by means of net. It is assumed that the post is rigidly connected (welded) to the base plate, consequently anchored to concrete retaining slab. Moreover, four different locations of barrier post end to end-plate connection are taken into considerations. All numerical calculations were performed with the use of SolidWorks software, with a simulation module fully based on finite element method.

2 Numerical model assumptions

For the numerical analysis purpose of the rockfall barrier post, a HEA240 profile made of S235JR steel grade class was chosen. The total length of mentioned post was assumed

2,50 m. Four different locations of barrier post end to end-plate connection measuring from the fixed support were considered i.e. 0,30; 0,40; 0,50 and 0,60 m, respectively. The analysed variants of barrier post are presented in Fig. 2, where the numbers in parentheses denote modifications shown via dashed lines i.e. in variant II all dimensions remain same as in variant I, whereas end to end-plate connection is distanced 400 mm from the fixed end and one more area for eccentric load case was taken into consideration (see section A-A in Fig. 2).

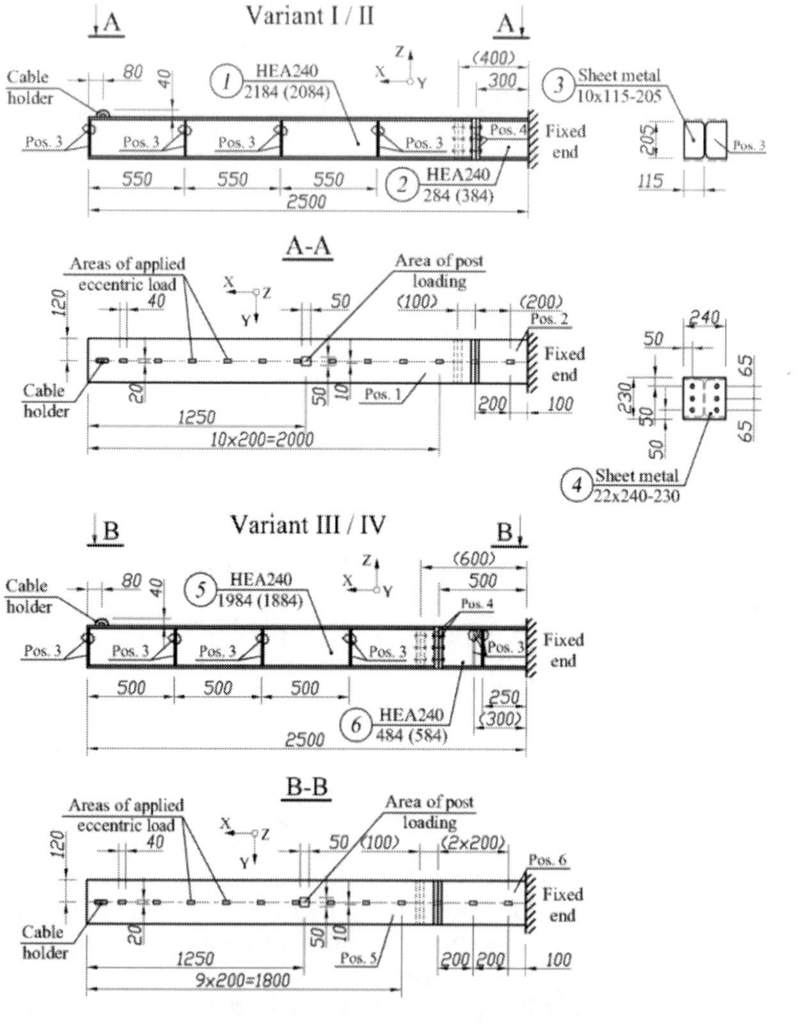

Fig. 2. Different variants of analysed fixed barrier post with variable location of end-plate connection.

According to the fact that the fixed type of rockfall barrier post was analysed, the end to end-plate connection was introduced to provide extra possibility of exchanging damaged via rockfall post element without the necessity of exchanging the whole post with also its foundation. In the end to end-plate connection six M24 12.9 grade high tensile bolts, electrically zinc plated were assumed. Connectors were arranged in two columns and three rows, where corner bolts were distanced from the metal sheet edges by 50 mm in the horizontal and vertical direction (see Fig. 2). Referring to the bolt manufacturer information [9] a tightening torque equal 1140 Nm was applied to each bolt. Additionally, for each connector two washers were provided, one under the bolt head and one under the nut. Flat rounded washers with external diameter Ø44 mm, internal diameter Ø26 mm and thickness

4 mm were introduced. On the basis of Polish national standard (PN-EN 1993-1-8:2006) connection meets all bearing capacity requirements with end-plate thickness assumed as 22 mm.

For the purpose of analysis two different types of loading were assumed. Additionally, it was preliminary assumed that the mountainside is inclined at 30 and analysed post at 60 degrees measuring to the horizontal terrain line. Hence the most unfavourable moment distribution in HEA profile was obtained for perpendicular impact direction to the XY-plane presented in Fig. 2. In the first case boulder impacts barrier post exactly in its mid-height (see area of post loading in Fig. 2). The boulder was simplified to the concentrated force of a 1000 kN value, distributed on the 50x50 mm area. The mentioned force reached its maximum value at 0.02 s from the start of analysis, after that load value tends to zero till 0.50 s. Such a simplification corresponds to the real case in which boulder hits the post, afterwards profile both elastically and plastically deforms under applied load and depending on the amount of accumulated elastic energy that energy is transferred backwards to the boulder. In the second loading case it was assumed that the boulder was caught via net in the middle point between two consecutive post and their mid-height. According to that only half of first case force value – 500kN was applied on the eccentricity. The posts were assumed to be spaced in the Y-axis direction by 3,00 m. Semi-rigid net fixing system was assumed with intermediate net supporting points distanced by 200 mm. In order to simplify the numerical model a catching net was neglected in the analysis. Hence concentrated force was applied on the eccentricity. The point of action was located 1,50 m in the Y-axis direction and 1,25 m in the X-axis direction measuring from the point where longitudinal profile axis crosses YZ-plane with an applied fixed boundary condition. In the opposite Z-axis direction the mentioned force point of action was distanced by 0,25 m measuring from top surface of HEA beam flange. The force acting direction was also assumed as the most unfavourable – perpendicular to the XY-plane and opposite to the Z-axis. On the basis of performed numerical calculations and considerations concluded in [10, 11], maximum energy in the barriers is accumulated at approximately 0.15 up to 0.8 s depending on net rigidity and location of boulder impact. In the presented paper maximum energy accumulation was assumed at 0.5 s, furtherly load was removed – force value equal 0 (boulder moving backwards under the influence of net elastic rigidity). The second and further boulder impacts under the gravity were neglected in the performed analysis, since maximum energy is transferred to the barrier via the first impact.

In the presented barrier posts in Fig. 2, following boundary conditions were adopted with respect to the Cartesian coordinate system: X, Y, Z-axis displacements and rotations fixed on the external YZ surface of HEA profile and Z-axis displacement fixed onto the bottom cylindrical surface of cable holder. Between the end-plates and washers/end-plates contact without penetration was assumed. Hex connectors were represented by SolidWorks bolt elements with specified alloy steel material and appropriate tightening torque equal 1140 Nm. Similar analysis with tightening torque equal 0 Nm had to be performed in order to find the influence of this parameter on the obtained stress results. Moreover, the direction of force applied exactly on the HEA profile follows the post deformation, whereas concentrated force applied on eccentricity have constant direction. It is worth noting that the amount of energy transferred onto the catching net supporting areas in the case of eccentrically applied load depends on distance and deformation of host post.

In the SolidWorks software discretization of barrier posts have been performed with 3D-Solid 4-node tetrahedron finite elements. For S235JR steel grade elements the homogeneous, elastic and isotropic material was chosen with Young's modulus $E = 210$ GPa, Poisson's ratio $\nu = 0.28$ and mass equal 7800 kg/m^3, respectively. Fine mesh with element size around ~ 3mm was used for washers, area under and near the washers' elements and near the applied boundary conditions. For the other part of the model an adaptive mesh from 3 up to 50 mm was used.

3 Numerical results and discussion

In this section numerical results concerning HMH stress results obtained in the end-plate connection under bolts washers are presented. The following nomenclature is used to identify appropriate model: Variant I denotes beam, where welded post to the steel foundation plate have 0,30 m and the assembled post height is 2,20 m, Variant II – 0,40 m of welded post and 2,10 m of assembled post, Variant III – 0,50 m / 2,00 m and Variant IV – 0,60 m / 1,90 m, respectively. For the sake of comparative analysis in which the influence of bolts tightening torque onto stress obtained values in end-plate connection is going to be discussed, absolute maximum HMH stress values were chosen from the areas located under corner hex bolts washers. The obtained HMH stress values for load applied directly to the HEA post at time $t = 0.02$ s were presented in Tab. 1. Values for load applied on eccentricity at time $t = 0.50$ s are shown in Tab. 2.

Table 1. Absolute maximum HMH stress values under corner hex bolts washers for different end-plate connection location and two different values of tightening torque at $t = 0,02$ s. Impact load applied directly to the HEA post.

End-plate belonging to:	Tightening torque [Nm]	Variant I (30 / 220) [Pa]	Variant II (40 / 210) [Pa]	Variant III (50 / 200) [Pa]	Variant IV (60 / 190) [Pa]
Assembled post	0	1.115E+09	9.122E+08	7.274E+08	5.203E+08
	1140	1.298E+09	1.026E+09	7.754E+08	5.385E+08
Welded post	0	1.189E+09	9.160E+08	7.335E+08	5.401E+08
	1140	1.336E+09	9.966E+08	7.893E+08	5.513E+08

Table 2. Absolute maximum HMH stress values under corner hex bolts washers for different end-plate connection location and two different values of tightening torque at $t = 0,50$ s. Barrier post subjected to the eccentrically applied load.

End-plate belonging to:	Tightening torque [Nm]	Variant I (30 / 220) [Pa]	Variant II (40 / 210) [Pa]	Variant III (50 / 200) [Pa]	Variant IV (60 / 190) [Pa]
Assembled post	0	1.222E+09	9.624E+08	7.990E+08	6.424E+08
	1140	1.258E+09	9.844E+08	8.360E+08	6.954E+08
Welded post	0	1.262E+09	9.680E+08	8.170E+08	6.648E+08
	1140	1.295E+09	1.030E+09	8.751E+08	7.396E+08

In both tables the welded post denotes the post fixed to the foundation plate, whereas assembled post denotes the post supported with the retaining rope (see Fig. 2).

On the basis of HMH stress values presented in Tab. 1 and Tab. 2 it is clearly visible that the time to make the impact load reaches its maximum value, also makes the stress under the washer edges significantly exceed the steel yield stress. It should be noted here, that the stress concentration covered very small area and material under the washers was not ruptured. Comparison of the obtained results shows that the longer distance of end to end-plate connection from the foundation, the lower stress values are obtained, which is linked with lower value of bending moment. With a closer distance to the fixed support, moment value rapidly increases, thus the combination of shear forces and moments forces led to relatively higher values of stress in variant I – III in comparison to variant IV. The analysis of the obtained stress results, makes us notice that in both loading cases minimum HMH stress values were obtained in variant IV, which was consequently chosen as best solution. The percentage increase in HMH stress due to introduced tightening torque in hex bolts for each model variant for two different types of loading were presented in Fig. 3.

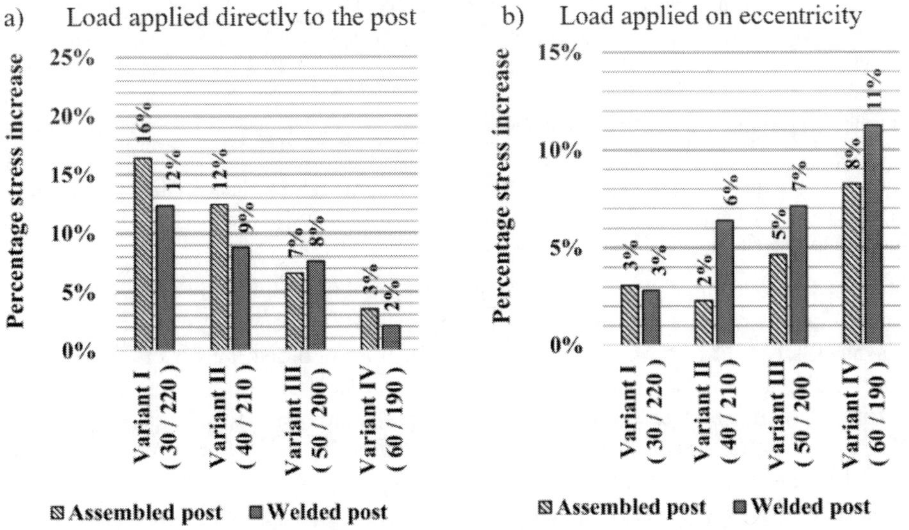

Fig. 3. Percentage increase of HMH stress under bolts washers in assembled post and welded post end-plate due to introduced tightening torque into the bolts. Two types of loading: a) load applied directly to the post at $t = 0.02$ s, b) load applied on eccentricity at time $t = 0.50$ s.

On the basis of Fig. 3 one may notice that introduced tightening torque into the bolts in case of load applied directly to the post profile, maximum percentage increase of stress is equal 16% for variant I, whereas in the case of load applied on eccentricity 11% maximum increase was obtained in variant IV. It is worth noting that the case of a force applied directly to the post (see Fig. 3a) the influence of tightening torque increases in welded post with closer location of end-plate to the fixed support. In the case of load applied on eccentricity (see Fig. 3b) the influence of tightening torque increases with an increased dimension of end-plate location in relation to the support due to the beam torsion. Given an information that the bolted connection should have been designed in the areas with possibly the lowest bending momentum forces as well as considering HMH stress values under the corner hex bolts washers, variant IV was chosen as the best from analysed solutions.

Despite the fact that Tab. 1 and Tab. 2 show the stress values correspond to the load reach its maximum value and each value exceeded adopted steel yield stress, the graph of stress-time function should also be discussed. In variant IV, where load was applied directly to the HEA profile, values of absolute maximum HMH stress under high tensile bolts with introduced tightening torque in the function of time was presented in Fig. 4. The location of bolts was presented with respect to the YZ-plane in the view of fixed support (see Fig. 2).

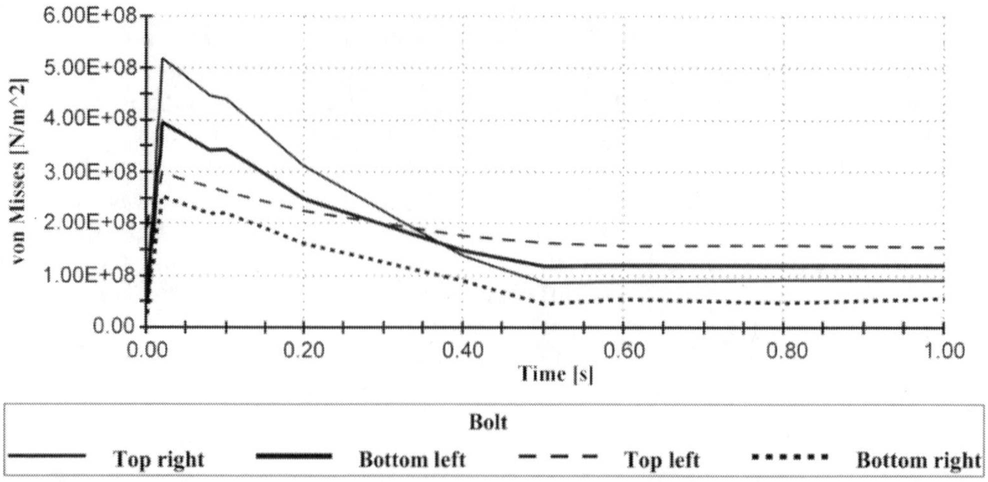

Fig. 4. Absolute maximum Mises stress in the time function for areas located under high tensile bolts washers in variant IV model. HEA profile directly subjected to the impact load. Other parameters: tightening torque equal 1140 Nm.

On the basis of Fig. 4 one can state that the applied impact load in time $t = 0.02$ s, where force reached its maximum value, plastic areas in end-plate connection under bolts washers could be observed. Moreover, plastic work of beam could also be observed near the fixed support and also near the connection of retaining cable. It is worth noting that plastic work of beam elements and end-plate connection was relatively short, approximately around 0.25 s and also the areas of plastic work were relatively small. Structure material was not ruptured. Considering the curve courses from Fig. 4, one may notice, that after approximately 0.50 s, stresses under hex bolt washers had increased in comparison to the original state, at which only initial stress due to introduced tightening torque had occurred. The mentioned stress increase was linked with the beam deformed shape arisen due to applied impact load – beam had not returned to the original rectilinear shape due to plastic work. In consecutive time steps values of HMH stress slightly fluctuated which was the result of beam vibrations as an effect of applied impact load, however it should be noted that further load impacts of boulder were neglected in the analysis. The plastic areas and scaled deformation for barrier post subjected to the eccentrically applied load at $t = 0.50$ s were presented in Fig. 5.

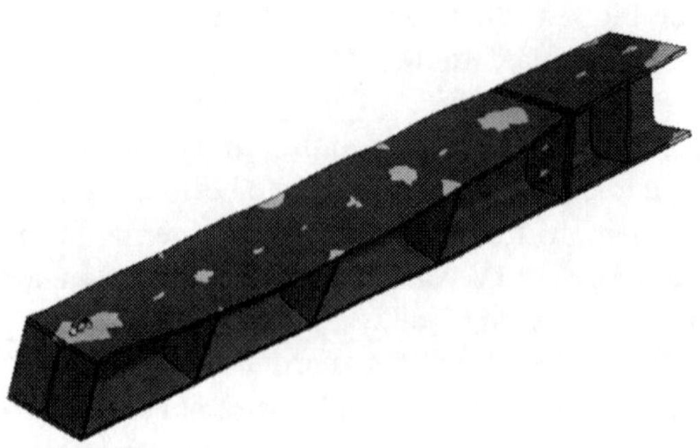

Fig. 5. Plastic areas and scaled deformation in the HEA beam subjected to the eccentrically applied load at $t = 0.50$ s.

4 Conclusion

The paper presents numerical analysis concerning the influence of high tensile bolts tightening torque on stress distribution in fixed rockfall barrier post with four different locations of end to end-plate connection. Two types of loading were taken into consideration i.e. impact load applied directly onto the HEA post profile and impact load applied on the eccentricity – boulder caught via net.

The analysis shows that the longer distance of end-plate connection from the fixed support the lower HMH stress values under the bolts were obtained. Moreover, it was shown that introduced maximum tightening torque, depending on the end-plate location increased HMH stress by 2 up to 16% under high tensile bolts in case of load applied exactly on barrier post, whereas in eccentrically loaded barrier by 2 to 11%, respectively. Despite that in both loading cases, when load reached its maximum value stress results exceeded the steel yield strength, the observed plastic areas were relatively small and material structure was not ruptured in the chosen model variant, where end-plate connection was distanced by 0,60 m from the fixed support. The numerical models presented in the paper can be only regarded as prototypes. Hence further analysis and model modifications should be considered.

References

1. N. Matsuokaa, H. Sakaib, Geomorphology **28**, 309-328 (1999)
2. J. Gardner, Albertan Geographer **6**, 15–20 (1977)
3. F. Letto, F. Perri, F. Cella, Catena **145**, 301-315 (2016)
4. Y. Kobayashi, E. L. Harp, T. Kagawa, Rock Mechanics and Rock Engineering **23**, 1-20 (1990)
5. G. B. Crosta, F. Agliardi, Natural Hazards and Earth System Sciences **4**, 583-598 (2004)
6. H. Aksoy, M. Ercanoglu, Natural Hazards and Earth System Sciences **6**, 941-954 (2006)
7. K. Thoeni, A. Giacomini, C. Lambert, S. W. Sloan, J. P. Carter, International Journal of Rock Mechanics and Mining Sciences **68**, 107-119 (2014)
8. F.Agliardi, G.B.Crosta, International Journal of Rock Mechanics and Mining Sciences **40**, 455-471 (2003)
9. https://www.fastenermart.com/files/metric_tighten_torques.pdf (date accessed: 20th December 2017)
10. T. Moon, J. Oh, B. Mun, Engineering Geology **172**, 41-56 (2014)
11. S. de Miranda, C. Gentilini, G. Gottardi, L. Govoni, A. Mentani, F. Ubertini, Engineering Structures **85**, 83-94 (2015)

Critical load of steel trusses - graphical methods

Piątkowski Michał[1,*]

[1]Koszalin University of Technology, Faculty of Civil Engineering, Environmental and Geodetic Sciences, Śniadeckich 2, 75-453 Koszalin, Poland

Abstract. The article presents graphical methods for determine critical loads of column and beam elements based on experimental results. The author presents the principles of using each method and the results of author's experimental tests on the instability phenomenon of planar steel truss with imperfections. The discussed methods were used to determine critical load of the tested truss, next compared with the results of numerical analysis. The validity of methods for determining the critical moment in the truss analysis has been confirmed.

1 Introduction

Knowledge of the critical load of compressive and flexural elements is important in the case of design considering global stability of structures: buckling and lateral-torsional buckling [1]. This load can be determined using empirical formulas, e.g. Euler formula, or computer programs, e.g. LT-Beam.

The critical load can also be determined experimentally [2]. However, experimental models always are *imperfect* structures whose equilibrium paths are characterized by limit points rather than bifurcation points [3]. The equilibrium path of the imperfect systems (formula 1a) are characterized by the critical load, which is the horizontal asymptotes and the initial imperfections that affect of the non-linear relationships between load and displacements [4] (figure 1c).

Experimental determination of the critical load of compressed elements is related to the transformation of Southwell, which converts the formula (1a) to the linear form (formula 1b). This function, in the coordinate system (δ / P; δ), intersects the vertical axis in the point, which is the initial imperfection (- e_0), while the tangent of its α angle is the critical force of the elements (figure 2.).

$$\delta = \frac{e_0}{\frac{P_{cr}}{P} - 1}\ ;\ \delta = \frac{\delta}{P} P_{cr} - e_0 \qquad \text{(1a); (1b)}$$

where:

δ transverse displacement of the compressed element,

* Corresponding author: michal.piatkowski@tu.koszalin.pl

e_0	initial imperfection,
P_{cr}	critical load of a perfect element,
P	real compressive load.

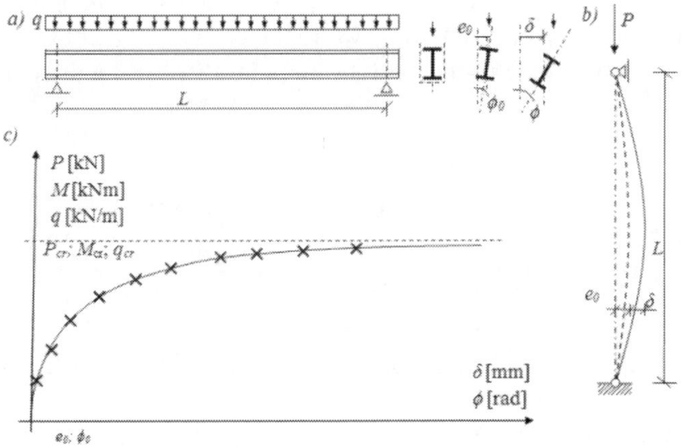

Fig. 1. Imperfect elements: a) beam; b) column; c) equilibrium path.

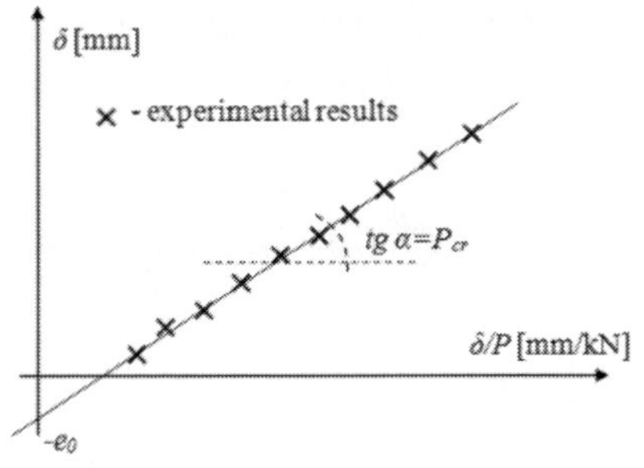

Fig. 2. Southwell's Method.

Similar rules relate to the experimental determination of the critical moment of bending elements. These elements are assumed with initial "out-of-straightness" imperfections e_0 and initial torsional imperfection ϕ_0. The increment of the twist angle ϕ in the beam is analogous to the increase of deformation δ in the compression element and can be expressed by the formula [3]:

$$\phi = \frac{\phi}{M} M_{cr} - \phi_0 \tag{1c}$$

where:

ϕ	angle of torsional deformation,
ϕ_0	initial torsional deformation,
M_{cr}	critical moment of bending beam,
M	max. moment.

2 Methods to determine the critical moments

Initial attempts to use the Southwell method in the analysis of experimental data of bending beams showed a good agreement between the slope of the line and the empirically determined critical moment [5]. In the 20th century, many researchers proposed their own modifications of the graphic method, e.g.:

- Massey (1963),

- Trahair (*Modified Plot* - 1969),
- Meck (*Skewed Plot* - 1977).

These methods employ the different coordinate systems and different forms of equations (1b) and (1c). The Massey's method should be used for beams at bending with a single initial imperfection e_0 or ϕ_0 [5]. The Massey's plots are constructed in the coordinate system (δ/M^2; δ) or (ϕ/M^2; ϕ), while the critical moment (or critical load) is the square root of the slope angle of the straight line (2a) or (2b). This method involves linear equations in the form of:

$$\delta = \sqrt{M_{cr}}\,\frac{\delta}{M^2} - e_0 \,;\; \phi = \sqrt{M_{cr}}\,\frac{\phi}{M^2} - \phi_0 \qquad (2a);\ (2b)$$

The Trahair's modification rely on transforming the formulas (1b) and (1c) into the form:

$$\delta \cdot M = \delta \cdot M_{cr} - \delta_0 \cdot M \,;\; \phi \cdot M = \phi \cdot M_{cr} - \phi_0 \cdot M \qquad (3a);\ (3b)$$

This transformation, called *Modified Plot*, creates a straight line in the coordinate system (δ; δM), whose angle of inclination is a critical moment, unfortunately it doesn't give a graphical representation of the initial imperfections in an obvious way [6].

The last variant of the graphical method, discussed in this paper, is called "*Skewed Plot*" due to the specific coordinate system, in which the lateral displacement and the twisting of the cross-section occur simultaneously.

$$\delta = \alpha\frac{\phi}{M} - \delta_0 \,;\; \phi = \beta\frac{\delta}{M} - \phi_0 \qquad (3a);\ (3b)$$

where:

$$\alpha = \frac{\pi^2 \cdot EJ}{L^2} \,;\; \beta = G \cdot J_T$$

E	modulus of elasticity,
J	second moment of area - weak axis
L	length of elements,
G	shear modulus,
J_T	second polar moment of area.

In this method, the critical load is calculated on the basis of both plots according to the formula:

$$M_{cr} = \sqrt{\alpha \cdot \beta} \qquad (3c)$$

The work of planar trusses are resembles the work of beams at bending (truss is a kind of beams with a castellated web). The experimental determination of the critical load of the compression chord in this elements can be done by different ways: using the equilibrium path of a separate chord (applying the Southwell method); or using the equilibrium path of an entire truss (by the Southwell or Meck method). Typically, in experimental studies, the critical load is firstly determined and next the critical force or critical moment are calculated. It should be noted that the use of the equilibrium path of a separate compression chord does not allow to determine initial imperfections of an entire truss (e_0 i ϕ_0), instead a certain overall imperfection only.

In the experimental tests an external critical load is usually determined. Next the critical force of the compression element or the critical moment of the bending element is determined based on the known critical load value. The stability problem in beam elements is well recognized and described, e.g. in [7].The relationship between the critical moment M_{cr} of the bending I-beam and the critical force N_{cr} of the compression chord was presented in the paper [8]. The literature also includes information on equivalent truss models in the analysis of bending I-beams [9]. Thus there is an analogy between buckling of the upper truss chord and lateral-torsional buckling of the bending beam.

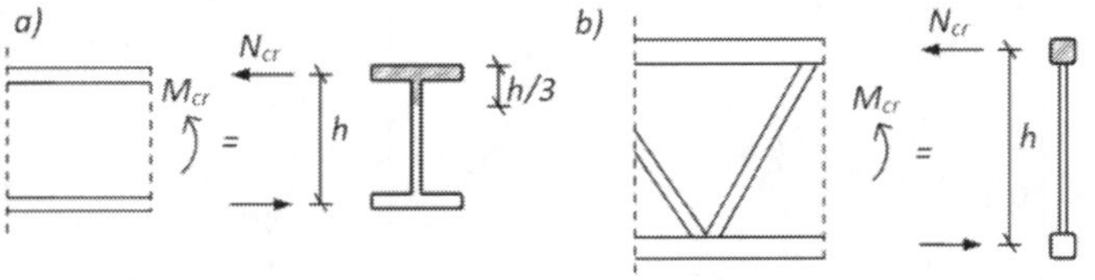

Fig. 3. Relation between the critical moment of the beam and the critical force of the compression chord: a) I-beam; b) truss.

3 Experimental research

Comparison of graphical methods for determining the critical moment in planar trusses was carried out on the results of author's experimental research. That research concerned seven meter long research models shown in Figure 4. The tests were carried out on simply supported single-span models with lateral, rigid support at the ends of the chords. At the beginning of the tests, the measurement was made of real, initial out-of-plane imperfections in the state without external load. During the study, the gravitational load was applied gradually in the nodes of the upper chord. The tests were carried out in the elastic range, limiting the possible deformation of chords up to 150 mm. The research models were made of steel with declared yield point of f_y = 235 MPa. The results of experimental tests, in the form of horizontal displacements of the upper and bottom chords, relative to the measured initial imperfections, are presented in Table 1. The displacements were measured at the chord nodes using inductive displacement sensors with a range of 300 mm and an accuracy of 0.001 mm.

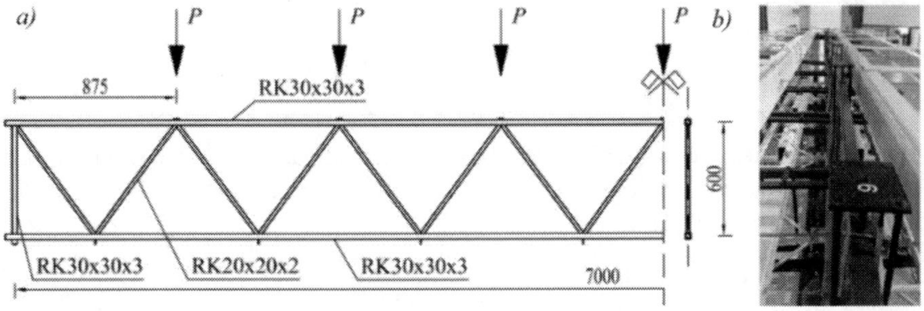

Fig. 4. Research model: a) scheme; b) Lateral-torsional buckling of a truss.

The research studies included also determination of the impact of various, simultaneous systems of chord imperfections on the equilibrium path of the truss. These configurations, e.g. opposite chord imperfections, were induced by means of an additional lateral load. The tests showed a significant effect of the imperfection of the tension chord on the shape of the equilibrium path of the compression chord and therefore also on the limit load on the truss [10]. The article presents only the basic results of experiment obtained for imperfect models without forcing any special imperfection configurations.

For the considered static scheme, the relationship between the load and the maximum bending moment of the equivalent beam can be expressed in the following form (this same is true for critical load and critical moment):

$$M_{max} = L \cdot P \tag{4}$$

Table 1. Experimental results.

Nodal load P [kN]	Model 1		Model 2		Model 3	
	upper chord (comp.) δ_u [mm]	bottom chord (tension) δ_b [mm]	upper chord (comp.) δ_u [mm]	bottom chord (tension) δ_b [mm]	upper chord (comp.) δ_u [mm]	bottom chord (tension) δ_b [mm]
$e_{0.u}$; $e_{0.b}$ initial imperfection (middle of chord)	4.803	1.203	5.399	3.304	10.412	9.711
0	0	0	0	0	0	0
0.20	0.024	0.060	0.438	0.223	0.101	0.015
0.40	3.853	1.050	2.871	1.731	1.444	-0.572
0.60	8.271	4.173	10.279	6.933	5.217	1.1815
0.65	11.605	6.582	12.957	8.835	6.717	2.072
0.70	14.623	8.500	17.410	11.947	10.071	4.379
0.75	18.284	11.261	24.741	17.170	14.722	7.524
0.80	25.281	16.249	34.758	24.100	20.057	11.085
0.82	31.504	20.558	39.746	27.424	23.577	13.754
0.84	37.410	24.740	45.217	31.086	28.369	16.692
0.86	42.855	28.413	53.474	36.752	33.364	19.332
0.88	53.036	35.408	60.566	41.475	39.842	23.957
0.90	61.637	41.264	77.964	53.598	46.099	28.441
0.92	69.766	46.502	96.275	66.020	57.653	35.975
0.94	89.142	59.994	132.429	91.225	69.688	43.888
0.96	-	-	-	-	106.986	68.567

On the basis of the obtained results, general deformation of the entire truss was determined: horizontal displacement δ_{truss} and rotation ϕ in accordance with Fig. 5b. These results are summarized in Table 2.

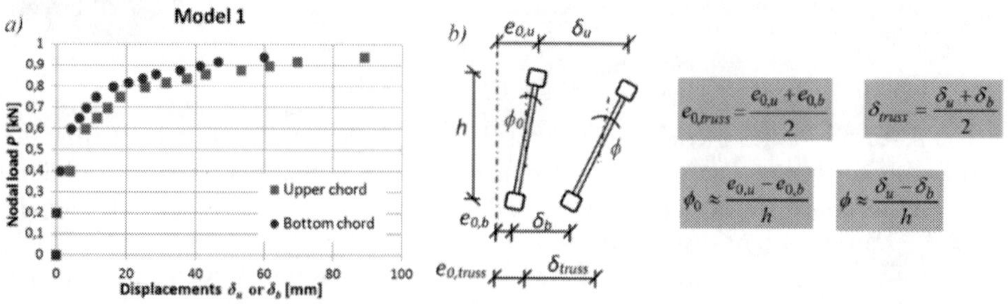

Fig. 5. Measured chord deformations: a) equilibrium path of the chords - Model 1; b) the calculating method of the truss deformation.

Table 2. Deformation of truss.

Nodal load P [kN]	Moment (form. 4) [kNm]	Model 1		Model 2		Model 3	
		Displacements δ_{truss} [mm]	Rotation ϕ [rad]	Displacements δ_{truss} [mm]	Rotation ϕ [rad]	Displacements δ_{truss} [mm]	Rotation ϕ [rad]
initial imperfection $e_{0,truss}$; ϕ_0		3.003	0.006	4.352	0.0035	10.062	0.0012
0	0.00	0.000	0.000	0.000	0.0000	0.000	0.0000
0.20	1.40	0.077	0.00016	0.331	0.0004	0.058	0.0001
0.40	2.80	2.452	0.00467	2.301	0.0019	0.436	0.0034
0.60	4.20	6.222	0.00683	8.606	0.0056	3.199	0.0067
0.65	4.55	9.094	0.00837	10.896	0.0069	4.395	0.0077
0.70	4.90	11.562	0.01021	14.679	0.0091	7.225	0.0095
0.75	5.25	14.773	0.01171	20.956	0.0126	11.123	0.0120
0.80	5.60	20.765	0.01505	29.429	0.0178	15.571	0.0150
0.82	5.74	26.031	0.01824	33.585	0.0205	18.666	0.0164
0.84	5.88	31.075	0.02112	38.152	0.0236	22.531	0.0195
0.86	6.02	35.634	0.02407	45.113	0.0279	26.348	0.0234
0.88	6.16	44.222	0.02938	51.021	0.0318	31.900	0.0265
0.90	6.30	51.451	0.03396	65.781	0.0406	37.270	0.0294
0.94	6.44	58.134	0.03877	81.148	0.0504	46.814	0.0361
0.94	6.58	74.568	0.04858	111.827	0.0687	56.788	0.0430
0.96	6.72	-	-	-	-	87.777	0.0640

4 Critical moment of a planar truss

The numerically determined critical loadvalue of the researched structure is 0.971 kN. This load was determined in the SOFiSTiK program by means of linear buckling analysis (LBA) [11]. Numerical analysis was performed on a perfect model with nominal material and cross-section characteristics. The critical force in the compression chord corresponding to the critical load is 11.3 kN, while the appropriate buckling length factor is $\mu = 0.374$.

Based on the values summarized in Tables 1st and 2nd selected plots were developed.

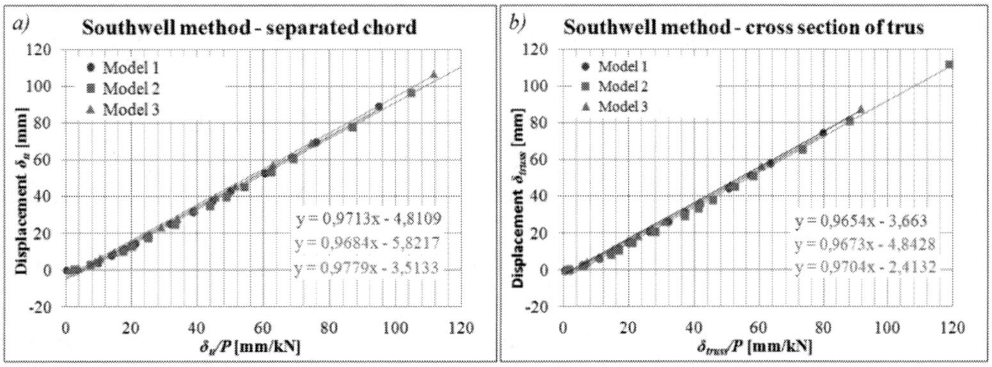

Fig. 6. Graphical results of the Southwell method (all models): a) for the separated upper chord; b) for the whole truss.

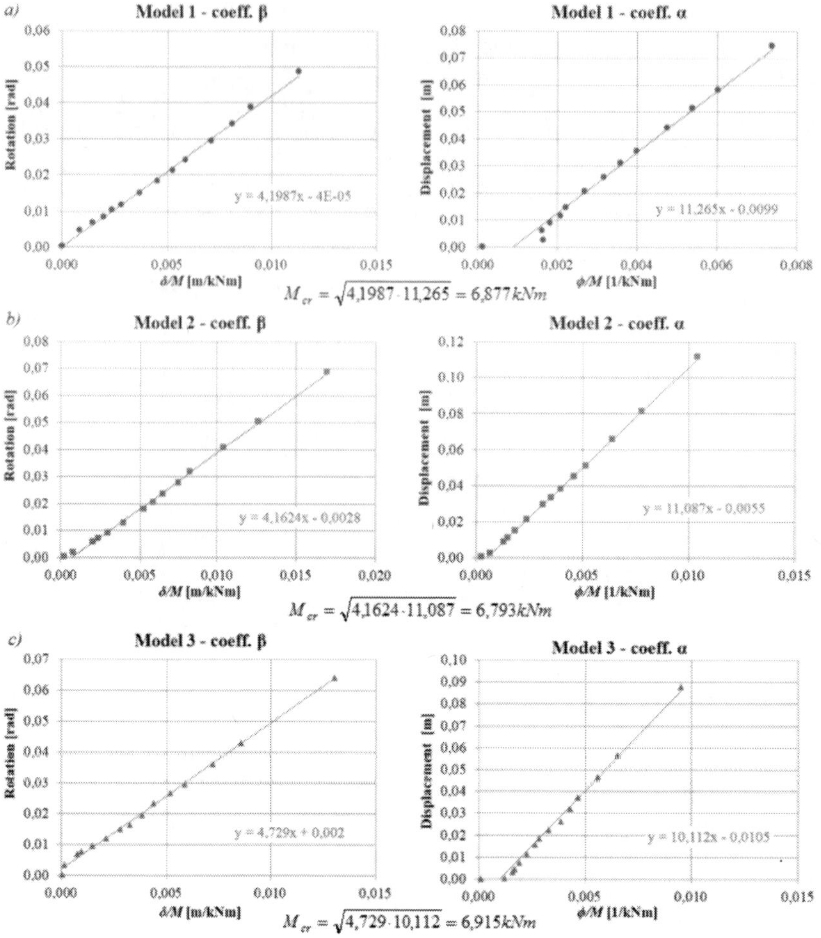

Fig. 7. Graphic results of the Meck method: a) Model 1; b) Model 2; c) Model 3.

The critical loads obtained on the basis of the values from the diagrams are shown in Table 3.

Table 3 Critical load of truss.

Method		Model 1		Model 2		Model 3	
Southwell (separated, upper chord)		0.9713		0.9684		0.9779	
Southwell (whole truss)		0.9654		0.9673		0.9704	
Massey							
($\delta/M^2;\delta$)	($\phi/M^2;\phi$)	0.979	1.011	0.987	1.003	0.976	**1.034**
Trahair							
($M\delta;\delta$)	($M\phi;\phi$)	0.950	1.005	0.953	0.974	0.965	1.012
Meck		0.983		0.970		0.988	

Comparing the obtained results, it should be recognized, that all the methods presented at the beginning of the paper allow to properly determine the critical load on the truss due to the lateral-torsional buckling (buckling of the compression belt). The differences between the respective average values and the value obtained on the basis of the numerical analysis do not exceed 10%. The average value obtained of all experimental results is 0.981 kN, while the maximum deviation from the average value is 5.4% (1.034 kN - Massey model 3 - coordinate system $\phi/M^2;\phi$).

5 Analysis of results and conclusions

The article compares the methods of graphical determination of the critical load on the basis of experimental data. The best-known methods and results of the author's experimental tests are presented. Based on the obtained results, the critical load was determined according to the rules each of the described method. Selected methods are presented in the form of graphical plots. The conducted analysis and obtained results allow for the following conclusions:

- The possibility of applying graphical methods, dedicated to beam elements (Massey, Trahair. Meck) to determine the critical load of a planar truss (in the case of the analyzed static scheme) confirms the validity of the use of the term "lateral-torsional buckling of the truss".
- The critical load values calculated on the basis of plots based on the torsion (Massey and Trahair method) are greater than their counterparts calculated on the basis of the horizontal displacement of the truss. This situation may result from neglecting the influence of the vertical displacement of each belt on twisting of the truss.
- Due to small differences between the numerical and experimental value of the critical load for all methods, it can be assumed that for the purposes of engineering practice, these methods can be used interchangeably.
- It can also be assumed that the analysis of the equilibrium path of a separate upper chord is sufficient to assess the critical load of an entire truss structure.

References

1. PN-EN 1993-1-1:2006 Eurocode 3: Design of steel structures. Part 1-1: General rules and rules for building [in Polish]
2. S.P. Timoshenko, J.M. Gere, *Theory of elastic stability*, Arkady, Warszawa (1963)
3. T.V. Galambos, A.E. Surovek, *Structural stability of steel: Concepts and applications for structural engineers*, John Wiley and son. New Jersey (2008)
4. W.F. Chen, E.M. Lui, Structural stability: Theory and implementation, Elsevier (1987)
5. P. Mandal, C.R. Calladine, Lateral-torsional buckling of beams and the Southwell plot, International Journal of Mechanical Sciences, **44**, 2557-2571 (2002)
6. I. Kalkan, Application of Southwell Method on the analysis of lateral torsional buckling tests on reinforced concrete beams, International Journal of Engineering Research and Development, **2**, 58-66 (2010)
7. M. Giżejowski, R. Szczerba, M. Gajewski, Influence of imperfections on LTB resistance of steel rolled and welded beams, Journal of Civil Engineering. Environment and Architecture, **3(1)**, 447-460 (2017)
8. H. Mehri, R. Crocetti, P.J. Gustafsson, Unequally spaced lateral bracings on compression flanges of steel girders, Structures, **3**, 236-243 (2015)
9. S.V. Smyrnaios, A. Iliopoulos, I. Vayas, Truss models for inelastic stability analysis and design of steel plate girders, Engineering Structures, **105**, 165-173 (2015)
10. M. Piątkowski, Elastic lateral buckling of steel truss with imperfections, Materiały Budowlane, **8**, 82-83 (2016) [in Polish]
11. SOFiSTiK - Structural Desktop: User interface of SOFiSTiK software. SOFiSTiK AG. (2008)

Analysis of load carrying capacity and destruction mechanism of an intermediate anchorage steel plate for pillar jib crane

Małgorzata Gordziej-Zagórowska[1,*], and *Wojciech* Migda[1]

[1]Gdańsk University of Technology, Faculty of Civil and Environmental Engineering, ul. Narutowicza 11/12, 80-233 Gdańsk, Poland

Abstract. Pillar jib cranes are often mounted to existing floors with the use of an intermediate anchorage steel plate instead of using independent foundations. The process of designing such an intermediate plate for a pillar jib crane and a destruction pattern for plates of insufficient thickness are presented in this article. The required steps of the designing process are introduced and also the required level of detail for the numerical analysis. The aim of this paper is to provide a secure and safe approach to the design of such intermediate anchorage plates.

1 Introduction

Pillar jib cranes find their way to all kind of different production facilities very often after the design and erection process is finished, therefore a way of secure mounting to existing floors has to be provided. The simplest and easiest way for this is the use of a so called intermediate steel anchorage plate (Fig. 1). The basic design as for the used shape is simple, a plate can be either circular or square (sometimes a rectangular shape might be useful). The size of the designed plate can be estimated by the maximum allowed tension force in the anchor bolts. Standards for steel design [1], [2] provide only basic rules for the design of such an intermediate plate and the design of (planar) dimensions and the thickness of the plate require additional calculations. For the calculations of the load carrying capacity of an intermediate steel plate the use of a FEM (finite element method) model is very helpful, several aspects, beside the dimension and thickness and material strength of the steel plate, need to been taken into consideration. These aspects are: stiffness of the floor the baseplate will be anchored to, the concrete strength of the floor, the stiffness of the soil underneath the floor, the placement and diameter of the anchor bolts and the tensile strength.

While the initial estimation of the proper placement of the anchors and the dimension of the plate is not very difficult due to existing information for calculations [2], [3], [5] the whole design process requires proper understanding of how the different parameters interact with each other and how they influence the final result – the correct dimensions of the plate.

* Corresponding author: malgor@pg.edu.pl

2 Steps of the design process

The design process of an intermediate anchorage steel plate for a jib pillar crane can be divided into the following steps:

- defining of the basic known parameters: design forces for the crane, dimension of the pillar base plate and placement of the bolts (fasteners between base plate and intermediate plate), thickness of the floor plate (slab), material properties of the floor and the soil parameters for the overall floor stiffness,
- initial definition of the intermediate plate planar dimensions and anchor placement based on the floor thickness,
- definition of the FEM model and the proper application of the supports and acting forces,
- analysis of the results.

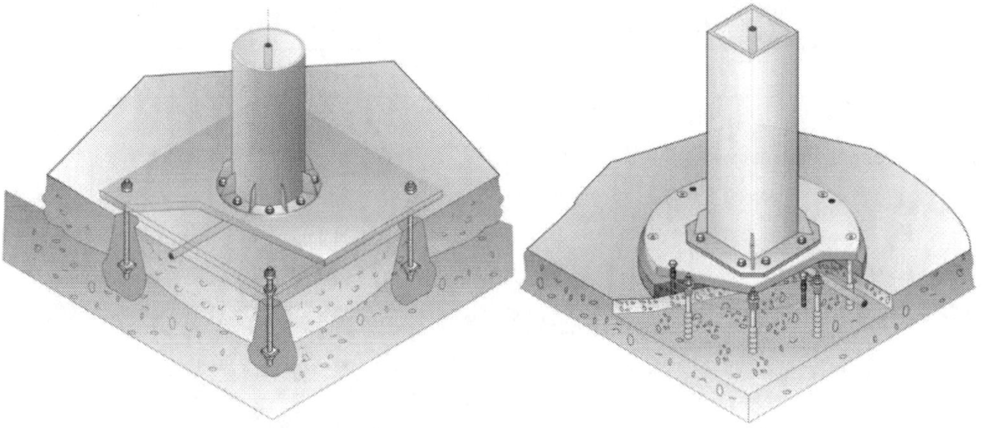

Fig. 1. Square and round intermediate anchorage plate [3].

3 Definition of the basic parameters

The basic parameters required for the proper design start with the design working loads and the geometry of the base plate anchorage for the chosen pillar crane. For the example used in this paper a jib pillar crane model LS 235 from ABUS with a span of 5000 mm and a work load of 1000 kg was used, as presented in Fig. 2. The characteristic loads are: axial load (vertical force at base plate) $N_F = 17.7$ kN and overturning moment (OTM) $M_F = 56.7$ kNm, these values have been used as these have been provided by the producer for the calculations of a intermediate plate.

In the next step the thickness of the concrete floor has to be determined, for this analysis a thickness of 300 mm was assumed. Further more it was assumed that the concrete is of grade C20/16. The effective maximum usable length for chemical (adhesive) anchoring bolts is less then the floor thickness, because while drilling through the whole floor at the bottom of the floor a cracking can occur as shown in Fig. 4. It is recommender to reduce the drill depth by 50 mm to avoid the cracking, it is also recommended to drill the last bits without hammering.

The following step requires few initial assumptions and simplified calculations as a starting point – these are the diameter of the intermediate plate (in case of a square plate the length of the sides should be equal to the diameter of a circular plate) and the distance between the anchors.

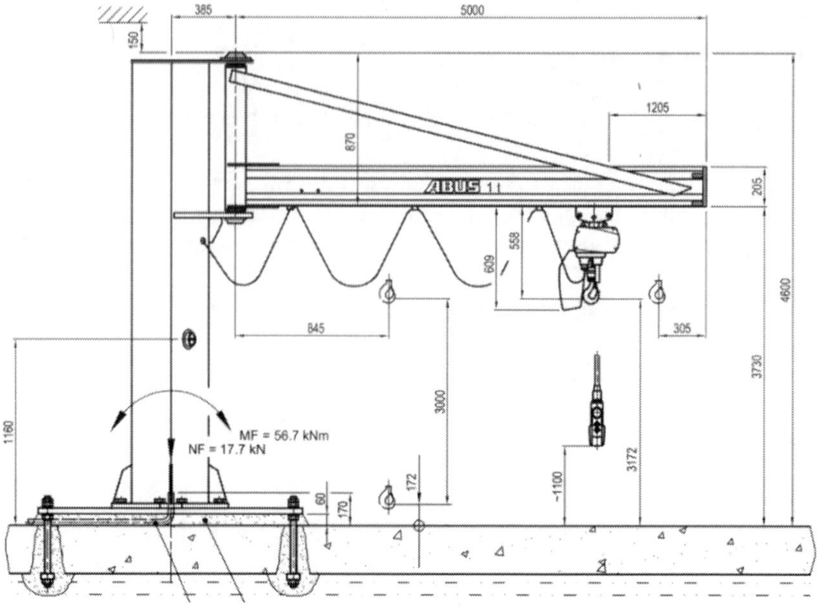

Fig. 2. General dimensions of the jib pillar crane.

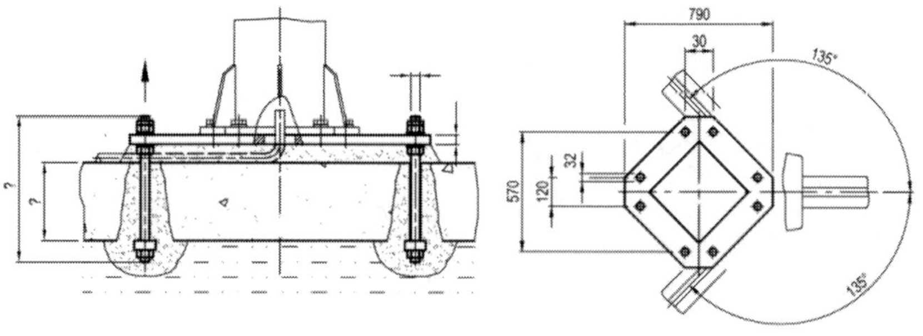

Fig. 3. Dimensions of the base plate and the hole for the base plate bolts.

The distance between the anchors can be calculated based on the number of anchors and basis of the length of the anchors assuming that the concrete cone of force dispersion around every anchor – assuming that chemical anchors provide higher tensile (anchorage) force than the concrete cone of force dispersion - has an angle of spread of 33.7° (according to [4] – 8.10.3 (5)), which is not only a safer approach (by reducing the size of the cone) then using the typical angle for the cone of spread for concrete that is around 55° but seems more accurate since these anchors are pre-tensioned (just as tendons, for which [4] suggests 33.7°) before the workload of the crane and correlating forces occur.

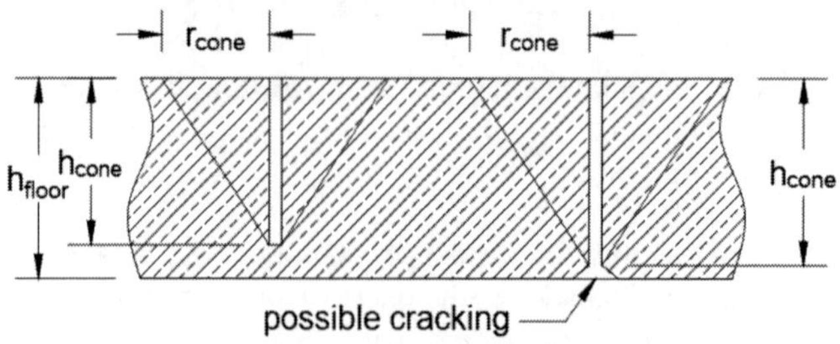

Fig. 4. Dimensions for the estimation of the concrete breakout cone.

Usually the use of 12 anchors as total provides a good distribution of forces leaving 30° between anchors. The distance between anchors should be calculated so that the cones of spread around the anchors don't interact. With this in mind the distance between the anchors d_{cone} should be not less as presented in equation (1):

$$d_{cone} = 2 * r_{cone} = 2 * h_{cone} * \tan 33.7 \quad (1)$$

where:

r_{cone}	radius of the cone, $r_{cone} = 167$ mm,
h_{cone}	effective length of the anchor bolt – a safe assumption is $h_{cone} = h_{floor} - 50$ mm,
h_{floor}	thickness of the concrete floor.

4 Initial calculations for the intermediate plate dimensions

Based on the distance between the anchors, the diameter D_{plate} of the intermediate plate can be calculated (2). A larger diameter of the intermediate plate will result in lower forces in the anchor bolts and also in lower compression force on top of the floor.

$$D_{plate} \cong \frac{n * d_{cone}}{\pi} + 2 * b \quad (2)$$

where:

n	number of anchors – for this study 12 anchor bolts have been used,
b	distance from centre of the anchor axis to the edge of the intermediate plate – 50 mm is usually a good value.

Using the above equations and assumptions (h_{cone} = 250 mm, n = 12, b = 50 mm) the calculated diameter of the intermediate plate is 1373 mm, for the further calculations a diameter of 1400 mm was used.

At this stage it is possible to check if the overturning moment of the crane, as a pair of forces will generate a force in the most exposed anchor bolt that is smaller than the resistance of the cone of force of dispersion (F_{cone}), for this the area of the cone (A_{cone}) has to be calculated and the total force of dispersion (3).

$$F_{cone} = A_{cone} * \tau_{Rd,C20/16} = \pi * r_{cone} * h_{cone} * \tau_{Rd,C20/16} \quad (3)$$

where:

A_{cone}	conical surface, $A_{cone} = 1311$ cm^2,
$\tau_{Rd,C20/16}$	shear resistance of concrete (according to [6]), $\tau_{Rd,C20/16} = 0.022$ kN/cm^2,
F_{cone}	force of dispersion of the concrete cone around the anchor bolt, $F_{cone} = 28.8$ kN.

With the force of dispersion calculated the class and diameter of the anchor bolt can be determined, for this example anchors of class 5.6 [2] and M20 have been used.

For the numerical analysis yet another parameter is required – the stiffness of the floor k. The stiffness can be either experimentally determined, which is the most preferred solution – for this a load should be placed at the location of the designed intermediate plate and the deflection of the floor under the load should be measured. If this is not possible the stiffness of the floor should be calculated (using computer software like Autodesk Robot Structural Analysis) or estimated, for instance using the procedure presented in [7]. A value of k = 0,03 MPa/mm = 30 000 kN/m^3 can be used for industrial floors with a thickness of 300 mm (including the underlying soil), this means that a force creating a pressure of 0,03 MPa will generate a displacement of the floor that is equal to 1 mm. By implementing the floor and soil stiffness in such a way takes the soil and floor in this calculations into account without the need of modelling them explicit. It is yet recommended to measure the stiffness of the floor if possible, because it will provide more accurate results.

5 Numerical model and analysis

With all the initial parameters established the numerical FEM model was created using the program RFEM by Dlubal. The following parameters have been incorporated:

- diameter of the intermediate plate: 1400 mm,
- radius of the axis of the anchor bolts: 650 mm (= 1400/2 -50),
- node supports for anchor bolts, supporting only displacements ux, uy and uz only against tension, compression is handled by surface support,
- surface support with a stiffness of 30 000 kN/m^3, and providing only support against compression,
- load transfer from crane pillar base plate to intermediate plate was handled with nodal forces located at the places of bolts (as shown in Fig. 3) – 8 nodes,
- the overturning moment was implemented by replacing a moment with a pair of forces (as shown in Fig. 6),
- the vertical force was equally applied to all 8 nodes (as shown in Fig. 6),
- two different cases have been analysed with different safety factors: one to estimate the forces in the anchor bolts and one for the contact compression for the floor – the load and combination cases are presented in Table 1 and 2,
- different thicknesses of the intermediate plate have been used, starting at 50 mm and reducing the thickness by 5 mm up to a thickness of 10 mm,
- as the material of the intermediate plate steel of grade S355 was used,
- the size of the FE was set to 1 cm and the calculation method that was used, was a postcritical analysis.

Table 1. Load cases used in the numerical analysis.

Load case	Type of load
LC 1	dead load
LC 2	vertical force
LC 3	overturning moment

Table 2. Load combinations with safety factors used in the numerical analysis.

Load combination	Loads and safety multipliers
CO 1	0.9*LC1 + 0.9*LC2 + 1.5*LC3
CO 2	1.35*LC1 + 1.5*LC2 + 1.5*LC3

In Fig. 5 and Fig. 6 the FEM model is presented showing the node supports (light blue) and surface supports (red), also showing the nodal forces representing the vertical force N_F and the overturning moment M_F.

6 Evaluation of the results

The results of the numerical analysis have been presented Table 3. For comparison the maximum equivalent von Mises stress that occurred in the intermediate plate, the maximum force (in form of Pz reaction of the nodal supports) in the bolts and the maximum contact stress (in form of surface support) have been taken into consideration. The comparison was carried out for the different thickness of the intermediate steel plate, starting at 50 mm finishing at 10 mm with a step of 5 mm. As can be seen from the results, not only did the values of the examined parameters alter but also the distribution of stress and deformation of the intermediate plate. In Fig. 7 to Fig 9. the overall deformation of the plate, the contact stress and the equivalent von Mises stress are presented, from load case 1 (CO1) for the thinnest and the thickest plate analysed.

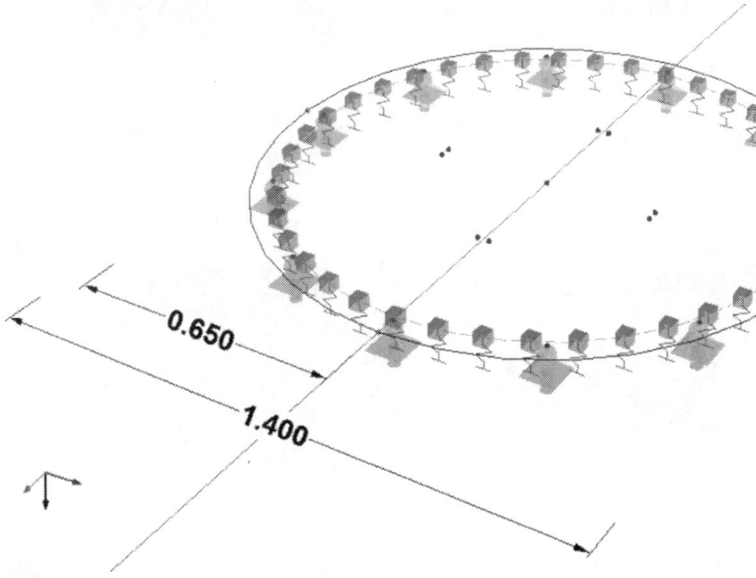

Fig. 5. FEM model of the base plate and the holes for the base plate bolts and supports.

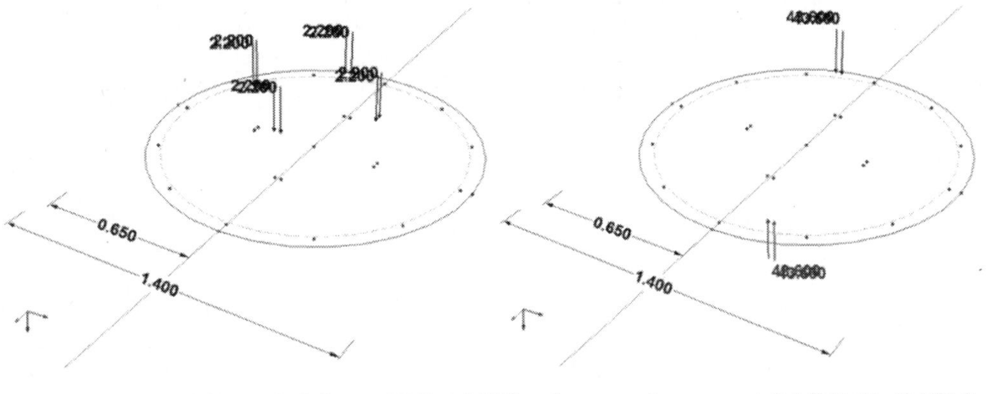

Fig. 6. Placement of the vertical forces ($N_F/8 \cong 2.2$ kN) and overturning moment ($M_F/2*0.57 \cong 49.7$ kN).

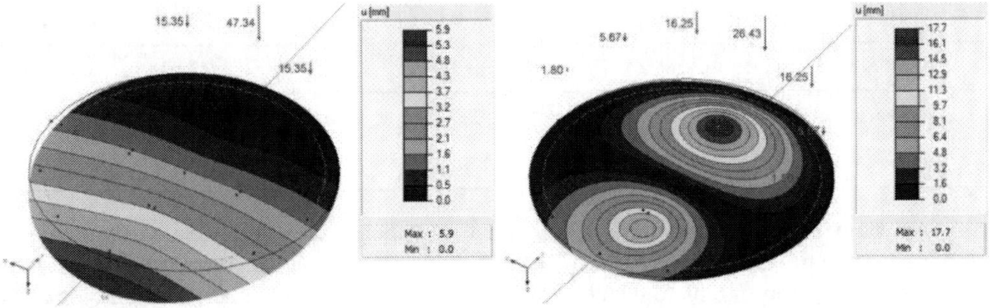

Fig. 7. Global deformation of the intermediate plate for 50 mm and 10 mm thickness (50 mm plate shown on the left).

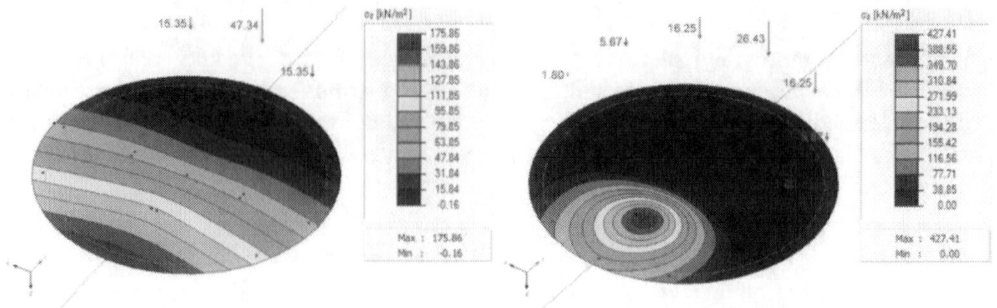

Fig. 8. Contact stress of the intermediate plate for 50 mm and 10 mm thickness (50 mm plate shown on the left).

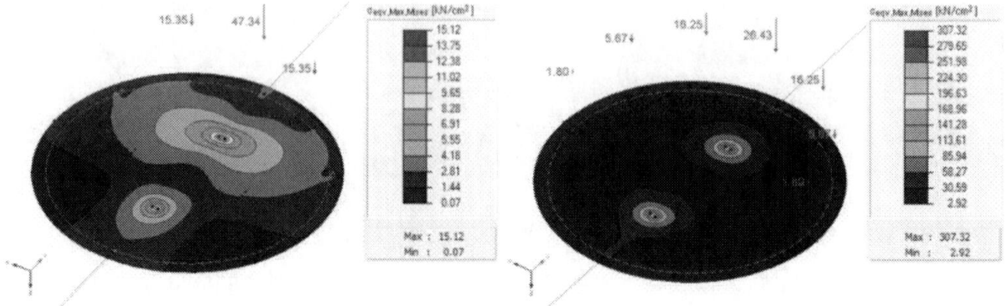

Fig. 9. Equivalent von Mises stress of the intermediate plate for 50 mm and 10 mm thickness (50 mm plate shown on the left).

Table 3. Comparison of the maximum results for contact stress, equivalent von Mises stress and axial force in anchor bolts for different intermediate plate thickness from the numerical analysis for both load combinations.

Load combination	**CO 1**			**CO 2**		
Plate thickness [mm]	$\sigma_{eqv,max}$ [kN/cm²]	$P_{z,max}$ [kN]	$\sigma_{contact,\,max}$ [kN/m²]	$\sigma_{eqv,\,max}$ [kN/cm²]	$P_{z,max}$ [kN]	$\sigma_{contact,\,max}$ [kN/m²]
50	15.12	47.34	175.86	14.80	50.69	187.85
45	18.46	37.52	182.46	18.05	39.93	193.80
40	23.12	29.76	190.58	22.61	31.36	201.19
35	29.92	24.24	200.43	29.30	25.16	210.01
30	40.44	23.73	210.42	39.57	22.67	219.18
25	58.01	24.30	217.84	56.71	23.32	224.81
20	90.03	24.77	220.22	88.09	23.83	231.36
15	151.29	25.47	292.50	150.79	24.58	307.39
10	307.32	26.43	427.41	297.38	25.56	444.38

As can be seen from the presented results the proper thickness of the intermediate plate is very important, because the use of a very thick plate results in increased force in the bolts, whereas a too thin plate results in a very large deformation and equivalent von Mises stress that would result in destruction of the intermediate plate and also increase in contact stress that could lead to local destruction of the concrete floor. The optimum thickness for the plate is 30 mm for the presented case, that is where the forces in the anchoring bolts are the lowest. The results in the bolt (first decreasing and then increasing, as shown in Fig. 10) indicate that a state of best (equal) force distribution is achieved by a plate of 30 mm thickness.

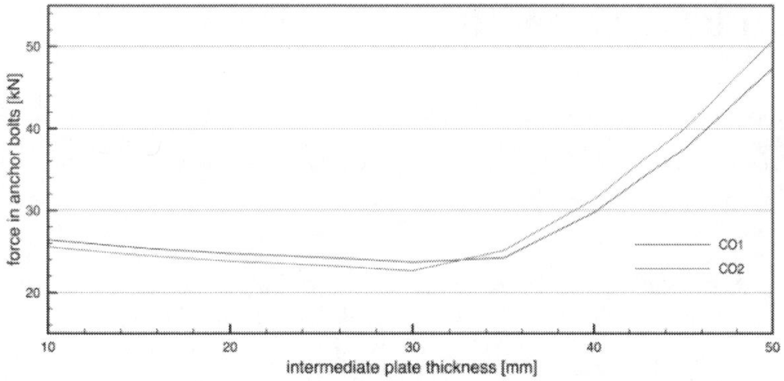

Fig. 10. Force in anchor bolts for both load cases in relation to the intermediate plate thickness.

7 Conclusions

A rather simple task as the design of an intermediate plate for a jib pillar crane requires a good understanding of the interaction of all the parameters that are involved in the design process of such a plate. This paper should give a fair overview of how these parameters influence one another and what aspects need to be considered. As can be seen from the analysis of the results, the most critical role plays the thickness of the intermediate plate. Furthermore it can be seen that the required FEM model for the calculations is very simple but delivers all needed information in order to verify the assumed geometrical and material properties of the designed plate.

References

1. EN 1993-1-5, Eurocode 3: Design of steel structures - Part 1-5: General rules - Plated structural elements, Eur. Comm. Stand. (2006)
2. EN 1993-1-8, Eurocode 3: Design of steel structures - Part 1-8: Design of joints, Eur. Comm. Stand. (2005)
3. ABUS Kransysteme GmbH, AN 12283 03.17, www.abuscranes.com
4. EN 1992-1-1, „Eurocode 2: Design of concrete structures - Part 1-1 : General rules and rules for buildings", Eur. Comm. Stand. (2004)
5. Model Code 1990: Design Code, CEB-FIB, Thomas Telford Services Ltd., Lausanne ISBN 0 7277 1696 4 (1993)
6. PN-B-03264:2002. Konstrukcje betonowe, żelbetowe i sprężone. Obliczenia statyczne i projektowanie
7. W. Ryżyński, B. Karczewski, Wyznaczanie parametrów obliczeniowych podłoża warstwowego pod posadzką przemysłową. Nowoczesne Hale, **5**, 14, 34-38 (2014)

Impact of single-layer dome modelling on the critical load capacity

Paweł Zabojszcza[1,*], *Urszula* Radoń[1], and *Paulina* Obara[1]

[1]Kielce University of Technology, Faculty of Civil Engineering and Architecture, al. Tysiąclecia Państwa Polskiego 7, 25-314 Kielce, Poland

Abstract. In the study an attempt was made to evaluate the effect produced by the modelling of the single-layer lattice dome on its critical load capacity. The modelling concerned the means of connecting bars in a node, bars of the lattice dome, and the effect of geometric imperfections. Taking steel covers, two basic means of modelling of how bars are connected in the node can be distinguished, namely pin and rigid joints. In the study, the pin joint was SBP-1 type connector, whereas the rigid joint - WABI-1 connector. In the description of bars, truss and frame elements were employed. Each element accounted for geometric nonlinearities in the Lagrange description. Regarding a frame element, the physical relationships represented the elastic behaviour of the structure with the use of the Hooke's law. With respect to the compression truss elements, a nonlinear relationship resulting from experimental investigations was additionally employed. Stability analysis of the structure was performed by means of the Finite Element Method using Abaqus and Robot Structural Analysis software. In order to obtain the load-displacement relations, the Riks arc length method was used. The analysis was focused on global modes of stability loss due to snap-through and bifurcation.

1 Introduction

Single-layer lattice domes are often used in many modern engineering structures. The key advantages offered by these structures include low dead load combined with relatively high critical load capacity. In the design of such structures stability analysis plays a vital role [1-5]. Stability loss in bar structures can be global or local in character. In the first case, bifurcation of the balance position or node snap-through can occur, the other case additionally involves buckling or lateral torsional buckling of individual bars of the system.

In literature, qualitatively different means of estimating the critical load capacity of lattice domes can be found. In the study [6], Głąbik gave formulas for the critical load capacity of a triangular mesh dome. In the formula, the slenderness of the bars and initial node deviation in assembly were accounted for. In the study [7], Bródka et al. estimated critical load capacity of a structure constructed from bars according to the Prandtl model using a method of successive approximations. Rządkowski, Kowal, Słowik, Wawszczak in

*Corresponding author: zabojszcza.pawel@gmail.com

studies [8,9] emphasised effect of geometric imperfections of bars, especially of their axes, on the critical load capacity. It can be even a few times lower than the critical load capacity of the dome with perfect geometry. Experimental investigations on the previously prepared models were frequently used [8,9].

In the paper an attempt was made to carry out numerical simulations by means of the commonly available software. The simulations concerned the assessment of the effect of single-layer lattice dome modelling on its critical load capacity. Different types of bar connection in a node, means of modelling dome bars and geometric imperfections were taken into account.

2 Means of modelling bars in the nodes of the single-layer lattice domes

Joints and nodes are necessary to transfer actions, to create constraints in the elements, or between them and to transfer external forces. Components of joints and nodes are susceptible to rupture, pressure, cut, local stability loss or shear [10]. In steel covers, basically two ways of bar connection in nodes can be differentiated, namely pin and rigid joints. In the study, the pin joint was SBP-1 type connector, whereas the rigid joint - WABI-1 connector.

In 1988, Schlaich-Bergermann & Partner, Stuttgart, Germany, published the basic principles of a reticulated structure with a splice connector, whose first implementation SBP-1 is shown in Figure 1. The node connector consists of two flat plates connected by a single central bolt. Simultaneously, a clamp for cable bracings can be connected to the node through the central bolt. Each structural member is connected to the horizontal splice plates by two or more bolts in single shear. The central bolt allows for an easy adjustment of the horizontal angle between the structural members. Vertical angles can be accommodated by means of folding the splice plates. Twist angles can be adjusted only in a highly limited range of imperfections. In consequence of the small section height of the splice plates, this node connector can transfer limited bending moments only [11].

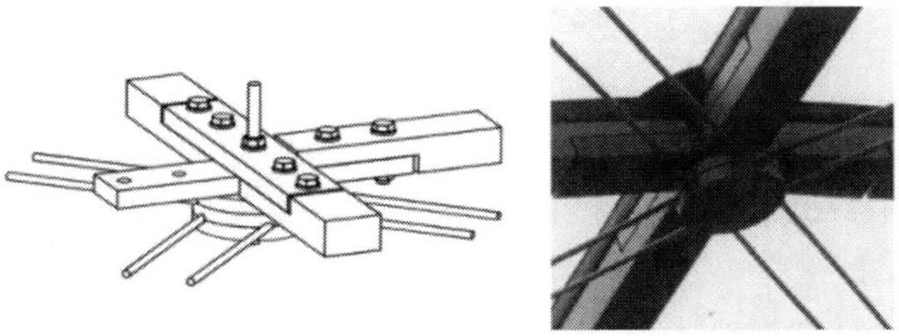

Fig. 1. Splice Connector SBP-1 [11].

The node, shown in Figure 2, is a welded end-face connector WABI-1, which was developed by Waagner-Biro AG, Vienna, Austria, for the courtyard roof of the British Museum in London. The node consists of a star-shaped plate with 5 or 6 arms. Each arm runs between adjacent structural members. These nodes are made of thick plates by cutting perpendicular to the plate surface. The end-faces of the structural members have a double mitre cut to match with the corresponding node gap between adjacent arms. The thickness of the node plate is less than the height of the connected structural members. Both, top and bottom surface of the node plates are connected to the members by fillet welds, the side surfaces are connected by butt welds. Horizontal, vertical and twist angles of structural members at this node can be accommodated by the geometry of the double mitre cuts at the end of the members. High bending moments, up to the full member strength, can be transferred [11].

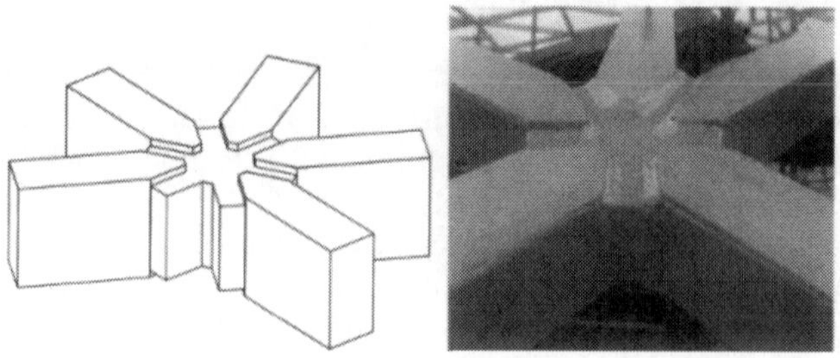

Fig. 2. End-Face Connector WABI-1 [11].

3 Means of modelling bars and the numerical techniques used

In the study, two attempts to bar modelling, by means of truss and frame elements were considered. Both models took into account geometric nonlinearities in the Lagrange description. Applying the frame element, the physical relationships represented the elastic behaviour of the structure with the use of Hooke's law. Regarding the compressive truss elements, a nonlinear relationship resulting from experimental investigations was additionally employed.

The sources of the nonlinear relations $\underline{\sigma} = f(\underline{\varepsilon})$ are, apart from strictly material features, also geometrical imperfections. In this analysis, differential constitutive relations were neglected, these were generated for an entire element (i.e. for individual bars). This approach does not allow to separate the material and geometrical imperfections. The experiments were carried out using bars 163 to 572 mm in length. Therefore, it was of utmost importance to properly interpret the scale effect. The solution was reached creating a dimensionless load-displacement relation (RLD) due to imperfect compressed bar (Fig. 3). The parameters of the RLD were derived using experimental test results presented by Kowal and Radoń [12].

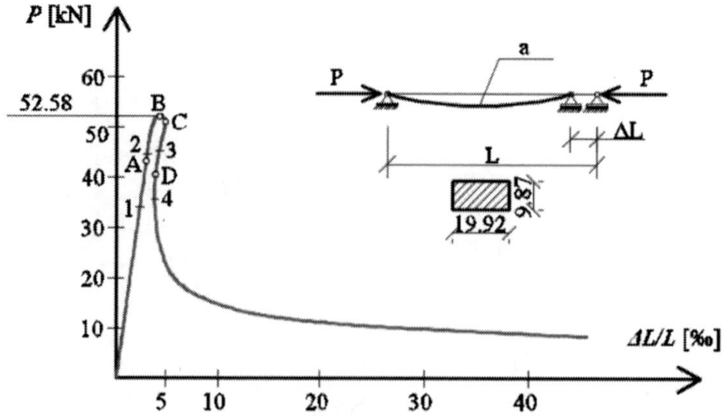

Fig. 3. Example of a RLD due to a bar with imperfections.

Stability analysis of the structure was performed with the FEM-based software: Abaqus and Robot Structural Analysis. In order to obtain the load-displacement relation, the Riks arc length method was used [13]. Four years after the arc length method was proposed by Riks, Crisfield came up with an alternative to that method. The latter was employed in many commercial FEM programs [14]. Another variant of the constant arc length method, similar to that developed by Crisfield, was put forward by Ramm. It was termed a modified Riks-Wempner method [15].

4 Numerical results

A lattice dome consisting of 49 nodes and 120 elements (Fig. 4) was analysed in the paper. The dome geometry is shown in Table 1. The elements of the structure made of RO101.6x6 steel pipes of yield point f_Y=235 MPa and Young's modulus E=210 GPa. One-parameter nodal load applied in every node was taken into account. In the study, the determination of the critical load capacity of the lattice dome was performed using the following variants:

- Variant I – geometrically non-linear truss model with linear physical relationships,
- Variant II – geometrically and physically non-linear truss model,
- Variant III –geometrically non-linear truss model with linear physical relationships, taking into account geometric imperfections of the bars,
- Variant IV –geometrically non-linear frame model with linear physical relationships,
- Variant V–geometrically non-linear frame model with linear physical relationships, taking into account geometric imperfections of the bars.

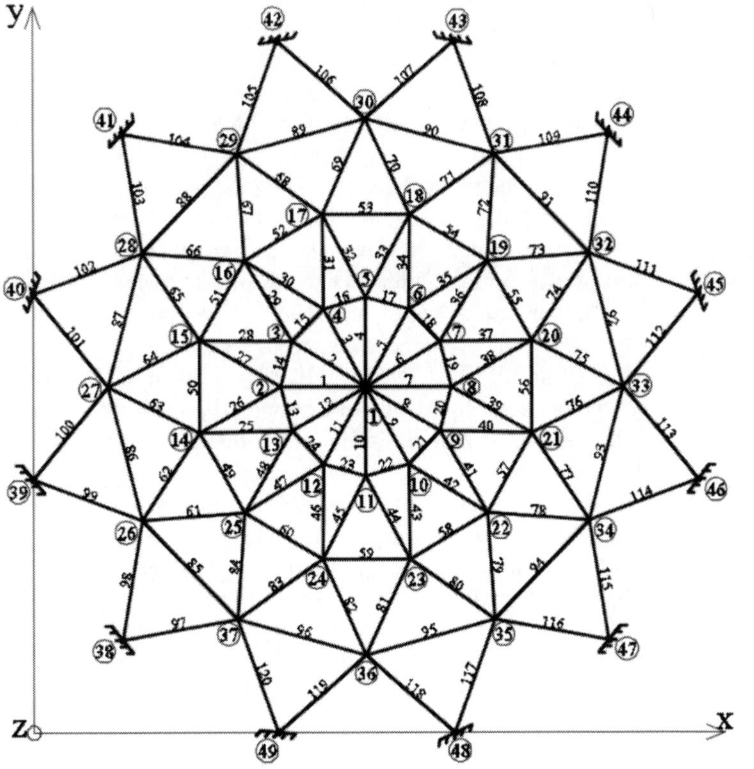

Fig. 4. Mesh of the lattice dome.

Table 1. Geometry of the lattice dome.

No. of node	x [m]	y [m]	z [m]	No. of node	x [m]	y [m]	z [m]	No. of node	x [m]	y [m]	z [m]
1	14.489	14.489	4.50	18	16.430	21.733	3.40	34	24.232	8.864	1.80
2	10.739	14.489	4.15	19	19.792	19.792	3.40	35	20.114	4.746	1.80
3	11.241	16.364	4.15	20	21.733	16.430	3.40	36	14.489	3.239	1.80
4	12.614	17.736	4.15	21	21.733	12.548	3.40	37	8.864	4.746	1.80
5	14.489	18.239	4.15	22	19.792	9.186	3.40	38	3.882	3.882	0.0
6	16.364	17.736	4.15	23	16.430	7.244	3.40	39	0.0	10.607	0.0

7	17.736	16.364	4.15	24	12.548	7.244	3.40	40	0.0	18.371	0.0
8	18.239	14.489	4.15	25	9.186	9.186	3.40	41	3.882	25.095	0.0
9	17.736	12.614	4.15	26	4.746	8.864	1.80	42	10.607	28.978	0.0
10	16.364	11.241	4.15	27	3.239	14.489	1.80	43	18.371	28.978	0.0
11	14.489	10.739	4.15	28	4.746	20.114	1.80	44	25.095	25.095	0.0
12	12.614	11.241	4.15	29	8.864	24.232	1.80	45	28.978	18.371	0.0
13	11.241	12.614	4.15	30	14.489	25.739	1.80	46	28.978	10.607	0.0
14	7.244	12.548	3.40	31	20.114	24.232	1.80	47	25.095	3.882	0.0
15	7.244	16.430	3.40	32	24.232	20.114	1.80	48	18.371	0.0	0.0
16	9.186	19.792	3.40	33	25.739	14.489	1.80	49	10.607	0.0	0.0
17	12.548	21.733	3.40								

In all the analysed cases the node snap-through was a decisive mode of stability loss. For example, in variant I the linear bifurcation load reached a value of 44.897 kN, while the limit load for this case is 33.35 kN. In variant IV the linear bifurcation load reached a value of 50.52 kN, while the limit load for this case is 49.83 kN.

In Variant I, as the load grew until the limit point was reached, the most stressed element was bar 22 (Fig. 4). In Variant II, a decrease in the critical load capacity was observed. Additionally, compared to Variant I, it was accompanied by a displacement change corresponding to the load. In the studies [7,8,9], the authors clearly indicated a massive impact of imperfections on the critical load capacity of the structure. That could be seen in the model studies conducted by those researchers. For Variants III and V, an attempt was made to carry out the same numerical simulations. In the Abaqus environment, *BUCKLE* analysis was launched. It produced the structure buckling modes and the corresponding values of the critical force. Then, the perfect geometry of the bars was disturbed by means of nodal displacements computed in the previous step with the use of *IMPERFECTION* command. Next, computations were run again using *STATIC RIKS*. In both Variant III and Variant V, 5% and 10% values of the eigenvector of the first buckling mode were taken into account. As an example, the maximum displacement of the node number 1 in the frame structure for the first buckling mode of the structure is Uz=0.32 mm which means the adoption of imperfections at the level of 0.016 mm for 5% and 0.032 mm for 10%. The character of the load-displacement relation that was obtained clearly indicates the occurrence of the limit point, which had a definitely lower value. In Variant IV, as the load increased until the limit point was reached, the numbers of the most stressed elements changed. Initially, that was bar 17, and when 95% of the critical load capacity was applied, it was bar 115 (Fig. 4).

The critical load capacity values for the variants analysed in the study are summarised in Table 2. The Figures 5 and 8 show the deformation of the lattice dome for the truss model (Variant I) and the frame model (Variant IV), respectively. The descriptions of internal forces and stresses were also provided. Load-displacement path for geometrically non-linear model with linear physical relationships for truss and frame model, are show on the Figures 6 and 9, respectively. Figure 7 and Figure 10 show load-arc length path for geometrically and physically non-linear model with 5% of the value of the eigenvector of the first buckling mode for truss (Variant III) and frame (Variant V) models, respectively.

Table 2. Summary of the critical load capacity results.

Variant I	Variant II	Variant III		Variant IV	Variant V	
		5%	10%		5%	10%
33.35 kN	20.34 kN	21.45 kN	17.01 kN	49.83 kN	46.33 kN	44.62 kN

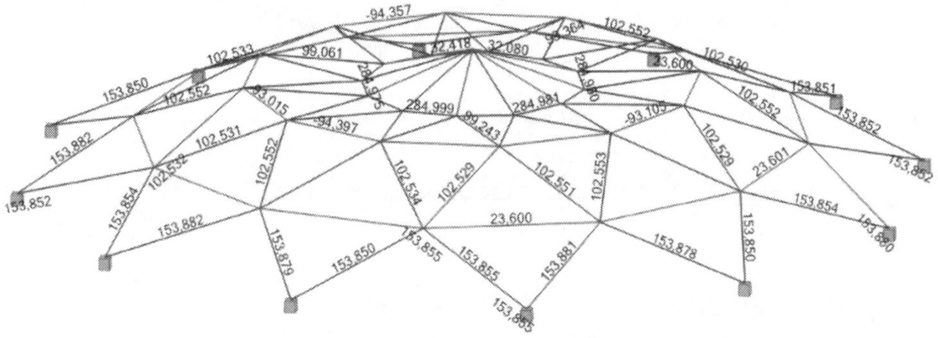

Fig. 5. Deformation of the lattice dome for the truss model with the description of axial forces in the bars at the limit point (Variant I).

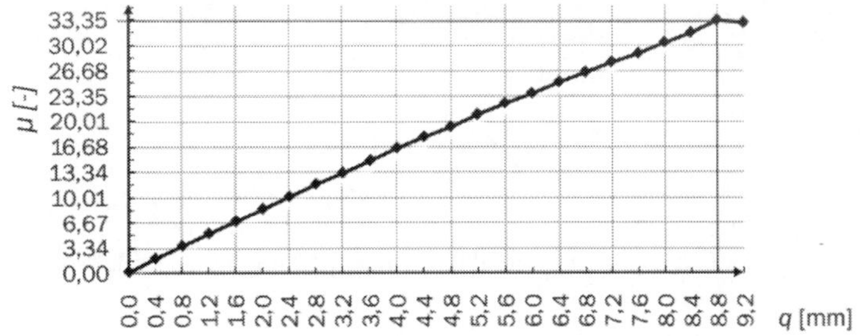

Fig. 6. Load-displacement path for geometrically non-linear truss model with linear physical relationships (Variant I).

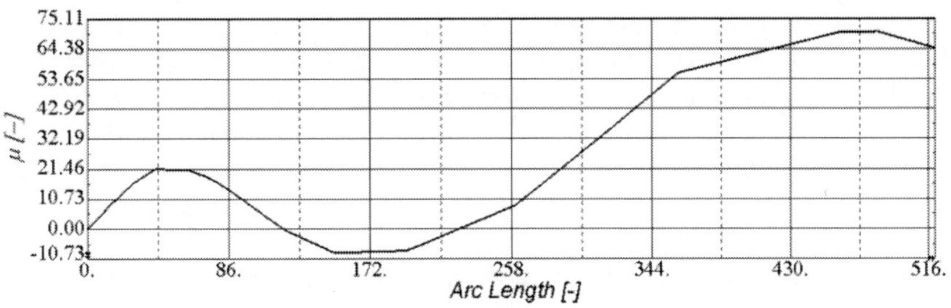

Fig. 7. Load-arc length path for geometrically non-linear truss model with 5% of the value of the eigenvector of the first buckling mode (Variant III).

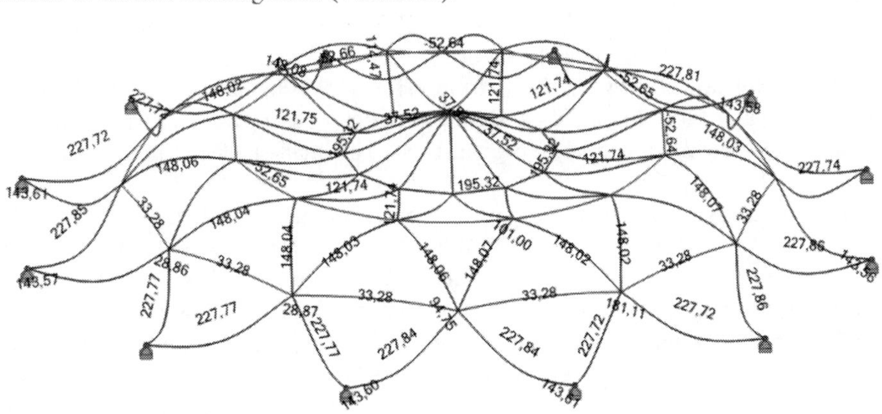

Fig. 8. Deformation of the lattice dome for the frame model with the description of the maximum stress in the bars at the limit point (Variant IV).

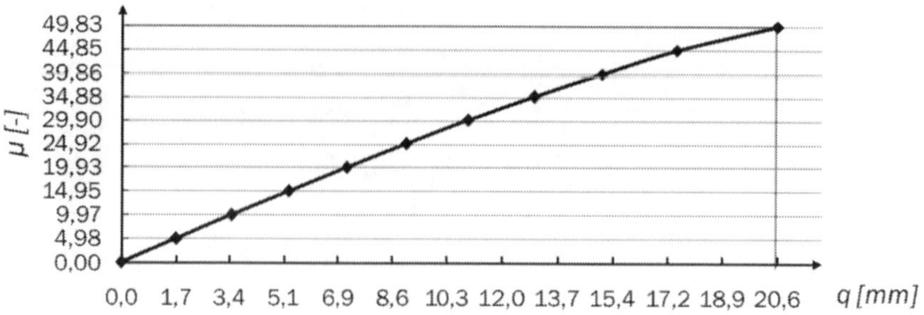

Fig. 9 . Load-displacement path for geometrically non-linear frame model with linear physical relationships (Variant IV).

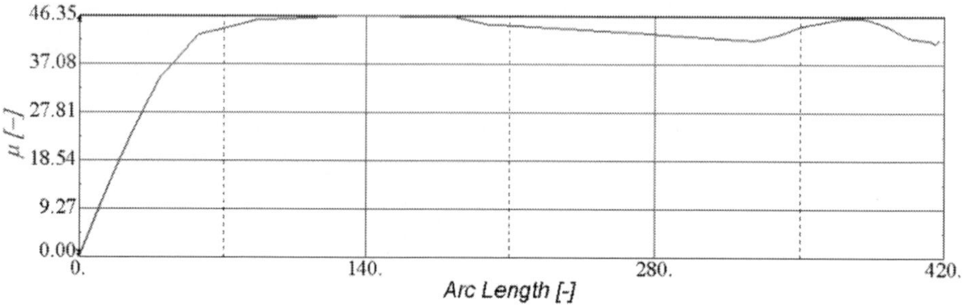

Fig. 10. Load-arc length path for geometrically non-linear frame model with 5% of the value of the eigenvector of the first buckling mode (Variant V).

5 Summary and conclusions

In the paper the effect of single-layer lattice dome modelling on the critical load capacity was discussed. Five variants were considered, the way of modelling was modified throughout. The modifications concerned the bar connections in nodes, bars of the lattice dome and the impact of geometric imperfections. In each case, geometric nonlinearities in the Lagrange description were taken into account. It was observed that the node snap-through was a decisive mode of the stability loss for all the cases. In Variant I, the value of the critical load capacity, equal to 33.35kN, was determined. In Variant II, non-linear physical relationships for the truss element were taken into account, which lowered the value of the critical load capacity to 20.34 kN. Variant III accounted for the disturbances to the perfect geometry of the bars caused by the displacement of nodes obtained from the structure buckling analysis with the use of truss elements. Numerical analysis performed with the ABAQUS program (using the *IMPERFECTION* command) showed a substantial decrease in the critical load capacity. The reductions, to the values of 21.45 kN and 17.01 kN, constituted 5% and 10%of the value of the eigenvector of the first buckling mode, respectively. In Variant IV, a frame structure was considered, in which a linear relationship between strain and stress was utilised. The analysis demonstrated that the value of the critical load capacity was 50.58 kN. Additionally, as the load increased, a change in the most stressed elements, namely from bar no.17 to the near-support bar no. 115, was observed. Finally, in Variant V, the geometry of the frame elements was disturbed by imperfections, as it was the case with Variant III. Again, the effect of imperfections on a decrease in the critical load capacity was observed, which was 5% to 46.33kN, and 10% to 44.62kN. Stability analysis of the structure was performed with the Finite Element Method using Abaqus and Robot Structural Analysis software. Differences in results between software were insignificant, less than 5%. In this paper B32 element was used for frame structure and T3D2 element for truss structure. The elements in the frame structure were additionally compacted 10 times.

Because of the height-to-span ratio of 0.155, the structure under analysis is a shallow dome. Structures of this type are characterised by strong nonlinearity, consequently they are extremely susceptible to stability loss resulting from the node snap-through. In such a case, it is not sufficient to carry out the analysis of initial stability. It is necessary to perform a nonlinear stability analysis that allows the determination of limit points related to that mode of stability loss.

References

1. J. Marcinowski, *Stability of elastic structures,* Dolnośląskie Wydawnictwo Edukacyjne (2017) [in Polish]
2. Z. Waszczyszyn, Cz. Cichoń, M. Radwańska, *Stability of Structures by Finite Element Methods*, Elsevier (1994)
3. U. Pawlak, M. Szczecina, *Dynamic eigenvalue of concrete slab road surface*, IOP Conf. Series: Materials Science and Engineering 245 (2017)
4. P. Obara, W. Gilewski, *Dynamic stability of moderately thick beams and frames with the use of harmonic balance and perturbation methods*, Bulletin of The Polish Academy of Sciences: Technical Sciences, **64,** 4,739-750 (2016)
5. W. Mochocki, P. Obara, U. Radoń, P. Zabojszcza, *Effect of single-layer truss dome geometry on critical load capacity*, Structure and Environmental, **93**, 152-164 (2017)
6. J. Głąbik, *Critical load in pin-jointed single-layer lattice domes,* Doctoral thesis, Politechnika Śląska (1970) [in Polish]
7. J. Bródka, I. Czmoch, M. Giżejowski, J. Marczewski, Ultimate capacity of steel lattice *dome*, Inżynieria i Budownictwo, **1**, 21-24 (1983) [in Polish]
8. Z. Kowal, J. Słowik, W. Wawszczak, *Principles of the design of bar-rib domes*, Inżynieria i Budownictwo, **7**, 247-250 (1992) [in Polish]
9. J. Rządkowski, *Limit load capacity in single-layer lattice domes loaded at nodes*, Inst. Bud. Pol. Wrocławskiej (1984) [in Polish]
10. J. Bródka, A. Kozłowski, *Design and computations of joints and nodes of steel structures*, PWT, **T.1** (2013)[in Polish]
11. S. Stephan, J. Sanchez-Alvarez, K. Knebel, *Reticulated Structures on Free-Form Surfaces,* Stahlbau (2004)
12. Z. Kowal, U. Radoń, *Proposition of a constitutive model of compressed rods for calculating the limit bearing capacity of real rod constructions*, Inżynieria i Budownictwo, **12**, 679-683 (1995)[in Polish]
13. E. Riks, *An incremental approach to the solution of snapping and buckling problems*, Int. J. Solids Struct., **15**, 529-551 (1979)
14. M. A. Crisfield, *A fast incremental/iterative solution procedure that handles snap-through*, Comput. Struct.,**13**, 55-62 (1981)
15. E. Ramm, *Strategies for tracing the non-linear response near limit points.* Nonlinear Finite Element Analysis in Structural Mechanics, 68-89, Springer (1981)

Investigation of natural frequencies of axially loaded thin-walled columns

Łukasz Żmuda-Trzebiatowski[1,*]

[1]Gdańsk University of Technology, Faculty of Civil and Environmental Engineering, ul. Narutowicza 11/12, 80-233 Gdańsk, Poland

Abstract. The paper deals with correlation between natural frequencies of two steel thin-walled columns and the corresponding applied load. The structures are made of cold-formed lipped channel sections. The columns lengths were assumed to follow two buckling patterns – global flexural and flexural-torsional buckling. In the thicker structure two material models were considered – linearly-elastic and elastic-perfectly plastic. Numerical computations cover dynamic eigenvalue problem, linear buckling and geometrically (and materially) non-linear analysis. The correlation between squares of natural frequencies and the applied load is linear in both columns. The first natural frequencies drop to zero due to structural buckling. This method, called the Vibration Correlation Technique, allows to predict buckling loads on the basis of measured vibration frequencies of the structures. Plasticity does not affect the corresponding curves – the use of the presented technique is limited to the structures exhibiting elastic buckling behaviour.

1 Introduction

Cold-formed thin-walled profiles are widely used in civil engineering, e.g. as elements strengthening silos or girts or purlins in roofs and wall structures. They exhibit three kinds of stability loss – local, distortional and global [1]. The latter can be flexural, torsional or a combination of both. The actual buckling mode is affected by material data and a slenderness ratio. The local buckling behaviour is linked with a small slenderness ratio. The slenderness ratio increase makes the structures exhibit distortional stability loss, next triggering global buckling modes. The buckling load in thick compressed elements is relatively large, therefore the limit load exceeds an elastic-range stability problem. Elastic buckling is observable in elements of a high slenderness ratio.

The prediction of realistic buckling loads of compressed thin-walled elements by means of experimental tests can be useful for the validation of numerical models of new or existing structures. One of the non-destructive methods allowing to determine the buckling load of structures is the Vibration Correlation Technique (VCT) [2]. It assumes linear correlation between compressive load and squares of natural frequencies for a majority of steel structures:

* Corresponding author: luktrzeb@pg.edu.pl

$$(f_n / f_{n0})^2 = 1 - (P / P_n) \quad (1)$$

where:

f_n	nth natural frequency of the loaded structure;
f_{n0}	nth natural frequency of the unloaded structure;
P	applied load;
P_n	buckling load corresponding to nth vibration mode.

While a structure experiences buckling, the first natural frequency tends to zero. Experimental tests show that this tendency is not always apparent – sometimes it reaches its minimum, next rises again [3].

Thus, recording acceleration signals of structures at two loading stages, next making a time-to-frequency domain the Fourier transform, it is possible to predict primary buckling loads. Such a technique can improve the safety of structural steel and composite buckling-prone structures.

Nevertheless, some shell structures, e.g. column-strengthened silos, are characterised by a non-linear load–frequency squared relationship [4]. The relation is linear only at the start; the second half shows it curvilinear leading to the predicted buckling greater than the realistic one.

In the literature the VCT has been mainly used in panels and cylindrical shells [5, 6]. The paper covers an application of the Vibration Correlation Technique to cold-formed thin-walled profiles exhibiting global buckling.

2 Investigated structures

2.1 Description of the columns

The effectiveness of the presented method was verified on two examples of thin-walled columns. The height of the first structure (A) was 9000 mm, the height of the second (B) was 2500 mm. In both cases the length was assumed to follow different buckling modes – both global, A – flexural, B – flexural-torsional. The columns were made of the same cold-formed lipped 4 mm thick channel section. The dimensions of the cross-section are given in Fig. 1a. The material is steel S235 with following parameters: Poisson's ratio υ = 0.3, elasticity modulus E = 210 GPa, specific weight ρ = 7850 kg/m^3 and yield stress f_y = 235 MPa.

2.2 Numerical models

The columns were modelled in the FEM-based ABAQUS software [7]. In both (A) and (B) models 4-node shell elements with reduced integration were applied. The FEM model of the column A consists of 41'152 elements, while the column B model consists of 22'000 elements.

The material model in the higher column exhibiting global flexural buckling (A) was set linearly-elastic due to the limit load much higher than the minimum buckling load. The second column height to trigger the influence of plasticity greater compared to stability impact. Therefore, the structure B covers two material models – linearly-elastic and elastic-perfectly plastic.

The bottom edges were fixed at both columns. Translations in all directions and rotations about all axes at top edges were constrained, except the vertical deflections (corresponding to the column longitudinal axes).

The load was applied vertically to the upper structural edges in order to provide axial compression. The column model view is shown in Fig. 1b.

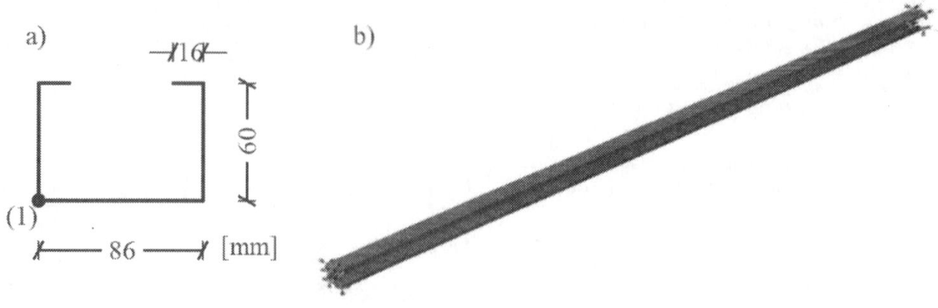

Fig. 1. The column cross-section (a) and the numerical model (b).

Numerical investigation of thin-walled columns covers dynamic eigenvalue problem, linear buckling and geometrically (and materially) non-linear analysis. Linear buckling analysis was performed in order to obtain the primary five buckling loads and the corresponding modes. The static, general method with a load as a control parameter was used to consider the cases of non-linear analysis. The initial, minimum and maximum increment sizes were equal to 0.001, 10^{-30} and 0.003, respectively. Geometrically non-linear analysis led to full equilibrium paths showing column load versus vertical displacement of a node 1 in a corner of the channel section in the upper edges of the structures (see Fig. 1a). Additionally, automatic stabilization in a form of a damping factor was introduced to provide solution convergence. The damping factor was equal to 10^{-8} in the column A and 10^{-7} in the column B. Dynamic eigenvalue analysis was conducted to find the primary natural frequencies and the corresponding vibration modes.

Geometrically (and materially) non-linear analysis and dynamic eigenvalue analysis were performed repeatedly. In each course the load increased from zero to a prescribed level in the non-linear analysis case in order to pre-stress the columns. Next, dynamic eigenvalue analysis was conducted. On the basis of the performed analysis the relationships between the squares of natural frequencies and the applied compressive loads were plotted, illustrating the conducted experiments.

3 Results

3.1 Column A

The evolution of maximum compressive forces in the column A against vertical displacement of the reference node resulting from geometrically non-linear analysis was plotted in Fig. 2a. The maximum load carrying-capacity was equal to 48.5 kN. The structure deformation produced by the applied load at its limit point exhibits global, flexural deformation shape (Fig. 2b). The limit load equals 212 kN, thus it does not act on stability loss in any way.

The first buckling load in the column A equals 43 kN. The corresponding buckling mode shows a global, flexural deformation shape including a single semi-wave. The deformation patterns in both static non-linear and buckling analysis converge. The second buckling load equals 59.8 kN, the corresponding buckling mode shows a flexural-torsional mode. The primary five buckling modes and their corresponding buckling loads are given in Fig. 3a. The Euler's buckling load with respect to the weak axis equals 43 kN.

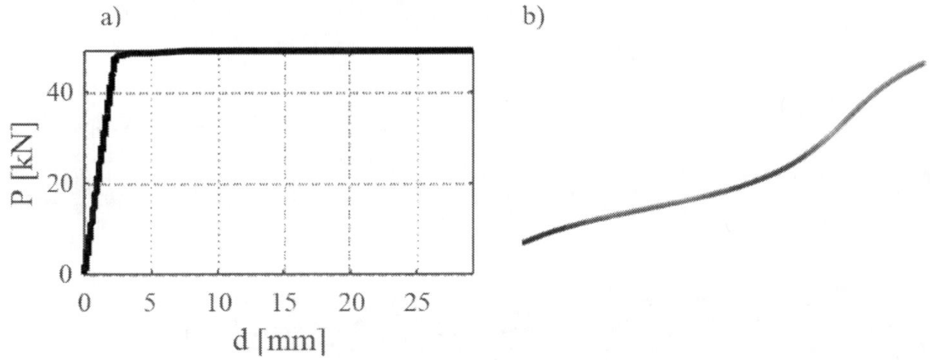

Fig. 2. The evolution of maximum compressive forces and vertical displacement of the node 1 in the column A obtained from static non-linear analysis (a) and the structure deformation (b).

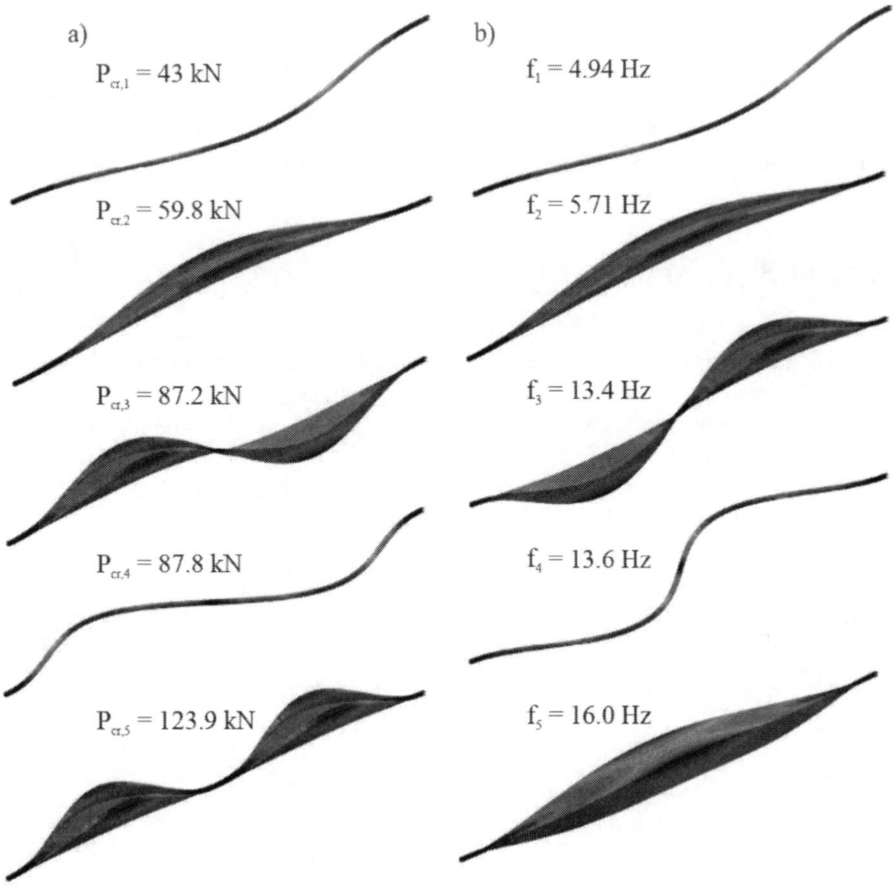

Fig. 3. The primary modes in the column A: linear buckling analysis (a) and dynamic eigenvalue analysis (b).

The first natural frequency equals 4.94 Hz, the second – 5.71 Hz. The primary five vibration modes and their corresponding natural frequencies are presented in Fig. 3b. The buckling and vibration one-to-four modes resemble, the fifth modes diverge.

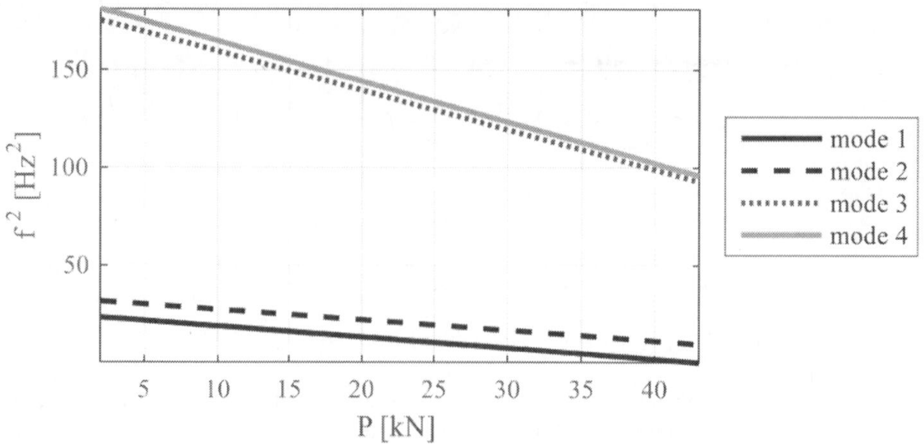

Fig. 4. The correlation between compressive load and squares of natural frequencies in the column A.

The correlation between the compressive forces and the squares of the first four natural frequencies was plotted in Fig. 4. Linear relations occur in each case – the lowest square of a correlation coefficient is 0.9999. The first natural frequency drops to zero when the load equals 43 kN – the exact value of the first buckling load. The Vibration Correlation Technique seems to work sufficiently well in the range of thin-walled compressed elements.

3.2 Column B

The evolution of maximum compressive forces in the column B against vertical displacement of the reference node resulting from the geometrically non-linear analysis was plotted in Fig. 5a. The maximum compressive load carrying-capacity is equal to 430.2 kN. The structure deformation caused by the applied load at its limit point shows a global, flexural-torsional deformation (Fig. 6a). Application of the damping factor 10^{-8} or omitting the solution stabilization result in a different deformation shape – still global, but flexural. Thus the maximum compressive carrying-capacity equals 555 kN, showing a 29% increase.

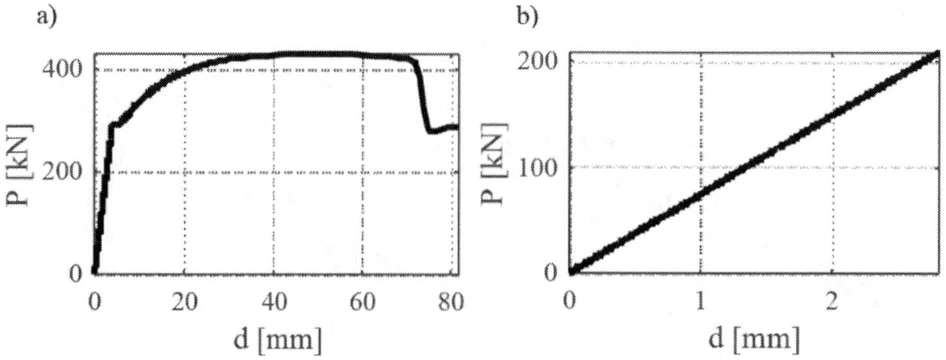

Fig. 5. The evolution of maximum compressive forces and vertical displacement of the node 1 in the column B obtained from static non-linear analysis with linearly-elastic (a) and elastic-perfectly plastic material models (b).

Numerical model of column B, employing the elastic-perfectly plastic material model shows the maximum limit load equal 212 kN. The corresponding equilibrium path and deformation shape are shown in Figs. 5b and 6b.

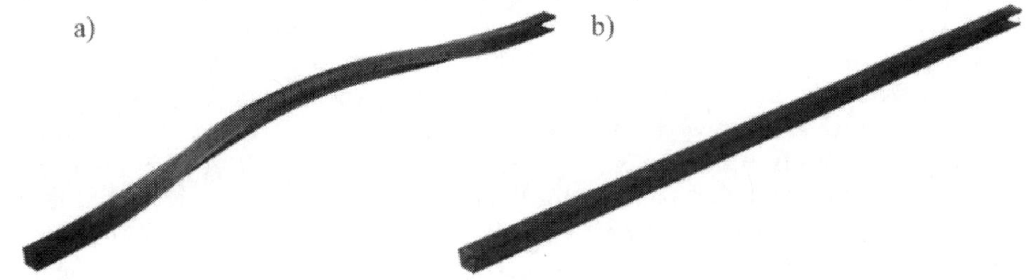

Fig. 6. The column B deformation at the time of the loss of stability with linearly-elastic (a) and elastic-perfectly plastic material models (b).

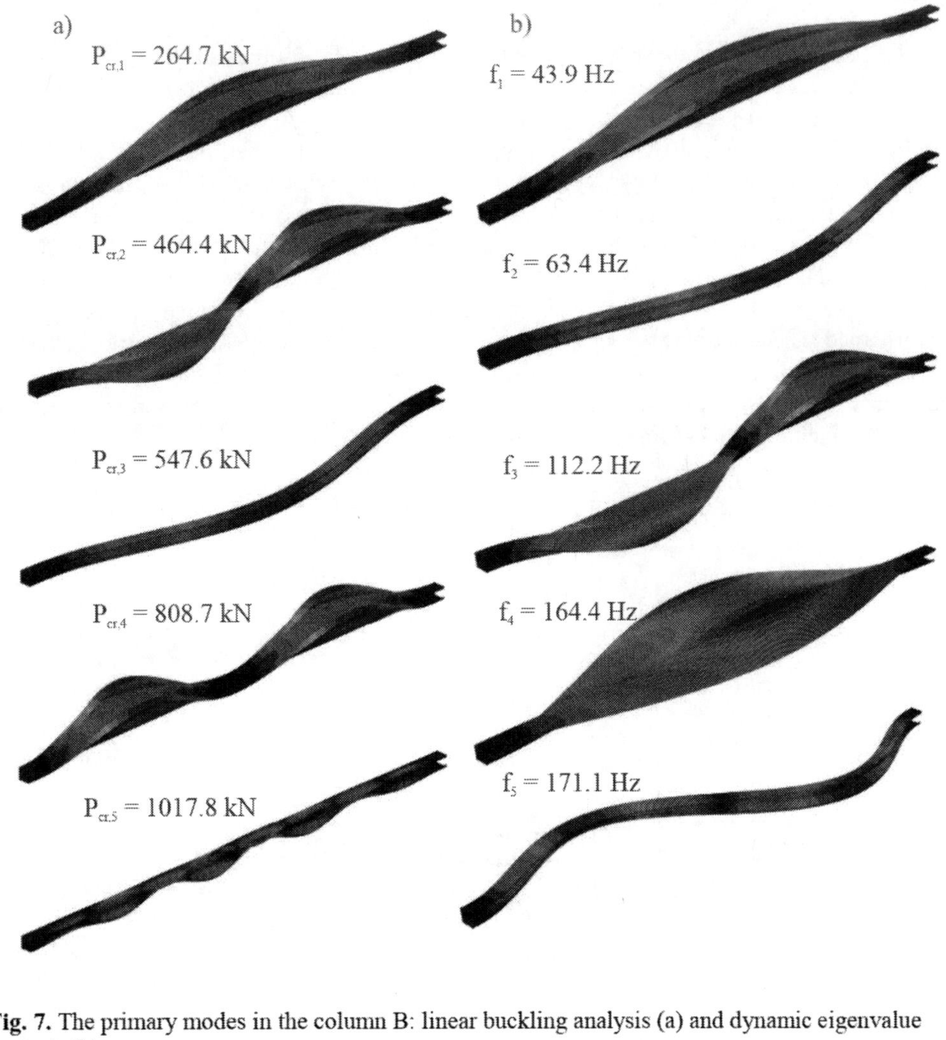

Fig. 7. The primary modes in the column B: linear buckling analysis (a) and dynamic eigenvalue analysis (b).

The first buckling load in the column B equals to 264.7 kN. The corresponding buckling mode shows a global, flexural-torsional deformation including a single semi-wave. The second and third buckling loads equal 464.4 and 547.6 kN, respectively. The corresponding buckling modes show a flexural-torsional shape with two semi-waves and a global, flexural shape, respectively. The first buckling load constitutes 61.5% of the maximum load-carrying capacity. It is possibly caused by difficulties to find a limit point because any intended imperfections of the investigated structures are not introduced. The introduction of a slight geometrical imperfection is intended to fix the problem, but the author wanted to analyse perfect column geometries. The deformations received in static non-linear analysis and buckling analysis converge. The primary five buckling modes and their corresponding buckling loads are given in Fig. 7a. The analytical flexural-torsional buckling load equals 266.8 kN. The difference between numerical and analytical results is lower than 1%.

The first natural frequency equals 43.86 Hz, the second – 63.37 Hz, the third – 112.2 Hz. The primary five vibration modes and their corresponding natural frequencies values are given in Fig. 7b. Both first buckling and vibration modes coincide, but the second and third modes interchange, making it still possible to find buckling–vibration mode equivalents. It is a key issue due to linear relations between the squares of the natural frequencies and the applied load only in the case of convergent buckling and vibration modes.

The correlation between the compressive forces and the squares of the first five natural frequencies was plotted in Fig. 8. Linear relations occur in each case to the point when the applied load becomes the first buckling load. The lowest square of a correlation coefficient is 0.9997. The first natural frequency drops to zero when the load equals 266 kN. While compared to the first buckling load the difference is lower than 0.5%. It is caused by a greater maximum compressive load-carrying capacity compared to the minimum buckling load. Here the post-buckling behaviour is analysed. When the applied load reaches 285 kN, the first natural frequency rapidly increases to the new maximum, next declines again. After reaching the first buckling load level the higher order natural frequencies change their performance. The square of the second frequency changes the slope of the curve. The curve corresponding to the square of the third frequency reaches its minimum and rises, whereas the fourth one starts to decrease rapidly. The Vibration Correlation Technique predicts the calculated buckling load sufficiently.

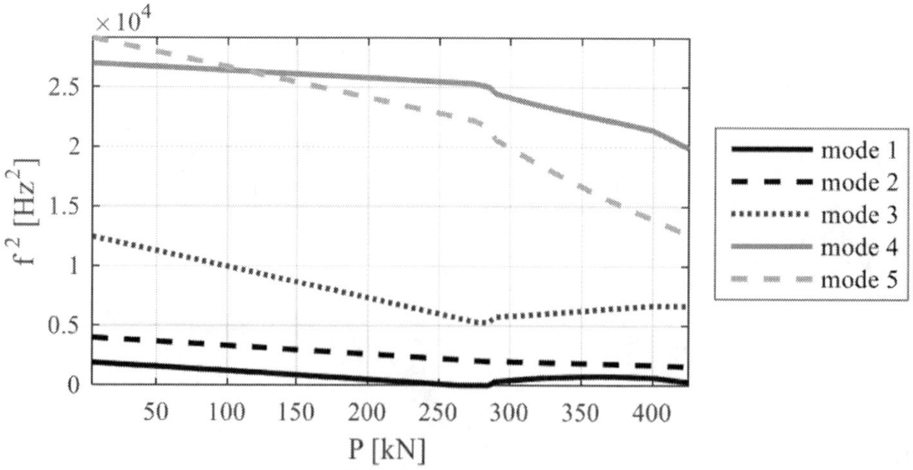

Fig. 8. The relationship between compressive load and squares of natural frequencies in the column B with linearly-elastic material model.

The introduction of plasticity into the analysis did not affect the vibration frequencies and the corresponding modes. The correlation between squares of the natural frequencies and applied load is linear, but it varies when the stresses in the column achieve the yield stress level (see Fig. 9). Therefore, the presented method to predict the buckling load limits

its use only to the elastic buckling range. While plasticity is the major structural account, the Vibration Correlation Technique fails.

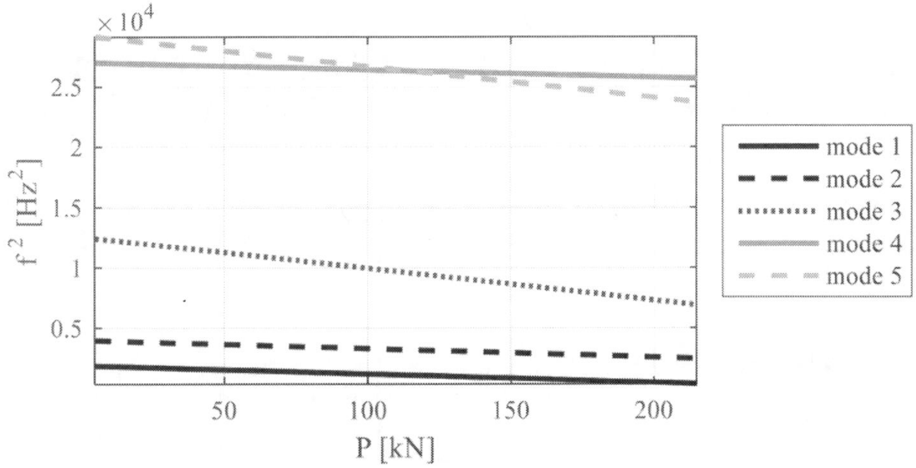

Fig. 9. The relationship between compressive load and squares of natural frequencies in the column B with elastic-perfectly plastic material model.

4 Conclusions

The Vibration Correlation Technique can be successfully applied in thin-walled columns exhibiting global – flexural-torsional and flexural – buckling. The effectiveness of the method to estimate the realistic buckling loads is high. Unfortunately, its range is limited to the buckling-prone structures. The method cannot be applied to predict the limit load of the structures; it is limited to elastic buckling loads only, thus the inelastic buckling cases are not dedicated for such estimations.

Local and distortional buckling patterns have not been investigated so far, thus the author is planning further research on such columns. Additionally, physical tests are planned on real thin-walled elements compressed in a testing machine, in order to compare numerical and experimental results.

References

1. *EN 1993-1-3* (2006).
2. J. Singer, J. Arbocz, T. Weller, *Buckling experiments: experimental methods in buckling of thin-walled structures: shells, built-up structures and additional topics. Vibration Correlation Techniques (VCT)*, **2** (John Wiley & Sons, New York, 2002)
3. M.A. Arbelo, S.F.M. de Almeida, V. Donadon, S.R. Rett, R. Degenhardt, S.G.P. Castro, K. Kalnins, O. Ozolins, THIN WALL STRUCT, **79**, 119-128 (2014)
4. Ł. Żmuda-Trzebiatowski, P. Iwicki, *Shell Structures: Theory and Applications*, **4**, 481-484 (CRC Press/Balkema, London, 2018)
5. H. Abramovich, D. Govich, A. Grunwald, PROG AEROSP SCI, **78**, 62-73 (2015)
6. E. Skukis, O. Ozolins, K. Kalnins, M.A. Arbelo, PROCEDIA ENGINEER, **172**, 1023-1030 (2017)
7. *Abaqus 6.12. Theory Manual* (Dassault Systemes, 2012)

Design procedure for steel column bases with stiffeners

Marcin Górski[1,*]

[1]Rzeszów University of Technology, Department of Building Structures, ul. Poznańska 2, 35-959 Rzeszów, Poland

Abstract. The procedure of calculation of steel column bases with stiffeners including plastic distribution of stresses in concrete has been shown. The strain model from section analysis, similar to model used in calculations of reinforced concrete columns has been used in this purpose. The assumption of linear strain of base plate under stiffeners and column has been made, according to Bernoulli-Euler Theory. The displacements of compressed part of columns bases have been associated with deformations of concrete foundation. Limitations of dimensions of compression zone have been defined based on results of FEM analysis. Ultimate compressive strain in the concrete under compression and bending, according to Eurocode 2 has been taken into consideration in the procedure. Deformation of base plate based on stiffness coefficients used in determining of rotational stiffness of steel joints has been assumed in the tensioned part. Incomplete utilisation of bolts resistance due to insufficient deformation has been also included. All the mentioned assumptions results in universal procedure that allows calculation of resistance of column bases under any combination of axial force and bending moment. Finally design example with an entire proceeding has been presented.

1 Introduction

Columns are elements of steel structures often subjected not only to axial force, but also to bending moment, which are transferred into foundations. It requires proper design of a column base. In the case of small values of bending moments, usually a simple base plate with suitable thickness is sufficient. Calculation of column bases with such geometry is shown in Eurocode 3 [1]. When the force transferred into foundation grows significantly, the single base plate becomes insufficient, so vertical stiffeners of base plate are required. The actual code [1] does not provide guidelines for calculations of column bases with such complex geometry. There are only procedures (e.g. Fisher's formula [2]) to analyse rigid column bases, based on assumption of elastic analysis and linear distribution of stress in concrete under column base, while Eurocode allows plastic analysis. The paper proposes a universal calculation procedure of steel column bases with stiffeners considering plastic stress distribution.

* Corresponding author: mgorski@prz.edu.pl

2 Assumptions of procedure

2.1 General

The assumption of rigid column core and stiffeners has been made. It results in linear deformation of part of base plate under these elements. Therefore the model according to Bernoulli-Euler Theory can be used in the procedure. However, nonlinear behavior of concrete results in different centroid of foundation compared to steel column [3]. As a result the base plate is situated in discontinuity point of a structure, so the strain equation cannot be used directly. Instead of strain, total vertical displacements of distinctive points of base plate in compression and tension zone have been determined.

2.2 Compression zone

The column cross-section together with vertical stiffeners, enlarged on both sides by additional bearing width *c* determines the available effective area of concrete (Fig. 1).

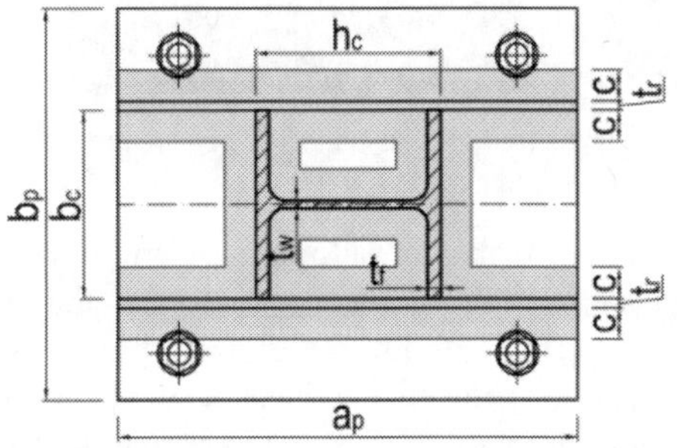

Fig. 1. Available effective part of concrete taken into account in calculations.

A simplified, rectangular distribution of stresses in compression zone of concrete has been assumed, which results in Eq. (1).

$$x_{eff} = \lambda \cdot x \tag{1}$$

where:

$\lambda = 0{,}8$ for concrete up to strength class C50/60 according to [4].

x the distance from the ends of stiffeners, where the compressive strain is maximum, to the neutral axis.

The value of x_{eff} should be set to achieve the equilibrium of vertical forces (resultant force in concrete should counterbalance the resistance of tension zone increased by the acting axial force).

The range of compression zone depends on the eccentricity of the compressive force (bending moment to axial force ratio). In the case of pure bending compressive stress in concrete occurs only in the zone located at the ends of stiffeners. Along with the increase of the axial force, the position of neutral axis moves towards the plate end in tension [5]. The FEM analysis by means of Autodesk Simulation Mechanical software has been conducted to verify these assumptions. The model of exemplary steel column base made of 8-nodes brick and 4-nodes tetrahedral elements has been analyzed under various loads. The computations consider surface-to-surface contact, as well as nonlinear material models. Selected results of analysis are shown in Fig. 2 – 3.

Fig. 2. The range of compressed zone (grey colour) for eccentricity $e = 2{,}5$ m.

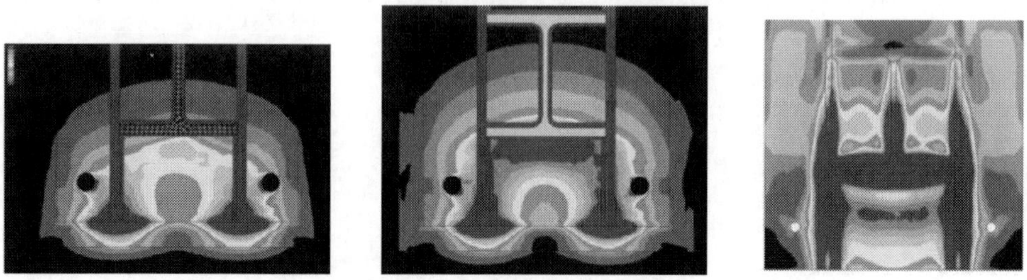

Fig. 3. Maps of compression **stress** under base plate for eccentricity e = 0,8 m (at the left), e = 0,4 m (in the middle) and e = 0,125 m (at the right).

The performed analysis confirms the adopted assumptions concerning the shape of a compression zone. Furthermore, the analysis reveals that the stress distribution under the column flange is not symmetrical about the flange, but it is concentrated on the side subjected to higher compression. Based on these results, the simplification rely on using only necessary part of area of equivalent T-Stubs in compression has been made (Fig. 4).

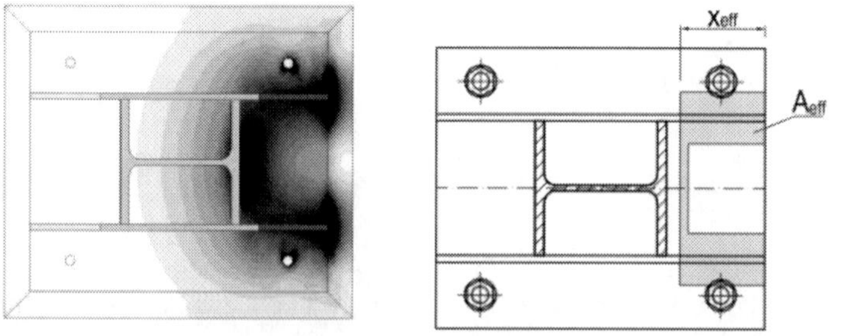

Fig. 4. Distribution of compressive stress under base plate. At the left: from FEM analysis. At the right: simplified, adopted in procedure.

According to [4] the ultimate strains in concrete $\varepsilon_{cu,3}$ = 3,5‰ (under pure bending) and $\varepsilon_{cu,2}$ = 2,0‰ (under pure compression) have been assumed in the procedure. The strain $\varepsilon_{cu,2}$ detects more versatile character, because in ULS it always occurs at a distance 3/7 x from the ends of stiffeners, regardless of the actual value of eccentricity [6]. Therefore this point has been assumed a characteristic one for a base plate in a compressed part. The total vertical displacement of a base plate in this point (ΔC_{cu2}) is equal the total vertical displacement of foundation directly under it. The vertical displacement of any point of the top surface of foundation is given by the expression (2):

$$\Delta C(x, y) = \int_0^{h_f} \varepsilon_z(x, y, z)dz \quad (2)$$

where:

$\varepsilon_z(x,y,z)$	the function of distribution of vertical strain in foundation under pressure from the base plate,
h_f	height of foundation,
x,y	coordinates of a particular point of the top surface (in horizontal plane),
z	coordinate on vertical axis.

While the coordinates (x,y) are known, the formula (2) can be written as (3):

$$\Delta C = \int_0^{h_f} \varepsilon_z(z)dz \tag{3}$$

where:

$\varepsilon_z(z)$	the function of distribution of vertical strain in foundation through the vertical axis under characteristic point of base plate in compressed part.

Finding the mathematic formula of the mentioned function is a troublesome task. Therefore simplification of ΔC_{cu2} formula was made in the procedure. The strain distribution through the depth of foundation has been assumed as the same as the stress distribution under partially loaded areas, according to [4]. Additionally, according to [7] the ratio between loaded area A_{c0} and design area A_{c1} can be assumed as (4):

$$\sqrt{\frac{A_{c1}}{A_{c0}}} = 1{,}5 \tag{4}$$

Thus the average strain through the height of the foundation equals (5):

$$\varepsilon_{cu2,mean} \approx 0{,}7 \cdot \varepsilon_{cu2} = 0{,}7 \cdot 0{,}002 = 0{,}0014 \tag{5}$$

An approximate coefficient equal 0,7 has been calculated from geometric relations.

Finally, the total vertical displacement of the base plate due to concrete deformation at a distance 3/7 x from the ends of stiffeners (ΔC_{cu2}) is given by the formula (6):

$$\Delta C_{cu2} = h_f \cdot \varepsilon_{cu2,sr} = 0{,}0014 \cdot h_f \tag{6}$$

2.3 Tension zone

The tension zone consists of a base plate in bending and anchor bolts in tension. The resistance of this zone ($F_{T,Rd}$) depends on failure mode of equivalent T-Stub in tension – complete yielding of the flange (mode 1-2) or anchor bolt failure (mode 3). In both cases reaching a full resistance of tension zone needs necessary vertical displacement of base plate at the location of anchor bolts. This displacement depends on vertical displacement of stiffeners and the stiffness of tension zone.

Vertical displacement of stiffeners in the line of anchor bolts (ΔL_T) is proportional to displacement in compression zone ΔC_{cu2}, according to the formula (7):

$$\Delta L_T = \frac{7}{4} \cdot \Delta C_{cu2} \cdot \left(\frac{a_p - e_1}{x} - 1 \right) \tag{7}$$

where:

a_p	length of base plate,
e_1	distance from the tension edge of base plate to the centre of anchor bolts (Fig. 5).

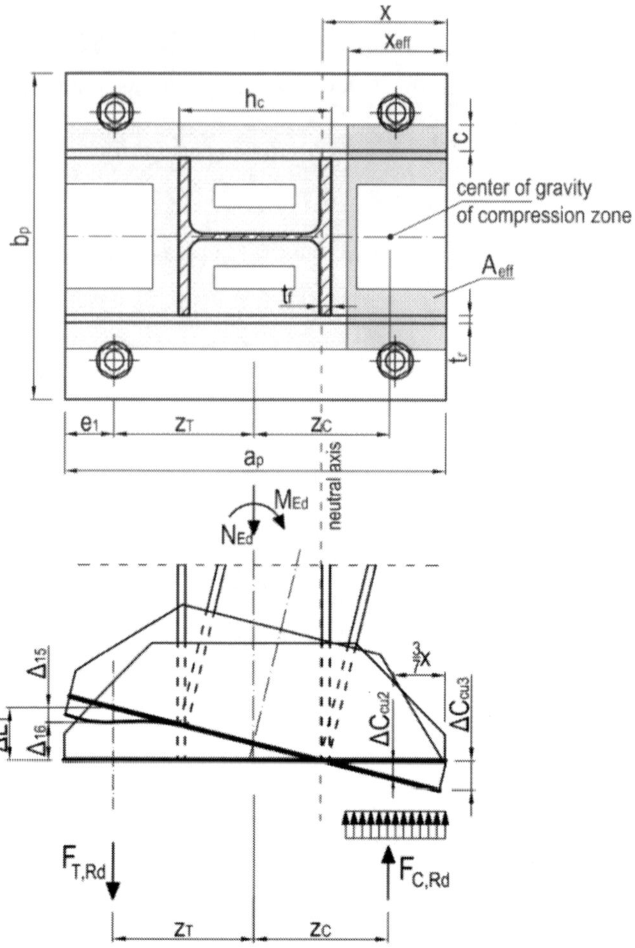

Fig. 5. Scheme of distribution of forces and displacements in column base.

According to [1] the stiffness of tension zone of column base can be written as (8):

$$S_T = \frac{E}{\mu\left(\frac{1}{k_{15}} + \frac{1}{k_{16}}\right)} \tag{8}$$

where:

k_{15}	stiffness coefficient of base plate in bending under tension
k_{16}	stiffness coefficient of anchor bolts in tension
μ	stiffness ratio $S_{j,ini}/S_j$
E	modulus of elasticity

The value of stiffness ratio μ depends on the ratio $M_{j,Ed}$ / $M_{j,Rd}$, in the case of column bases can be written as (9):

$$\mu = \begin{cases} 1{,}0 & \text{for} \quad M_{j,Ed} \le \frac{2}{3} M_{j,Rd} \\ \left(1{,}5 \cdot \frac{M_{j,Ed}}{M_{j,Rd}}\right)^{2,7} & \text{for} \quad M_{j,Ed} > \frac{2}{3} M_{j,Rd} \end{cases} \tag{9}$$

Therefore, if $M_{j,Ed} = M_{j,Rd}$ then $\mu \approx 3{,}0$ and the required vertical displacement of stiffeners in the line of anchor bolts to achieve the full resistance of tension zone ($\Delta L_{T,Rd}$) equals (10):

$$\Delta L_{T,Rd} = \frac{F_{T,Rd}}{S_T} = \frac{F_{T,Rd} \cdot \mu}{E} \cdot \left(\frac{1}{k_{15}} + \frac{1}{k_{16}} \right) = \frac{3 \cdot F_{T,Rd}}{E} \cdot \left(\frac{1}{k_{15}} + \frac{1}{k_{16}} \right) \tag{10}$$

While $\Delta L_T \geq \Delta L_{T,Rd}$ the resistance of tension zone is full, equal $F_{T,Rd}$. Otherwise this resistance will be reduced to a certain value $F_{T,Rd,red}$. The dependency between $F_{T,Rd,red}$ and ΔL_T has been assumed as the same as between bending moment in joint $M_{j,Ed}$ and the rotational stiffness ϕ, according to [1], which results in formula (11):

$$F_{T,Rd,red} = \begin{cases} \dfrac{\Delta L_T \cdot E}{\dfrac{1}{k_{15}} + \dfrac{1}{k_{16}}} & \text{for} \quad F_{T,Rd,red} \leq \dfrac{2}{3} F_{T,Rd} \\[2ex] \left(\dfrac{\Delta L_T \cdot E \cdot F_{T,Rd}^{\;2,7}}{3 \cdot \left(\dfrac{1}{k_{15}} + \dfrac{1}{k_{16}} \right)} \right)^{0,27} & \text{for} \quad F_{T,Rd,red} > \dfrac{2}{3} F_{T,Rd} \end{cases} \tag{11}$$

If $\Delta L_T \leq 0$ there is no tension zone in column base and $F_{T,Rd,red} = 0$.

3 Steps of the procedure

The procedure to compute bending resistance of the column base consists of the steps:
Step 1: Calculation of the resistance of tension zone $F_{T,Rd}$ and the associated necessary displacement $\Delta L_{T,Rd}$ (10).
Step 2: If necessary, assumption the reduced resistance of tension zone $F_{T,Rd,red}$ due to insufficient displacement in this zone.
Step 3: Calculation of the required resistance of compression zone using formulas (12a) or (12b):

$$F_{C,Rd} = F_{T,Rd} + N_{Ed} \tag{12a}$$

$$F_{C,Rd} = F_{T,Rd,red} + N_{Ed} \tag{12b}$$

Step 4: Determining the range of compression zone x_{eff} and x.
Step 5: Calculation of the vertical displacement of tension zone ΔL_T (7) and comparison with $\Delta L_{T,Rd}$ (10).
Step 6: If necessary, calculation of the actual resistance of tension zone $F_{T,Rd,red}$ (11) and comparison with assumed in step 2. In the case of discrepancies repeating steps 2 – 6 until satisfactory compatibility is reached.
Step 7: Calculation of the bending resistance using formula (13a) or (13b):

$$M_{j,Rd} = F_{C,Rd} \cdot (z_C + z_T) - N_{Ed} \cdot z_T \tag{13a}$$

$$M_{j,Rd} = F_{C,Rd} \cdot z_C + F_{T,Rd,red} \cdot z_T \tag{13b}$$

4 Design example

The calculation of bending resistance of an exemplary column base under N_{Ed} = 200 kN (Fig. 6) using proposed procedure has been presented below.

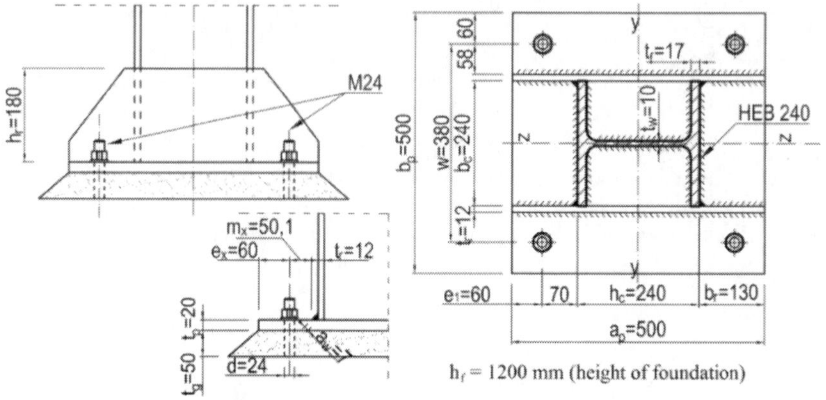

Fig. 6. Dimensions of analyzed column base.

Step 1: Calculation the resistance of tension zone and necessary displacement to achieve the full resistance of this zone:

$$F_{T,Rd} = 185{,}6kN$$

$$\Delta L_{T,Rd} = \frac{3 \cdot F_{T,Rd}}{E} \cdot \left(\frac{1}{k_{15}} + \frac{1}{k_{16}} \right) = \frac{3 \cdot 185600}{210000} \cdot \left(\frac{1}{5{,}34} + \frac{1}{2{,}57} \right) = 1{,}53mm$$

Step 2: Assumption the reduced resistance of tension zone due to insufficient displacement in this zone:

Assumption of sufficient displacement has been made, therefore: $F_{T,Rd,red} = F_{T,Rd}$

Step 3: Calculation of the required resistance of compression zone:

$$F_{C,Rd} = F_{T,Rd} + N_{Ed} = 185{,}6 + 200 = 385{,}6kN$$

Step 4: Determining the range of compression zone (Fig. 7):

$$F_{C,Rd,1} = A_{eff,1} \cdot f_{cd} = 16880 \cdot 17{,}8 = 300{,}5kN$$

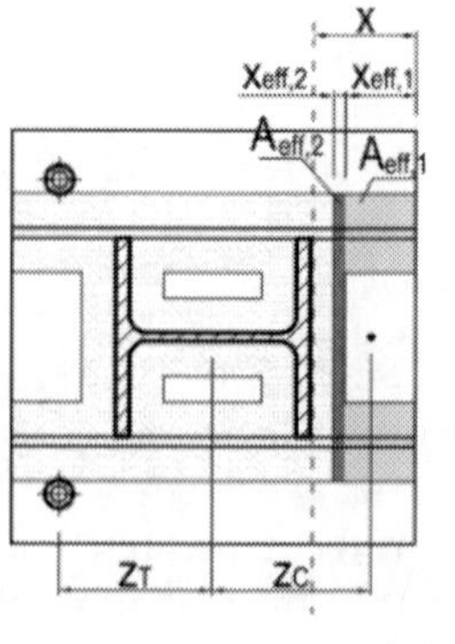

Fig. 7. The range of compression zone of analyzed column base for N_{Ed} = 200 kN.

$$F_{C,Rd,2} = F_{C,Rd} - F_{C,Rd,1} = 385{,}6 - 300{,}5 = 85{,}1kN$$

$$x_{eff,2} = 13{,}8mm$$

$$x = \frac{x_{eff}}{\lambda} = \frac{101{,}9}{0{,}8} = 127{,}4mm$$

Step 5: Calculation of the vertical displacement of tension zone:

$$\Delta L_T = \frac{7}{4}\cdot\Delta C_{cu2}\cdot\left(\frac{a_p - e_1}{x} - 1\right) = \frac{7}{4}\cdot 1{,}68\cdot\left(\frac{500-60}{127{,}4} - 1\right) = 7{,}2mm$$

Step 6: Calculation of the actual resistance of tension zone:

$$\Delta L_T = 7{,}2mm > \Delta L_{T,Rd} = 1{,}53mm \rightarrow F_{T,Rd,red} = F_{T,Rd}$$

Step 7: Calculation of bending resistance:

$$M_{j,Rd} = 385{,}6\cdot(194{,}7 + 190)\cdot 10^{-3} - 200\cdot 190\cdot 10^{-3} = 110{,}5kNm$$

The diagram with the M – N curve for analysed geometry of column base is shown in Figure 8. The shape of the curve is similar to the diagram for reinforced concrete columns.

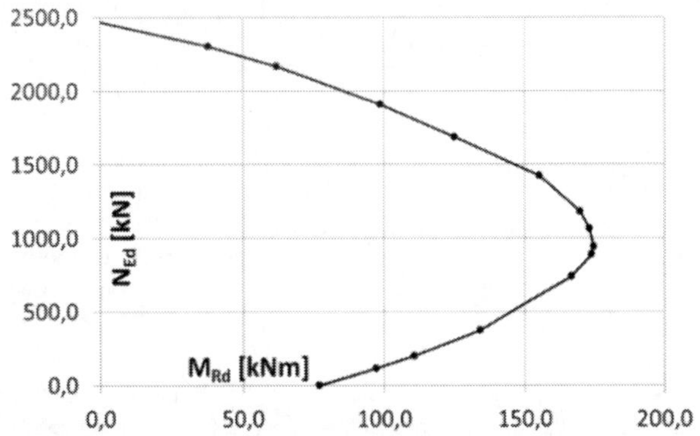

Fig. 8. The M – N curve for analyzed geometry of column base.

5 Conclusions

The proposed procedure allows to easily calculate the resistance of column bases with stiffeners under any combination of loads. This universal calculation method based on Eurocode assumptions is an interesting alternative to existing procedures with many restrictions in the usage. It can be used not only for column base with stiffeners, but also for other problematic cases with complex geometry. Application of the deformation analysis allows to include together the properties of steel base plate and the concrete foundation.

References

1. EN 1993-1-8 (2005): Eurocode 3: Design of steel structures - Part 1-8: Design of joints
2. M. Łubiński, A. Filipowicz, W. Żółtowski, *Konstrukcje metalowe, część I: podstawy projektowania* (Arkady, Warszawa, 2000)
3. A. Matuszak, Czasopismo Techniczne, **107**, 4-B, 75-86 (2010)
4. EN 1992-1-1 (2004): Eurocode 2: Design of concrete structures - Part 1-1: General rules and rules for buildings
5. J. Żmuda, *Projektowanie Konstrukcji Stalowych, cz. 1 Dźwigary kratownicowe, słupy, ramownice* (PWN, Warszawa, 2016)
6. A. Łapko, B. Jensen, *Podstawy projektowania i algorytmy obliczeń konstrukcji żelbetowych* (Arkady, Warszawa, 2015)
7. Access Steel, SN037a-PL-EU

Dynamic analysis of temporary steel grandstand equipped with different types of bracing system

Natalia Lasowicz[1,*], and *Tomasz* Falborski[1]

[1]Gdańsk University of Technology, Faculty of Civil and Environmental Engineering, ul. Narutowicza 11/12, 80-233 Gdańsk, Poland

Abstract. In the paper, behaviour of a temporary steel grandstand equipped with two different types of bracing system has been analysed through the numerical study. A typical solution concerning application of a diagonal tubular members has been compared with elements proposed by authors and called polymer dampers. A polymer element consists of two L-shape steel members bonded with polymer mass. The aim of the paper is to verify the effectiveness of polymer dampers in reduction of vibrations of temporary metal grandstands subjected to dynamic loads induced by spectators. The results of the study show that the response of structure equipped with two analysed bracing systems is different variable. The application of the polymer dampers results in substantial reduction in structural vibrations. Moreover, the decrease in values of accelerations of the grandstand with polymer dampers allows to maintain comfort of spectators at an acceptable level.

1 Introduction

Dynamic loads are often considered the most important ones at the design stage of civil engineering structures [1]. Newly designed and erected metal structures are often composed of very slender and light components that are more easily excitable by different types of dynamic load [2]. Wind, earthquake or crowd load effects determine the design procedure of metal structures (that are) regularly subjected to such significant dynamic loads (see, for e.g. [3-8]). Vibrations of civil engineering structures, generated by dynamic loads, may be very dangerous to the structural integrity as well as may act strongly human perception and comfort [9, 10].

In the past, metal grandstands were mainly used during sport events. Nowadays, these types of structures are commonly applied for music concerts, festivals or other events with large number of participants [2]. Moreover, the use of lighter and slender structural members leads to the reduction in values of natural frequencies and that is why such structures are more easily induced to vibrations. As a result of this situation, resonance may occur in the case of dynamic loads due to behaviour of spectators (e.g. jumping) which may even cause severe damages to the grandstand [11-14]. Structural vibrations also lead to the

* Corresponding author: natmajew@pg.edu.pl

reduction in comfort of spectators [9,10]. A number of different methods have been proposed in order to minimize such situations. One of the methods is to increase stiffness of the structure [15]. Another solution assumes application of additional dampers so as to increase dissipation of energy during vibrations [16].

In the paper, behaviour of a temporary steel grandstand equipped with two different types of bracing system has been analysed through the numerical study. A typical solution concerning application of a diagonal tubular member has been compared with an element proposed by authors and called polymer damper. It consists of two L-shape steel members L 50x50x5 mm bonded with polymer mass of thickness 5 mm. The applied polymer mass is a specially designed flexible two-component grout based on polyurethane resin [17, 18]. This kind of material has been already used for filling of cracks in brick walls [19, 20]. It has been proven, based on the previous experimental studies, that the material is characterized by excellent damping properties [21, 22].

The aim of the paper is to verify the effectiveness of polymer dampers in reduction of vibrations of temporary metal grandstands subjected to dynamic loads induced by spectators. The first stage of the investigation deals with modal analysis so as to determine dynamic parameters, such as modes of free vibrations and the corresponding natural frequencies for both structural models. The second stage of the numerical study concerns dynamic transient analysis. In the analysis, the dynamic load has been assumed a series of synchronous repetitive impacts, as expressed by Fourier series (see [23]).

2 Numerical analysis

Two different numerical models of a temporary metal grandstand erected using scaffolding system have been generated using a commercial programme MSC Marc. Total dimensions of the analysed structure are: length 16.8 m, width 10.8 m and height 7.5 m. The first describes the structure equipped with typical diagonal stiffener members of tubular cross-section, while the second one concerns the structure with polymer dampers. Both models have been presented in Fig.1 and Fig. 2, respectively. Scaffolding system and stiffeners consisting of slender tubular structural members have been modelled by standard two-node beam-column elements, while platforms and benches have been generated as standard four-node shell elements available in the program. In the case of polymer, standard eight-node (six degrees of freedom for each node) solid elements have been used. The base of all models has assumed fixed only in translational directions. Steel, as a structural material, has been described by the following material properties: E=210 GPa, v=0.3, ρ=7850 kg/m3. The behaviour of polymer has been simulated with the use of the Mooney-Rivlin material model that is the most frequently adopted method for modelling complex mechanical behaviour of elastomers and rubber-like solids [24, 25]. The following material constants for the five-parameter Mooney-Rivlin model have been applied: C_{10}=889.490 kPa, C_{01}=-245.840 kPa, C_{20}=-155.310 kPa, C_{11}=93.786 kPa, C_{30}=11.148 kPa. The bulk modulus has been set to be 2.5 GPa, as commonly used for elastomers and rubber-like materials (see [17, 21]). Structural damping was considered in the analysis.

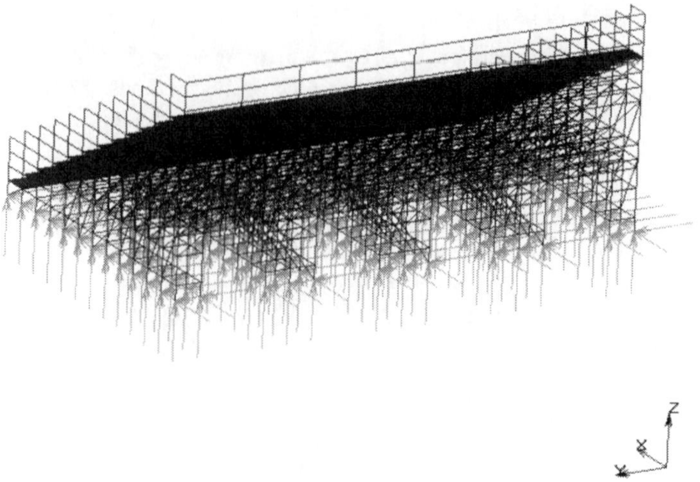

Fig. 1. FE model of temporary grandstand equipped with typical tubular diagonal members.

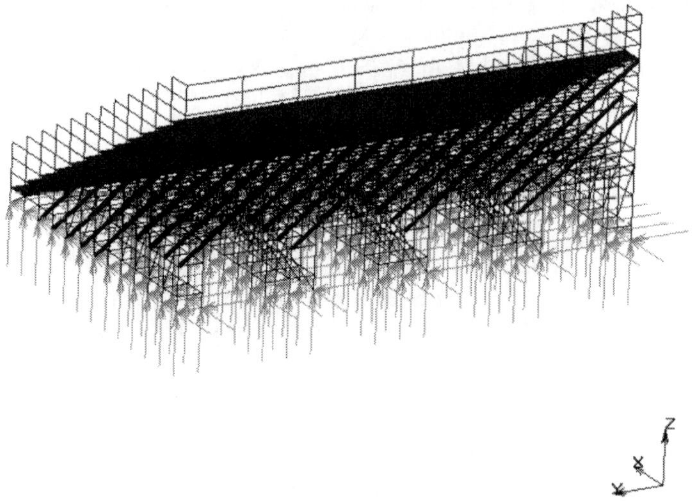

Fig. 2. FE model of temporary grandstand equipped with polymer dampers.

2.1 Modal analysis

The first stage of numerical analysis is focused on modal analysis in order to determine horizontal (left-to-right) modes of free vibrations and corresponding natural frequencies of temporary grandstand equipped with two different types of bracing system. In the case of such structures horizontal vibrations are found to be the most important ones [12, 23]. An empty and occupied structure has been analyzed numerically. In the modal analysis the mass of about 300 spectators has been assumed. First horizontal (Y direction) mode of free vibrations of temporary grandstand equipped with tubular elements and polymer dampers have been presented in Figure 3 and Figure 4, respectively. The values of natural frequencies corresponding to horizontal modes of free vibrations for empty and occupied temporary grandstand are collected in Table 1.

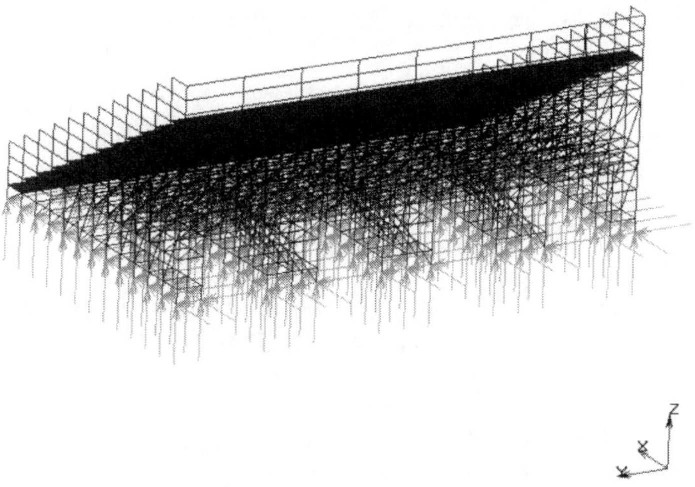

Fig. 3. 1st horizontal mode of free vibrations of temporary grandstand equipped with typical tubular diagonal members (Model 1).

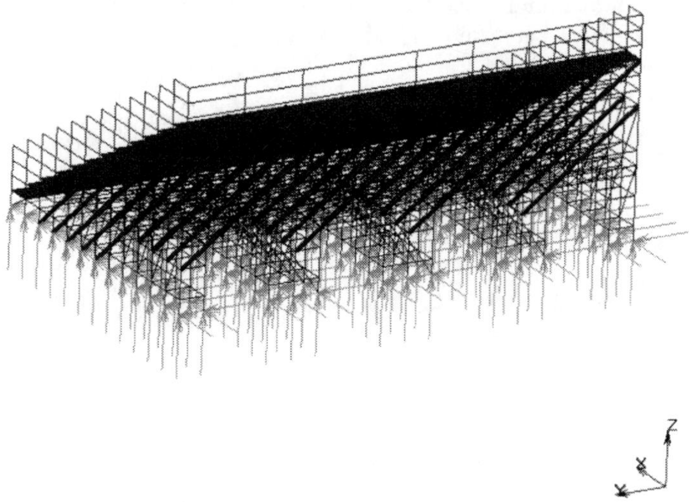

Fig. 4. 1st horizontal mode of free vibrations of temporary grandstand equipped with polymer dampers (Model 2).

Table 1. Natural frequencies determined for an empty and occupied structure equipped with two different types of bracing system.

Model	**Natural frequency [Hz]**	
	Empty structure	Occupied structure
Model 1	6.43	4.00
Model 2	7.00	4.55

It can be seen from Table 1 that mass of the spectators significantly decreases values of natural frequencies (by as much as 38% in the case of Model 1, and 30% in the case of Model 2). However, there is no significant difference between values of natural frequencies of structure equipped with tubular diagonal elements and polymer dampers. The difference between determined values of natural frequencies are not greater than 8% and 9% for an empty and occupied structure, respectively. Based on the previous study [12, 26, 27] it should be underlined that the estimated natural frequency values of an empty and occupied structure are in the safe range.

2.2 Dynamic transient analysis

The second stage of the investigation is focused on dynamic transient analysis. The behaviour of temporary metal grandstand equipped with two different bracing systems under human-induced excitation due to jumping has been analysed. In the analysis, the dynamic load has been assumed to be consisted of synchronous repetitive impacts, as expressed by Fourier series (see [23]). According to the Polish Standards [28], additional load in horizontal (left-to-right) direction of (6% of total value of load acting in vertical direction) has been assumed in dynamic transient analysis.

The aim of the dynamic analysis has been conducted to determine peak values of accelerations and displacements of the structure equipped with tubular and polymer elements (compare also [29]). Acceleration time histories for both models obtained at the level of the highest steel platform are presented in Figure 5 and Figure 6. The displacement time histories are presented in Figure 7 and Figure 8, respectively. The peak values of acceleration and displacement of temporary grandstand equipped with two different types of bracing system under human-induced excitation due to jumping at 2 Hz estimated from dynamic transient analysis are summarized in Table 2.

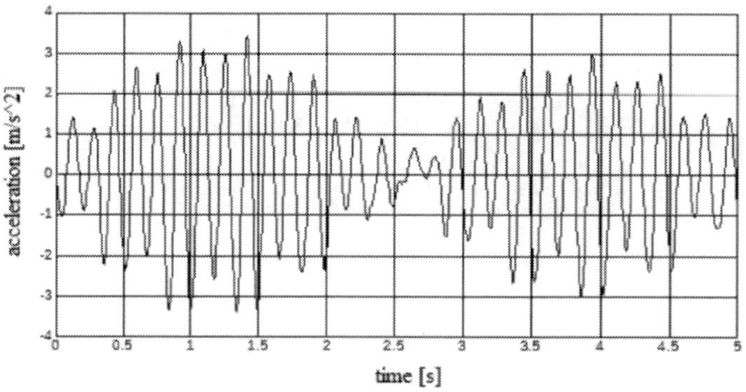

Fig. 5. Acceleration time history for temporary steel grandstand equipped with tubular elements (Y direction).

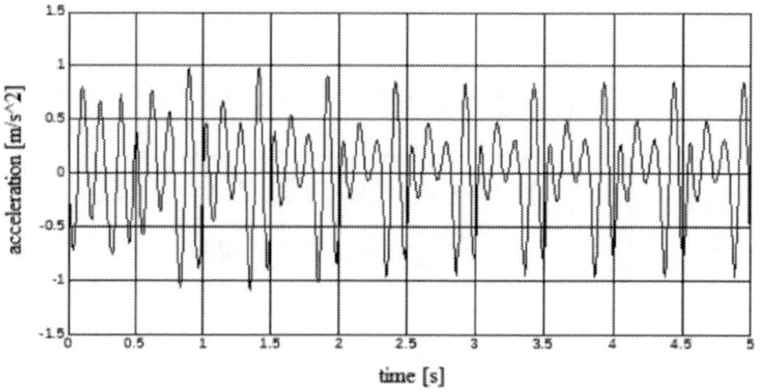

Fig. 6. Acceleration time history for temporary steel grandstand equipped with polymer dampers (Y direction).

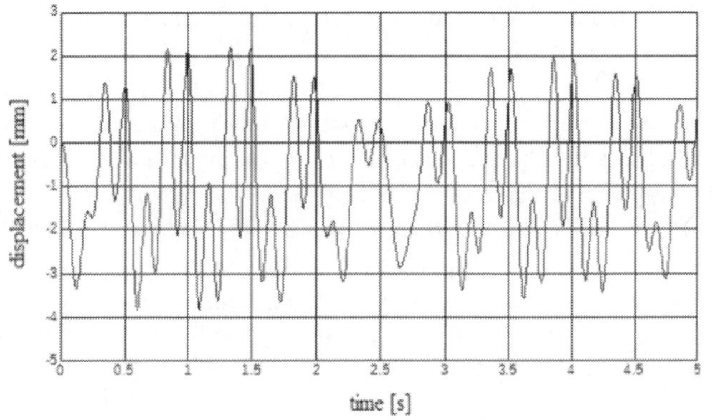

Fig. 7. Displacement time history for temporary steel grandstand equipped with tubular elements (Y direction).

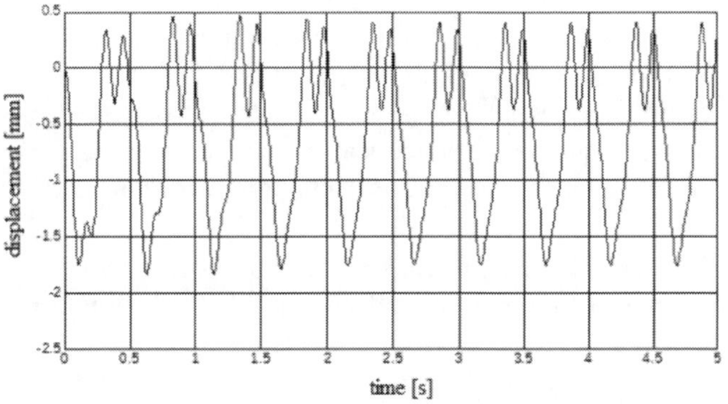

Fig. 8. Displacement time history for temporary steel grandstand equipped with polymer dampers (Y direction).

Table 2. Peak left-to-right horizontal accelerations and displacements for Model 1 and Model 2.

Model	Acceleration [m/s^2]	Displacement [mm]
Model 1	3.42	3.83
Model 2	1.10	1.84

The lowest recommended acceleration value to avoid panic among spectators, is equal to 0.35g (where g is the acceleration of gravity - see [30]). It can be seen from Table 2 that the peak value of acceleration estimated for Model 1 is almost equal to 0.35g and it can be considered as unacceptable while the peak value of 0.11g has been reached in the case of Model 2 and can be considered disturbing. It should be pointed out that the peak value of acceleration estimated for the structure equipped with polymer dampers is more than three times lower than the value determined for grandstand with typical tubular member. The peak displacement estimated from the dynamic transient analysis for the structure with polymer dampers is more than two times lower than the value obtained for grandstand with typical tubular elements.

3 Conclusions

The paper deals with effectiveness of polymer dampers in reduction of temporary steel grandstand vibrations. Typical solution using diagonal tubular elements to improve stiffness of such structures has been compared with the application of polymer dampers consisting of two L-shape steel members bonded with polymer mass. In the first stage of the study modal analysis has been conducted in order to determine dynamic parameters, such as modes of free vibrations and corresponding natural frequencies, of temporary steel scaffolding grandstand equipped with two different types of bracing system. Next, dynamic transient analysis has been conducted and behaviour of the structure under human-induced excitation due to jumping has been determined.

The results of the study show that the response of structure equipped with two analysed bracing systems is different. The application of the polymer dampers results in substantial reduction in structural vibrations. Moreover, the decrease in values of accelerations of the grandstand with polymer dampers allows us to maintain comfort of spectators at an acceptable level.

References

1. A.K. Chopra, *Dynamics of Structures* (Prentice-Hall, USA, 2010)
2. P. Reynolds, P. Mohanty, A. Pavic, Use of operational modal analysis on empty and occupied stadia structures. *Proceedings of 1st International Operational Modal Analysis Conference*, Copenhagen 2005
3. T. Falborski, R. Jankowski, Numerical evaluation of dynamic response of a steel structure model under various seismic excitations, Procedia Engineering, **172**, 277-283 (2017)
4. H. Elwardany, A. Seleemah, R. Jankowski, Seismic pounding behavior of multi-story buildings in series considering the effect of infill panels, Engineering Structures, **144**, 139-150 (2017)
5. R. Jankowski, S. Mahmoud, *Earthquake-Induced Structural Pounding* (Springer, Cham, Switzerland, 2015)
6. R. Jankowski, S. Mahmoud, Linking of adjacent three-storey buildings for mitigation of structural pounding during earthquakes, Bulletin of Earthquake Engineering, **14**, 3075-3097 (2016)
7. H. Naderpour, R. C. Barros, S.M. Khatami, R. Jankowski, Numerical study on pounding between two adjacent buildings under earthquake excitation, Shock and Vibration, article ID 1504783 (2016)
8. E. Simiu, R. Scanlan, *Wind Effects on Structures, Third edition* (John Wiley and Sons, USA, 1996)
9. S. P. Nhleko, M. S. Williams, A. Blackeborough, Vibration perception and comfort levels for an audience occupying a grandstand with perceivable motion, *Proceedings of IMAC XXVII: A Conference and Exposition on Structural Dynamics*, Orlando 2009.
10. N. Lasowicz, R. Jankowski, The effectiveness of polymer damper in damage reduction of temporary steel grandstand, Key Engineering Materials, **713**, 171-174 (2016)
11. V. Brito, R. L. Pimentel, Cases of collapse of demountable grandstands, Journal of Performance of Constructed Facilities, **23(3)**, 151-159 (2009)
12. D. Crick, G. Y. Grondin, Monitoring and analysis of a temporary grandstands, Structural Engineering Report No 275, (2008)
13. N. Jacobs, Temporary Demountable Structures: the need for guidance, The Structural Engineer, **74(5)**, (1996)

14. N. Noss, *Investigation of human-structure interaction through experimental and analitycal studies* (Bucknell University, USA, 2012)
15. T. Ji, B. R. Ellis, Effective bracing system for temporary grandstand, The Structural Engineer, **75(6)**, 95-100 (1997)
16. N. Noormohammadi, P. Reynolds, Control of human induced vibrations in stadia using a hybrid tuned mass damper, *Proceedings of International Conference on Noise and Vibration Engineering*, 1119-1132 (2012)
17. T. Falborski, R. Jankowski, Advanced Hysteretic Model of a Prototype Seismic Isolation System Made of Polymeric Bearings, Applied Sciences **8(3)**, 400 (2018)
18. 18. A. Kwiecień, Polimerowe złącza podatne w konstrukcjach murowych i betonowych (Politechnika Krakowska, Poland, 2012)
19. A. Kwiecień, Flexible polymer adhesives versus stiff mineral and epoxy adhesives tested dynamically on masonry columns strengthened using of bonded GFRP mesh, *Proceedings of 8th International Conference on Structural Dynamics*, Belgium 2011
20. A. Kwiecień, B. Zając, Energy approach in analysis of masonry columns confined with GFRP mesh bonded on stiff and flexible adhesives, Structural Analysis of Historical Constructions, 1864-1872 (2012)
21. T. Falborski, R. Jankowski, Experimental Study on Effectiveness of a Prototype Seismic Isolation System Made of Polymeric Bearings, Applied Sciences **7(8)**, 808 (2017)
22. N. Lasowicz, A. Kwiecień, R. Jankowski, Enhancing the seismic resistance of columns using polymer adhesive - experimental study, Key Engineering Materials **624**, 478-485 (2015)
23. B. R. Ellis, T. Ji, J. D. Littler, The response of grandstands to dynamic crowd loads, Structures and Buildings,**140(4)**, 355-365 (2000)
24. R. H. Finney, *Finite Element Analysis, Engineering with Rubber: How to design Rubber Components*, (Gent A. N., Germany, 2001)
25. R. H. Finney, A. Kumar, Development of material constants for nonlinear finite-element analysis, Rubber Chemistry and Technology, **61(5)**, 879-891 (1988)
26. J. D. Littler, Temporary demountable grandstand: dynamic response, Building Research Establishment, **3**, (2000)
27. BRE Digest, The response of structures to dynamic crowd loads, **426** (1997)
28. PN-EN 13200-6:2013-02 Obiekty widowiskowe - Część 6: Trybuny demontowalne (tymczasowe)
29. N. Lasowicz, A. Kwiecień, R. Jankowski, Experimental study on the effectiveness of polymer damper in damage reduction of temporary steel grandstand, Journal of Physics: Conference Series, **628**, 1-7 (2015)
30. Kasperski M., Actual problems with stand structures due to spectator-induced vibrations, *Proceeding of the 3rd European Conference on Structural Dynamics*, 455-461 (1996)

MATEC Web of Conferences **219**, 03001 (2018) https://doi.org/10.1051/matecconf/201821903001
BalCon 2018

Soil-structure interaction effects on modal parameters of office buildings with different number of stories

Natalia Lasowicz[1,*] and *Tomasz* Falborski[1]

[1]Gdansk University of Technology, Faculty of Civil and Environmental Engineering, Narutowicza 11/12, 80-233 Gdańsk, Poland

Abstract. The paper summarizes the results of a numerical investigation designed to study the soil-structure interaction effects on modal parameters of three office buildings. The reinforced-concrete 4-storey, 8-storey, and 12-storey office buildings, each with additional two levels of embedded basements, represent low, medium, and high-rise structures, respectively. In order to conduct this research, detailed finite-element structure models were prepared. Soil-foundation flexibility was represented with the use of spring-based solutions, incorporating foundation springs and dashpots. The influence of diverse soil conditions (represented by their average effective profile velocities and shear moduli) on the dynamic characteristics of the analyzed three office buildings (e.g. fundamental vibration periods) was investigated and discussed.

1 Introduction and motivation

Structural response of a building subjected to various dynamic excitations, such as earthquake ground motions, mining-induced tremors, impact loads, or jumping crowd loads, may be affected by many different factors including base isolation (see, for example, [1,2]), structural dampers (see, for example, [3-5]), pounding effects (see, for example, [6-11]) etc. Among a number of factors, which may considerably modify the dynamic characteristics, and thus alter the seismic response of a structure, the interaction between the structure foundation and the underlying soil is considered to be one of the most important contributors (see, for example, [12-16]). Even though the soil-foundation flexibility may have a significant impact on dynamic behaviour, many studies either do not incorporate soil-structure interaction, which may result in misrepresentation of the actual building's response, or are performed for simplified stick models, according to which buildings are idealized as multi-degree-of-freedom systems (see, for example, [17-19]), which, on the other hand, may not be fully appropriate for irregular structures.

Motivated by the preceding discussion, the present study was designed to investigate the soil-structure interaction effects on modal parameters of three reinforced-concrete office buildings. The 4-storey, 8-storey, and 12-storey office buildings represent low, medium, and high-rise structures, respectively. In order to conduct this research, detailed finite-

[*] Corresponding author: natmajew@pg.edu.pl

element structure models were firstly prepared. In the next step, the soil-foundation flexibility was represented with the use of spring-based solutions, incorporating foundation springs and dashpots, as it is the most commonly adopted approach for idealizing the soil-foundation interface in current engineering practice. The influence of diverse soil conditions (represented by their average effective profile velocities and shear moduli) on the modal characteristics of the analyzed structures (e.g. fundamental vibration periods) was investigated and discussed.

2 Numerical models

Three office buildings considered in the present study are irregular reinforced-concrete structures made in a slab-column system with an interior shear wall core. The 4-storey, 8-storey, and 12-storey structures have the same floor plans, meaning that only the total number of stories is different. Additionally, each building has two levels of fully embedded basements with total height of 2×4.35 m. The height of the typical above-grade storey is 3.75 m, whereas the height of the emergency exit located at the rooftop is 2.5 m. Accordingly, the overall heights of the 4-storey, 8-storey, and 12-storey structures are 26.2 m, 41.2 m, and 56.2 m, respectively. The plan dimensions of each structure are 27 m wide by 42 m long, whereas the gross dimensions of the foundation mat are 28 m wide by 43 m long. Reinforced-concrete exterior basement walls, interior staircase walls, floor slabs, and mat foundations were developed using 4-node shell elements, whereas reinforced-concrete columns were modelled with the use of beam elements available in an educational version of Autodesk Robot Structural Analysis Professional 2017. As an example, the numerical model of the 12-storey office building is presented in Fig. 1.

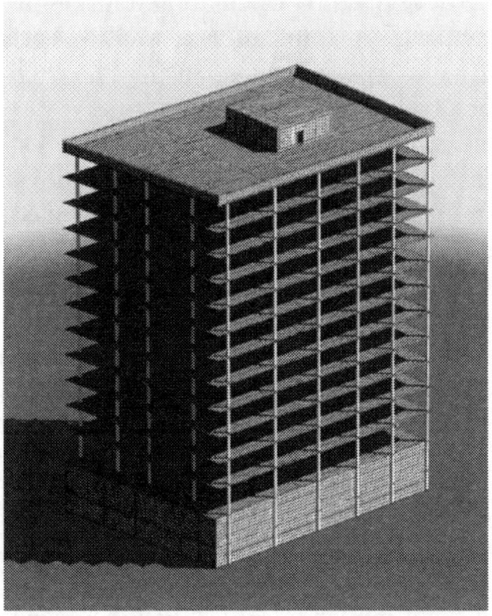

Fig. 1. Numerical model of the 12-storey office building.

3 Soil characteristics

Detailed description of soil conditions is an important component in every dynamic analysis incorporating soil-structure flexibility (see, for example, [20-22]). More specific procedures and guidelines for estimation of shear wave velocities (including calculation of overburden-corrected shear wave velocities) may be found, for example, in [23,24]. In order to investigate the soil-structure interaction effects on the modal parameters of the office buildings considered in the present study, three different site conditions were utilized. Soil properties were assumed in accordance to [25]. Geotechnical data represented by shear wave velocities, mass densities, and Poisson's ratios are briefly summarized in Table 1.

Table 1. Characteristics of the soil types considered in the study.

Soil type	Shear wave velocity (m/s)	Mass density (kN/m^3)	Poisson's ratio (-)
Dense soil	500	20	0.40
Stiff soil	300	18	0.35
Soft soil	150	16	0.30

4 Soil-structure flexibility

Soil-structure interaction may be implemented into numerical analysis through many different approaches, either by incorporating foundation springs and dashpots or modelling the soil beneath the structure as a solid continuum with finite elements (see, for example, [26-28]). In the present study soil-foundation flexibility was implemented using spring-based solutions proposed in [29], which are identified among the most commonly adopted equations in current engineering practice. This approach includes calculations of soil springs to capture translational and rotational degrees of freedom, and dashpots to address soil damping effects. For each site condition considered in this study, characteristics of springs and dashpots were developed by calculating translational (k_x, k_y) and rotational (k_{xx}, k_{yy}) stiffnesses for rectangular rigid foundation as well as dashpot coefficients (c_x, c_y, c_{xx}, c_{yy}). The foundation stiffness was calculated using the static stiffness as well as the embedment correction factors η. The base spring stiffness was subtracted from the overall horizontal stiffness to determine the portion of the horizontal stiffness attributed to passive pressure resistance against basement walls (total translational stiffness is larger due to embedment). Vertical springs and dashpots were distributed over the footprint of the foundation, allowing the foundation to deform in a natural manner. Edge intensities were adjusted to match the overall rocking stiffness values, as the vertical soil reaction is not uniform, and tends to increase near the edges of the foundation. Correction factors were determined using the equations presented in [30]. Corner intensities ware calculated as the average of edge intensities in both directions. It should be also underlined that both horizontal and vertical springs are elastic with no compression capacity limit and zero tension capacity.

5 Modal analysis

In the final step of the current numerical investigation, modal analysis was conducted to investigate the influence of soil-structure flexibility (i.e. represented by three soil conditions presented in Section 3) on modal parameters of the analyzed office buildings. The results obtained for the 4-storey, 8-storey, and 12-storey office buildings were compared with those obtained for fixed-base conditions. As an example, fundamental modes of vibration, in both transverse and longitudinal directions, of the 12-storey fixed-base building are presented in Fig. 2. Results obtained for all site conditions considered in the study are briefly summarized in Table 2 and also presented in Fig. 3.

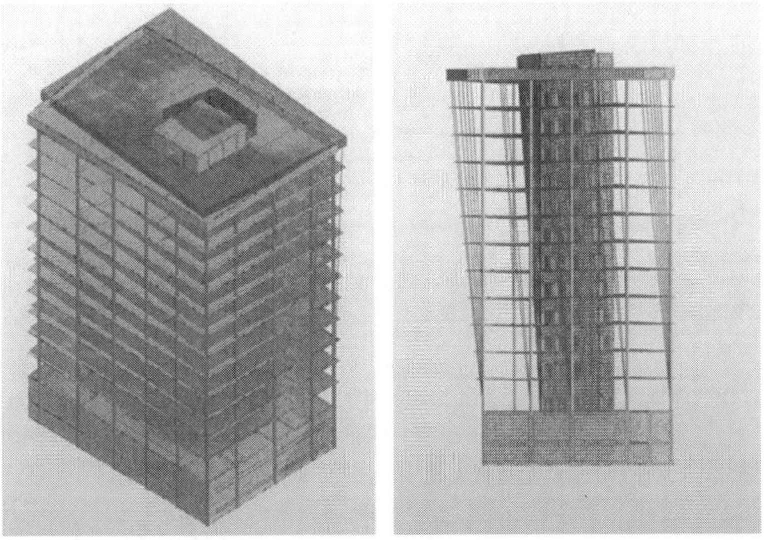

a)

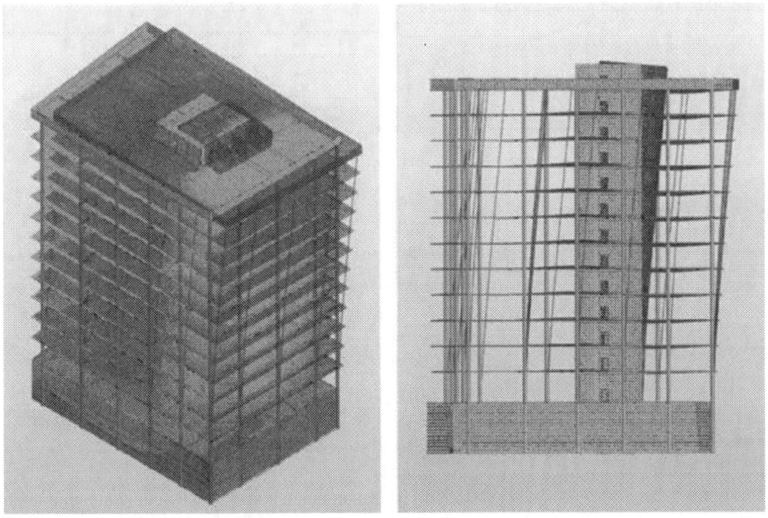

b)

Fig. 2. Transverse: (**a**) and longitudinal, (**b**) modes of vibrations for the fixed-base 12-storey office building.

Table 2. Results of modal analysis.

Site conditions	Frequency (Hz)					
	4-storey building		8-storey building		12-storey building	
	Transverse mode of vibration	Longitudinal mode of vibration	Transverse mode of vibration	Longitudinal mode of vibration	Transverse mode of vibration	Longitudinal mode of vibration
Fixed-base	3.64	4.51	1.89	2.15	1.15	1.20
Dense soil	3.39	4.19	1.69	1.86	1.01	1.02
Stiff soil	3.28	3.95	1.64	1.78	0.96	0.98
Soft soil	3.21	3.67	1.54	1.71	0.91	0.92

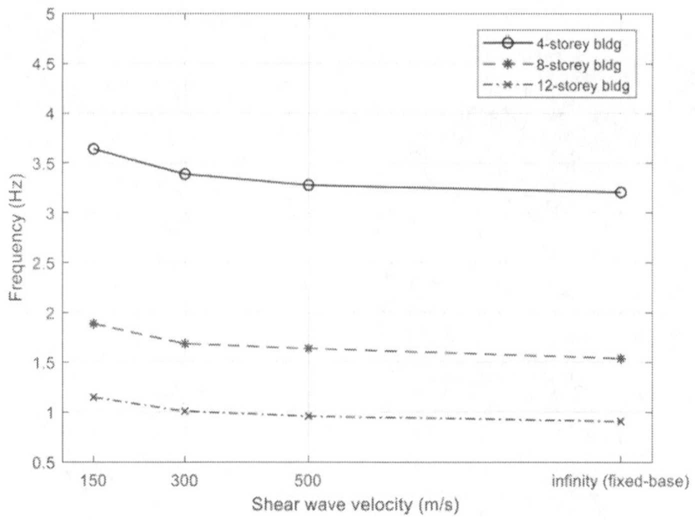

a)

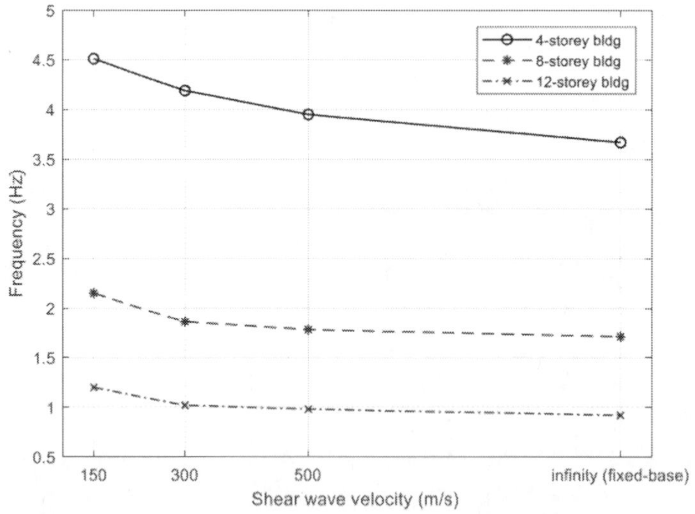

b)

Fig. 3. Frequencies for transverse: (**a**) and longitudinal, (**b**) modes of vibration for the analyzed buildings.

5 Results and conclusions

The present study was designed to examine the soil-structure interaction effects on the modal parameters of reinforced-concrete office buildings with different number of stories. Three different site conditions were utilized into numerical analysis to investigate the importance of the soil flexibility. Next, the results were compared with those obtained for the fixed-base structures. The numerical investigation resulted in the following conclusions:

1. Close inspection of Table 2 and Fig. 3 clearly shows that soil flexibility can considerably modify the modal parameters of the building structure by lengthening its fundamental period (the lower the fundamental frequency, the longer the fundamental period of vibration). As expected, the effects of soil-foundation flexibility become

more significant as the stiffness of the underlying soil decreases. The present investigation confirmed that for stiff and dense soils modal parameters were only modestly affected. For the soft soil the reduction levels in the fundamental frequencies in transverse direction for the 4-storey, 8-storey, and 12-storey buildings are 11.8%, 18.5%, and 20.8%, respectively, when compared to the fixed-base models. The reduction levels in the fundamental frequencies in longitudinal direction for the 4-storey, 8-storey, and 12-storey buildings are 18.6%, 20.4%, and 23.3%, respectively, when compared to fixed-base models.

2. Results obtained exhibits that the soil-structure interaction effects seem more profound for taller buildings as the reduction levels in fundamental frequencies in both transverse and longitudinal directions are the highest for the 12-storey building.

Conducted numerical investigation, even though utilizing simple engineering approach of incorporating the soil-foundation flexibility with springs and dashpots, clearly demonstrates the significance of soil-structure interaction. Ignoring the site conditions, especially in case of soft soils, may not provide accurate results in dynamic analysis.

References

1. T. Falborski, R. Jankowski, Experimental Study on Effectiveness of a Prototype Seismic Isolation System Made of Polymeric Bearings, Applied Sciences **7(8)**, 808 (2017)
2. T. Falborski, R. Jankowski, Advanced Hysteretic Model of a Prototype Seismic Isolation System Made of Polymeric Bearings, Applied Sciences **8(3)**, 400 (2018)
3. N. Lasowicz, R. Jankowski, The effectiveness of polymer damper in damage reduction of temporary steel grandstand, Key Engineering Materials, **713**, 171-174 (2016)
4. N. Lasowicz, A. Kwiecień, R. Jankowski, Experimental study on the effectiveness of polymer damper in damage reduction of temporary steel grandstand, Journal of Physics: Conference Series, **628**, 1-7 (2015)
5. N. Lasowicz, A. Kwiecień, R. Jankowski, Enhancing the seismic resistance of columns using polymer adhesive - experimental study, Key Engineering Materials **624**, 478-485 (2015)
6. R. Jankowski, S. Mahmoud, *Earthquake-Induced Structural Pounding* (Springer, Cham, Switzerland, 2015)
7. H. Elwardany, A. Seleemah, R. Jankowski, Seismic pounding behavior of multi-story buildings in series considering the effect of infill panels, Engineering Structures, **144**, 139-150 (2017)
8. R. Jankowski, S. Mahmoud, Linking of adjacent three-storey buildings for mitigation of structural pounding during earthquakes, Bulletin of Earthquake Engineering, **14**, 3075-3097 (2016)
9. H. Naderpour, R. C. Barros, S.M. Khatami, R. Jankowski, Numerical study on pounding between two adjacent buildings under earthquake excitation, Shock and Vibration, **2016**, article ID 1504783 (2016)
10. R. Jankowski, Pounding between superstructure segments in multi-supported elevated bridge with three-span continuous deck under 3D non-uniform earthquake excitation, Journal of Earthquake and Tsunami, **9**, Paper no. 1550012 (2015)
11. R. Jankowski, Impact force spectrum for damage assessment of earthquake-induced structural pounding, Key Engineering Materials, **293-294**, 711-718 (2005)
12. G. Gazetas, *Foundation Vibrations in: Foundation Engineering Handbook* (Springer, USA, 1991)

13. J.P. Wolf, *Dynamic Soil-Structure Interaction* (Prentice-Hall, USA, 1985)
14. G. Mylonakis, G. Gazetas G., Seismic soil-structure interaction: beneficial or detrimental? Journal of Earthquake Engineering **4**, 277-301 (2000)
15. J.P. Stewart, G.L. Fenves, R.B. Seed, Seismic soil-structure interaction in buildings I: Analytical methods, Journal of Geotechnical and Geoenvironmental Engineering **125(1)**, 26-37 (1999)
16. J.P. Stewart, G.L. Fenves, R.B. Seed, Seismic soil-structure interaction in buildings II: Empirical findings, Journal of Geotechnical and Geoenvironmental Engineering **125(1)**, 38-48 (1999)
17. T. Falborski, R. Jankowski, Numerical evaluation of dynamic response of a steel structure model under various seismic excitations, Procedia Engineering, **172**, 277-283 (2017)
18. A.K. Chopra, *Dynamics of Structures* (Prentice-Hall, USA, 2010)
19. R.W. Clough, J. Penzien, *Dynamics of Structures* (McGraw-Hill, USA, 1993)
20. A.S. Veletsos, A.M. Prasad, Seismic Interaction of Structures and Soils: Stochastic Approach, Journal of Structural Engineering **115(4)**, 935-956 (1989)
21. J.P. Stewart, G.L. Fenves, System identification for evaluating soil–structure interaction effects in buildings from strong motion recordings, Earthquake Engineering & Structural Dynamics **27(8)**, 869-885 (1998)
22. H.L. Wong, M.D. Trifunac, J.E. Luco, A comparison of soil-structure interaction calculations with results of full-scale forced vibration tests, Soil Dynamics and Earthquake Engineering **7(1)**, 22-31 (1998)
23. National Earthquake Hazards Reduction Program: Soil-Structure Interaction for Building Structures (2012)
24. Applied Technology Council: Improvement of Nonlinear Static Seismic Analysis Procedures (2005)
25. Pacific Earthquake Engineering Research Center: Guidelines for Estimation of Shear Wave Velocity Profiles (2012)
26. J.P. Stewart, S. Kim, J. Bielak, R. Dobry, M.S. Power, Revisions to soil-structure interaction procedures in NEHRP design provisions, Earthquake Spectra **19(3)**, 677-696 (2003)
27. H.L. Wong, J.E. Luco, Dynamic interaction between rigid foundations in a layered half-space, Soil Dynamics and Earthquake Engineering **5(3)**, 149-158 (1986)
28. G. Mylonakis, S. Nikolaou, G. Gazetas, Footings under seismic loading: Analysis and design issues with emphasis on bridge foundations, Soil Dynamics and Earthquake Engineering **26(9)**, 824-853 (2006)
29. A. Pais A., E. Kausel, Approximate formulas for dynamic stiffness of rigid foundations, Soil Dynamics and Earthquake Engineering **7(4)**, 213-227 (1988)
30. C.W. Harden, T.C. Hutchinson, Beam-on-Nonlinear-Winkler-Foundation Modeling of Shallow, Rocking-Dominated Footings, Earthquake Spectra **25**, 277-300 (2009)

Simplified method of applying loads to flat slab floor structural model

Maciej Tomasz Solarczyk[1,*], and *Andrzej* Ambroziak[1]

[1]Gdańsk University of Technology, Faculty of Civil and Environmental Engineering, Narutowicza 11/12, 80-233 Gdańsk, Poland

Abstract. The article analyses the impact of the live load position on the surface of a reinforced concrete flat slab floor of 32.0 m × 28.8 m. Four variants of a live load position are investigated: located on the entire concrete slab, set in a chessboard pattern, applied by bands and imposed separately in each of the slab panels. Conclusions are drawn upon differences in bending moments, the time of calculation and the size of output files. The problems in the interpretation of results are presented too. A procedure is presented to model the reinforced concrete structures in computational programs. The recommendations of the Eurocodes are presented regarding to load combinations in the Ultimate Limit State (ULS). Convergence analysis of the finite element mesh is carried out to verify the obtained results. The law status on the implementation of the Building Information Modelling (BIM) technology in Poland points out significant time savings in the application of this technology.

1 Introduction

Structural design is generally based on virtual mapping of a real object. The designer is supported by a number of computational tools to complete the task. A widespread automation of design works makes it available for the engineers with little experience. It should be emphasized that the results of numerical analysis should be permanently and critically assessed by the designer after computations. While a designer is not experienced the assessment may not be sufficiently reliable.

In the design process the task of a structural designer is to determine the static model and the load layout, finally, to verify the results. These tasks include: definition of the structure (dimensions, types of finite elements, material data, support types), determining of finite element mesh (FE), load application, load combination statement, verification of the FE mesh and the final internal forces. These steps are necessary for the designer of reinforced concrete structures to determine the required reinforcement due to an entirety of Ultimate Limit States (ULS) and Serviceability Limit States (SLS).

Each of the modelling steps is important due to real structural analysis. The most difficult task is a proper representation of subsequent elements included in structural system (columns, beams, slabs, etc.) together with their assembly. There is a number of modelling possibilities,

* Corresponding author: maciej.solarczyk@pg.edu.pl

each one detecting advantages and disadvantages. Some important information about slab-beam structures is given in articles [1, 2] and in the book [3].

It should be emphasized that even a structure accurately mapped in a computational model but subjected incorrectly defined loads will not be analysed properly. While the loads are overestimated the designed structure does not pose a threat to the safety and life of people at operation but it is uneconomic. When the loads are underestimated the result may trigger a structural disaster, to possibly bring fatalities.

The article assesses the impact of the load action to the structural response, i.e. internal forces. The factors affecting the design efficiency are the calculation time, simplicity in the interpretation, accuracy of the solution.

2 Combinations of actions in ULS in terms of Eurocode

Due to Ultimate Limit States, the design load values should be applied, forming possible combinations of actions. According to the standard [4] the partial critical factors $\gamma_G = 1.35$ and $\gamma_Q = 1.5$ are assumed for permanent and variable actions, respectively.

According to the principle given in 5.1.3 (1)P in Eurocode 2 [5]: *In considering the combinations of actions (...) the relevant cases shall be considered to enable the critical design conditions to be established at all sections, with the structure or part the structure considered.*

In addition, the note to point 5.1.3 (1)P in [5] states: *(...) The following simplified load arrangements are recommended for buildings:*

(a) Alternate spans carrying the design variable and permanent load (...); other spans carrying only the design permanent load (...) and

(b) Any two adjacent spans carrying the design variable and permanent loads (...). All others spans carrying only the design permanent load (...).

It should be noted that standard recommendations [5] do not always converge the principles of structural mechanics. If the recommendation from subsection (a) is intended to the determine the maximum bending moments in the span, the subsection (b) does not results in a predetermined maximum moment at the support. While determining the influence line of bending moment at the support of a continuous beam, two adjacent spans and every second subsequent span should be loaded with alternating load. Nevertheless, the difference in the results is considered insignificant.

3 Description of flat slab floor structural model

The investigated flat slab floor is a component of an office building. One of the storeys has been separated to form a numerical model. The reinforced concrete slab of 20 cm thickness is made of C20/25 concrete. The height of the storey is 3.0 m. The slab supports are reinforced concrete columns 30 cm × 30 cm made of C20/25 concrete. The numerical model assumes that the concrete columns are situated above and below the considered slab. In the x-direction the slab shows five spans of 6.4 m, while in the y-direction there are four spans of 7.2 m. The assumed actions are: dead structural load (taken from the software), complementary dead load of a finished floor, equal 2 kN/m^2 (estimation) and imposed uniformly distributed load equal 3.0 kN/m^2. The value of imposed load is taken according to Tables 6.1 and 6.2 (see PN-EN 1991-1-1:2004 standard [6]). The B category of office areas are subjected to a load equal 3.0 kN/m^2.

Computations were conducted by means of the Robot Structural Analysis Professional (RSAP) system, version 2015. The software is integrated with the graphically environment servicing for modelling, analysis and dimensioning of various structural element types. It provides: generating FEM meshes, performing calculations and verifying the results, dimensioning individual elements and preparing documentation of the project. There are national versions including national standards with design specifications, see [7,8].

The calculations were carried out on computer with processor Intel Core i7 with RAM of 4,0 GB.

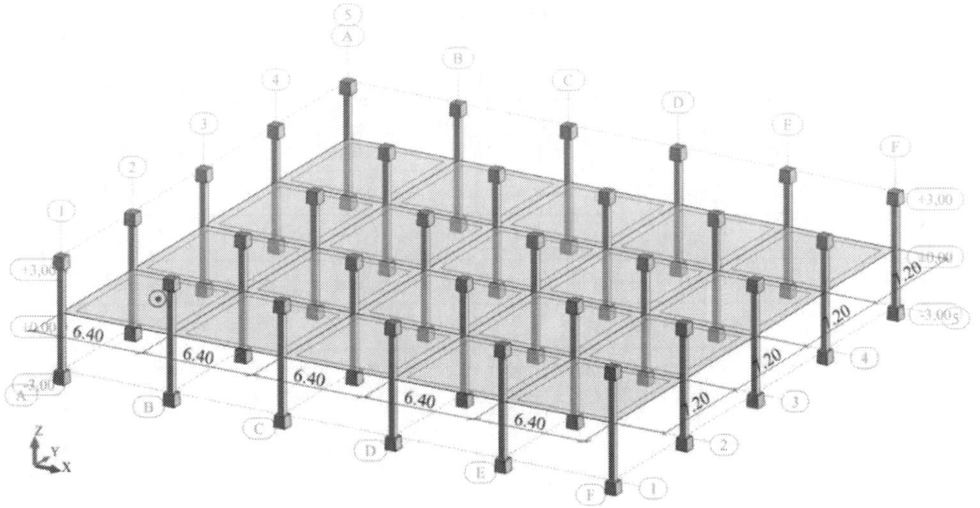

Fig. 1. Geometry of the analysed flat slab floor with the point designation in which the analysis of the results of internal forces is made.

For the sake of comparative analysis four variants were considered to locate the load on a slab-column structure:

1) **variant 1** - live load located on an entire surface of the slab (no combinations for this type of load, thus an increase in the variable action partial factor by additional 25% -30% was assumed to obtain a global coefficient $\gamma_Q = 1.9$);
2) **variant 2** - live load set in a chessboard pattern load (in this variant three combinations in the Ultimate Limit State are specified);
3) **variant 3** - live load set by bands (three combinations are determined in ULS);
4) **variant 4** - live load applied separately in each panel of slab (combinations in ULS are generated automatically by the RSAP program).

4 Finite Element Mesh refinement

An inherent step of the FEM computational procedure is a FE mesh convergence analysis in the RASAP case, described in detail in the books [7, 8]. This is necessary check of the method correctness due to its approximate character. A characteristic point is chosen to check the convergence. Comparing the internal force values in a selected point the convergence of the solution is assessed with respect to mesh refinement.

In the worked case, the FE mesh was doubled to achieve satisfactory results. In the investigated model the mid-point of A-B and 1-2 section were taken into account for convergence check (see Fig. 1). The FE mesh was verified in numerically due to load variant 1. In the presented calculations 1% difference in bending moments is taken as a limit for the subdivision of FE mesh (due to a regular FE mesh and the simplified live load assessment in variant 1).

Table 1. Analysis of finite element mesh convergence.

No.	Mesh density (size of FE)	M_{xx} [kNm/m]	Difference [%]
1.	80 cm × 80 cm	33.30	-
2.	40 cm × 40 cm	34.02	2.16
3.	20 cm × 20 cm	34.66	1.88
4.	10 cm × 10 cm	34.86	0.58

Table 2. Computation time and output file size under convergence of the FE mesh.

No.	Mesh density (size of FE)	Time	Output file size
1.	80 cm × 80 cm	5 s	4.2 MB
2.	40 cm × 40 cm	20 s	7.5 MB
3.	20 cm × 20 cm	2 min	24.7 MB
4.	10 cm × 10 cm	25 min	92.8 MB

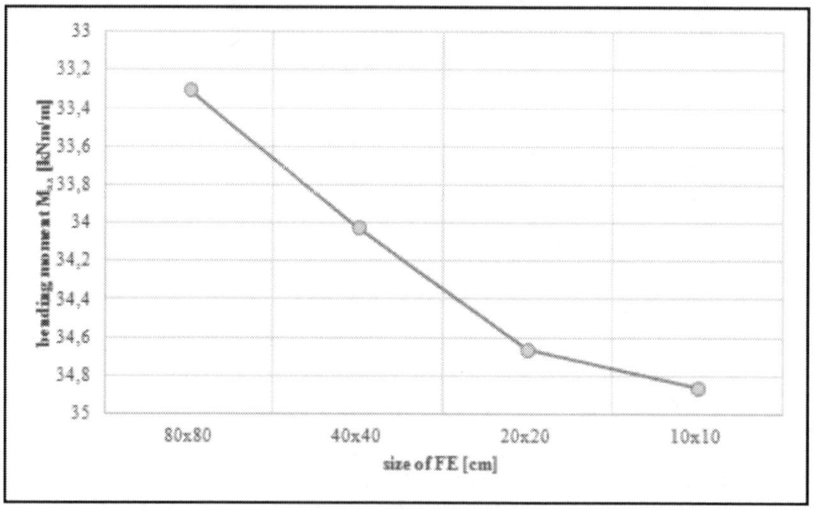

Fig. 2. Finite Elements mesh convergence.

The results in Tables 1 and 2 point out the FE size of 20 cm × 20 cm to give satisfactory internal forces results, this mesh is accepted for final calculations. Additionally, this mesh density provides an acceptable calculation time equal 2 minutes and a reasonable output file size with calculation results (about 25 MB).

5 Analysis of bending moments

While the numerical calculations are performed a detailed comparison of bending moments for a concrete slab is done. Figs. 3-4 present a bending moment M_{xx} map [kNm/m] for the ULS.

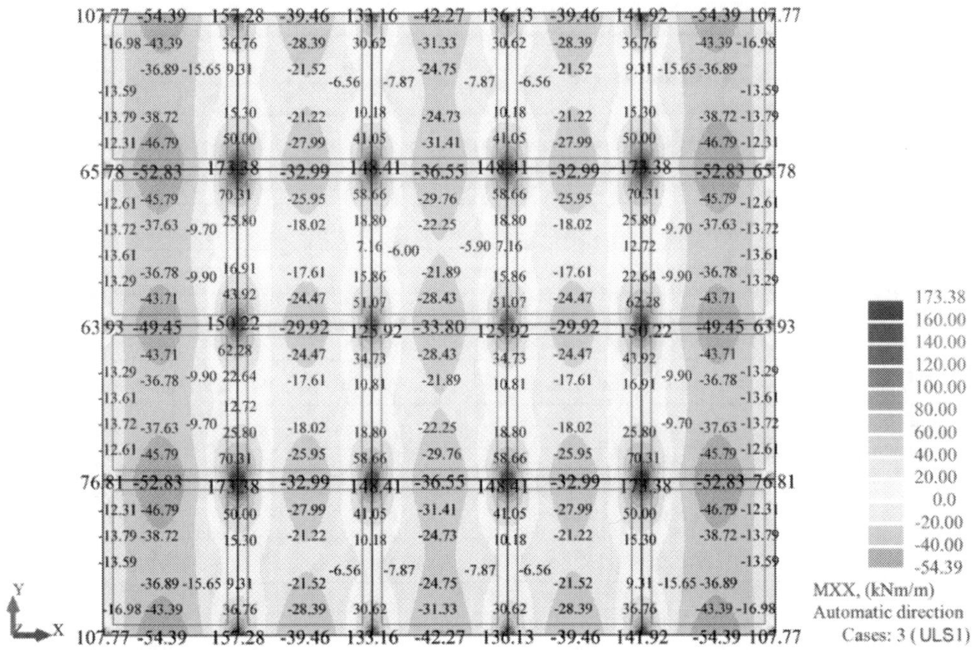

Fig. 3. Map of bending moment M_{xx} – variant 1 for ULS1.

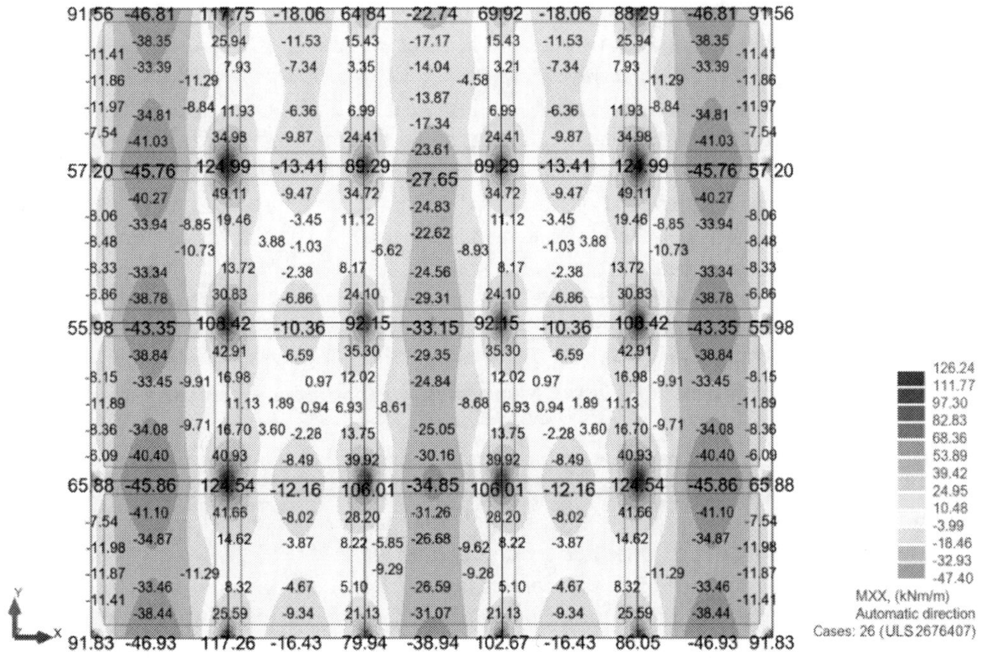

Fig. 4. Map of bending moment M_{xx} – variant 4 for ULS 2676507/4194304.

Table 3. Results of bending moments - differences in results compared to the accurate solution (variant 4).

No.	Variant of calculations	M_{xx} [kNm/m]	Difference M_{xx} [%]	M_{yy} [kNm/m]	Difference M_{yy} [%]
1.	variant 1	34.66	7.87	40.62	6.78
2.	variant 2	32.34	0.65	37.98	-1.43
3.	variant 3	35.36	10.05	37.38	-3.30
4.	variant 4	32.13	-	38.44	-

Table 4. The calculation time and the size of output files.

No.	Variant of calculations	Time	File size
1.	variant 1	2 min	24.7 MB
2.	variant 2	2 min 30 s	55.0 MB
3.	variant 3	2 min 30 s	55.3 MB
4.	variant 4	5 min 50 s	268.7 MB

6 Discussion

The calculation results in variant 1 are safe, overestimating the bending moments M_{xx} and M_{yy} by about 7%. It can therefore be concluded that the adopted global coefficient $\gamma_Q = 1.9$ has been slightly scaled. The imposed live load applied in the chessboard pattern gives similar results for M_{xx} (0.65%), it underestimates the bending moment M_{yy} (by 1.43%) as well as the load applied by bands (3.30%) in the variant 3.

The calculation time due to different loading variants is also compared. The calculation time for variants: 1, 2 and 3 is almost constant, equal to approx. 2 minutes. The calculations time for the variant 4 takes it higher - about 6 minutes. Additionally, the variant 4 results is a more time-consuming presentation in RSAP program, compared to others variants. The least size of output file is detected for the variant 1, equal to 10% of variant 4 capacity.

The calculations for variant 1 may be acceptable for the ULS, however this variant underestimates the deflection values due to SLS. In the present calculations the problem of scratched concrete is neglected. Nevertheless, in real structural calculations the decrease of stiffness of the reinforced concrete structure after scratching must be taken into account.

The load application and CPU time of calculations to asses a given variant of numerical calculations of a structural model. An efficient loading pattern is a constant uniformly distributed load on a whole surface of a concrete slab. Applying a load on the slab structure with bands (variant 3) or on a chessboard (variant 2) pattern are slightly more time consuming but may be acceptable. Imposing a load to each slab panel separately (variant 4) is hard and time-consuming, thus it is disqualified from practice.

From the viewpoint of numerical correctness the choice of a finite element mesh size (convergence analysis) is more considerable than an accurate definition of imposed live loads.

An important factor to act upon the solution effectiveness is a direct interpretation of numerical results. Live load applied on an entire slab surface makes a single load combination thus no mistake occurs here. The extreme internal forces due to chessboard and band load patterns appropriate number of combinations of actions exists. Extreme bending moments in support areas can be assessed from the combination with the loads added from both chessboards or bands patterns, whereas in the span area for separate chessboards or bands. It is quite complicated in this respect to interpret the results for the load applied separately in each slab panels. In this case tedious and hard investigation occurs for the values and location of maximum internal forces distribution, because the combinations in this variant are generated automatically by the RSAP program.

Due to the simplified method of deflection calculation (e.g. used the formulas given in paragraph 16.5.1 in [9]) the problem arises to determine the value of the α_M coefficient depending on the shape of the bending moment diagram. For the most commonly considered cases of continuously loaded components with bending moments at the ends of the span (according to Fig. 16.4 d) in [9]) it is necessary to determine the values of bending moments at the edges of the element from the ultimate combination of actions for the maximum moment in the span. Thus it can be problematic to consider the results from the variant 4 with a load applied in each panel separately.

Determination of internal forces in a structural model is necessary for reinforcement design. In reinforced concrete structures the amount of required reinforcement is specified under ULS and SLS combinations. The designer of reinforcement deals with the economics of the solution (approximation of the adopted reinforcement to the necessary due to structural calculations) and by facilitating the execution of reinforcement skeleton on the construction site. For this purpose, the designer should select the diameters, spacing and dimensions of reinforcement properly. The calculated reinforcement is almost greater than (or rarely equal to) the reinforcement adopted on construction drawings. In a majority of cases there is an additional load capacity reserve of reinforced concrete structures. Thus it can be concluded that slight differences in the results of internal forces obtained from the variants (precisely and approximate) of imposed loads can be negligible from a practical point of view.

In addition, it is noticeable the challenging architectural ideas corresponding to novel, slender structures trigger the Serviceability Limit State (deflection or scratch) as decisive. In order to meet these requirements, the designer must apply additional reinforcement to the structure raising the load bearing capacity again.

Finally, in the design process of flat slab floor it is a key issue to assure the punching resistance. Punching strength is a critical point in flat slabs design due to the lack of its theoretical description, empirical code-based formulations are still the most widespread to deal with the slab-column connections. Recommendation in point 5.6.1 of the standard [10] says: *The degree of longitudinal reinforcement of elements checked for puncture (...) cannot be less than 0.5% in both directions*. According to this requirement, the designer of flat slab floor has an additional reserve of load bearing capacity, thus possible differences may be omitted between the exact and simplified determination of internal forces out of analysed loading variants.

7 Conclusions

Nowadays the building industry in Poland tends to decrease time and costs allocated to structural design. The economic account becomes dominant in the buildings engineering process. The analysis of entire 3D structures becomes a widespread design practice, instead of a selected research upon e.g. a selected storey only.

A three-dimensional model was applied in the stadium structure in Gdansk for the European Football Championship in 2012 [11] - according to the designer, this 3D solution reduced the design time of the facility by as much as 85%. In accordance with the European Parliament and Council Directives adopted in 2014: classic [12] and sectoral [13] - Building Information Modelling (BIM) technology should be implemented in the public construction process by all member states of the European Union. Poland introduced the directives in the act [14]. Consequently, all public investments must implemented the BIM technology, allowing for design based on 3D models.

The investigation performed in the paper can be perspective regarding the application of loads to the flat slab floor structural model. The proposed simplification is intendent to reduce design errors, save the CPU time and make a more comprehensive result analysis.

It can be concluded that in the case of Ultimate Limit State calculations the most effective way of applying loads to flat slab floor structural model is variant 1 (live load located on an entire surface of the slab with variable action partial factor $\gamma_Q = 1.9$). It allows to save the computational time, leads to a small size of output file and provides a straightforward and direct interpretation of numerical results.

References

1. P. Kossakowski, Inżynierski problem komputerowego modelowania pracy żelbetowej płyty dwuprzęsłowej z uwzględnieniem sprężystej podatności belki, Przegląd Budowlany, **10**, 19-24 (2012)
2. P. Kossakowski, Uwzględnienie wpływu sprężystej podatności belek w numerycznym modelowaniu stropów żelbetowych, Przegląd Budowlany, **11**, 24-31 (2014)
3. W. Starosolski, *Komputerowe modelowanie betonowych ustrojów inżynierskich wybrane zagadnienia*, edition V, volumes 1 and 2, Gliwice: Wydawnictwo Politechniki Śląskiej (2013)
4. PN-EN 1990:2004, Eurocode – Basis of structural design
5. PN-EN 1992-1-1:2008, Eurocode 2: Design of concrete structures. Part 1-1: General rules and rules for buildings
6. PN-EN 1991-1-1:2004, Eurocode 1: Actions on structures. Part 1-1: General actions – Densities, self-weight, imposed loads for buildings
7. A. Ambroziak, P. Kłosowski, *Autodesk Robot Strutural Analysis Podstawy Obliczeń*, Gdańsk: Wydawnictwo Politechniki Gdańskiej (2013)
8. A. Ambroziak, P. Kłosowski, *Autodesk Robot Structural Analysis Wymiarowanie konstrukcji stalowych i żelbetowych*, Gdańsk: Wydawnictwo Politechniki Gdańskiej (2016)
9. M. Knauff, *Obliczanie konstrukcji żelbetowych według Eurokodu 2*, Warszawa: Wydawnictwo Naukowe PWN (2012)
10. PN-B 03264:2002, Konstrukcje betonowe, żelbetowe i sprężone. Obliczenia statyczne i wymiarowanie
11. R. Wojdak, *Nietypowe autorskie konstrukcje budowlane. Stadion piłkarski na EURO 2012 w Gdańsku i inne obiekty*, Gdańsk: Wydawnictwo Politechniki Gdańskiej (2010)
12. Directive of the European Parliament and of the Council 2014/24/EU of 26 February 2014 on public procurement and repealing Directive 2004/18/EC
13. Directive 2014/25/EU of the European Parliament and of the Council of 26 February 2014 on procurement by entities operating in the water, energy, transport and postal services sectors and repealing Directive 2004/17/EC
14. Ustawa z dnia 22 czerwca 2016 r. o zmianie ustawy – Prawo zamówień publicznych oraz niektórych innych ustaw, Dz.U. 2016 poz. 1020

Acoustic non-destructive testing of high temperature degraded concrete with comparison of acoustic impedance

Richard Dvořák[1,*], *Zdeněk*Chobola[1], and *Ivo*Kusák[1]

[1]Brno University of Technology, Faculty of Civil Engineering, Veveří 331/95, Brno, Czech Republic

Abstract. The paper is focused on non-destructive measurement of high temperature degraded concrete test specimens of three mixtures different by the use of coarse aggregate. Testing is done by ultrasonicpulse velocitymethod and Impact-Echo method. Non-destructive results are compared with destructive tests. Ultrasonic pulse velocity, dominant resonance frequency and acoustic impedance are discussed and compared with changes in density, cubic compressive strength, and tensile strength of concrete. The paper suggests possible assessment of degraded concrete by the change in acoustic impedance dependent on residual tensile strength.

1 Introduction

Acoustic non-destructive methods like Impact-Echo or ultrasound pulse velocity method can be used to determine the location and extent of flaws such as cracks delamination, voids, honeycombing and debonding in plain, reinforced and post tensioned concrete structures. It can be also used to determine the condition of structure, which has been degraded by different factors[1].

During the measurement by Impact-Echo method, a frequency spectrum is evaluated. In terms of condition assessment of test specimens, a resonance frequency of longitudinal, transverse, and torsional wave propagation is determined. The change in frequency spectrum makes it possible to distinguish a change in physical-mechanical properties.

Ultrasonic velocity pulse method is based on measuring the travel time of ultrasonic pulse through the test specimen and acquiring the shortest travel time for chosen test position of exciter and receiver. In terms of degraded plain concrete, where cracks and microcracks occur, and the material begin to be more fragile, the ultrasonic pulse travels longer path, as it is avoiding the air voids, cracks, and inhomogeneity. This leads to longer travel time of ultrasonic pulse.

The principles behind these methods lay in the theory of stress wave propagation in heterogenous solid materials.

1.1 Stress wave propagation

The impact-echo method belongs to stress wave propagation non-destructive methods based on the use of transient stress waves. A stress wave pulse is excited by mechanical impact introduced into a structure of test specimen. This pulse travels into the structure as dilatational (P) and distortional (S) waves and along the surface as a Rayleigh (R) wave as is shown at Fig.2. The P- and S-waves propagate into the structure along spherical waveforms and are reflected by internal cracks, voids, interface between different mediums and by external boundaries of the structure. A P-wave is associated with normal stress, while an S-wave is associated with shear stiffness. A Rayleigh wave, also called a surface wave or R-wave, propagates along the surface of a solid, and particle motion is retrograde elliptical[2].

Fig 1shows a situation when a P- or S-wave front is incident on a boundary between dissimilar media, "specular" reflection occurs. At a boundary between two different media only a portion of a stress wave is reflected. The remainder penetrates the underlying medium (wave refraction). The angle of refraction, β, is a function of the angle of incidence, θ, and the ratio of wave speed, C_2/C_1, in the different media, and is given by Snell's Law. Depending on the angle of incidence, a P-wave can be partially reflected as both P- and S-waves and can be refracted as both P- and S- waves[3].

$$R_n = \frac{Z_2 - Z_1}{Z_2 + Z_1} \tag{1}$$

The portion of an incident ray of a P-wave that is reflected at an interface between two media depends on the specific acoustic impedances of each medium (equation 1).

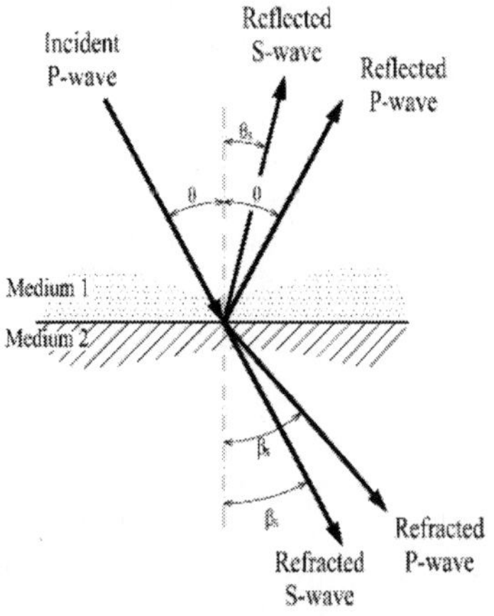

Fig 1. Reflection and refraction of P-wave [3].

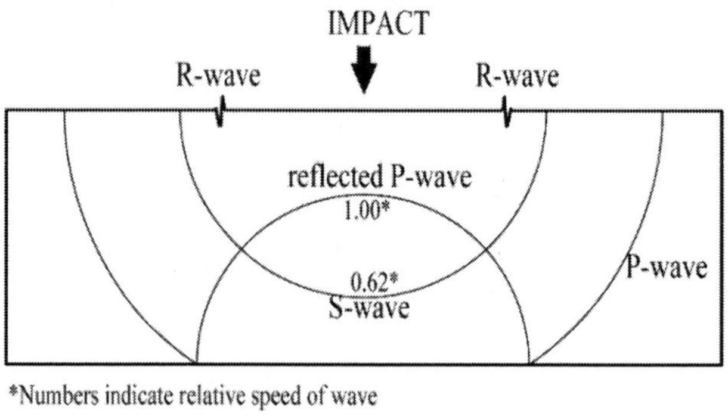

Fig.2. Different types of waves [3].

If mechanical wave propagating through concrete (concrete have acoustic impedance 9-10·10^6kg·m^{-2}·s) hit interface with air void (air have acoustic impedance 0.413 kg·m^{-2}·s), mechanical wave is reflected and serve as new source of longitudinal, transverse andRayleigh waves.

$$Z = D \cdot v_L \tag{2}$$

The acoustic impedance of a material *Z*is density *D*multiplied by acoustic velocity $v_L$in material. This parameter therefore combines the change in two material attributes, which are genuinely affected by high temperature.

1.2 Degradation of concrete

Concrete is a heterogenous material, which consists of fine and coarse aggregate, cement adhesive, additives, and mix water. Each compound reacts to elevated temperature differently, therefore the changes in physical-mechanical and physical-chemical properties cause a synergistic effect.

Consequently,we can divide the process of concrete degradation by high temperature by reaction of aggregate, cement matrix and free and bounded water content.

1.2.1 Aggregate

Aggregates with a high content of silica suffers to phase change of quartz at 574°C when β-quartz transform to α-quartz. Approximately at 300 – 400°C silica aggregate begins to form microcracks at interface with cement matrix, which leads to increased porosity and absorption[4].

Limestone aggregate is slightly less resistant to high temperatures, but the decrease in strength for silica and limestone aggregate is more alike up to 700°C. From this temperature a limestone aggregates have shown better performance, thanks to decomposition of calcite $CaCO_3 \rightarrow CaO + CO_2\uparrow$. Exceeding a threshold of 900°C both limestone and silica aggregate perform similar in term of residual strength[4].

1.2.2 Cement

The reaction of hydraulic cement binder to elevated temperature was thoroughly researched in past decades. The degradation process of cement matrix consists of two main compounds, CSH gels and portlandite $Ca(OH)_2$. Dehydration of CSH gels take place in temperature range of 105 to 1000°C and consists of multi-step reaction. Portlandite dehydrate in temperature range from 400 to 550°C[5]. Due to volume change during dehydration, the cement tends to shrink and form microcracks.

1.3 Applied materials

For the sake of an experiment a set of 3 mixtures A, B and C was designed. A number of 63 test specimens of diameter 100 × 100 × 400 mm were manufactured from each mixture. All teste specimens were stored in water containment for 28 days. Test specimens were afterwards preheated in laboratory oven at 110°C for 78 hours. By this, we remove residual free water content to prevent concrete from explosive spalling[6].

The mixture was prepared with use of fine aggregate of fraction 0/4 mm from local quarry Žabčice was and coarse aggregate of fraction 8/16 mm and 11/22 mm from local quarry Olbramovice. The applied fine aggregate mainly consist of quartz and coarse aggregate mainly consist of granodiorite, where the main compounds are feldspar, quartz, and biotite.Portland cement used in experiment was CEM I 42.5 R from local cement works Mokrá. Specific amounts of each compounds are shown inTable 1.

Table 1. Mixture designs.

Compounds	**Amount of each compound for 1 m^3 [kg]**		
	Mixture A	**Mixture B**	**Mixture C**
Cement CEM I 42.5 R	345	345	345
Fine aggregate Žabčice 0/4 mm	848	896	813
Coarse aggregate Olbramovice 4/8 mm	-	-	1010
Coarse aggregate Olbramovice 8/16 mm	980	521	-
Coarse aggregate Olbramovice 11/22 mm	-	391	-
SuperplastizerSicaViscocrete 2030	2.8	2.5	3.1
Mix water	160	173	176

Test specimens were then fired in ceramics furnace Rhode KE 130B (shown at Fig.3)with temperature gain of 5°C/min. The aimed temperatures were maintained for 1 hour, the test specimens were spontaneously cooled to room temperature inside the furnace. Each mixture was divided to 1 reference set and 6 temperature fired sets. The reference set was kept at laboratory conditions at relative humidity of 50-60% and air temperature 19-21°C. After firing in the furnace, specimens were moved to laboratory conditions resembling the ones for a reference set.

Fig.3. Ceramic furnance Rhode KE 130 B.

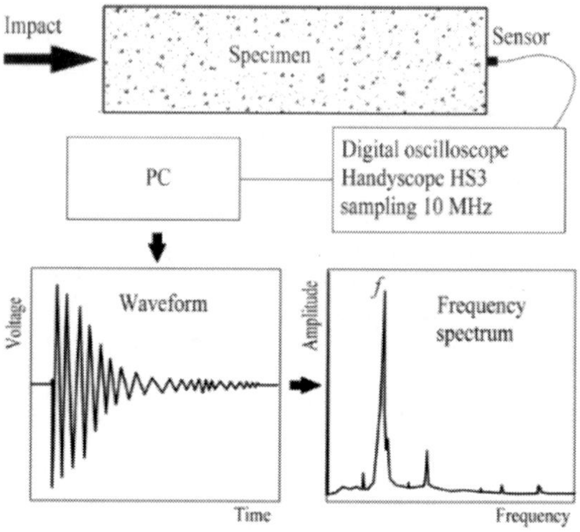

Fig.4. Impact-Echo test setup.

1.4 Used methods

1.4.1 Ultrasound pulse velocity method

The ultrasonic pulse analyser 58-E4900 was applied for the measurement. Measurement by ultrasonic pulse method was conducted in longitudinal way of test specimens laying on rubber pads. In order to determine the shortest time needed for ultrasonic pulse to travel the length of test specimen each test consists of three test positions of exciter and receiver. The first position is at centre of upper surface, second at the middle and third at centre of the specimen base. Three different travel speeds were acquired and the shortest was picked asv_L[7].

1.4.2 Impact-Echo

Measurement by Impact echo was conducted on test specimens placed on rubber pads. As exciter of mechanical stress waves a spherical hammer of a total mass 70.5 g was used. In order to perform vibration a piezoelectric sensor MIDI 446s12 was used, signal processing was performed by the use of digital oscilloscope Handyscope HS3 with sampling speed set to0.2MHz.In order to record the longitudinal waves, the receiver was placed in the front of test beam and exciter hit the test specimen at the opposite side [8].The measured signal was transformed to frequency spectrom by Fourier fast transformation [9].The measurement procedure is shown on Fig.4. Each recorded signal lasted 335 ms and was made up by 65.5 thousands records. Due toRAM memory instalation in the oscyloscope, each signal also contained first 3 ms of signal before the impact of mechanical impuls.

1.4.3 Destructive tests

In order to compare the non-destructive results obtained by ultrasound pulse velocity method and Impact-Echo method destructive tests were conducted. The test specimens were tested by four-point bending test on flexural tensile strength[10] on each end of broken specimens a cubic compressive strength test was made[11].

2 Results of measurement

2.1 Density and absorbance

Test specimens have shown a slight loss of mass for thermally degraded test specimens as is shown atFig.5. The average density of non-degraded concrete specimens was for mixture A 2360 kg·m^{-3}, for mixture B 2380 kg·m^{-3} and for mixture C 2210 kg·m^{-3}. Density of mixture A lost 7.6% for set degraded by 1000°C. This decrease in density was for mixture A lowest from all sets, mixture B lost 8.8% and mixture C nearly 9%. Highest decrease of mass appeared between 20 and 200°C for mixture B and C. This leap is due to pre-drying in laboratory oven of mixture B and C. Temperature sets of 200 and 400°C of Mixture A has not been pre-dried in the beginning of the experiment.

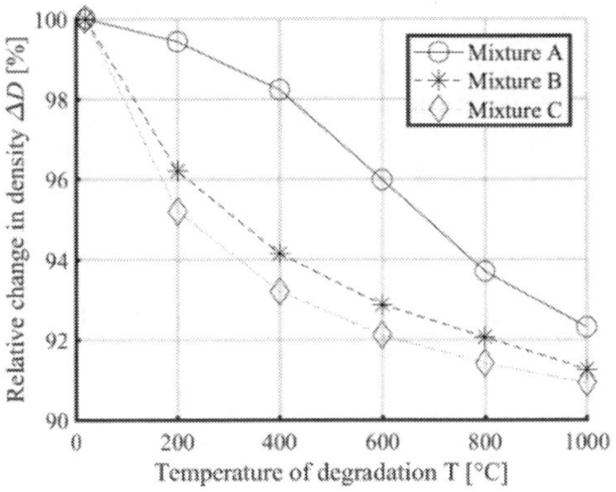

Fig.5. Relative change in density.

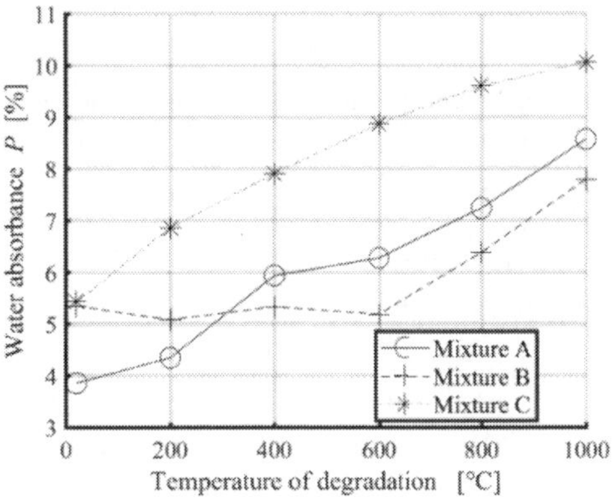

Fig.6. Change of water absorbance.

As mentioned in chapter 1.2, thermal degradation of concrete leads to the increase in porosity.Fig.6shows the results of water absorbance based on Czech norm for measurement of water absorbance and leakage of concrete[12]. Water absorbance of non-degraded specimens was equal 3.8% for mixture A, 5.2% for mixture B and 5.3 for mixture C. The highest water absorbance 10% was detected for mixture C at a 1000°C set.

2.2 Impact echo and ultrasonic pulse velocity measurement

The measurement by Impact-Echo method has shown a change in dominant frequency spectrum (Fig.7.). The reference set showed frequency 5.22 kHz for mixture A, 5.41 kHz for mixture B and 4.92 kHz for mixture C. The first decrease in dominant frequency has appeared at temperature 400°C for mixture A. This change is due to the pre-drying of test specimens of mixture B and C. The highest decrease in longitudinal frequency for all mixtures appeared between temperature 400 and 600°C. I this temperature range also occurs phase change of β-quartz to α-quartz and decomposition of portlandite (1.2.2). Above the temperature of 600°C the rate of decrease in resonance frequency falls. The lowest measured frequencies occurred for 1000°C temperature sets. Resonance frequency for mixture A was 0.67 kHz, for mixture B 0.63 kHz and for mixture C 0.78 kHz.

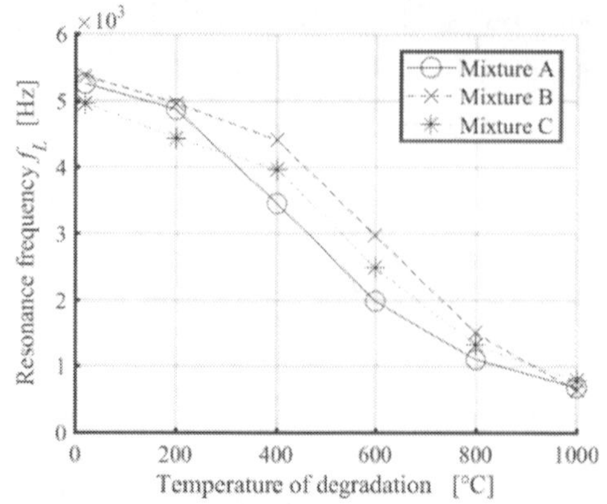

Fig.7. Measurement of longitudinal resonance frequency by IE method.

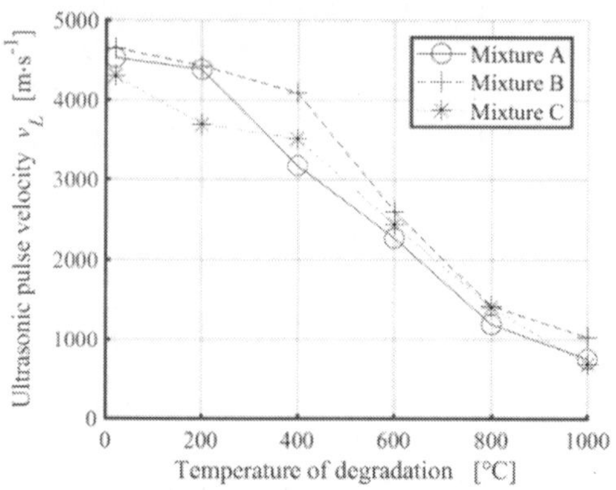

Fig.8. Ultrasounic pulse velocity measurement.

The results of pltrasonic pulse velocity measurement have shown a similar trend in fall of pulse velocity for each temperature sets as is shown on Fig.8. The non degraded mixtures reached velocity from 4350 $m \cdot s^{-1}$to 4620 $m \cdot s^{-1}$. A major decrease occured between temperature 400 and 600°C, which was most significant for mixture B, where pulse velocity change from 4090 $m \cdot s^{-1}$ to 2600 $m \cdot s^{-1}$. This mixture have shown highest resistance to degradation in term of ultrasonic pulse velocity mesurement. At 1000°C, the measured value value for mixture B is 1026 $m \cdot s^{-1}$, the higher than velocity of mixture A 747 $m \cdot s^{-1}$ and mixture C 678 $m \cdot s^{-1}$. This is quite an oposite result than in the Impact-Echo measurement.

2.3 Destructive tests

The results of destructive tests are shown at Fig.9 and Fig.10. For each temperature and reference set a flexural tensile strengthtest and cubic compressive strength test have been conducted. Compressive strength for reference set for mixture A was 66 MPa, for mixture B 74 MPa and for mixture C 50 MPa.

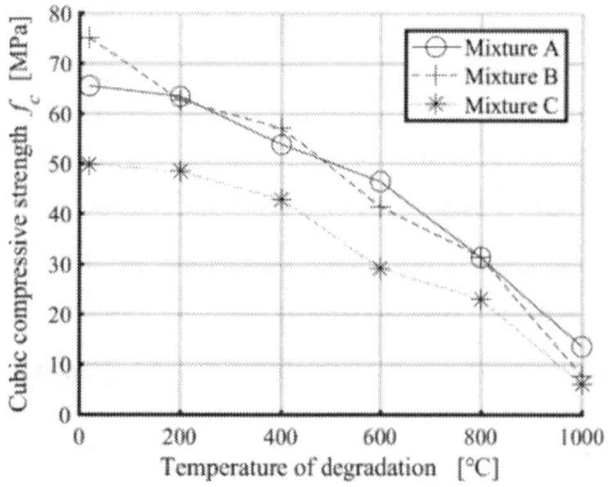

Fig.9. Cubic compressive strength test.

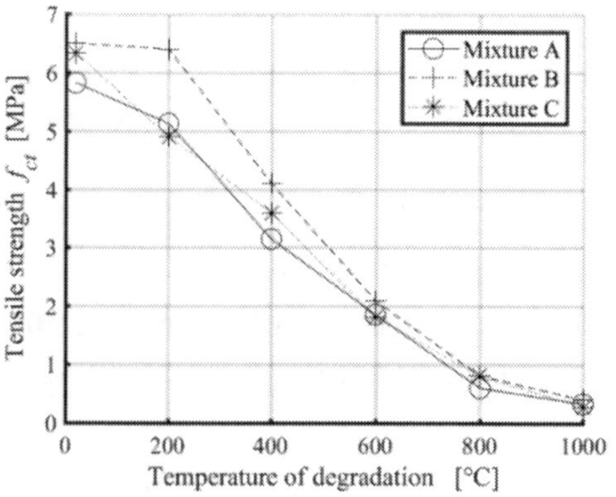

Fig.10. Flexural tensile strength test.

The first decrease in strength appeared at temperature 200°C, which was most significant for mixture B, although in Impact-Echo and Ultrasonic pulse velocity measurement this mixture seemed most resistant to temperature degradation.

2.4 Change in acoustic impedance

The measured values of density and acoustic velocity in specimens, make it possible to compute acoustic impedance for each temperature set. Fig.11 presentsrelation of acoustic impedance with the change of flexural tensile strength. We can easily recognize, although reference sets show different flexural tensile strength and acoustic impedance, after 40°C the measured values match and dispersion of acoustic resistance for 600, 800 and 1000°C corelate with polynomial trendline of order 2. The trendline reliability value R is 0.977.

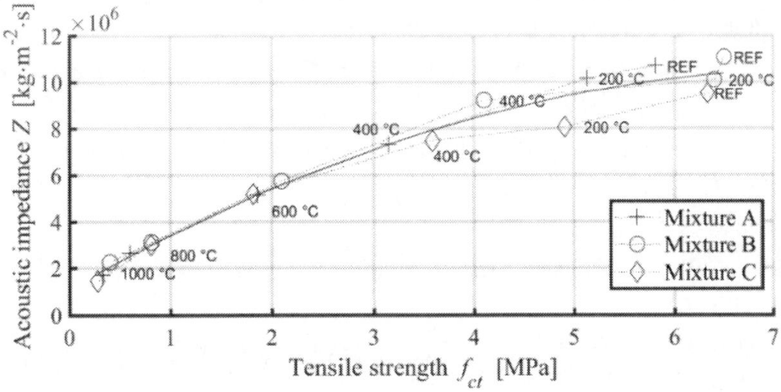

Fig.11. Change of acoustic impedance Z to flexural tensile strength.

It seems that, at high temperatures, above 400°C the rate of acoustic impedance on tensile strength properties of test specimens normalize to narrow values.

3 Conclusions

The presented paper aims at non-destructive testing of high temperature degraded concrete test specimens of three mixtures, different by used coarse aggregate. It compares the measured results of 6 temperature sets and 1 reference set for each mixture. A change in measured resonance frequency by impact-echo testing proves a correlation between change of resonance frequency and physical-mechanical state which is represented by destructive tests. Change of acoustic impedance on change of flexural tensile strength is observed. Non-degraded reference sets of all three mixtures have shown different initial values of measured acoustic impedance and tensile strength. At the temperature degraded sets above the 400°C the measured values are more intersected by the trend line. From 600 to 1000°C reliability R of a polynomial trendline of order 2 is 0.977. It seems that the measurement of acoustic impedance for tested degraded specimens may be used as a reliable tool to assessits residual mechanical properties. In the past studies of vibroacoustic methods used on high-temperature degraded concrete has proposed [13]combined application with destructive tests [14]. The approach of the presented paper suggests a process to accurate the condition of thermally degraded concrete via its acoustic impedance obtained from ultrasonic pulse velocity method, or resonance method and the density of concrete. This would be in practice conducted on core bore specimens or insitu. The temperatures during fire accidents in average reach from 400 to 650°C, these temperatures and the highest ones show a strong dependence between acoustic impedance and tensile strength. This means that most of concrete structures severely damaged by fire, could be assessed by this approach. This measurement can be also understood as a possible initial classification of thermally degraded concrete.

This publication was developed under the project No. LO1408 "AdMaS Up – Advanced Materials, Structures and Technologies" supported by Ministry of Education, Youth and Sports under the „National Sustainability Programme I" and under the project GAČR No.16-02261S supported by Czech Science Foundation and under the project FAST-J-18-5275 supported by Faculty of Civil Engineering of VUT.

References

[1] G.Epasto, E. Proverbio, V. Venturi,Materials and Structures**43** (2010)

[2] N.Carino, Proceedings of the 2001 Structures Congress & Exposition**15** (2001)

[3] V.Malhotra, N. Carino, CRC Press **2** (2004)

[4] A.Savva, , P. Manita, K.K. Sideris,Cement and Concrete Composites**27** (2005)

[5] Q. Zhang, Journal of Thermal Analysis and Calorimetry**6** (2012)

[6] J. Zhao, J. Zheng, G. Peng, K. V. Breguel,Cement and Concrete Research**65** (2014)

[7] ČSN 73 1371 (1983)

[8] ASTM C1383-15 (2015)

[9] R. Bracewell,*The Fourier Transform & Its Applications***3** (USA: McGraw-Hill Science/Engineering/Math, 1999)

[10] ČSN EN 12 390-5. (2009)

[11] ČSN EN 12 390-3 (2009)

[12] ČSN 73 1316 (1990)

[13] A. Ghorbanpoor, A. Benish, N. Benish,*Non-destructive testing of wisconsin highway bridge: Wisconsin highway research program—final report*(Wisconsin Department of Transportation, 2003)

[14] K. Krzemien, I. Hager,Construction and Building Materials**96** (2015)

MATEC Web of Conferences **219**, 03004 (2018) https://doi.org/10.1051/matecconf/201821903004
BalCon 2018

The effect of macro polymer fibres length and content on the fibre reinforced concrete

Aleksandra Mariak[1,*], and *Marzena* Kurpińska[1]

[1] Gdańsk University of Technology, Faculty of Civil and Environmental Engineering, Narutowicza 11/12, 80-233 Gdańsk, Poland

Abstract. The paper presents studies of a ready-mix concrete containing polymer fibres of three different lengths: 24, 38 and 54 mm. The performed tests allowed to determine the effect of fibre volume fraction and length on the concrete strength. The basic parameters of concrete mixture (consistency, air content and bulk density) were identified. Fibre reinforced concrete belongs to a group of composite materials. The polymer fibres are applied in the concrete in structures where the reduction of shrinkage cracking as well as corrosion resistance and fire temperatures are required. It is widely known, that the cracking behaviour of concrete structures depends on flexural tensile strength of concrete. The addition of fibres significantly improves the tensile strength. The experimental study, including axial compressive strength and center-point loading flexural tensile strength, was carried out. The scope of the research was also expanded by the usage of a scanning microscope. The test results showed the effect of fibre length and fibre combinations on mechanical properties of concrete. The effect of the research is to formulate guidelines due to the quantity of macro polymer fibres. In general, appropriate fibre content brings a beneficial effect e.g. improves better workability of a concrete mixture.

1 Introduction

Fibre reinforced concrete is composite material with cement matrix and short-cut, dispersed steel, glass, carbon or plastic fibres [1]. It differs from ordinary concrete with a change of plastic behaviour, thus it is a very interesting material of multiple applications. It is commonly used for industrial floors. However, for the production of typical concrete elements steel reinforcement bars are more and more often being replaced by dispersed reinforcement. Additional protection of the reinforced concrete surface by impregnation can significantly reduce the process of gas and liquid permeability. Modification of physical and mechanical properties of concrete makes the structure more resistant to abrasion, erosion and corrosion [2-4] and shows an economical reason too [5-6].

* Corresponding author: aleksandra.mariak@pg.edu.pl

2 Fibres characteristics

The matrix of fibre reinforced concrete is a brittle material, but limit deformability of dispersed fibres is significantly higher than the deformability of the matrix. As compared to high-strength composites, the percentage of dispersed fibre reinforcement is approximately 1 - 2% by volume. Even just a small portion of fibres brings a significant modification of concrete properties. Bulk density of cement grout ranges between 2,0 - 2,25 g/cm^3, Young's modulus is 10 - 30 GPa and tensile strength 3 - 6 MPa. The addition of fibres significantly changes its performance according to the characteristics of the fibre type used (Table 1). Fibres can be dosed in different configurations and intermediate solutions can be created in the form of MFRP, GFRP, GFRC, MFRC fibres (Fig. 1).

Table 1. Properties of fibres [7].

Fibre type	ϕ [μm]	ρ [g / cm^3]	E [GPa]	f_{ctm} [MPa]	V [%]
steel	5-500	7,84	200	500-2000	0,5-3,5
glass	9-15	2,60	70-80	2000-4000	2-3,5
asbestos	0,02-04	3,00	180	3300	2-3
polypropylene	20-200	0,90	5-7	500-750	8
nylon	-	1,10	4	900	13-15
polyethylene	-	0,95	0,30	0,7	10
carbon	9	1,90	230	260	1
kevlar	10	1,45	65-133	360	2-4
acrylic	18	1,18	14-19,5	400 1000	3

The question of the choice of the appropriate type and amount of fibres and their influence on mechanical properties is the subject of a multiple scientific research [8-14]. Many research results have been published on fibre reinforced concrete, but a large number of issues have not been investigated yet. Much more research work has been published on steel fibre concrete than on polymer fibre concrete. It is well known that the introduction of high strength thin polymer fibres into the fragile matrix results in a new building material, not comparable to concrete or steel reinforced concrete. The presence of fibres modifies the properties of the surrounding matrix.

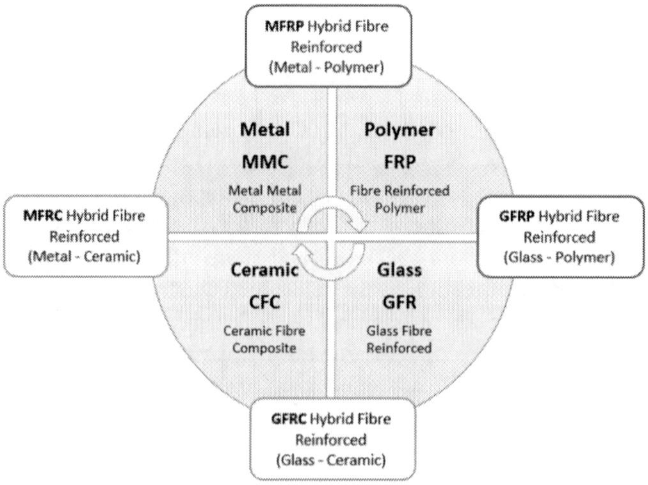

Fig. 1. Types of fibres used in concrete.

3 Macro polymer fibres

In accordance with EN 14889-2, polymer fibres are classified as II class fibres - macro-fibres with a diameter of > 0,30 mm. They are usually used to increase so called residual flexural strength. They have structural applications and influence the load capacity of the elements, and thus their strength. Polymeric fibres inhibit crack propagation [15]. In this respect, observation of the material's behaviour in the post-cracking phase is particularly important. Such tests should only be carried out when the stress condition is as homogeneous as possible. Conventional strength tests known from concrete technology, such as compression or splitting tests, are not suitable because the actual stress distribution is not known. Relatively uniform state of deformation occurs in axial tensile test. One of the innovative methods of testing fibre reinforced concrete is X-Ray tomography [16]. Another way is to monitor concrete elements implemented in many objects by research teams [17-19].

The effectiveness of the added fibres depends e.g. on: fibres content, their shape, dimensions and adhesion to the cement matrix. The adhesion of the polymer fibre to the cement matrix has a significant impact on the mechanical properties of the composite, and in particular on the amount of energy accumulated during the destruction process (Fig. 2). During specimen destruction process, at the beginning of the elastic deformation range, longitudinal crack appears at a certain distance from the fibre surface.

Fig. 2. Adhesion of Astra Polyex Mesh 2000 fibre to cement matrix. Photo SEM Hitachi TM3030.

However, not all phenomena occurring in the contact zone between grout and fibre are noticeable in the diagram of the pulling force as a function of displacement. It may be explained with local extinction of mutual displacements due to friction between fibre and matrix [13]. Improved adhesion enhances fibre resistance to pulling force.

Based on scientific research, many empirical formulae describing the influence of factors on the mechanical properties of fibre reinforced concrete are derived. High flexural tensile strength and high flexural cracking resistance are the basic characteristics of concrete with fibres added. With low fibres content, the flexural tensile strength does not change as rapidly as in case of non-reinforced concrete. After exceeding the limit of proportional deformation, the behaviour of concrete changes suddenly, both before and after exceeding the destructive load. The specimen does not break and exhibits a significant continuous resistance while further deformed. At the same time, the nature of destruction varies depending on the type and amount of fibre used. Cracking resistance refers to the ability of fibre reinforced concrete to absorb loads in a defined deformation range. It is a measure of stress changes in concrete during cracking.

Numerous studies are being carried out in scientific centres around the world to determine the relationship between the adhesion characteristics of the matrix to polymer fibres of different shapes and the mechanical properties of fibre reinforced concrete.

Unfortunately, most of the publications concern fibre reinforced concretes containing steel fibre [9, 20] while much less scientific research is devoted to cement composites modified with polymer, basalt, glass or kevlar fibres.

4 Experimental studies

The main aim of the study was to determine the effect of polymer fibre volume fraction and length on the improvements of mechanical concrete properties. Experimental studies were conducted using copolymer fibres (polypropylene ethylene) Astra Polyex Mesh 2000. The considered fibres are filament fibres twisted into bundles and available in three lengths 24, 38 and 54 mm (Fig. 3). Fibres parameters were given in Table 2. The paper presents the results of preliminary studies aimed at determining the influence of fibre length on the change of concrete properties. The tests were carried out on concrete mixture, modified with fifteen compositions of different fibre lengths. A fixed total fibre mass of 4 kg/m^3 has been adopted. Individual compositions were marked with the following symbols MFLC I - MFLC XV.

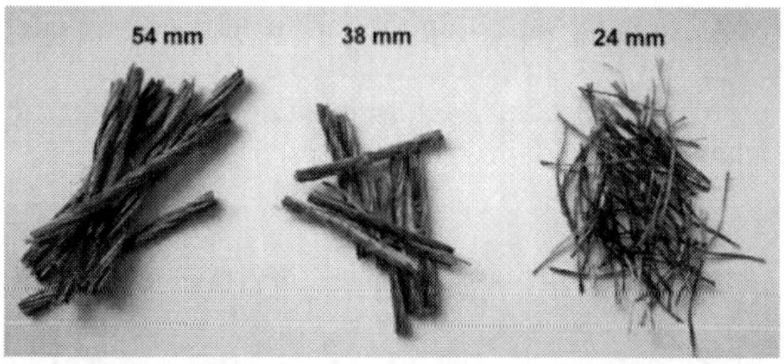

Fig. 3. ASTRA PolyexMesh 2000 fibres in lengths of 54, 38, 24 mm.

In all tests, the same composition of concrete matrix class C 30/37 was used: CEM I 42,5 R - 380 kg, water - 196 kg, sand 0/2 - 760 kg, gravel 2/8 – 1050 kg, PCE super-plasticizer – 0,65% of cement mass, Astra Polyex Mesh 2000 fibres - 4 kg, w/c = 0,52.

Table 2. Parameters of Astra Polyex Mesh 2000 polymer fibres.

length [mm]	54	38	24
fibre diameter [mm]	0,45		
tensile strength [MPa]	650-750		
slenderness l/d [-]	120	84	63
number of fibres per 1 kg [pcs.]	11000	156000	250000
total length of reinforcement [mb/kg]	5940	5928	13500
recommended dose [kg/m^3]	1,5 - 5		
Young modulus E of a fibre [GPa]	5,9		

The choice of the content and type of dosed fibres can be presented graphically using a triangle Fig. 4. The sides of the triangle are the shares of the individual fibre lengths: 24, 38 and 54 mm and their percentage content: 0%, 25%, 50%, 75%, 100%. The composition of fifteen different concrete mixtures is characterised by a varying combination of different length fibre content. The table of quantities of the selected fibre for each tested mixture is presented in Table 3.

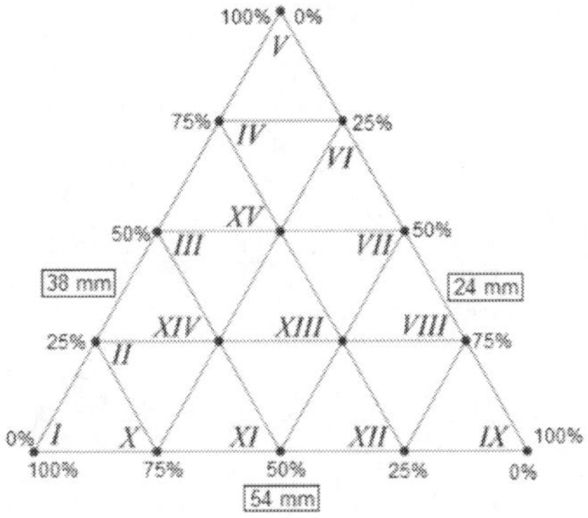

Fig. 4. Triangle of fibre lengths selection.

Table 3. Percentage content of fibres for every mixture.

Mixture no. MFLC	**Percentage content of fibres [%]**		
	54 mm	**38 mm**	**24 mm**
I	100	0	0
II	75	25	0
III	50	50	0
IV	25	75	0
V	0	100	0
VI	0	75	25
VII	0	50	50
VIII	0	25	75
IX	0	0	100
X	75	0	25
XI	50	0	50
XII	25	0	75
XIII	25	25	50
XIV	50	25	25
XV	25	50	25

Prism specimens of the dimensions of 40 x 40 x 160 mm were prepared from these concrete mixtures in the quantity of 9 pcs. for each variant as well as 100 mm cubic samples were prepared from these concrete mixtures in the quantity 12 pcs. for each variant. The specimens were compacted using a vibrating table, with the same vibration times for a specific series of specimens. Before the specimens were produced, the consistency and air content in the concrete mixture were examined.

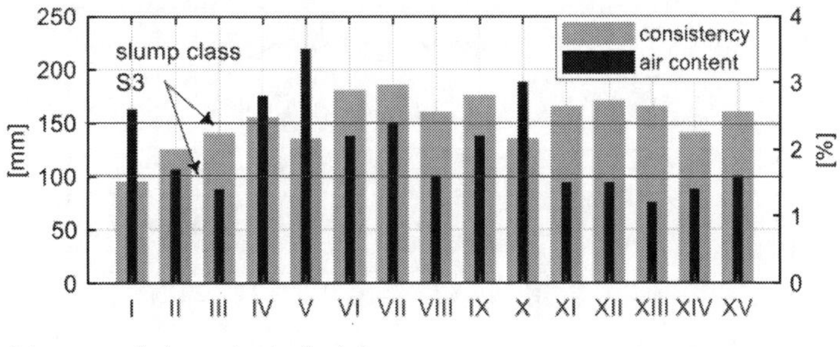

Fig. 5. The consistency and air content of mixtures.

The consistency measurement results were presented in Fig. 5, measured using Abrams cone's slump after 45 minutes from mixing cement with water. Fifteen mixtures with a total volume of 35 dm^3 were made. It was assumed that after 45 min the consistency of concrete mixture should reach class S3 (100 ÷ 150 mm). The first acceptance criterion was a visual evaluation of the concrete mixture. Mixture should not separate and there should be no bleeding. During the study it was observed that concrete mixtures marked with the symbols MFLC II, III, V, X and XIV (consistency 125 - 140 mm) featured a very good workability. Consistency of mixtures of MFLC IV, VI, VII, VIII, IX, XI, XII, XIII and XV where higher than 150 mm. Bleeding symptoms were noticed in mixtures marked with MFLC VI and VII. Air content of the concrete mixtures ranged from 1,2% to 3,5%, with the highest air content in mixtures marked MFLC V, X, IV and I. Bulk density of all mixtures ranged from 2306 to 2368 kg/m^3.

The prepared specimens were cured in water at a temperature of 20 ± 2 °C for 28 days (for beams) and 90 days (for cubes). Then, strength tests were carried out on pre-determined dates. Beams (40 x 40 x 40 x 160 mm) were subjected to the flexural tensile strength test (Fig. 6), and then compressive strength test were carried out on half-beams aged 1, 7 and 28 days. In the case of flexural tensile strength test for all specimens the increase in tensile strength was recorded after 28 days in relation to the 7-day tensile strength (Fig. 7). The largest increase concerned specimens made from mixtures MFLC VII, VIII and IX. These mixtures did not contain 54 mm fibres but mainly had short fibres. The standard deviation for tensile strength for each set of samples in specified term did not exceed the following values: SD1 < 0,33, SD7 < 0,54, SD28 < 0,78 [MPa]. Based on the obtained results it can be concluded that the strength of the cement matrix is very important. Whereas fibre content specified by the manufacturer is between 1,5 kg and 5 kg, tests indicate that the recommended fibre content shall not be greater than 3 kg. The type and size of specimens tested are also of key significance. The aim of the study was to verify the influence of fibre content on the rheological changes of concrete mixtures and on the mechanical properties of concrete. Noticeable changes of strength indicate that it would be recommended to produce specimens in accordance with PN-EN 14889-2 for all variants.

Fig. 6. Flexural tensile strength test.

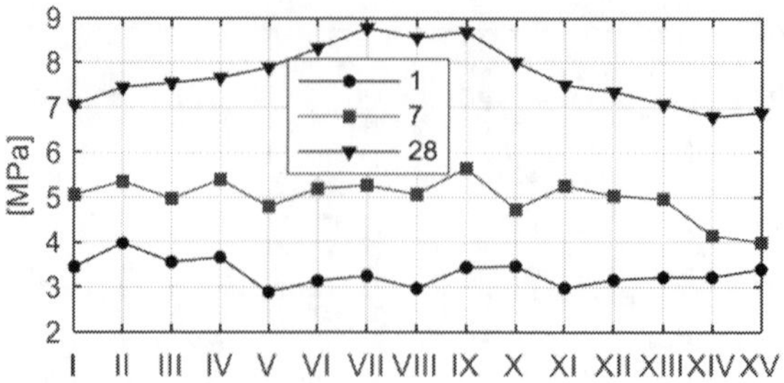

Fig. 7. Prism tensile strength after 1, 7, 28 days.

The results of compression strength tests carried out on half-beams were presented in Fig. 8a. The highest strength after 28 days was noted in case of MFLC VI (57,7 MPa), IV (56,2 MPa), VII (55,7 MPa), XII (54,3 MPa) and X (54,2 MPa). In all cases the compressive strength after 28 days was higher than 50 MPa. After 1 day of hardening, the highest compressive strength was obtained in case of MFLC IX (19,3 MPa) and X (18,8 MPa) compositions, while the lowest strength was obtained in case of MFLC XIV (13,6 MPa) and XV (13,9 MPa) compositions. The standard deviation for prism compressive strength for each set of samples in specified term did not exceed the following values: $SD1 < 1{,}55$, $SD7 < 4{,}20$, $SD28 < 3{,}13$ [MPa].

In order to determine the scale effect, compression strength tests were also carried out on 100 mm cubic specimens. For each grout, 6 specimens aged 28 and 90 days were tested. The results of compressive strength tests for cubic specimens after 28 and 90 days of curing are characterized by low variability (Fig. 8b). Only three mixtures show a decrease in long-term strength (MFLC II, V, VII). After 28 days, the highest compressive strength was obtained for mixtures marked MFLC VII (50,4 MPa), XI (50,0 MPa), VI (49,5 MPa) and XIV (49,4 MPa), while the lowest one for MFLC XIII (41,2 MPa). The standard deviation for prism compressive strength for each set of samples in specified term did not exceed the following values: $SD28 < 2{,}54$, $SD90 < 4{,}63$ [MPa].

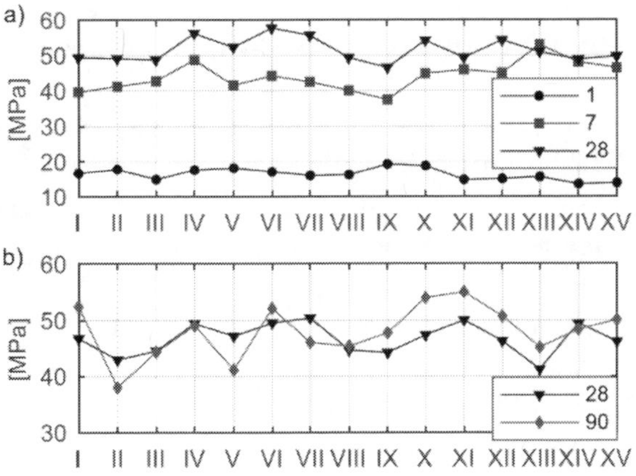

Fig. 8. a) Prism compressive strength, **b)** Compressive strength of 100 mm cube.

5 Conclusions

Based on the conducted experimental studies, the following conclusions were formulated:

- All mixtures featured good workability, with the best workability and rheology was noted in case of fibre reinforced concrete mixtures marked with MFLC VI, VII (mixture of 38 mm and 24 mm fibres) and IV (mix of 54 mm and 38 mm fibres). While using the mixture containing 100 % of 54 mm long fibres, the concrete mixture revealed slight bleeding. The recipe needs correction and application for example some addition of zeolite. This additive improves the workability of the concrete mixture and completely eliminates disadvantageous bleeding. It causes strength parameters increase, as it was confirmed by numerous tests.
- The results of the compression strength test after 28 days determined on specimens 40 x 40 x 160 mm are approximately 15 % - 20 % higher than those obtained on cubic samples with a side of 100 mm.
- In the case of the flexural tensile strength test, tensile strength results obtained after 28 days are approximately 10 % - 30 % higher than results obtained after 7 days.

- The test results are influenced by the specimen size, especially if more than 2 kg of fibres were used and the fibre length exceeds 24 mm.
- Authors noted that the amount of cement grout should be adjusted so as to good use the adhesive properties of polymer fibres to the cement matrix and thus increase the cohesion of the concrete composite. Some mixtures show a decrease in long-term strength (after 90 days) which may be the result of less fibre adhesion and dispersion.
- Contact surface of polymer fibres to the cement matrix should be taken into account at the phase of designing the fibre reinforced concrete mixture. For this purpose it is possible to modify the cement matrix with selected admixtures and additives.

References

1. A. M. Neville, *Properties of Concrete, fourth edition*, London, (1996)
2. M. Kurpińska, T. Ferenc, *II International Conference of Computational Methods in Engineering Science*, (2017)
3. M. Kurpińska, T. Ferenc, Shell Structures: Theory and Applications, **4**, 549-552 (2018)
4. N. Ciak, M. Ciak, J. Harasymiuk, *IX Konferencja Dni Betonu*, **2**, 499-511, (2016)
5. B. Grzyl, E. Miszewska-Urbańska, M. Apollo, *E3S Web of Conferences*, EDP Sciences, (2017)
6. D. Kowalski, B. Grzyl, A. Kristowski, Civil and Environmental Engineering Reports **26.3**, 5-13 (2017)
7. A. Gallowicz-Sarzalejo, Fibres as structural element for reinforcement of concrete, Technical Manualy, Maccaferri, (2015)
8. M. A. Glinicki, A. Wysokowski, A. Żurawicka, Mechanical properties of fibreced concrete for thin repairs concrete pavements. Brittle Matrix Composities **VII**, 75-84 (2005)
9. J. P. Romualdi, J. A. Mandel, J. Amer. Concr. Inst., **61**, 657-670 (1964)
10. A. Bentur, S. Mindes, *Fibre reinforced cementious composities*, Elsevier Applied Science, London and New York, (1990)
11. A. M. Brandt, *Cement based composities: materials, mechanical and properties and performance*, E&FN SPON, London, (1995)
12. A. Katz, A. Bentur, Advanced Cement Based Materials, **3**, 1-13 (1996)
13. A. M. Brandt, M. A. Glinicki, Archives of Civil Engineering, **3**, 399-426 (1999)
14. J. Kasperkiewicz, *Konferencja Naukowa – Jabłonna 1979*, Wrocław, (1982)
15. F. Grzymski, Builder, **90**, 90-93 (2017)
16. Ł. Skarżyński, J. Tejchman, Strain, **52** (1), 26-45 (2016)
17. J. Chróścielewski, A. Mariak, A. Sabik, B. Meronk, K. Wilde, Adv. Sci. Technol. Res. J., **10**, 254-262 (2016)
18. M. Miśkiewicz, Ł. Pyrzowski, K. Wilde, O. Mitrosz, Polish Maritime Research, **24**, 149-155 (2017)
19. A. Mariak, J. Chróścielewski, K. Wilde, Shell Structures: Theory and Applications, **4**, 557-560 (2018)
20. M. A. Glinicki, Drogi i Mosty, **3**, 5-36 (2002)

The experimental comparison of parameters signals of acoustic emission during destructive test on composites based on cement

Michaela Hoduláková[1,*], *Dalibor* Kocáb[1], *Libor* Topolář[1], and *Barbara* Kucharczyková[1]

[1]Brno University of Technology, Faculty of Civil Engineering, Veveří 331/95, Brno, Czech Republic

Abstract. The paper reports the analysis of acoustic emission signals captured during compressive strength tests of fine-grained cement-based composites of various composition. Acoustic emission monitoring belongs to a non-invasive and passive non-destructive testing technique. It is one of the most sensitive experimental techniques to monitor cracking of materials. While the acoustic emission phenomenon is directly associated with nucleation of cracks in materials based on cement, the measurement of these kinds of materials is readily applied to detect cracking activity. An understanding of microstructure–performance relationships is the key to true understanding of material behaviour.

1 Introduction

The concept of the water-cement ratio (w/c) is the basis of concrete technology because this ratio directly impacts the ultimate density, porosity, and mechanical properties of materials based on cement see Figure 1 [1].

The lower this ratio is, the stronger, more sustainable, less permeable and more durable the materials based on cement will be [1, 2]. This is the reason, why superplasticizers, which allow the production of concrete (and other materials based on cement) with lower water content without compromising workability, are very popular and widely applied. In a short term, the early mechanical properties are linked primarily to the w/c ratio, because the first hydrates that give the cement paste its strength are the ones that develop on the surface of the cement particles. In spite of the fact, that in materials with superplasticizers, there is not enough water for complete hydration, compressive strength continues to increase as the water-cement ratio (w/c) decreases. However, it was proved that it is not the hydration degree of cement particles that is determining compressive strength of the material, but their closeness in the cement paste [2].

This paper addresses to testing of several types of composites based on cement during the destructive test by one of the most interesting evaluation and diagnostic techniques for concrete structures, an acoustic emission (AE) technique (Figure 2). This non-destructive method is recently in great demand for maintenance purposes [3].

* Corresponding author: hodulakova.m@fce.vutbr.cz

The AE method shows a significant potential to be applied for in-situ monitoring and evaluation of the current state of structures. It is applied during loading and it is often used to detect a failure at a very early stage, long before a structure completely fails [4]. This method is commonly used in bridge loading tests (e. g. the measurements of the bridge pillars) [5].

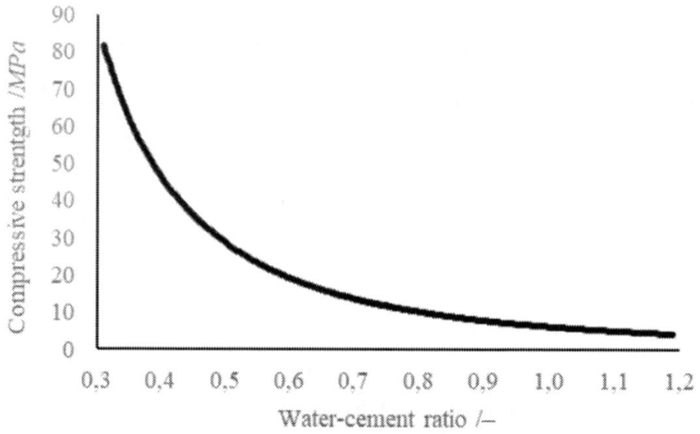

Fig. 1. The relationship between water-cement ratio and the compressive strength of concrete [6].

This method is based on detection of elastic waves by sensors directly attached (except for contactless sensors) to the surface of the specimen (Figure 3 and 4). These waves are defined as AE waves and generated due to the rapid release of stored strain energy from a localised source within a material. The sources of AE are deformation processes such as crack growth and plastic deformation. Propagations and patterns of cracks are mainly affected by loading types and conditions [7]. The initial cracking position may be dependent on the internal cracks and flaws during the loading, and the mechanical behaviour of specimens [8, 9].

After the detection of AE signals, as dynamic motions at the surface of a material, these signals are converted into the electric signal that is further amplified and filtered [4].

The measurement sensors are possible to calibrate after the sensors were stuck to the specimens (before loading tests) by means of Pen-test (Hsu-Nielsen source) [10].

The AE method only tracks the active actions (which take place during the measurement) in the material. The shape of the structure and the previous defects do not affect the measurement. The using of the AE method itself does not act upon the test specimen and the results [4].

There are some signal parameters which are useful to characterise the source of the AE event and are influenced by the crack mode. There is, for example, the amplitude, the duration, the rise time, AE energy and the cumulative number of AE counts.

Four parameters of each signal are recorded, calculated from the wave-forms (energy, duration, amplitude, the cumulative number of AE counts). These parameters are collected as the components of a vector associated with the signal they derive from.

The advantage of the AE technique, compare to other non-destructive testing method, is that AE method uses only few sensors in order to monitor the AE activity on entire structure. Another advantage is that the AE method does not require stopping during the loading, compared with the ultrasound method (which can be used for the measurement of defects too). In our experiment it is important not to stop the press during the compressive strength test because every stopping action could influence the value of the modulus of elasticity [4].

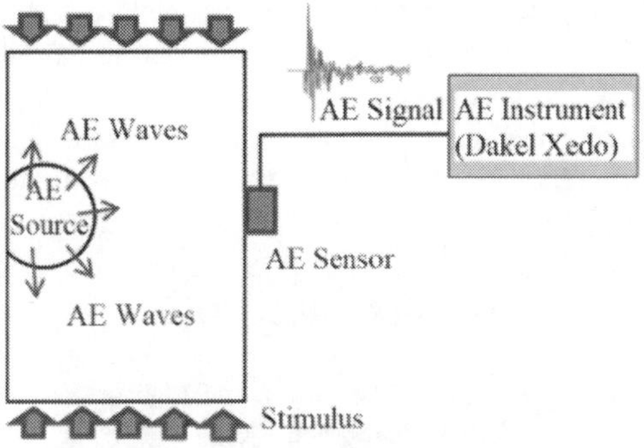

Fig. 2. Principle of acoustic emission measurement.

There are some limitations of the AE method too. In the case of a quick loading during compressive strength test, the AE method is less useful because there is a large number of signals that may not be recorded by the hardware of the AE measuring device.

Determining the behaviour of different kind of the material on the damage behaviour under compression loading using acoustic emission is quite common and is popular among other researchers [11, 12, 13, 14].

2 Experimental setup

The main aim of the performed experiments was to compare the acoustic responses during compressive strength test performed on the specimens at the age of three days. For this purpose, fine-grained cement composites were made, each with a different water-cement ratio and plasticizer content. The fresh composite formula is based on the ČSN EN 196-1 standard [15] for the testing of cement strength characteristics. Composites "M" were made of quartztite sand with the aggregate grain size distribution of 0–2 *mm*, CEM I 42.5 R Portland cement (Cement Plant Mokrá, Czech Republic) and water in ratio of 3:1:0.4 and 3:1:0.5 respectively (s:c:w). The composition of composites "MP" was modified by adding a Sika Viscocrete polycarboxylate-based super-plasticizer (Sika Group, Czech Republic) in amount of 1% by cement mass. The rest of components were the same. Table 1 shows the composition of composites. From each mixture, three test specimens with dimensions of 40 × 40 × 160 *mm* were manufactured and tested [16].

Table 1. Fresh mortar composition.

Identification	Components	
	s:c:w	Superplasticizer by cement mass [%]
42_M_040	3:1:0.40	-
42_M_050	3:1:0.50	-
42_MP_040	3:1:0.40	1
42_MP_050	3:1:0.50	1

All specimens were coated with the PE foil during the hardening in laboratory conditions at a temperature of (21 ± 2)*°C* and relative humidity of (60 ± 10)% until the time of testing. While testing compressive strength each specimen was placed in a testing FORM+TEST DELTA 6-300 hydraulic press with a range of 0 – 300 *kN* as shown in Figure 3. Two acoustic emission sensors (type-IDK 09) with 35 *dB* preamplifiers were fixed to each specimen using beeswax. Acoustic emission signals were taken by measuring equipment DAKEL XEDO. Universal measurement and diagnostic system DAKEL-XEDO was developed by ZD Rpety-Dakel company. XEDO is a modular system. One communication card and up to 15 various input cards can be located in a metal box. Communication between cards within a box is performed by a hi-speed bus. It allows sampling of the signal from one sensor (speed up to 8 *MSamples/sec*), enumerates standard acoustic emission parameters, process emission events parameters for possible emission source localization.

Fig. 3. Specimen location in the hydraulic press.

Fig. 4. The specimen 42_MP_050 after the test.

3 Experimental results

In order to describe acoustic emission signals formed during the compressive strength test in the prismatic beam, the attention was focused on the cumulative number of AE overshoots, duration, amplitude and energy of these signals. Different types of cracks generate different AE signals, these differences can be related to the properties of the material [17, 18]. Duration of acoustic emission signal is the time difference between the first and last threshold crossings. The amplitude is the greatest measured voltage in a waveform. This is an important parameter in AE inspection because it determines detectability of the signal. The signals with amplitudes below the operator-defined minimum threshold are not recorded. The last monitored parameter acoustic emission energy is directly proportional to the area under the acoustic emission waveform.

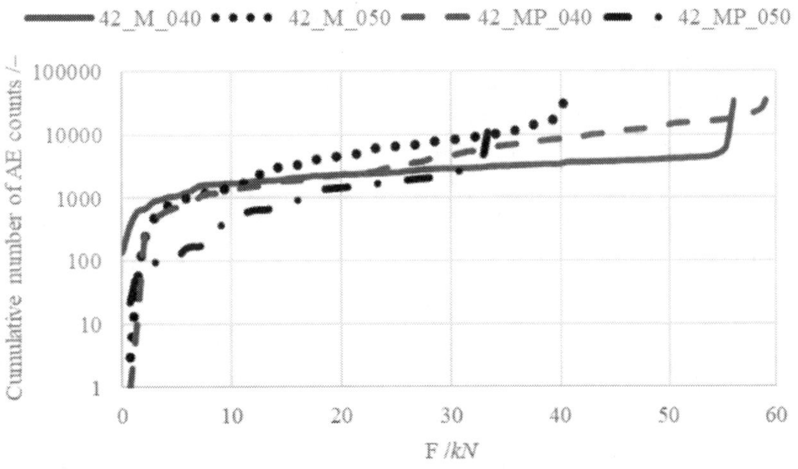

Fig. 5. The dependence of the cumulative number of AE overshoots on the loading force.

Figure 5 presents the loading process in relation with the AE events for all of the investigated composites. Based on the maximum value of a loading force, it can be estimated that the highest compressive strength was achieved in the case of the specimens made of mixture with the w/c of 0.40 with the addition of plasticizer. On the other hand, the mixture with the w/c of 0.50 with plasticizer exhibited the lowest estimated compressive strength. In relation to the AE method, supposedly the least amount of cracks (within the interval of 10-30 *kN*) were formed in the specimens with the w/c of 0.50 with plasticizer. In contrast, the greatest amount of cracks was detected in the case of specimens with the w/c of 0.50 without plasticizer. The above described data are supposedly connected to the strength of the bond of the cementitious matrix of the material, where the mixtures with the lower w/c have stronger bonds, more brittle at the same time. This effect is particularly evident for the mixture with the w/c of 0.40 without a plasticizer, exhibiting a very sharp increase of AE counts at the moment of failure.

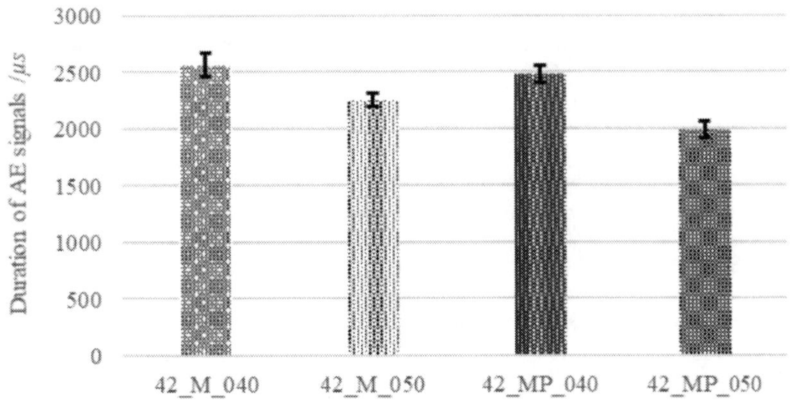

Fig. 6. The duration of AE signals for each specimen.

Figure 6 shows average duration of the AE signals for individual specimens during loading. It is evident that the specimens with a lower w/c show lower AE signal attenuation than the specimens with a higher water-cement ratio. This is probably due to the strength of the cementitious matrix and the fact that the specimens with the higher w/c exhibit the formation of air pores causing increased attenuation of the AE signal.

Figure 7 presents average amplitudes of the recorded AE signals. A higher amplitude value for specimens with the higher w/c indicates the formation of larger cracks during loading. This is caused again by the fact that the bond of the cementitious matrix of the specimens with the higher w/c is more damage-prone.

The calculated values of the energy of the captured AE signals are shown in Figure 8. Obviously, the highest value in the case of specimens with the w/c of 0.40 and plasticizer was detected. This energy is predominantly related to the type of destruction of the specimen, where the specimen with the highest energy failed "explosively" into multiple pieces. The other specimens are more or less crumbled, see Figure 4.

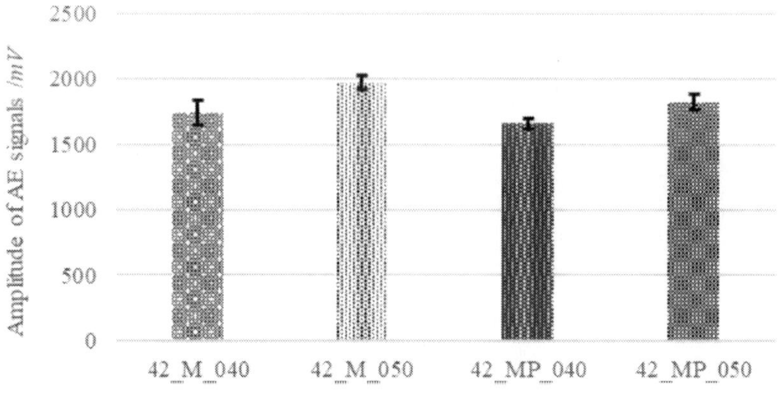

Fig. 7. The amplitude of AE signals for each specimen.

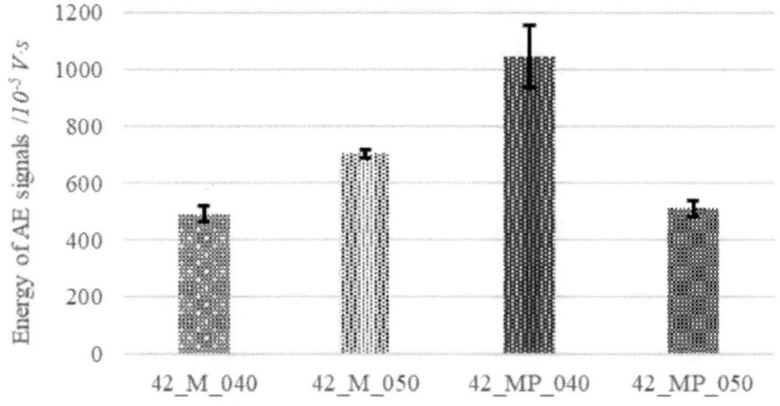

Fig. 8. The energy of AE signals for each specimen.

4 Conclusion

The experiments indicate that the acoustic emission method can significantly contribute to a more detailed insight into the behaviour of composite materials and structures during their lifespan. The acoustic emission method appears to be an interesting supplementary method for measuring the behaviour of composite constructions, structures and materials, during their production or under loading.

Based on the performed experiments, it can be stated that:

- a higher number of acoustic emission signal counts correspond to the formation of a higher number of new microcracks,

- a higher amplitude value of the acoustic emission signals or released energy of acoustic emission signals correspond to more significant structural changes, which occur in the structure of the material.

Larger structural changes during the test were performed in specimens with a higher water-cement ratio. The plasticizer reduced structural damages.

This research was supported by the Czech Science Foundation under the project no. 17-14302S "Experimental analysis of the early-age volume changes in cement-based composites".

References

1. H. W. Reinhardt, C. U. Grosse, *Advanced testing of cement based materials during setting and hardening* (RILEM Publications S.A.R.L., 2005)
2. P.C. Aitcin, R. J. Flatt, *Science and technology of concrete admixtures* (Woodhead Publishing, 2015)
3. M. Ohtsu, T. Suzuki, Journal of Acoustic Emission, **22** (2005)
4. Ch. U. Grosse, M. Ohtsu, *Acoustic Emission Testing* (Heidelberg: Springer, 2008)
5. P. Olaszek, J. R. Casas, G. Świt, *On-site assessment of bridges supported by acoustic emission*, in: Proceedings of the Institution of Civil Engineers-Bridge Engineering, **169(2)**, 81-92 Thomas Telford Ltd (2015)
6. Journal of the Network of African Countries on Local Building Materials and Technologies, **3(2)** (HABITAT, 1994, 42 p.): Technology Profile No. 1 - Fibre-concrete Roofing**. New Zealand Digital Library [online]. University of Waikato [cit. 2018-01-17]. Available on: http://www.nzdl.org/gsdlmod?e=d-00000-00---off-0cdl--00-0----0-10-0---0---0direct-10---4-------0-1l--11-en-50---20-about---00-0-1-00-0-0-11-1-0utfZz-8-00&a=d&cl=CL1.137&d=HASH01f1813030e134a3d7292eca.5
7. E. K. Tschegg, A. Schneemayer, I. Merta, K. A. Rieder, Cement and Concrete Composites, **62** (2015)
8. M. A. A. Aldahdooh, N. M. Bunnori, Construction and Building Materials, **45** (2013)
9. R. V. Sagar, B. R. Prasad, Nondestructive Testing and Evaluation, **27** (2012)
10. Z. Cáb, *Using of non-destructive test of accoustic emission for prolonging the operation life of composite pressure cylinders*. TRANSACTIONS of the VŠB–Technical University of Ostrava, Safety Engineering Series, **11**(2), 51-55 (2016)
11. N. Kosmann, J. M. Karsten, M. Schuett, K. Schulte, B. Fiedler, *Determining the effect of voids in GFRP on the damage behaviour under compression loading using acoustic emission*. Composites Part B: Engineering, 70, 184-188 (2015)
12. L. Topolář, L. Pazdera, P. Cikrle, *Acoustic emission monitoring during static modulus elasticity test of concrete specimen*. In Applied Mechanics and Materials (Vol. 486, pp. 267-272). Trans Tech Publications (2014)
13. E. D. Pasiou, D. Triantis, *Correlation between the electric and acoustic signals emitted during compression of brittle materials*. Frattura ed Integrità Strutturale, (40), **41** (2017)
14. Q. Meng, M. Zhang, L. Han, H. Pu, T. Nie, *Effects of acoustic emission and energy evolution of rock specimens under the uniaxial cyclic loading and unloading compression*. Rock Mechanics and Rock Engineering, **49**(10), 3873-3886, (2016)
15. EN 196 – 1: *Methods of testing cement* (2005)
16. L. Topolář, L. Pazdera, B. Kucharczyková, J. Smutný, K. Mikulášek, Applied Sciences, **7** (2017)
17. K. Ohno, M. Ohtsu, Construction and Building Materials, **24** (2010)
18. Z. Li, Y. Xi, Non-Destructive Testing in Civil Engineering (1995)

Assessment of sulfur concrete properties for use in civil engineering

Michał Lewandowski[1,*], and *Renata* Kotynia[1]

[1]Lodz University of Technology, Department of Concrete Structures, al. Politechniki 6, 90-942 Lodz, Poland

Abstract. Sulfur concrete is a material that was introduced to the civil engineering industry as a substitute of a typical concrete in which cement is a binder. Sulfur concrete properties are provided by a sulfur polymer produced in the process of synthesis of the ordinary sulfur with modifiers that prevent a process of crystallization of sulfur what results in stabilizing sulfur polymer structure. The purpose of this investigation was to evaluate characteristics of sulfur concrete, which was designed with combustion products. It was established that mechanical parameters of sulfur concrete should correspond with properties of a cement concrete used in production processes of prefabricated elements of road and construction industry. Accurateness of parameters of a designed mixture was confirmed and rated by a series of tests of hardened concrete – physical, mechanical (compression strength), durability in harsh environment. The test results confirmed low water absorption, high resistance to wear abrasion, high freeze-thaw scaling resistance (surface weathering) in a salty environment and high compressive and tensile strength of individual compositions. However, due to the internal structural damage, the freeze-thaw-attack cube tests did not prove the freeze-thaw resistance of the sulfur concrete specimens. The reason was low freeze-thaw-attack resistance of sulfur binders.

1 Introduction

Sulfur, used in this research, is the so-called "recycled sulfur" due to fact that it is obtained in action of purifying fuel – natural gas, oil, coal, however it is characterized with high purity: 99,95% content of elementary sulfur. Although its satisfactory quality, most of the produced sulfur is placed at stockpiles.

Natural result of sulfur resources growth is the increase of research on a new application of this product. Replacement of a cement with a sulfur polymer as a binder in concrete compounds, is one of the analyzed directions of solving a problem of a sulfur waste. The first attempts of utilizing sulfur as a binder for concrete compounds date back to the 1920s. The first results have shown that the new material showed a tendency to lose its mechanical properties in time. It was probably caused by disintegration of internal sulfur polymer bonds. Additionally, first sulfur concrete samples were susceptible to freeze-thaw cycles

[*] Corresponding author: michal.lewandowski@p.lodz.pl

and humid or water environment [1]. In order to solve this problem, several modifiers of standard properties of sulfur binder were prospected.

Modifiers, such as dicyclopentadien, styrene, turpentine, furfural, are substances that added to the liquid sulfur polymer, change its physio-chemical properties by impeding process of crystallization of sulfur, it results in stabilization of polymer structure with physical parameters irrespective of a passing time. The advantage of using the modified sulfur polymer as a binder in manufacturing sulfur concrete is the ability of fast compressive strength gaining. It is estimated that full compressive strength of produced elements is reached after 24 hours from its forming. This is a result of the incubating process, directly affected by solidification of the product.

Another benefit of sulfur concrete products is the possibility of utilization wastes, whose amount at stockpiles is still increasing. According to GUS [2] (central statistical office) data from 2014, in Poland it was produced 131,3 million tons of industrial waste materials from which only 27,6 % was recycled. The total mass of wastes kept at stockpiles (data from the end of 2014) has reached 1683,5 million tons. Materials that are most frequently stored at stockpiles nearby power plants and electrical power and heating stations are slags and fly ashes – by-products of coal combustion.

A beneficial feature of a sulfur concrete is a possibility of applying all kind of wastes regardless to their quality – those types of additives do not come into reaction with a binder and might be used as fillers. Using wastes, which without additional refinement do not comply with requirements of additives used in civil engineering and despite this might be used as fillers of sulfur concrete products of small architecture without any risk, resulting in economic and ecological benefits.

Based on literature, sulfur concrete is characterized by high durability in harsh environments such as salt or acidic solutions, is unresponsive to biodegradation, radiation activity and is proof to freeze – thaw cycles, is material hardly absorbing liquids, has high compression strength, is easily recyclable [3-7]. Due to all of mentioned advantages of sulfur concrete, there is a wide range of potential applications of this product.

2 Component materials and production methods

2.1 Sulfur polymer

Sulfur polymers are materials produced from natural sulfur modified by chemical substances that are added at strictly defined conditions and proportions. The following modifiers: dicyclopentadiene (DCPD), styrene, turpentine and furfural are added to the liquefied sulfur to transform properties of this material by inhibiting its crystallization - that stabilizes the structure enabling preservation of consistent physical properties irrespectively the time. All of them are high-boiling liquids or low setting solids imparting the binder with plasticity. They undergo co-polymerization with particles of a sulfur in a temperature of 140°C, creating linear (styrene) or cross-linked (DCPD) structures – depending on a type of modifier [8]. The influence of produced sulfur polymers on people's health was not tested yet, but tests conducted by National Institute of Hygiene of Poland (No. HK/W/1008/01/2009) on other sulfur mortar have confirmed safety of use of tested sulfur polymer.

2.2 Aggregate

An aggregate used in the project was selected with a special emphasis on achieving satisfactory mechanical parameters of the product. The secondary criterion of an aggregate selection was the economical aspect. The following types of an aggregate commonly used in a typical cement concrete were applied in the project: sand, gravel, granite and dolomite.

Additives in the sulfur concrete (fly ash, phosphogypsum) were selected to obtain intended grain curve consistent with ACI 5482 code [1]. Two main aims were considered in additives assortment. A fine-grained dust sealing the aggregate matrix was used as the first one, and cheaper and easily accessible waste materials, stored in huge amounts at stockpiles, were utilized in the second. At the same time it favors competitiveness of the product considering the economic aspect as additives fly ash and phosphogypsum have been used.

2.3 Production process

Production process of a sulfur concrete based on melting of an elementary sulfur and simultaneous mixing it with modifiers at the temperature in a range between 130 and 140°C. Heated up aggregate and other additives are added to a liquidized sulfur polymer binder and mixed all together. Finally, after obtaining homogeneous compound, it is ready to be poured into molds. In the research, cubic and cuboid forms were used. From one batch, usually about 16 cubic and 9 cuboid samples were performed. The details of batch compositions are presented in Table 1.

Table 1. Selected compositions of batches.

No.	Batch No.	Sulfur	Modifier		Sand	Coarse aggregate 2-8 mm		Additive	
		%	Type	%	%	Type	%	Type	%
1	3	30	-	-	60	-	-	fly ash	10,0
2	6	22,5	str	2,5	62,5	-	-	fly ash	12,5
3	8	22,5	str	2,5	61,0	-	-	fly ash	14,0
4	10	22,5	str tur	1,25 1,25	62,5	-	-	fly ash	12,5
5	12	19,0	dcpd	1,0	20,0	gravel	60,0	-	-
6	15	19,0	dcpd	1,0	20,0	granite	60,0	-	-
7	16	19,0	dcpd	1,0	20,0	dolomite	60,0	-	-
8	20	21,1	dcpd	1,1	15,6	gravel	54,4	fly ash	7,8
9	22	24,5	dcpd	1,3	6,4	gravel	58,7	ph	9,1
10	24	24,5	dcpd fur	0,65 0,65	13,1	gravel	52,0	fly ash	9,1
11	25	24,5	dcpd fur	0,65 0,65	13,1	gravel	52,0	ph	9,1
12	26	21,1	dcpd	1,1	20,0	gravel	57,8	-	-
13	27	24,5	dcpd	1,3	13,1	gravel	52,0	fly ash	9,1
14	28	24,5	dcdp	1,3	13,1	gravel	52,0	ph	9,1
15	30	21,1	dcpd	1,1	20,0	gravel	57,8	-	-

Abbreviations:
dcpd – dicyclopentadiene
ph- phosphogypsum
str – styrene
tur – turpentine
fur – furfural

2.4 Experimental tests

The main aim of the batch composition was to produce durable sulfur concrete products, able to withstand environmental conditions, their exposure should be highly durable and immune to weather conditions. The following mechanical and durability tests were carried out:

- Water absorption [9]
- Bulk density [9]
- Compressive strength [10]
- Bending tensile strength [11]
- Determination of ultrasonic pulse velocity [12]
- Freeze-thaw resistance [9]
- Freeze-thaw resistance in salty environment – scaling [13]
- Abrasion resistance [13]

All of experimental tests were carried out according to the European or National Standards listed in references.

Additionally, after performing series of freeze – thaw cycles to sulfur concrete specimens, it was decided to test freeze – thaw resistance of sulfur polymer itself. Tests were conducted according to the Polish Standard PKN-CEN/TS 12390-9:2007 [14].

3 Test results

Cuboid polymer specimens of sulfur polymer (40x40x160 mm) were subjected to the freeze – thaw cycle test in a temperature ranged between -20°C and +20°C within one cycle. All specimens, irrespectively of the modifier, cracked in a short time after freeze – thaw cycles have begun (Fig.1). First cracks have appeared on a surface of polymer specimens within first 72 hours, full cracks (when a specimen breaks) were observed after 7 days from beginning of the tests.

Fig. 1. Cuboid samples of sulfur polymers after 75 freeze – thaw cycles.

The polymeric binders were not affected by additives such as flying ash and phosphogypsum. Application of additives in dust form and aggregate of a good quality guarantees obtaining product of low water absorption and high compressive strength. Application of coarse aggregate with 2-8 mm graining in the mixture resulted with a significant improvement of strength parameters of sulfur concrete (Table 2, Fig. 2).

Table 2. Mechanical parameters of selected sulfur concrete batches.

No.	Batch No.	Density	Water absorption	Compressive strength [MPa]		Bending strength [MPa]		Ultrasonic wave velocity [km/s]
		[kg/m³]	[%]	Early	Late	Early	Late	
1	3	2285	-	41,0	41,0	4,3	-	-
2	6	2160	0,23	14,5	18,0	4,2	4,2	3,44
3	8	2053	0,37	12,1	14,2	4,1	4,1	3,15
4	10	2029	0,36	11,2	14,1	3,4	3,8	3,23
5	12	2351	0,35	57,3	56,1	8,7	7,7	3,91
6	15	2296	0,31	65,6	53,4	5,2	-	3,82
7	16	2387	0,47	57,0	51,7	7,3	-	4,15
8	20	2235	0,07	50,6	68,7	13,3	10,4	4,07
9	22	2300	0,10	39,7	45,8	7,4	2,3	3,97
10	24	2241	0,06	38,4	41,9	4,9	5,3	3,94
11	25	2256	0,04	49,6	50,5	4,5	3,7	3,79
12	26	2420	0,25	54,5	54,6	5,5	5,8	3,89
13	27	2341	0,04	58,7	58,2	10,1	10,0	3,82
14	28	2390	0,04	61,8	51,3	8,0	5,7	3,91
15	30	2300	0,13	55,4	44,9	7,0	5,5	3,76

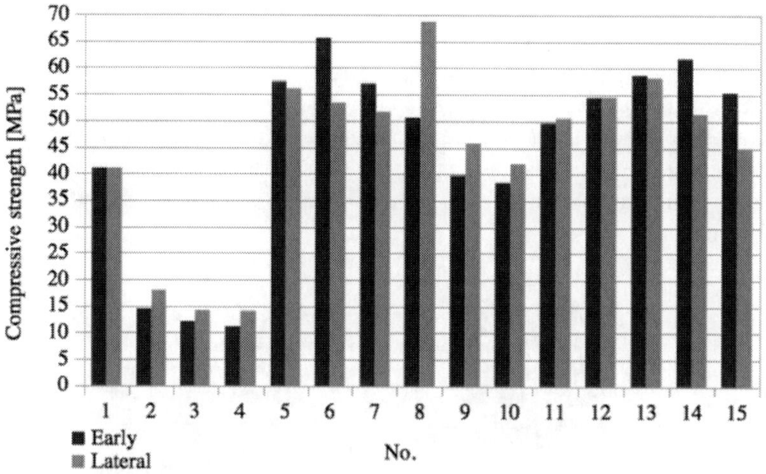

Fig. 2. Comparison of early and late strength of sulfur concrete samples.

Modification of sulfur polymer with styrene (STR) significantly decreased strength of the sulfurous mortar from 41,0 MPa to approximately 14,5 MPa. Natural aggregate (gravel, granite, dolomite) did not influence the compound strength. The curing time of samples did not affect the compressive or bending tensile strength. Early strength (1-4 days of age) and late strength after 28 days were similar (Fig. 2). The differences between the investigated parameters were caused by heterogeneity of the samples. The main reason of the results dispersion was varied conditions of samples casting. The time of casting is strictly limited, so the temperature changes significantly affected the final strength of sulfur concrete specimens. Additionally, differences in bending tensile strength observed on specimens tested in early and late stage of incubation might be connected with excessive maximum aggregate dimension contained in sulfur concrete (8 mm) for that kind of form (40x40x160 mm), which is designed to test mortar parameters (maximum aggregate dimension of 4 mm).

The freeze-thaw resistance of sulfur concrete was conducted using volumetric and surface method. In the first one the strength loss limit is 20% of reference specimen strength and cracks on the surface of samples subjected to a specific number of freezing-thawing cycles are not allowed. According to the Polish standard [14], concrete members used in the road infrastructure require the freeze-thaw resistance of F150, which corresponds to 150 freeze-thaw cycles. Several test specimens did not meet those requirements (Table 3).

Table 3. Durability test results of selected sulfur concrete batches.

No.	Batch No.	Freeze-thaw resistance, samples compressive strength [MPa]			
		Reference	Tested	Strength loose	Number of cycles
1	3	-	-	-	-
2	6	16,6	12,0	28%	200
3	8	11,4	10,2	11%	200
4	10	12,5	9,1	27%	200
5	12	61,9	26,2	58%	200
6	15	56,1	36,3	35%	200
7	16	35,8	31,2	13%	200
8	20	65,5	46,8	29%	100
9	22	46,9	14,7	69%	200
10	24	42,7	29,3	31%	200
11	25	50,4	7,7	85%	100
12	26	54,2	8,1	85%	150
13	27	55,4	31,5	43%	150
14	28	58,1	42,6	27%	150
15	30	59,1	43,6	26%	150

At the same time, frost resistance evaluated with the second method, by measuring the surface exfoliation of samples in water solution of de-icing salt (NaCl), proved sulfur concrete very resistant (Table 4). The summary amount of the exfoliated material after 28 freeze-thaw cycles expressed in kg/m^2 was exiguous, and after 56 cycles did not exceed 0,3 kg/m^2.

The obtained abrasion test results affirmed that selected compositions of sulfur concrete are highly resistant to abrasion. The tested samples qualifies for the highest, 4th class of abrasion resistance according to [13] (Table 5).

Table 4. Durability tests results - freeze-thaw resistance in 3% salt solution (scaling).

Specimen No.	Mass of scaled material after 28 cycles (g)	Mass of scaled material after 56 cycles (g)	Mass of scaled material after 28 cycles (kg/m^2)	Mass of scaled material after 56 cycles (kg/m^2)
1A	1,50	2,51	0,16	0,26
2A	1,05	1,78	0,11	0,19
3A	0,21	0,64	0,02	0,07
4A	0,24	1,36	0,03	0,14
5B	0,03	0,11	0,00	0,01
6B	0,07	0,36	0,01	0,04
7B	0,01	0,10	0,00	0,01
8B	0,01	0,13	0,00	0,01

Table 5. Durability test results of selected sulfur concrete batches.

Specimen No.	width of groove (mm)		Requirements
	mean value	results in correlation to calibration sample	
2	15,1	20,0	Class 4 ≤ 20 mm
3	14,4	19,0	
4	15,0	19,5	

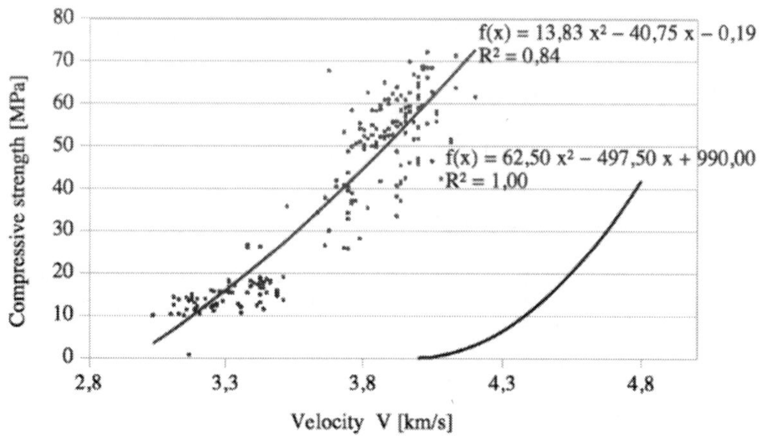

Fig. 3. Correlation curve of Compressive Strength - Velocity for sulfur concrete in comparison to standard curve for cement concrete according to PN-EN 13791.

In order to determine the compressive strength of sulfur concrete an ultrasonic pulse velocity method was used as well. The speed of wave for particular batches were repeatable and fall into between 24,1 and 31,0 μs, with an average speed value of 26,6 μs, which corresponded to a wave velocity of 3,76 km/s. The correlation between the velocity of an ultrasonic pulse in concrete and its compressive strength was estimated basing on the hypothetical curves or a reference tests. Acoustic parameters significantly depend on the tested material. For a typical, well condensed cement concrete with a density between 1800 and 2500 kg/m^3, the velocity varies between 3,0 to 4,8 km/s. The correlation between compressive strength and wave velocity for sulfur concrete in reference to the normalized curve for cement concrete is presented in Fig 3.

4 Discussion

The presented test results of sulfur concrete confirm low water absorption and high resistance to grating in dry conditions. The composite proved high freeze-thaw resistance in salty environment and can be recommended for production of small-sized prefabricated elements. Furthermore, the compressive and tensile strength of individual compositions allowed to design products of a high strength class. However, the sulfur concrete did not prove freeze-thaw resistance with cyclical freezing in the air and the thawing in a water. The probable reason of the vulnerability was the sulfur binder. Regarding this, the products made of the sulfur concrete cannot be exposed to severe conditions in Polish climate (XF class of environment exposition), as their durability does not meet the requirements of the Standards.

The research results clearly indicate that the material can be applied in the construction industry mostly for indoor and underwater or underground applications, below the freezing zone (for example underground cesspits).

The additional restraint are sulfur's flammable properties, which are usually omitted, and relatively low melting temperature - around 130°C.

Apart from further freeze-thaw resistance development, the evaluation of abrasion in water conditions is highly recommended due to possible application in underwater environment and all types of piping. Production of larger, reinforced sulfur concrete's members or even structures also requires further testing, but the starting point is the application of reinforcement that is immune to or properly protected from a corrosion. The composite bars seem to be a natural solution as the sulfur concrete occurs to be highly aggressive environment for conventional steel reinforcement.

In conclusion, despite the limitations of the obtained sulfur concrete composition, its potential applications are numerous and also economically justified.

The authors acknowledge financial support from the National Center for Research and Development and National Fund for Environmental Protection and Water Management for the applied research project GEKON No1/O5/213122/26/2015.

References

1. ACI Committee 548, ACI 548.2R-93 (Reapproved 1998) - *Guide for Mixing and Placing Sulfur Concrete in Construction* (1998)
2. Central Statistical Office – *Statistical Information and Elaborations* (2015)
3. M. Abdel-Mohsen, M. El Gama, Cement and Concrete Composites, **31,** 186-194, (2009)
4. M.M. Vlahovic, Martinovic, Construction and Building Materials, **25,** 3926-3934, (2011)
5. S. Myoungsu, K. Kyuhun, Construction and Building Materials, **69**, 167-176 (2014)
6. M. M. Vlahovic, M. M. Savic, Materials and Design, **34**, 346-354 (2011)
7. M. Abdel-Mohsen, D. Amr El, Journal of Environmental Geotechnics, **2(2)**, 95-103 (2014)
8. I. Błasiak, Ł. Łuszczyk, Chemik **41(9)**, 243 (1988)
9. PN-88 B-06250 – Ordinary Concrete
10. PN-EN 12390-3:2011 - Testing hardened concrete - Part 3: Compressive strength of test specimens
11. PN-EN 196-1:2016-07 - Methods of testing cement - Part 1: Determination of strength
12. PN-EN 12504-4:2005-09 Testing concrete. Determination of ultrasonic pulse velocity
13. PN-EN 1338:2005 - Concrete paving blocks - Requirements and test methods
14. PKN-CEN/TS 12390-9:2007 – Testing hardened concrete – Part 9: Freeze-thaw resistance – Scaling

and humid or water environment [1]. In order to solve this problem, several modifiers of standard properties of sulfur binder were prospected.

Modifiers, such as dicyclopentadien, styrene, turpentine, furfural, are substances that added to the liquid sulfur polymer, change its physio-chemical properties by impeding process of crystallization of sulfur, it results in stabilization of polymer structure with physical parameters irrespective of a passing time. The advantage of using the modified sulfur polymer as a binder in manufacturing sulfur concrete is the ability of fast compressive strength gaining. It is estimated that full compressive strength of produced elements is reached after 24 hours from its forming. This is a result of the incubating process, directly affected by solidification of the product.

Another benefit of sulfur concrete products is the possibility of utilization wastes, whose amount at stockpiles is still increasing. According to GUS [2] (central statistical office) data from 2014, in Poland it was produced 131,3 million tons of industrial waste materials from which only 27,6 % was recycled. The total mass of wastes kept at stockpiles (data from the end of 2014) has reached 1683,5 million tons. Materials that are most frequently stored at stockpiles nearby power plants and electrical power and heating stations are slags and fly ashes – by-products of coal combustion.

A beneficial feature of a sulfur concrete is a possibility of applying all kind of wastes regardless to their quality – those types of additives do not come into reaction with a binder and might be used as fillers. Using wastes, which without additional refinement do not comply with requirements of additives used in civil engineering and despite this might be used as fillers of sulfur concrete products of small architecture without any risk, resulting in economic and ecological benefits.

Based on literature, sulfur concrete is characterized by high durability in harsh environments such as salt or acidic solutions, is unresponsive to biodegradation, radiation activity and is proof to freeze – thaw cycles, is material hardly absorbing liquids, has high compression strength, is easily recyclable [3-7]. Due to all of mentioned advantages of sulfur concrete, there is a wide range of potential applications of this product.

2 Component materials and production methods

2.1 Sulfur polymer

Sulfur polymers are materials produced from natural sulfur modified by chemical substances that are added at strictly defined conditions and proportions. The following modifiers: dicyclopentadiene (DCPD), styrene, turpentine and furfural are added to the liquefied sulfur to transform properties of this material by inhibiting its crystallization - that stabilizes the structure enabling preservation of consistent physical properties irrespectively the time. All of them are high-boiling liquids or low setting solids imparting the binder with plasticity. They undergo co-polymerization with particles of a sulfur in a temperature of 140°C, creating linear (styrene) or cross-linked (DCPD) structures – depending on a type of modifier [8]. The influence of produced sulfur polymers on people's health was not tested yet, but tests conducted by National Institute of Hygiene of Poland (No. HK/W/1008/01/2009) on other sulfur mortar have confirmed safety of use of tested sulfur polymer.

2.2 Aggregate

An aggregate used in the project was selected with a special emphasis on achieving satisfactory mechanical parameters of the product. The secondary criterion of an aggregate selection was the economical aspect. The following types of an aggregate commonly used in a typical cement concrete were applied in the project: sand, gravel, granite and dolomite.

Additives in the sulfur concrete (fly ash, phosphogypsum) were selected to obtain intended grain curve consistent with ACI 5482 code [1]. Two main aims were considered in additives assortment. A fine-grained dust sealing the aggregate matrix was used as the first one, and cheaper and easily accessible waste materials, stored in huge amounts at stockpiles, were utilized in the second. At the same time it favors competitiveness of the product considering the economic aspect as additives fly ash and phosphogypsum have been used.

2.3 Production process

Production process of a sulfur concrete based on melting of an elementary sulfur and simultaneous mixing it with modifiers at the temperature in a range between 130 and 140°C. Heated up aggregate and other additives are added to a liquidized sulfur polymer binder and mixed all together. Finally, after obtaining homogeneous compound, it is ready to be poured into molds. In the research, cubic and cuboid forms were used. From one batch, usually about 16 cubic and 9 cuboid samples were performed. The details of batch compositions are presented in Table 1.

Table 1. Selected compositions of batches.

No.	Batch No.	Sulfur	Modifier		Sand	Coarse aggregate 2-8 mm		Additive	
		%	Type	%	%	Type	%	Type	%
1	3	30	-	-	60	-	-	fly ash	10,0
2	6	22,5	str	2,5	62,5	-	-	fly ash	12,5
3	8	22,5	str	2,5	61,0	-	-	fly ash	14,0
4	10	22,5	str tur	1,25 1,25	62,5	-	-	fly ash	12,5
5	12	19,0	dcpd	1,0	20,0	gravel	60,0	-	-
6	15	19,0	dcpd	1,0	20,0	granite	60,0	-	-
7	16	19,0	dcpd	1,0	20,0	dolomite	60,0	-	-
8	20	21,1	dcpd	1,1	15,6	gravel	54,4	fly ash	7,8
9	22	24,5	dcpd	1,3	6,4	gravel	58,7	ph	9,1
10	24	24,5	dcpd fur	0,65 0,65	13,1	gravel	52,0	fly ash	9,1
11	25	24,5	dcpd fur	0,65 0,65	13,1	gravel	52,0	ph	9,1
12	26	21,1	dcpd	1,1	20,0	gravel	57,8	-	-
13	27	24,5	dcpd	1,3	13,1	gravel	52,0	fly ash	9,1
14	28	24,5	dcdp	1,3	13,1	gravel	52,0	ph	9,1
15	30	21,1	dcpd	1,1	20,0	gravel	57,8	-	-

Abbreviations:
dcpd – dicyclopentadiene
ph- phosphogypsum
str – styrene
tur – turpentine
fur – furfural

2.4 Experimental tests

The main aim of the batch composition was to produce durable sulfur concrete products, able to withstand environmental conditions, their exposure should be highly durable and immune to weather conditions. The following mechanical and durability tests were carried out:

- Water absorption [9]
- Bulk density [9]
- Compressive strength [10]
- Bending tensile strength [11]
- Determination of ultrasonic pulse velocity [12]
- Freeze-thaw resistance [9]
- Freeze-thaw resistance in salty environment – scaling [13]
- Abrasion resistance [13]

All of experimental tests were carried out according to the European or National Standards listed in references.

Additionally, after performing series of freeze – thaw cycles to sulfur concrete specimens, it was decided to test freeze – thaw resistance of sulfur polymer itself. Tests were conducted according to the Polish Standard PKN-CEN/TS 12390-9:2007 [14].

3 Test results

Cuboid polymer specimens of sulfur polymer (40x40x160 mm) were subjected to the freeze – thaw cycle test in a temperature ranged between -20°C and +20°C within one cycle. All specimens, irrespectively of the modifier, cracked in a short time after freeze – thaw cycles have begun (Fig.1). First cracks have appeared on a surface of polymer specimens within first 72 hours, full cracks (when a specimen breaks) were observed after 7 days from beginning of the tests.

Fig. 1. Cuboid samples of sulfur polymers after 75 freeze – thaw cycles.

The polymeric binders were not affected by additives such as flying ash and phosphogypsum. Application of additives in dust form and aggregate of a good quality guarantees obtaining product of low water absorption and high compressive strength. Application of coarse aggregate with 2-8 mm graining in the mixture resulted with a significant improvement of strength parameters of sulfur concrete (Table 2, Fig. 2).

Table 2. Mechanical parameters of selected sulfur concrete batches.

No.	Batch No.	Density	Water absorption	Compressive strength [MPa]		Bending strength [MPa]		Ultrasonic wave velocity [km/s]
		[kg/m³]	[%]	Early	Late	Early	Late	
1	3	2285	-	41,0	41,0	4,3	-	-
2	6	2160	0,23	14,5	18,0	4,2	4,2	3,44
3	8	2053	0,37	12,1	14,2	4,1	4,1	3,15
4	10	2029	0,36	11,2	14,1	3,4	3,8	3,23
5	12	2351	0,35	57,3	56,1	8,7	7,7	3,91
6	15	2296	0,31	65,6	53,4	5,2	-	3,82
7	16	2387	0,47	57,0	51,7	7,3	-	4,15
8	20	2235	0,07	50,6	68,7	13,3	10,4	4,07
9	22	2300	0,10	39,7	45,8	7,4	2,3	3,97
10	24	2241	0,06	38,4	41,9	4,9	5,3	3,94
11	25	2256	0,04	49,6	50,5	4,5	3,7	3,79
12	26	2420	0,25	54,5	54,6	5,5	5,8	3,89
13	27	2341	0,04	58,7	58,2	10,1	10,0	3,82
14	28	2390	0,04	61,8	51,3	8,0	5,7	3,91
15	30	2300	0,13	55,4	44,9	7,0	5,5	3,76

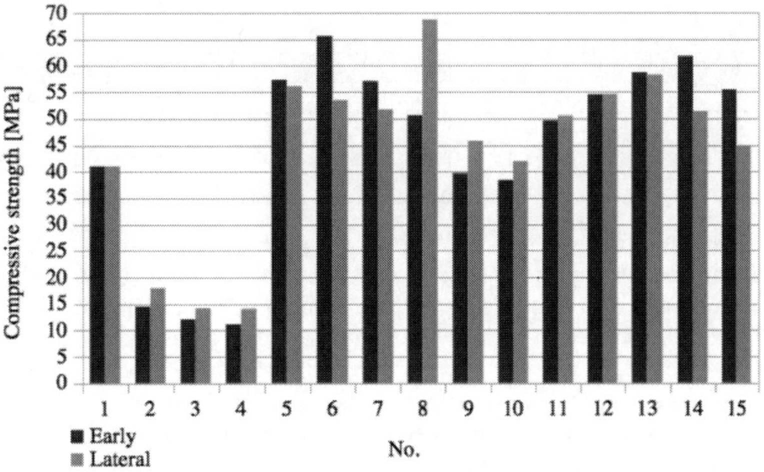

Fig. 2. Comparison of early and late strength of sulfur concrete samples.

Modification of sulfur polymer with styrene (STR) significantly decreased strength of the sulfurous mortar from 41,0 MPa to approximately 14,5 MPa. Natural aggregate (gravel, granite, dolomite) did not influence the compound strength. The curing time of samples did not affect the compressive or bending tensile strength. Early strength (1-4 days of age) and late strength after 28 days were similar (Fig. 2). The differences between the investigated parameters were caused by heterogeneity of the samples. The main reason of the results dispersion was varied conditions of samples casting. The time of casting is strictly limited, so the temperature changes significantly affected the final strength of sulfur concrete specimens. Additionally, differences in bending tensile strength observed on specimens tested in early and late stage of incubation might be connected with excessive maximum aggregate dimension contained in sulfur concrete (8 mm) for that kind of form (40x40x160 mm), which is designed to test mortar parameters (maximum aggregate dimension of 4 mm).

The freeze-thaw resistance of sulfur concrete was conducted using volumetric and surface method. In the first one the strength loss limit is 20% of reference specimen strength and cracks on the surface of samples subjected to a specific number of freezing-thawing cycles are not allowed. According to the Polish standard [14], concrete members used in the road infrastructure require the freeze-thaw resistance of F150, which corresponds to 150 freeze-thaw cycles. Several test specimens did not meet those requirements (Table 3).

Table 3. Durability test results of selected sulfur concrete batches.

No.	Batch No.	Freeze-thaw resistance, samples compressive strength [MPa]			
		Reference	Tested	Strength loose	Number of cycles
1	3	-	-	-	-
2	6	16,6	12,0	28%	200
3	8	11,4	10,2	11%	200
4	10	12,5	9,1	27%	200
5	12	61,9	26,2	58%	200
6	15	56,1	36,3	35%	200
7	16	35,8	31,2	13%	200
8	20	65,5	46,8	29%	100
9	22	46,9	14,7	69%	200
10	24	42,7	29,3	31%	200
11	25	50,4	7,7	85%	100
12	26	54,2	8,1	85%	150
13	27	55,4	31,5	43%	150
14	28	58,1	42,6	27%	150
15	30	59,1	43,6	26%	150

At the same time, frost resistance evaluated with the second method, by measuring the surface exfoliation of samples in water solution of de-icing salt (NaCl), proved sulfur concrete very resistant (Table 4). The summary amount of the exfoliated material after 28 freeze-thaw cycles expressed in kg/m^2 was exiguous, and after 56 cycles did not exceed 0,3 kg/m^2.

The obtained abrasion test results affirmed that selected compositions of sulfur concrete are highly resistant to abrasion. The tested samples qualifies for the highest, 4th class of abrasion resistance according to [13] (Table 5).

Table 4. Durability tests results - freeze-thaw resistance in 3% salt solution (scaling).

Specimen No.	Mass of scaled material after 28 cycles (g)	Mass of scaled material after 56 cycles (g)	Mass of scaled material after 28 cycles (kg/m^2)	Mass of scaled material after 56 cycles (kg/m^2)
1A	1,50	2,51	0,16	0,26
2A	1,05	1,78	0,11	0,19
3A	0,21	0,64	0,02	0,07
4A	0,24	1,36	0,03	0,14
5B	0,03	0,11	0,00	0,01
6B	0,07	0,36	0,01	0,04
7B	0,01	0,10	0,00	0,01
8B	0,01	0,13	0,00	0,01

Table 5. Durability test results of selected sulfur concrete batches.

Specimen No.	width of groove (mm)		Requirements
	mean value	results in correlation to calibration sample	
2	15,1	20,0	Class 4 ≤ 20 mm
3	14,4	19,0	
4	15,0	19,5	

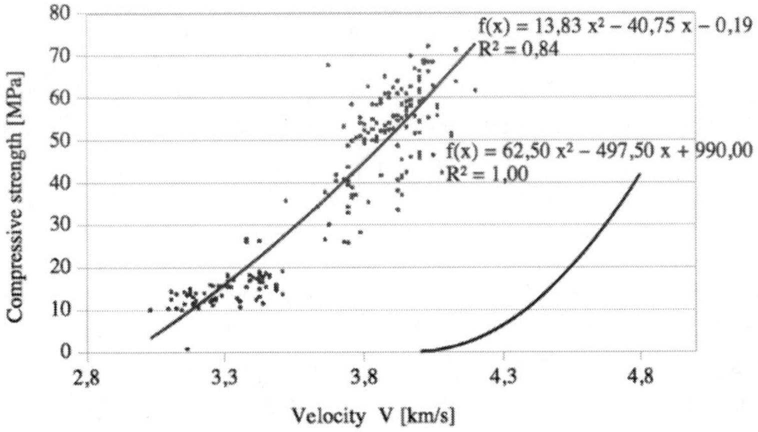

Fig. 3. Correlation curve of Compressive Strength - Velocity for sulfur concrete in comparison to standard curve for cement concrete according to PN-EN 13791.

In order to determine the compressive strength of sulfur concrete an ultrasonic pulse velocity method was used as well. The speed of wave for particular batches were repeatable and fall into between 24,1 and 31,0 μs, with an average speed value of 26,6 μs, which corresponded to a wave velocity of 3,76 km/s. The correlation between the velocity of an ultrasonic pulse in concrete and its compressive strength was estimated basing on the hypothetical curves or a reference tests. Acoustic parameters significantly depend on the tested material. For a typical, well condensed cement concrete with a density between 1800 and 2500 kg/m^3, the velocity varies between 3,0 to 4,8 km/s. The correlation between compressive strength and wave velocity for sulfur concrete in reference to the normalized curve for cement concrete is presented in Fig 3.

4 Discussion

The presented test results of sulfur concrete confirm low water absorption and high resistance to grating in dry conditions. The composite proved high freeze-thaw resistance in salty environment and can be recommended for production of small-sized prefabricated elements. Furthermore, the compressive and tensile strength of individual compositions allowed to design products of a high strength class. However, the sulfur concrete did not prove freeze-thaw resistance with cyclical freezing in the air and the thawing in a water. The probable reason of the vulnerability was the sulfur binder. Regarding this, the products made of the sulfur concrete cannot be exposed to severe conditions in Polish climate (XF class of environment exposition), as their durability does not meet the requirements of the Standards.

The research results clearly indicate that the material can be applied in the construction industry mostly for indoor and underwater or underground applications, below the freezing zone (for example underground cesspits).

The additional restraint are sulfur's flammable properties, which are usually omitted, and relatively low melting temperature - around 130°C.

Apart from further freeze-thaw resistance development, the evaluation of abrasion in water conditions is highly recommended due to possible application in underwater environment and all types of piping. Production of larger, reinforced sulfur concrete's members or even structures also requires further testing, but the starting point is the application of reinforcement that is immune to or properly protected from a corrosion. The composite bars seem to be a natural solution as the sulfur concrete occurs to be highly aggressive environment for conventional steel reinforcement.

In conclusion, despite the limitations of the obtained sulfur concrete composition, its potential applications are numerous and also economically justified.

The authors acknowledge financial support from the National Center for Research and Development and National Fund for Environmental Protection and Water Management for the applied research project GEKON No1/O5/213122/26/2015.

References

1. ACI Committee 548, ACI 548.2R-93 (Reapproved 1998) - *Guide for Mixing and Placing Sulfur Concrete in Construction* (1998)
2. Central Statistical Office – *Statistical Information and Elaborations* (2015)
3. M. Abdel-Mohsen, M. El Gama, Cement and Concrete Composites, **31,** 186-194, (2009)
4. M.M. Vlahovic, Martinovic, Construction and Building Materials, **25,** 3926-3934, (2011)
5. S. Myoungsu, K. Kyuhun, Construction and Building Materials, **69**, 167-176 (2014)
6. M. M. Vlahovic, M. M. Savic, Materials and Design, **34**, 346-354 (2011)
7. M. Abdel-Mohsen, D. Amr El, Journal of Environmental Geotechnics, **2(2)**, 95-103 (2014)
8. I. Błasiak, Ł. Łuszczyk, Chemik **41(9)**, 243 (1988)
9. PN-88 B-06250 – Ordinary Concrete
10. PN-EN 12390-3:2011 - Testing hardened concrete - Part 3: Compressive strength of test specimens
11. PN-EN 196-1:2016-07 - Methods of testing cement - Part 1: Determination of strength
12. PN-EN 12504-4:2005-09 Testing concrete. Determination of ultrasonic pulse velocity
13. PN-EN 1338:2005 - Concrete paving blocks - Requirements and test methods
14. PKN-CEN/TS 12390-9:2007 – Testing hardened concrete – Part 9: Freeze-thaw resistance – Scaling

Computational simulations of the cable-stayed bridge over the Vistula River in Krakow

Michał Topolewicz[1,*], and *Krzysztof* Topolewicz[2]

[1] Gdańsk University of Technology, Faculty of Civil and Environmental Engineering, Narutowicza 11/12, 80-233 Gdańsk, Poland
[2] TOP PROJEKT

Abstract. This article presents the execution method of a cable stayed bridge over the Vistula River in Krakow together with FEM-based computational simulations carried out during the design process. The performed calculations concern the global system in its final stage and in the construction stages as well as local analysis of construction details. The obtained results made it possible to work out efficient solutions that contribute to a significant reduction of the project cost and execution time shortening.

1 Introduction

The cable-stayed bridge across the Vistula River in Krakow, designated as MD-7 [1], is located along the important expressway S7. The main span, 200,0m long is one of the longest spans in Poland. The main contractor STRABAG chose the TOP PROJEKT team led by an experienced bridge designer Krzysztof Topolewicz to optimize the execution design. The aim of the optimization was to reduce the overall investment costs adjusted to the available technological solutions and the completion date.

The FEM computational analysis was performed during the design process to prove the fulfillment of a load capacity condition in final state, construction stages and to check the critical details of the superstructure. The experience gained in the course if analysis, similarly to the cases presented in [2-8], indicates the importance of the method to computationally simulate the structural behaviour. A key problem is the selection of an appropriate theoretical description due to an analysed issue in order to yield correct conclusions. A detailed analysis of construction stages contributes to a significant reduction of costs and time, additionally, it leads to effective structural solutions.

2 Description of the bridge

The MD-7 bridge consists of two parallel structures for two roadways of the S7 expressway (Fig. 1). Each structure consists of three superstructures: two three-span prestressed concrete box girder access viaducts over the floodplain and the main bridge with a cable-stayed span. The total length of the bridge is 706,0m a 390 m section (45+50+200+50 +45) is the main cable-stayed structure.

* Corresponding author: michal.topolewicz@pg.edu.pl

Fig. 1. The bridge MD-7 under construction.

The main superstructure consists of two subsections of different cross-section. In the main span a ladder deck steel-concrete composite section occurs with two steel box girders, I-beam cross girders and a reinforced concrete slab 280 mm thick (Fig. 2). In the access spans both longitudinal and cross girders are made of prestressed concrete. This distinction adopted at the optimization stage brought several economic and structural benefits such as: reduction of the span weight, elimination of negative reaction forces on the ballast spans supports, additionally it allowed for parallel execution of a number of elements to accelerate the working schedule. A normal C50/60 class concrete was used to cast the deck slab in the main span, however the possibility of using light-weight concrete was analysed in other bridges in future as part of the academic work [9].

The deck is suspended from two-armed 66,0m height towers by parallel strand cable-stays. The active cable anchorages are located under the deck in both concrete and steel sections. The strands pass continuously through the saddles on the pylons. The number of strands in the cable-stays varies from 24 to 37. The spacing of the anchorages in the deck is 10.0 m.

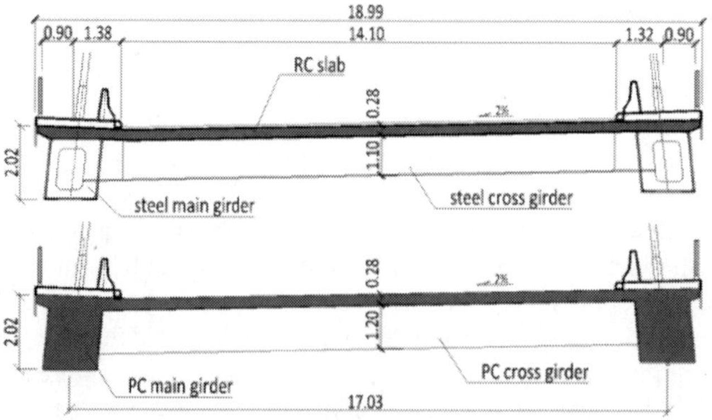

Fig. 2. Cross section of the main span (steel-concrete) and of the access span (prestressed concrete).

The bridge bearing system is a floating system with virtual fixed point in the middle of the main span. The deck is supported on the towers by neoprene bearings of relatively low rigidity, the supports on other piers are pot bearings. The function of a longitudinal fixed bearing is taken by cable-stays.

There are four concrete pylons (two for each structure). Each pylon includes a footing, a cylindrical column, a transverse girder, two arms and top bracing deep beam. The cylindrical column diameter is 9m, its height is 4,8 m. A transverse girder of trapezoidal shape, located directly on the column, is the most massive element of the bridge. It is post-tensioned with 24 19-strand cables. The pylon arm is a hollow reinforced concrete section up to 30 m height above the deck. The higher part is a solid section with cable-stay saddles installed inside. Each saddle is adjusted in space to the angle of cable-stays.

The foundations of the four pylons have been designed as square footings, their dimensions are 20,0x20,0x4,1m. Their bearing capacity is ensured by precast reinforced concrete piles of a 40x40 cm section. The length of piles is 9,0m, their bases are embedded in a layer of compacted fine sand.

The bridge has been designed according to both Eurocodes and former Polish Standards (although the investor only required the latter). The bridge was tested under the load model LM1 (α=1,0) and the A-class bridge load model from PN-S. The fatigue strength check was based on the FLM3 model.

3 Structural analysis

3.1 Global static analysis

Both static analysis and dimensioning were performed in SOFiSTiK the FEM-based computational environment [10-12]. The global three-dimensional computational model of the bridge (Fig. 3) was made of: 3890 Timoshenko beam elements with 6 nodal degrees of freedom (main girders, cross beams, piers, pylons), 28480 four-node Mindlin-Reissner shell elements (deck slab) and 64 cable elements (cable-stays). The number of nodes was 51 989, the number of dofs was 126 484. The support conditions were defined in the form of elastic springs with appropriate stiffness, assessed upon geotechnical calculation.

Fig. 3. Visualization of the global finite element model of the MD-7 bridge.

The global model was applied for static calculations in the final stage and in the construction stages. It was used to determine tensile forces introduced to cable-stays and deck displacements in the execution stages.

Additionally, local partial models were created to analyse structural details, e.g. cable-stay anchorages, post-tensioning cable anchorages, joints of steel and concrete sections, precast deck planks. A detailed model of a pylon was analyzed to investigate the erection process too.

3.2 Construction stage analysis

The concrete towers were erected using sliding scaffolding. The erection process was analysed by means of a detailed model incorporating shell finite elements (Fig. 4).

The construction principle was analogous to the cantilever method. Each segment introduced a load on the previous ones. The inclination of the pylon arms caused their bending. In order to minimize additional bending moments in the arms temporary struts were applied, but nevertheless it was necessary to calculate the arm deflection to obtain the designed geometry, introducing a precamber. The influence of concrete age on its elastic modulus was included in the analysis due to time of execution of a single segment lower than 3 days.

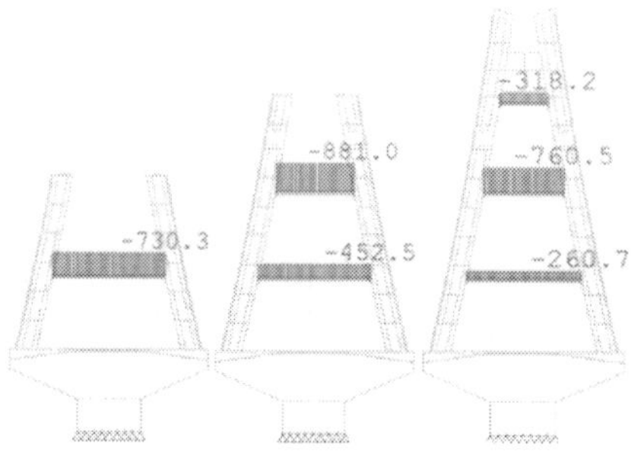

Fig. 4. Forces in temporary struts during in different construction stages of pylon.

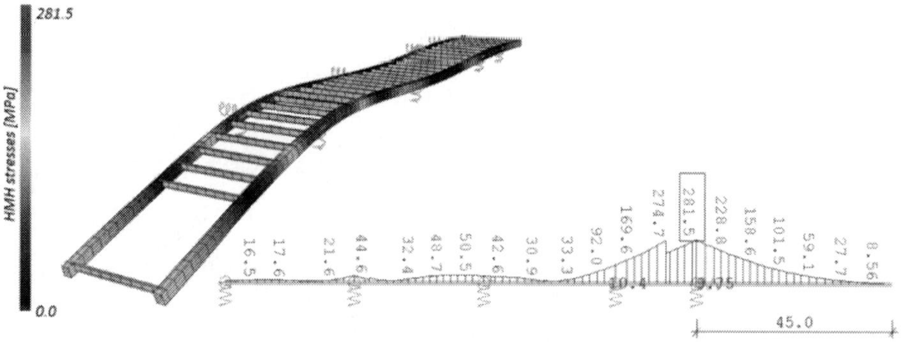

Fig. 5. Computational model for analysis of incremental launching of steel structure and a graph of HMH stresses in a steel box girder at the maximum cantilever overhang length.

The main span of the bridge is a steel-concrete composite ladder deck structure. The execution of steel girders was performed by means of incremental launching. In order to analyse it during this process the steel structure was isolated from the global bridge model and analysed with variable support conditions encountered during the launching (Fig. 5).

In this case, the concrete span had to be checked as well, because the assembly of steel segments was carried out on an access span already built. The steel structure weight and the working equipment load, e.g. cranes were to be considered in the global bridge analysis.

The access spans were erected on stationary scaffolds, span-by-span with the use of additional temporary supports. After their execution, on the deck of the bridge on one of the banks the steel structure was assembled and launched over the river. The incremental launching was carried out with the help of four temporary supports in the river. After joining the concrete and steel sections the process of concreting the deck slab in the main span was started. It was executed together with stay assembling and tensioning. The introduction tensile forces into the cable-stays caused progressive lifting of the deck from temporary supports. As a result, the performed calculations were non-linear due to boundary non-linearity. There was also geometrical non-linearity caused by cable sag due to its low tensile forces during construction stages.

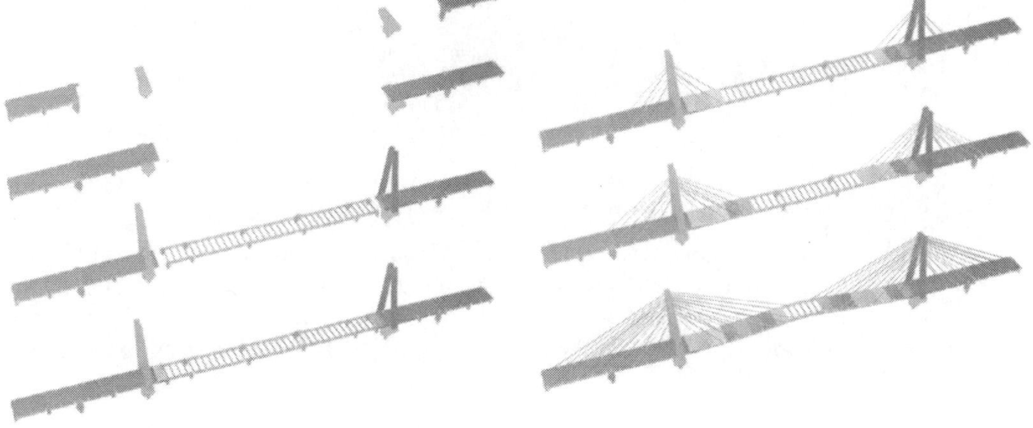

Fig. 6. Several of the modelled construction stages.

All erection stages have been considered in the global bridge finite element model (Fig. 6). The main advantage of a complex modelling, introducing shell elements for a deck slab is inherent consideration of a shear lag effect. Thus there is no need to use the effective width approach. This is important, because in every construction stage the effective width of a slab may be variable, to act strongly on the accuracy of the detection of displacements and stress distribution in the deck slab, especially when the main girders are located at the cross section edges. In this case it should be noted that the anchorage zone of cable-stays is not terminated in the steel girder, but it extends to the slab where tensile stresses are introduced due to point load character, to be considered during the slab reinforcement design

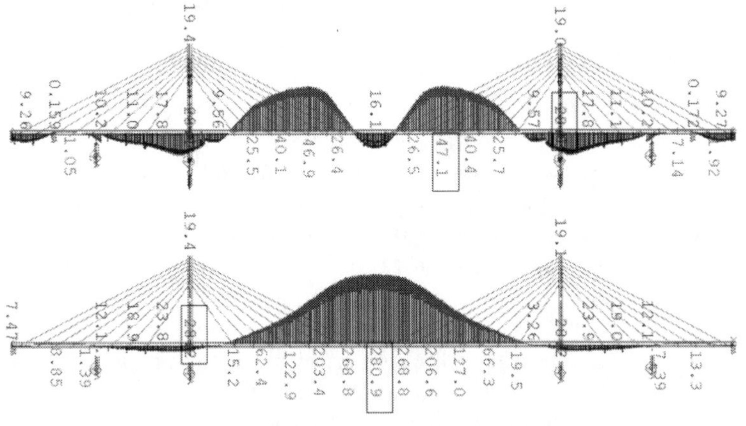

Fig. 7. Displacements of the deck before and after tensioning of cable stay no 7.

In the middle of the main span the vertical displacement amplitudes between the stage of casting concrete slab 10 m segment and tensioning of a cable-stay were approximately 300 mm (Fig. 7). Geodetic measurements were carried out in real-time during execution. They allowed to check the structural geometry, taking into account temperature influence. The measured deviations from the theoretical altitude in construction stages did not exceed 40 mm and final deviations after closing the execution works were less than 20mm.

3.3 Analysis of construction details

3.3.1 Joint between steel and concrete section

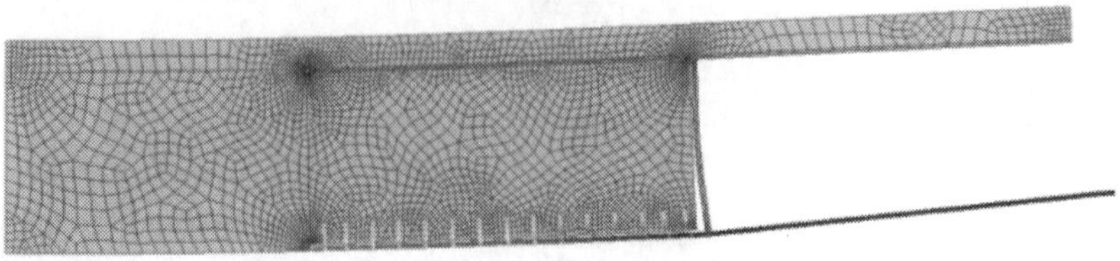

Fig. 8. Local plane model of joint between steel and concrete section.

A joint between steel and concrete sections was analysed in a plane stress model, focusing on shear studs flexibility. There are dead anchorages of post-tensioning cables located in the concrete part. Each strand was formed into an "onion" to smoothly develop the prestressing force into a section. In the plane model (Fig. 8) the prestressing force has been modelled as a surface load.

3.3.2 Cable-stays anchorage

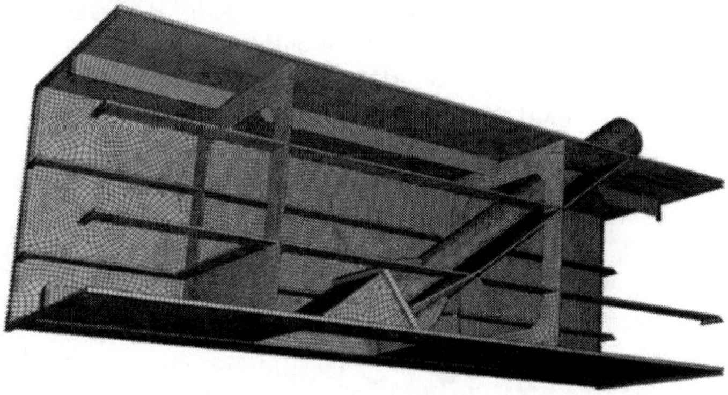

Fig. 9. Spatial local FEM model of cable-stay anchorage in steel structure (one web removed to improve visibility).

The analytical details, e.g. the anchorage of cable-stays in steel structure form a challenge to the designer. It is necessary to combine all design aspects, like ensuring load-bearing capacity, durability and maintenance. In order to meet all of these requirements, detailed models were created of cable-stays anchorages in steel structure (Fig. 9).

In the concrete section the anchorages of cable-stays have to be designed individually due to lack of systemic reinforcement solutions available, the aim was to avoid the problems similar to [13]. A confined concrete theory was applied to design a spiral reinforcement [14]. All anchorages were dimensioned for the breaking load of tendons, in accordance, assuming that the anchorage should not be the weakest element of a cable-stay. In the case of fatigue strength anchorages are designed to last longer than the cable itself. In order to conduct the fatigue checks in those untypical elements the hot-spot method was applied [15].

3.3.3 Reinforced concrete precast formwork planks

The use of precast formwork planks allows to efficiently design the steel structure cross girder. The planks were both slab formwork and lateral-torsional bracing of the cross beam top flange. The use of planks accelerated the execution time of the slab too. A single segment of 10,0m length was executed in less than 3 days. The precast filigree planks had a record breaking length of 5,0m. For this reason, a detailed finite element model was created. The plank was modelled by four-node shell elements, the rebars were beam elements. The concrete stiffness in the longitudinal direction was reduced due to cracking. Geometry of rebars was defined by means of imperfections corresponding to their buckling modes resulting from the eigenproblem of stability. The accidental state was also investigated, detecting a single welded joint of the truss and a top rebar damaged.

Before applying the precast planks in the construction, a test load was carried out. It confirmed bearing capacity calculated previously.

4 Concluding remarks

The design and construction project of the bridge MD7 was an interesting challenge to the entire design and executive team. The use of advanced finite element method modelling for global and local design cases allowed to obtain an efficient constriction and a precise assessment of displacements and forces in cable-stays. Thus it was made possible to execute the bridge on time encountering no major problems.

The applied FEM-based solutions are particularly useful in the case of non-typical structures with unusual solutions, not covered by any guidelines. A detailed local analysis allows to design efficient solutions characterized by high durability. In the case of MD7 bridge, measurements made during construction, acceptance tests in the form of trial load test similar to [16-17] and the first results read from the installed monitoring system resembling [18-26] confirm the assumptions made in structural design and modelling.

However it should be noted that the analysis complexity also leads to difficulties at the stage of interpreting the results and verifying the correctness of the assumed model. The experience in the design and modelling of other large and smaller bridge structures is invaluable while analysing a complex bridge.

References

1. K. Topolewicz, *Executive detailed design (replacement design) of MD-7 bridge*, (TOP PROJEKT, STRABAG 2015)
2. M. Miśkiewicz, Ł. Pyrzowski, K. Wilde, J. Chróścielewski: Numerical analysis and in situ tests of the Grot Rowecki Bridge in Warsaw. Adv. in Mech., 405-408 (2016)
3. J. Chróścielewski, M. Miśkiewicz, Ł. Pyrzowski, K. Wilde, Assessment of tensile forces in Sopot Forest Opera membrane by in situ measurements and iterative numerical strategy for inverse problem, Shell Str.: Th. & Appl., **3**, 499-502 (2014)
4. M. Miśkiewicz, K. Daszkiewicz, T. Ferenc, W. Witkowski, J. Chróścielewski, Experimental tests and numerical simulations of full scale composite sandwich segment of a foot-and cycle- bridge, Adv. in Mech, 401-404 (2016)
5. Ł. Pyrzowski, M. Miśkiewicz, J. Chróścielewski, "The effect of fishing basin construction on the behavior of a footbridge over the port channel," PMR 2017 S1 (93), **24**, 182-187 (2017)
6. M. Miśkiewicz, Ł. Pyrzowski, Influence of an applied bearing system on behavior of multi-span footbridge, Shell Str.: Th. & Appl., **4**, 433-436 (2018)
7. J. Chróścielewski, M. Miśkiewicz, Ł. Pyrzowski, B. Sobczyk, K. Wilde, A novel sandwich footbridge - Practical application of laminated composites in bridge design and in situ measurements of static response, Comp. Eng., **126**, 153-161 (2017)

8. A. Mariak, M. Miśkiewicz, B. Meronk, Ł. Pyrzowski, K. Wilde, Reference FEM model for SHM system of cable-stayed bridge in Rzeszów, Adv. in Mech. – Kleiber et al. (Eds), 383-387, (2016),
9. M. Kurpińska, T. Ferenc: Application of lightweight cement composite with foamed glass aggregate in shell structures, Shell Str.: Th. & Appl., **4**, 549-552 (2018)
10. J. Chróścielewski, W. Witkowski, Four-node semi-EAS element in six-field nonlinear theory of shells. Inter. J. for Num. Meth. In Eng., **68**, 1137-1179 (2006)
11. J. Chroscielewski, A. Sabik, B. Sobczyk, et al. Nonlinear FEM 2D failure onset prediction of composite shells based on 6-parameter shell theory, Thin-walled structures, **105**, 207-219
12. L. Mazurkiewicz, J. Malachowski, P. Baranowski, et al., Comparison of numerical testing methods in terms of impulse loading applied to structural elements, J. of Th. and Appl. Mech., **51** ,615-625 (2013)
13. J. Chróścielewski , M. Miśkiewicz , Ł. Pyrzowski , B. Sobczyk, Analysis of damage in the zone of tensioning cables anchorages of a bridge superstructure, using CDP ABAQUS material model, Arch. of Civil Eng., **63**, 85-97 (2017)
14. J.B Mander, M. J. N. Priestley, R. Park, Theoretical stress-strain model for confined concrete. J. of Str. Eng., **114**, 1804-1825 (1998)
15. A. Nussbaumer, L. Borges, L. Davaine, *Fatigue design of steel and composite structures* (Wiley & Sons, 2012)
16. M. Miśkiewicz, K. Makowska, Displacement measurements during load testing of railway arch bridge, SGEM2017, **17**, 257-264 (2017)
17. I-K. Fang, C-R. Chen, I-S. Chang, Field Static Load Test on Kao-Ping-Hsi Cable-Stayed Bridge, J. of Bridge Eng., **9**, 531-540 (2004)
18. A. Janowski, K. Nagrodzka-Godycka, J. Szulwic, P. Ziolkowski, Remote sensing and photogrammetry techniques in diagnostics of concrete structures, Comput. Concr., **18**, 405–420 (2016)
19. M. Miśkiewicz, B. Meronk, T. Brzozowski, K. Wilde, Monitoring system of the road embankment, Baltic J. of Roads and Bridge Eng., **12**, 218-224 (2018)
20. M. Miśkiewicz, O. Mitrosz, T. Brzozowski Preliminary field tests and long-term monitoring as a method of design risk mitigation: a case study of Gdańsk Deepwater Container Terminal, PMR **24**: 106-114 (2017)
21. M. Miśkiewicz, Ł. Pyrzowski, J. Chróścielewski, K. Wilde, Structural Health Monitoring of Composite Shell Footbridge for Its Design Validation, BGC Geomatics 228-233 (2016)
22. M. Miskiewicz, L. Pyrzowski, K. Wilde, Structural Health Monitoring System for Suspension Footbridge, BGC Geomatics, 321-325 (2017)
23. K. Wilde, M. Miśkiewicz, J. Chróścielewski, SHM System of the Roof Structure of Sports Arena „Olivia". Stru. Health Mon. 2013, **2**, 1745-1752, (2013)
24. W. Kaminski., K. Makowska, M. Miśkiewicz, J. Szulwic, K. Wilde, System of monitoring of the Forest Opera in Sopot structure and roofing, SGEM 2015, **2**, 471-482 (2015)
25. J. Hou, L. Jankowski, J. Ou, An online substructure identification method for local structural health monitoring, Smart Mat. and Str., **22** (2013)
26. G. Suwala, L. Jankowski, A model-free method for identification of mass modifications, Str. Control and Health Mon., **19**, 216-230 (2012)

Maturity testing of 3D printing concrete with inert microfiller

Szymon Skibicki[1], *Maria* Kaszyńska[1], and *Mateusz* Techman[1,*]

[1]West Pomeranian University of Technology, Faculty of Civil Engineering and Architecture, Piastów 50, 70-311 Szczecin, Poland

Abstract. The 3D Printing of cement composites is one of the fastest developing technologies of modern concrete. The 3D printing involves concretes with high amounts of microfillers. The study analyses the influence of curing conditions on the development of strength of concretes applicable in 3D printing. Ten mixes were tested in the study. In the studied cases cement constituted to 50% of the mass, while microfillers such as lime stone powder, kaolin, quartz powder and sand (up to 2 mm) constituted to the rest of the mass. Samples were cured for 7 days in exothermic conditions at 5°C, 20°C, 35°C. Standard mortar samples of 4x4x16 cm and cylinders with 46.5 mm and height of 35 mm that simulate the printed 3D path were made. The compressive strength was tested after 12h, 24h, 48h, 72h, 168 h and 28 days. Based on the acquired results the temperature development function was formulated and activation energy was determined. The results showed that the proposed method is useful in evaluation of printed concrete curing. It can be also used to determine the time of loading the wall which can speed up the process of constructing while maintaining degree of safety.

1 Introduction

A strong focus placed by various research facilities on 3D concrete printing has resulted in a development of many new research methods of this novel constructing system [1–4]. However, there are still no established Standards defining requirements for this technology. The elements created by additive manufacturing show small cross-section, which causes them to adopt the external temperature rapidly. As a results the elements can be highly susceptible to changes in strength development caused by low or high temperatures.

The influence of temperature on the mechanical properties of ordinary concrete is well researched [5,6]. The correlation between temperature of curing and mechanical properties is used daily in civil engineering (maturity method). This correlation allows to predict the strength of concrete, which results in shorter striking time or faster loading. Premature striking of the formwork was a cause of numerous structural failures and collapses [5]. The 3D printing technology is issued with a similar problem of strength prediction in structures. Applying well known and widely used maturity method allows to determine the loading time (i.e. with upper floors) even in structures constructed by additive manufacturing. Currently only one research facility tries to address the issue [7].

The research was conducted on concretes with inert microfillers applicable in 3D printing due to their rheological properties and composition. For proposed mixes the apparent activation energy, coefficients of strength improvement for different temperatures, cross-over effect and equivalent time were determined. The study was conducted on standard samples and model samples similar to printed ones [4,8].

2 Background

2.1 Maturity method

The maturity method is directed to determine mechanical properties of concrete based on the measurements of temperature during curing [5,9]. The maturity method was firstly used in 1949 by Nurse [10] and MacIntosha [11]. The studies were continued by Saul [12] in 1951. The determined function was based on the so-called maturity index (1). Initially used function did not reflect correctly the reality. As soon as in 1977 a research team have proposed an Arhennius equation for better projection of concrete curing. To concur the research a so called equivalent time was proposed (2). Many researchers (inter alia Carino [13], Kaszyńska [14]) proved that the equation (2) better reflects the maturity of concrete in structures than the Saul's formula (1).

The maturity method evolved with following years and have derived into different functions and methods for determining the mechanical properties of concrete [13,15–18]. Today the most popular version is included in the American Standard ASTM C1074 [19], which was proposed by Carino [13]. The method has a wide practical application [20]. The aforementioned equation from ASTM C1074 is seen as (3).

$$M = \sum_0^t (T - T_0) \cdot \Delta t \tag{1}$$

where:

M maturity index [°C-hours or °C-days],
T average concrete temperature during the time interval Δt [°C],
T_0 datum temperature (usually assumed to be -10°C),
t elapsed time [hours or days],
Δt time interval [hours or days].

$$t_e = \sum_0^t exp\left[\frac{-E_a}{R} \cdot \left(\frac{1}{T} - \frac{1}{T_r}\right)\right] \cdot \Delta t \tag{2}$$

where:

t_e the equivalent age at the reference temperature,
E_a apparent activation energy $\left[\frac{J}{mol}\right]$,
R universal gas constant, 8.314 $\left[\frac{J}{mol \cdot K}\right]$,
T average absolute temperature of the concrete during interval Δt [K],
T_r absolute reference temperature [K].

$$S = S_u \cdot \frac{k \cdot (t - t_0)}{1 + k \cdot (t - t_0)} \tag{3}$$

where:

S strength at age t,
S_u limiting strength,
k rate constant, $[1/h]$ t_0 –age at start of strength development,
t elapsed time.

2.2 3D printing – size specimen problem

There is only a single study that tried to evaluate the applicability of maturity method for optimization of 3D printing [7]. However, it is an overview article, that does not present any computations of coefficients found in (1)-(3).

Currently there are no Standards for applicability of the mixes for 3D printing. Many research facilities [4,21–24] proposed own methods that require a printing device. The methods seem well-thought and clearly evaluate the applicability of the mix. However, they are costly and time consuming.

Several research facilities have proposed a framework for estimation of mix feasibility in 3D printing. A research team [8] proposed a test on a Φ60x35 mm samples in a specially designed test bench. The samples are loaded with several weights that simulate additional printed layers. A research team [4] proposed a cylinder stability test, which is conducted on a Φ60x40 mm samples with a 5.5 kg applied load. The second method also determines the deformation of the sample. The size of the deformation determines the readiness of the mix for printing. A research team [2] determines the usability of the mix based on a slump flow [21]. Other teams [8,25] determine the usability with standard rheological tests conducted in a set period of time.

3 Materials and experimental procedure

3.1 Mix compositions

The mixes were made using 50% of CEM I 52.5 R (Rapid Portland Cement, according to European Standard) and 50% consisting of various microfillers including quartz powder (QP), limestone powder (LP), kaolin (K) and fine aggregate (FA). SP was added in percentage of cement mass to reach required consistency according to [21,26]. Those 4 fillers were added in various combinations. Designed mixes are presented in Table 1. The water/cement ratio was set to 0.42. The mix consistency was regulated with a superplasticizer.

It should be noticed that these type of mixes are used in French research facilities [8,27] to 3D printing. The „LP/K" mix and its properties was presented in other studies [8] as a applicable 3D printing mix. The authors have confirmed the properties of the above mentioned mixes on their own.

Table 1. Mix compositions.

Mix	CEM I 52.5R	QP	LP	K	FA	W/C	%SP
	%					-	%
QP	X	X	-	-	-	0.42	0.80%
LP		-	X	-	-	0.42	0.68%
K		-	-	X	-	0.49	2.16%
FA		-	-	-	X	0.34	0.00%
QP/LP		X	X	-	-	0.42	0.80%
QP/K		X	-	X	-	0.42	1.08%
QP/FA		X	-	-	X	0.42	0.17%
LP/K		-	X	X	-	0.43	2.16%
LP/FA		-	X	-	X	0.42	0.00%
K/FA		-	-	X	X	0.42	0.34%

3.2 Experimental procedure

After mixing, the concrete samples were placed in a climatic chamber and stored for 7 days in 3 different temperatures of 5°C, 20°C, 35°C. After the first stage the samples were stored in a laboratory conditions at 20°C ± 2°. Two types of samples were prepared including 40x40x160 mm standard samples and cylindrical samples with diameter of 46.5 mm and height of 35 mm that simulated a shape of printed layers.

Table 2. Parameters of maturity function.

T [°C]	5	20	35	T_0	E/R	E	R^2
Mix	k [-]			[°C]	[°K]	[J/mol]	
LP_K	0.035	0.076	2.035	2.16	5231	43 492	0.98
LP_FA	0.036	0.065	2.144	1.19	4766	39 623	0.96
QP_LP	0.032	0.054	0.355	-2.21	3943	32 779	0.97
K_FA	0.026	0.054	0.144	-1.23	4373	36 356	1.00
QP_K	0.027	0.026	0.093	2.48	4465	37 120	0.70
QP_FA	0.026	0.067	0.304	-3.62	4206	34 967	0.98
FA	0.036	0.114	0.045	-0.42	5124	42 603	0.98
LP	0.029	0.059	0.315	-3.91	3903	32 448	1.00
K	0.021	0.024	0.047	-0.07	3993	33 202	0.80
QP	0.028	0.045	0.550	-1.20	4066	33 806	0.94

4 Test results

4.1 Compressive strength development

Compressive strength of concrete samples was tested after 12, 24, 48, 72, 168 and 672 hours. The test results are presented in Fig.1. There are the mean values calculated based on 4 measurements (cube specimens). Coefficient of Variation for the analysed specimen was lower than 4.58%. The highest compressive strength was reached by QP, QP/FA, QP/LP and LP/FA, while the lowest by K concrete.

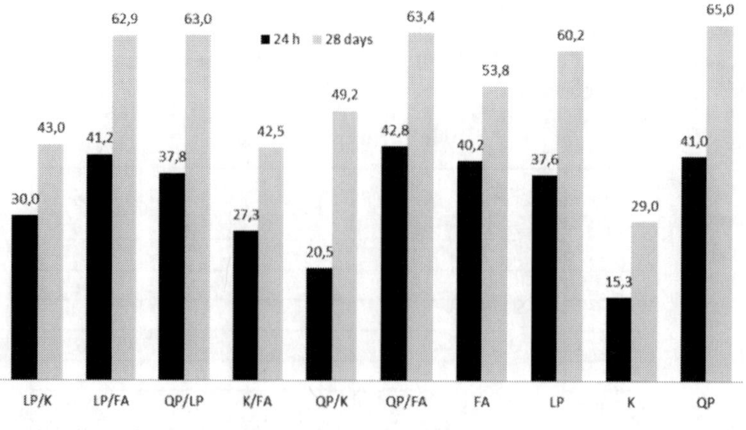

Fig. 1. Compressive strength of concretes after 24 h and 28 days.

The analysis determined the cube/cylinder compressive strength ratio ($f_{c.cube}/f_{c,cylinder}$). Values of compressive strength were obtained at 28 day-age after curing at 20°C. Mean value of $f_{c.cube}/f_{c,cylinder}$ is 1.74 with coefficient of variation of 3.93 %. Differences between cube and cylinder specimen are noticeable. Cylindrical specimens exhibit lower compressive strength.

The study focuses on 4 mixes (LP/K, QP/FA, QP/LP and LP/FA). Figure 2 and 3 present the strength development curves of selected mixes.

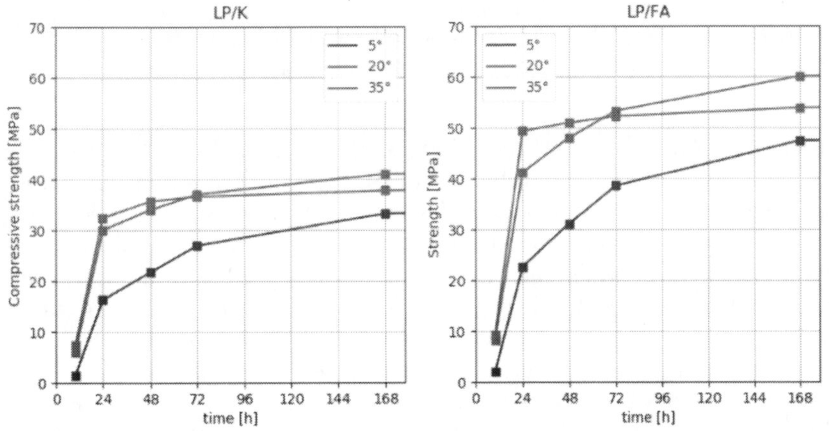

Fig. 2. Compressive strength development of LP/K and LP/FA cured in different temperatures.

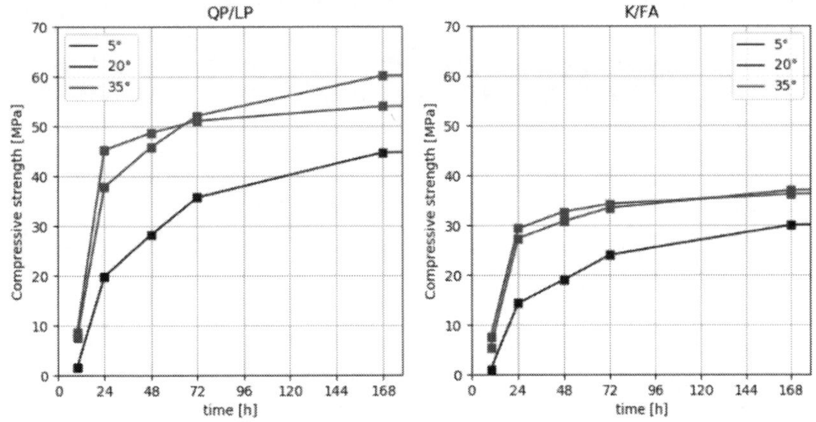

Fig. 3. Compressive strength development of QP/LP and K/FA cured in different temperatures.

4.3 Apparent activation energy

Apparent activation energy (E_a) was established in accordance with A1.1.8.2 of ASTM C1074-11. Table 2 contains the values of k (rate constant) coefficient, T_0 (datum temperature), E_a and R^2 (determination coefficient). The results show that E_a ranges from 34 to 44kJ/mol. The values of E_a between 34-44 kJ/mol are corresponding for this type of cement binder [28].

4.4 Equivalent time

For selected mixes an equivalent time was calculated based on the results in Table 2 and equation (4) (transformed formula (3)). Figure 4 shows equivalent time which is needed to reach 10%, 20%, 40%, 80% of ultimate compressive strength of concrete cured at 20°C ($S_{20°}/S_{u,20°}$). Figure 4 shows the results for selected mixes cured at different temperatures.

$$t_{e,5°/35°} = t_0 - \frac{S_{20°}}{k \cdot (S_{20°} - S_u)} \quad (4)$$

where:

$S_{20°}$ strength which respond 0,1 / 0.2 / 0.4 or 0.8 of ultimate compressive strength in 20°C,
S_u compressive strength in 5° or 35°,
k rate constant in 5° or 35°,
$[1/h]$ t_0 tiem at start of strength development in temperature 5° or 35°,
$t_{e,5°}$ / $t_{e,35°}$ equivalent time which is needed to reach strength threshold ($S/S_{u,20°}$ = *0.1 / 0.2 / 0.4 / 0.8*) in temperature 5°C or 35°C.

The difference in equivalent time is important in the early stages of concrete curing when the concrete has not yet developed full strength. In such a case, prematurely loaded structure might collapse [5]. During the first stage of curing up to S/S_u=0.4 the development rate of strength is slower in low temperature (5°C) causing significant differences in t_e. For example for mix K/FA difference at S/S_u=0.2 was 6.3h. This time seems short but in additive manufacturing even 6h is significant. In later stages (after S/S_u=0.4) the differences are lowered, but are not so important considering relatively high concrete strength. Another issue occurring during curing is a cross-over effect, described below.

Studied mixes have exhibited a visible cross-over effect (Fig.1, Fig.2). Strength development rate in different temperatures changes during curing which was already studied elsewhere [29,30]. Concretes with high amounts of cement or with rapid cements are strongly susceptible to the effect. Additionally, the higher temperature the more significant the differences in a final strength of concrete. In studied mixes the cross-over effect was visible between 30-90 hours. For mixes with fine aggregate (FA) and with kaolin (K) with different w/c ratio the effect has not occurred at all.

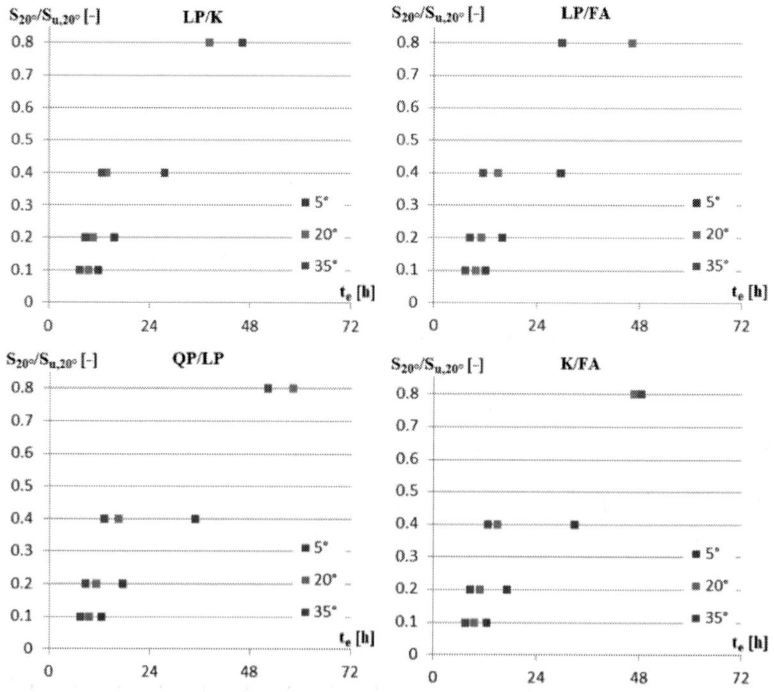

Fig. 4. Equivalent time for different compressive strength thresholds.

5 Conclusion

The conducted study has verified the composition of mixes with mineral additives suitable fo 3D printing. Highest compressive strength was obtained by QP, QP/FA, QP/LP and LP/FA concretes with addition of quartz and limestone powder, while the lowest by K concrete with kaolin.

The study focuses on the influence of temperature on the development of compressive strength. Two temperatures simulating conditions in spring-autumn (T=5°C) and summer (T=35°C) were taken into consideration. The influence of temperature was analyzed using maturity method. The study allowed to determine the values of activation energy for the mixes with mineral fillers and rapid Portland cement CEM I 52.5 R. This allowed to calculated the equivalent curing time and evaluation of compressive strength for different curing temperatures. Study has shown that in the case of 8 concretes the so-called cross-over effect occurred. The phenomenon occurs as decline in compressive strength development rate after 3 days in concretes cured in higher temperatures.

The study has shown that compressive strength of cylindrical samples that simulate printed layers is significantly lower than compressive strength of standard cubical samples 10x10x10 cm.

References

1. T. Wangler, E. Lloret, L. Reiter, N. Hack, F. Gramazio, M. Kohler, M. Bernhard, B. Dillenburger, J. Buchli, N. Roussel, *et al.*, Digital Concrete: Opportunities and Challenges. RILEM Letters, **1,** 67-75 (2016)
2. F. Bos, R. Wolfs, Z. Ahmed, T. Salet, Additive manufacturing of concrete in construction: Potentials and challenges of 3D concrete printing. Virtual and Physical Prototyping, **11**, 1-17 (2016)
3. Y. Tay, B. Panda, S. Paul, N. Noor Mohamed, M. Tan, K. Leong, 3D printing trends in building and construction industry: A review. Virtual and Physical Prototyping, **12**, 1-17 (2017)
4. A. Kazemian, X. Yuan, E. Cochran, B. Khoshnevis, Cementitious materials for construction-scale 3D printing: Laboratory testing of fresh printing mixture. Construction and Building Materials, **145**, 639-647 (2017)
5. N. Carino, H. Lew, The Maturity Method: From Theory to Application. In *Structures 2001*; Chang, P., Ed.: American Society of Civil Engineers: Reston, VA, 1-19 (2001)
6. S. Skibicki, Optimization of Cost of Building with Concrete Slabs Based on the Maturity Method. IOP Conf. Ser.: Mater. Sci. Eng., **245**, 1-11 (2017)
7. R.J.M. Wolfs. *3d printind of concrete structures (graduation thesis)*: Eindhoven University of Technology, 1-110 (2015)
8. A. Perrot, D. Rangeard, A. Pierre, Structural built-up of cement-based materials used for 3D-printing extrusion techniques. Mater Struct, **49**, 1213-1220 (2016)
9. Mariak A., Wilde K., Wyznaczanie wytrzymałości betonu na podstawie funkcji dojrzałości wg amerykańskiej normy ASTM C1074-11. Materiały Budowlane, **4,** 68-71 (2015)
10. R. Nurse, Steam curing of concrete. Mag Concrete Res, **2**, 79-88 (1949)
11. J. McIntosh, Electrical curing of concrete. Mag Concrete Res, **1**, 21-28 (1949)
12. A. Saul, Principles underlying the steam curing of concrete at atmospheric pressure. Mag Concrete Res, **2**, 127-140 (1951)

13. R. Tank, R. Carino, Rate Constant Functions for Strength Development of Concrete. ACI Materials Journal, **88 (1)**, 74-83 (1991)
14. M. Kaszynska, Early age properties of high-strength/high-performance concrete. Cement and Concrete Composites, **24**, 253-261 (2002)
15. H. Freiesleben, P. Pedersen, Curing of Concrete Structures. CEB Information Bulletin, **166**, 1-42 (1985)
16. J. Zhang, D. Cusson, P. Monteiro, J. Harvey, New perspectives on maturity method and approach for high performance concrete applications. Cement and Concrete Research, **38**, 1436-1446 (2008)
17. Knudsen T., The dispersion model for hydration of Portland cement: I. General concepts. Cement and Concrete Research, **14**, 622-630 (1984)
18. P. Fjellström, J.-E. Jonasson, M. Emborg, H. and Hedlund, Model for Concrete Strength Development Including Strength Reduction at Elevated Temperatures. Nordic Concrete Research, **45/1**, 25-44 (2012)
19. *ASTM C1074-11. Standard Practice for Estimating Concrete Strength by the Maturity Method.*
20. J. Myers. *The Use of Maturity Methods as a Quality Control Tool for High Performance Concrete Bridge Decks*: PCI / FHWA / FIB International Symposium on High Performance Concrete, 1-18 (2000)
21. G. Ma, Z. Li, L. Wang, Printable properties of cementitious material containing copper tailings for extrusion based 3D printing. Construction and Building Materials, **162**, 613-627 (2018)
22. T. Le, S. Austin, S. Lim, R. Buswell, R. Law, A.G.F. Gibb, T. Thorpe, Hardened properties of high-performance printing concrete. Cement and Concrete Research, **42**, 558-566 (2012)
23. E. Secrieru, S. Fataei, C. Schröfl, V. Mechtcherine, Study on concrete pumpability combining different laboratory tools and linkage to rheology. Construction and Building Materials, **144**, 451-461 (2017)
24. Z. Malaeb, H. Hachem, A. Tourbah, T. Maalouf, El Zarwi N., Hamzeh F., 3d concrete printing: Machine and mix design. International Journal of Civil Engineering and Technology (IJCIET), **6**, 14-22 (2015)
25. N. Roussel, F. Cussigh, Distinct-layer casting of SCC: The mechanical consequences of thixotropy. Cement and Concrete Research, **38**, 624-632 (2008)
26. G. Ma, L. Wang, A critical review of preparation design and workability measurement of concrete material for largescale 3D printing. Front. Struct. Civ. Eng., **22**, 382-400 (2017)
27. C. Gosselin, R. Duballet, P. Roux, N. Gaudillière, J. Dirrenberger, P. Morel, Large-scale 3D printing of ultra-high performance concrete – a new processing route for architects and builders. Materials & Design, **100**, 102-109 (2016)
28. Pinto R.C.A., Schindler A.K., Unified modeling of setting and strength development. Cement and Concrete Research, **40**, 58-65 (2010)
29. Wade, S.A.; Schindler, A.K.; Barnes, R.W.; Nixon, J.M., Eds. *Maturity Method Report No. 1*: Highway Research Center and Department of Civil Engineering at Auburn University, 1-307 (2006)
30. Brooks A.G., Schindler A.K., Barnes R.W., Maturity Method Evaluated for Various Cementitious Materials. Journal of Mat. in Civil Engineering, **19**, 1017-1025 (2007)

Comparison of calculation models' estimates with actual measured autogenous shrinkage in High-Performance Cement Composites

Adam Zieliński[1,*], *Karol* Federowicz[1], and *Maria* Kaszyńska[1]

[1]West Pomeranian University of Technology, Faculty of Civil Engineering and Architecture, al. Piastów 50, Szczecin, Poland

Abstract. There are several calculation models for estimation of shrinkage in concrete. The most popular ones are: ACI-209R-92 method, Bažant-Baweja model (B3), Gardner'a-Lockman'a model (GL 2000), CEB-FIP (CEB MC90-99), Spanish model (EHE), EuroCode method (EC-2) from 2004 and Tazawa–Miyazawa model (JSCE). Majority of aforementioned models do not include the deformation caused by the autogenous shrinkage. The methods assume the total deformation as a result of drying shrinkage action. Additionally, a non-standard method was proposed by Jonasson and Hedlund. However it only is used to determine the autogenous shrinkage of High-Performance Concretes (HPC). Tazawa-Miyazawa model and Jonasson-Hedlund method are based on the proportions and material properties such as water/cement ratio, water/binder ratio, type of the cement, or start of setting time. In contrary, the CEB MC90-99 and EC-2 models correlate the autogenous shrinkage with the overall compressive strength of concrete. The article presents the analysis of various calculation models in comparison to actual measurements. The autogenous shrinkage was measured in four cement composites starting at casting up till 28 days of curing.

1 Introduction

The first measurement of autogenous deformations have been made by La Chatelier in 1900. The researcher has described in his research a phenomenon occurring from the beginning of setting process up till later stages of cement hardening [1]. The research by Davis [2] and Lyman [3] in the 1940s have showed that a 5 year old concretes exhibit an autogenous deformation of 50-100 µm/m. Compared to thermal deformation and drying shrinkage the value of autogenous shrinkage appeared rather insignificant. Further studies of autogenous effects were omitted [4]. According to others researchers it can be assumed, that the total deformation of traditional concrete after 28 days is about 500 um/m, which of

* Corresponding author: adam.zielinski@zut.edu.pl

autogenous shrinkage is about 80 um/m. On the other hand we have high-performance concrete, where autogenous shrinkage can reach over 300 um/m when drying shrinkage is about 150 um/m. Total shrinkage of HPC is lower in comparison to traditional concrete. These analysis and shrinkage values are generalized and it should be analyzed due to composition of cement composites.

The breakthrough in the approach to autogenous shrinkage appeared after 1980s. The implementation of new generation of concretes and development of concrete technology created new issues that had to be addressed. Invention of chemical admixtures which reduce the water/cement ratio and implementation of reactive mineral additives, including silica fume launched the rise of High-Performance Concretes (HPC). Several factors have influenced the increase of the general value of shrinkage in concretes. For starters, use of reactive additives produced more compact cement matrix. With higher amounts of binder there was a noticeable increase of the chemical shrinkage, drying effect and autogenous shrinkage [5].

Dynamic implementation of HPC in real life has posed a necessity for scientific studies on this new technology. In mid-80s, despite strict control over water loss and reduction of temperature changes during curing there was a visible increase of cracking in constructed elements. Laboratory tests showed that rapidly developing autogenous deformations are the cause of this phenomenon [6].

There are numerous studies in the subject of autogenous deformations. Several methods were proposed to reduce the negative effect of autogenous shrinkage. Those include use of Shrinkage Reducing Admixtures (SRA), internal curing [7] or dispersed micro-reinforcement.

Several calculation models have been proposed that utilize the material proportions and compressive strength to predict the value of shrinkage.

In the paper the authors try to analyze the existing models and investigate the correctness of calculated estimations based on laboratory tests. The conducted tests proved some shortages in the models. It is necessary to include the amount of aggregate as a factor influencing autogenous shrinkage.

2 Mixture proportions and specimen

The study was performed on two High-Performance Self Consolidating Concretes with coarse, natural aggregate (C-3, C-4) and two cement composites with fine aggregate (C-1, C-2). The water/cement ratio was set to 0.28 (C-2, C-3), 0.34 (C-4) and 0.42 (C-1). C-1 and C-2 mixes contained high amount of cement and reduced amount of aggregate.

All of the mixes were made using Rapid Portland Cement CEM I 42.5 R, fly ash, silica fume and superplasticizer. The ratio of mineral additives to cement was constant. To obtain proper workability and self-consolidating properties the superplasticizer was dosed between 0.5% to 2.5% of cement mass. The mix design are visible in Table 1.

The autogenous deformation was measured on three 150x35x1150 mm samples. Displacement was measured from the casting time up till 28 days with 500 s intervals.

Compressive strength for the composites was different. After 28 days, the compressive strength on 10x10x10 cm specimens was: C-1: f_{cm} = 71 MPa; C-2: f_{cm} = 102,9 MPa; C-3: f_{cm} = 100,5 MPa and C-4: f_{cm} = 99,6 MPa.

Table 1. Mix design.

Mix	w/c	Cement [kg/m^3]	Mineral additives [kg/m^3]		Water [kg/m^3]	Superplasticizer [kg/m^3]	Aggregate [kg/m^3]	
							Natural	
			Fly Ash	Silica Fume			0 ÷ 2	2 ÷ 8
C-1	0,42	725	116	61	305	3,45	1004	-
C-2	0,28	794	127	67	222	18,0	1100	-
C-3	0,28	458	73,3	39	128	12	636	1092
C-4	0,34	450	72	38	155	8,0	624	1072

3 Calculation model characteristics

Four models were chosen for analysis, two based on component proportion and two based on compressive strength of concrete (the algorithms of all analyzed models are in Table 2).

Table 2. Autogenous shrinkage calculation models.

Tazawa and Miyazawa [8]

$$\varepsilon_c(t) = \gamma \cdot \varepsilon_{c0}(w/b) \cdot \beta(t) \cdot 10^{-6}$$

where:

$\varepsilon_c(t)$ – autogenous shrinkage at age „t",

γ – cement type coefficient:

$\gamma = 1{,}0$ – Portland cement; $\gamma = 0{,}6$ – belit cement

$\varepsilon_{c0}(w/b)$ – total autogenous shrinkage:

dla $0{,}2 \le w/b \le 0{,}50$ => $\varepsilon_{c0}(w/b) = 3070 \cdot exp\left[-7{,}2 \cdot \left(\frac{w}{b}\right)\right]$

dla $0{,}5 \le w/b$ => $\varepsilon_{c0}(w/b) = 80$

$\beta(t)$ – autogenous shrinkage development coefficient:

$$\beta(t) = 1 - exp[-a(t - t_0)^c]$$

where:

w/b – water/binder ratio,

t – concrete age [days],

t_0 – bonding initial time [days],

a, c – constants depending on w/c:

w/c	a	c
0,20	1,2	0,4
0,23	1,5	0,4
0,30	0,6	0,5
0,40	0,1	0,7
> 0,50	0,03	0,8

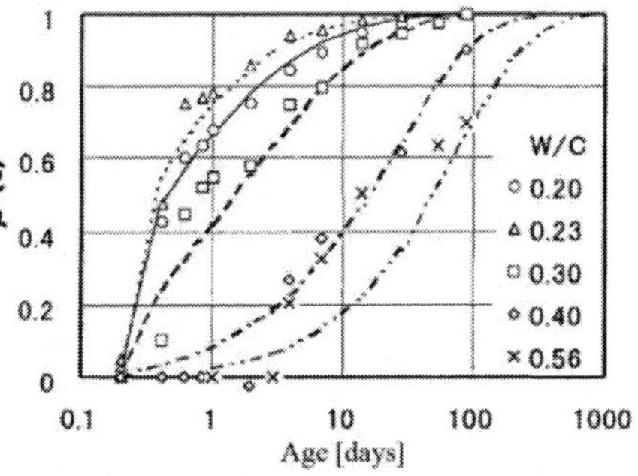

Jonasson-Hedlund [9]	$$\varepsilon_{as}(t) = [-0{,}65 + 1{,}3(w/b)] \cdot 10^{-3} \cdot exp\left[-\left(\frac{5}{t - t_{start}}\right)^{0,3}\right]$$ where: $\varepsilon_{as}(t)$ – autogenous shrinkage at age „t", t_{start} – starting age of autogenous shrinkage development (after 1 day)
Model CEB MC90-99 [10]	$$\varepsilon_{cas}(t) = -\alpha_{as}\left(\frac{f_{cm28}/10}{6 + f_{cm28}/10}\right)^{2,5} \cdot \left[1 - exp\left(-0{,}2\sqrt{t}\right)\right]$$ where: f_{cm28} – average compressive strength of concrete on day 28 (cylinder), [MPa], α_{as} – coefficient describing the type of cement: $\alpha_{as} = 800$ - CEM „L"; $\alpha_{as} = 700$ - CEM „N"/„R"; $\alpha_{as} = 600$ - CEM „R"
Model EC-2 (EN 1992-1-1) [11]	$$\varepsilon_{ca}(t) = \beta_{as}(t) \cdot \varepsilon_{ca}(\infty)$$ where: $\varepsilon_{ca}(\infty)$ - total autogenous shrinkage: $\varepsilon_{ca}(\infty) = 2{,}5 \cdot (f_{ck} - 10) \cdot 10^{-6}$ $\beta_{as}(t)$ – shrinkage development rate coefficient: $$\beta_{as}(t) = 1 - e^{(-0,2\sqrt{t})}$$

The Jonasson-Handlund method [9] that takes into account water/binder ratio can be only used for concretes with w/b≤ 0,4 and compressive strength of fcm ≥ 80 MPa. Tazawa-Miyazawa model can be used for concretes with w/b between 0,2 ≤ w/b ≤ 0,56, standard aggregate volume and surrounding temperature of t = 20 – 60°C.

The CEB MC90-99 and EC 2 models assume that the autogenous shrinkage depends mainly on concrete's compressive strength after 28 days.

4 Analysis of effectiveness of models for evaluation of autogenous shrinkage

Every empirical model determined the values of autogenous shrinkage in a set period. Figures 1 and 2 show the correlation of calculated values with actual measurement of autogenous shrinkage in C-1 and C-2 concretes.

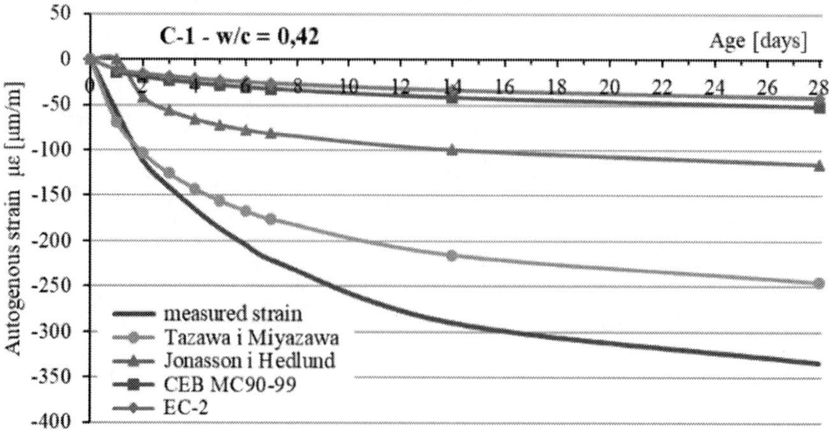

Fig. 1. Comparison of measured autogenous shrinkage of C-1 composite with calculated values.

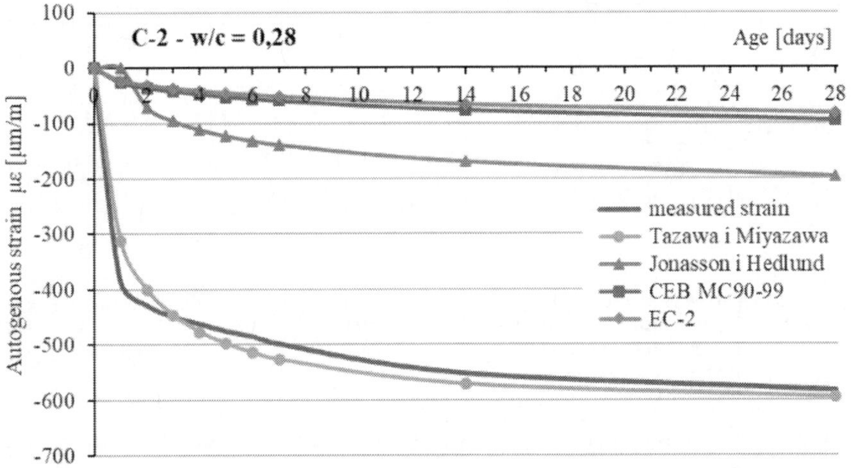

Fig. 2. Comparison of measured autogenous shrinkage of C-2 composite with calculated values.

Most approximate results were acquired by Tazawa-Miyazawa model. The comparison was most compatible for C-2 concrete with the lowest w/c ratio 0.28 and highest cement content. At higher water/cement ratio the model prognoses lower values of autogenous shrinkage. Model proposed by Jonnason and Hedlund understates the autogenous shrinkage due to its applicability only after a whole day. Compressive strength based models of CEB MC90-99 (CEB-FIP model) and EC-2 strongly understate the values of autogenous shrinkage. In case of the C-2 composite with fine aggregate, the calculated value constitutes to only 20% of measured value.

Figure 3 and 4 show the correlation of calculation models with measures of autogenous shrinkage in C-3 and C-4 composites. Figure 5 shows the comparison between model for all studied mixes.

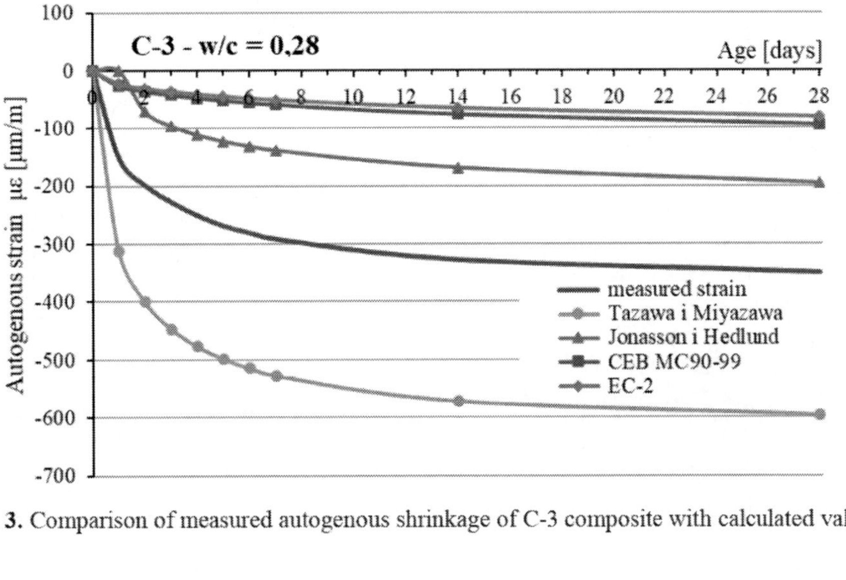

Fig. 3. Comparison of measured autogenous shrinkage of C-3 composite with calculated values.

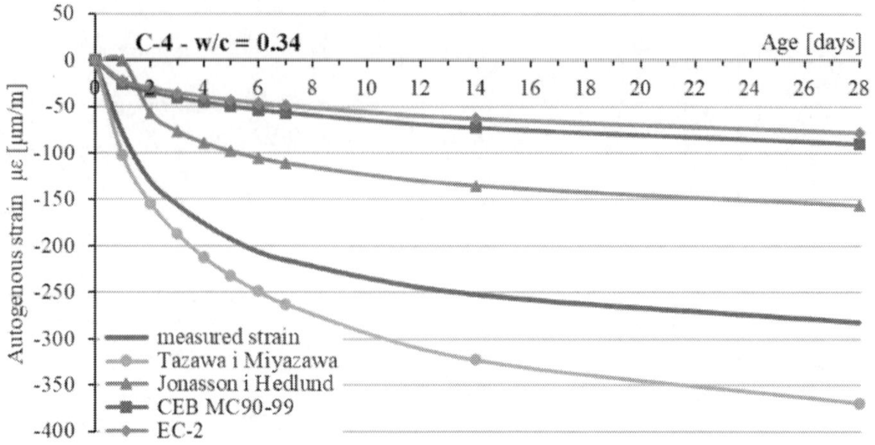

Fig. 4. Comparison of measured autogenous shrinkage of C-4 with calculated values.

The Tazawa-Miyazawa model does not account the amount and type of aggregate. This is visible considering the values calculated for C-2 and C-3 composites. The calculations for both mixes resulted in exact values of autogenous shrinkage. However, increasing the amount of aggregate reduced the overall autogenous shrinkage in C-3 composite.

The Jonnason-Hedlund model understates the values of autogenous shrinkage for C-2 and C-3. The model takes into account only shrinkage that occurs after a day. For C-2 and C-3 composites, the values of measured shrinkage in the first 24 h reached 50% of overall value, proving the method to be non-appropriable. Only after adding the values from the first 24 h, the model provides a correct representation of autogenous shrinkage.

The increase of the w/c ratio while retaining constant aggregate volume (C-3 and C-4) causes reduction of autogenous shrinkage and increase the correlation between CEB MC and EC-2 and measured values. With a lowered w/c ratio of 0.34 the compressive strength models understate the shrinkage almost three-folds.

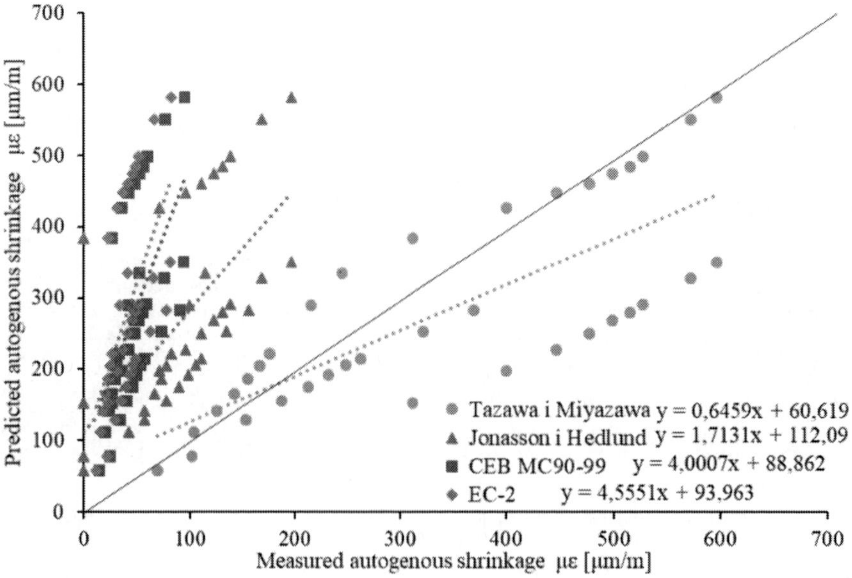

Fig. 5. Correlation of calculated and measured autogenous shrinkage after 1 - 7, 14, 28 days.

The plots of analyzed development of autogenous shrinkage in cement composites are arithmetic mean of 3 test specimens. On 28 day, the standard deviation was: for C-1: σ = 6,2 μm/m, for C-2: σ = 30 μm/m, for C-3: σ = 15 μm/m, for C-4: σ = 12 μm/m. In this situation, we can observe a large convergence of the registered autogenous strain on each of the three individual concrete specimens.

The increase of the w/c ratio above 0.34 increases the models-to-measurement correlation. However, the Tazaway-Miyazawa model overstates the values while other model understate them. In case of the C-1 mix, with the highest w/c ratio, high cement content and lowered volume of aggregate, rapidly developing shrinkage was understated by all models.

Highest models-to-measurement correlation was acquired for C-2 composite with lowest w/c. The cement composite C-2 has obtained the highest autogenous shrinkage, which was about 580 um/m after 28 day. Such a large strain was imposed by: low water/cement ratio, high content of cement and silica fume, lack of thick non-deformation natural aggregate.

The presented analysis shows that the value of autogenous shrinkage evaluated by CEM-MC and EC 2 methods is significantly understated. This is concurring with other studies [8]. Long et al. noticed that concretes with higher w/c ratio, thus lower autogenous shrinkage exhibit higher correlation with calculation models, especially the Japanese model [12].

5 Conclusions

Following conclusions can be drawn:

- The highest correlation was acquired for Tazawa-Miyazawa model. It was particularly precise for concretes with lower w/c and high content of cement. The model does not account for amount and type of aggregate;
- The Jonnason-Hedlund model strongly understates the autogenous shrinkage, due to being applied only after first 24 hours;
- The CEB MC90-99 and EC-2 models understate the actual value of autogenous shrinkage the lower the water/cement ratio;
- None of the models takes into account the influence of cement/aggregate ratio on predicted values of autogenous shrinkage.

References

1. Le Chatelier H., Sur les changements de volume qui accompagnent le durcissement des ciments, Bulletin de la Société pour l'Encouragement Industriel Nationale, **5** (1900)
2. Davis H.E. (ed.). *Autogenous volume change of concrete,* proc. ASTM **40**, 1103-1110 (1940)
3. Lyman C.G. *Growth and movement in Portland cement concrete,* Oxfort University Press (1934)
4. Jensen O.M. Hansen P.F., Cement and Concrete Research, **31(12)**, 1859–1865 (2001)
5. Aïtcin P.C. *High-performance concrete,* E & FN SPON (2007)
6. Palliere A.M., Buil M., Serrano J.J. Materials Science to Construction Materials Engineering, **3**, 990–997 (1987)

7. Kaszynska M., Zielinski A., Procedia Engineering, **108**, 608–615 (2015)
8. Tazawa E., Miyazawa S. Concrete Library of JSCE, **31**, 371–380 (1997)
9. Jonasson J.E., Hedlund H. (ed.). An engineering model for creep and shrinkage in high performance concrete, RILEM Proceedings PRO 17, Shrinkage of Concete, 507-529 (2000)
10. ACI 209R-92: Prediction of Creep, Shrinkage, and Temperature Effects in Concrete Structures
11. EN 1992-1-1:2008 Eurocode 2: Design of concrete structures - Part 1- General rules and rules for buildings
12. Long W.J., Khayat K.H., Xing F. The Open Civil Engineering Journal, 5, pp.116–123 (2011)

On the influence of the acceleration recording time on the calculation of impact severity indexes

Łukasz Pachocki[1,*], *Dawid* Bruski[1], *Stanisław* Burzyński[1], *Jacek* Chróścielewski[1], *Krzysztof* Wilde[1], and *Wojciech* Witkowski[1]

[1] Gdańsk University of Technology, Faculty of Civil and Environmental Engineering, ul. Narutowicza 11/12, 80-233 Gdańsk, Poland

Abstract. The paper concerns with the analysis of normative requirements pertaining to experimental setup of a crash test and its numerical modelling. An overview of parameters describing the collision of a vehicle with a road restraining system is presented. A short description of a concrete road safety barrier is presented. A brief description of numerical modelling procedures for crash tests is given as well. The parametric influence analysis is performed of the acceleration recording time on various crash test functionality parameters The simulations are carried out using LS-DYNA finite element code with a solver version R.8.1

1 Introduction

In January 2011 European Standards EN 1317 became mandatory in all countries within members of CEN (European Committee for Standardization) [1, 2]. Next year, in January 2012, guidelines for computational mechanics of crash testing were published, where basic validation procedures were specified [3, 4]. These documents defined several levels of performance for the barriers. There are three main criteria: the containment level, the impact severity levels and the deformation expressed by appropriate parameters. To qualify road safety barrier to an appropriate containment level a corresponding crash test must be performed on the basis of European Standards [2]. Another criterion is impact severity level which is described by two parameters: ASI (*Acceleration Severity Index*) and THIV (*Theoretical Head Impact Velocity*). The last criterion is the deformation of road restrained system. This paper is focused on the applicability of the data collection requirement for the calculation of ASI and THIV parameters for the numerical model post-processing.

The study presents a shortened description of characteristic parameters for crash tests specification in section 2. Section 3 contains general information of the vehicle and concrete barrier used in full-scale experimental crash test. Furthermore, a brief description is given of the procedure of crash test numerical modelling using LS-Dyna [5-6] finite element code. Parametric analysis was performed to check the influence of the travel time of the vehicle before impact on the calculation of ASI, THIV parameters. The results are presented in section 4.

* Corresponding author: lukpacho@pg.edu.pl

2 Characteristic parameters for the crash tests specification

2.1 Acceleration Severity Index (ASI)

According to European Standards [1] every vehicle, used for the calculation of ASI and THIV, shall contain at least three accelerometers for measurement in three global directions and optionally an angular velocity sensor. The accelerometers shall be mounted at a single point placed in the nearby of the vehicle's centre of mass. Calculation of ASI parameter is not required for Heavy Goods Vehicles (HGV) or buses, although, during full-scale crash test such data can be collected for the purpose of validation of the numerical model. The idea of ASI parameter is to bring the measure of motion severity for a person inside the vehicle during collision with road restrained system. Acceleration components are filtered with four-pole phase less Butterworth low-pass digital filter, detecting a cut-off frequency of 13 Hz. The ASI parameter is assumed as a single, maximum measure of the data collected during collision, according to the formula $ASI = \max[ASI(t)]$.

2.2 Theoretical Head Impact Velocity (THIV)

Another functional parameter calculated in crash test description is the Theoretical Head Impact Velocity (THIV). The concept of this parameter was to assess the possible amount of damage dealt to occupant inside the vehicle during collision with road vehicle restraint systems. It is assumed that at the moment of impact both, the vehicle and occupant's theoretical head show an identical horizontal velocity. After impact, the velocity of this theoretical cabin is calculated integrating accelerations in x and y directions. The vehicle's yaw angle can be measured using angular velocity collected by an appropriate sensor. Technically (in numerical simulations), accelerometers are modelled by rigid parts collecting all the required nodal data. According to European Standards [1] the limit value for the THIV parameter for all acceptable impact severity levels is 33 km/h.

2.3 Deformation

There are couple measures for the description of the road safety barrier deformation. They vary depending on a type of the vehicle. The first functional parameter is working width. It describes the maximum lateral distance between undeformed guideway and maximum dynamic position of any part of the barrier which was not detached during collision. If a vehicle deforms around barrier and it is not possible to interpret the working width correctly it is allowed to take the maximum lateral position of any part of the vehicle as an alternative. Another parameter is dynamic deflection which shows the maximum lateral dynamic displacement of any point of the traffic face (guideway) of the road restraint system. The last parameter is used only for certain types of vehicles i.e. HGVs and buses. This parameter is the vehicle intrusion. It shows maximum dynamic, lateral position of the vehicle from the undeformed traffic side. Besides those parameters, the general view of the permanent deformation should be shown.

In the upcoming part of the paper working width and dynamic deflection will be referred to as the performance parameters.

3 Description of the concrete road safety barrier

3.1 Geometry and specification

Road safety barriers are used to re-direct vehicles deviating from the road or to prevent dangerous breakthrough events. In the places where significant damage cannot be allowed road restrained systems with higher containment levels are used. As an example high containment level concrete barrier protecting bridge column is presented in fig. 1. Concrete barriers were developed to produce stiff response of the road restrained system i.e., energy dissipation with relatively small transverse displacement. The main concept is to convert energy of the impact into frictional sliding energy due to high friction coefficient between the barrier and the basis. The analysed barrier is composed of single elements, called segments. It weighs 4 tonnes and is 8 m long. The segments are connected using coupling elements which work as an interlock construction. Ultimately, the barrier works as a single chain, restraining vehicles during collision. The minimal length of permanent barrier installation is 80 m. This requirement does not apply to the temporary barrier which is a subject of the work.

Fig. 1. Concrete barrier protecting the bridge column (source: www.debonte.com).

3.2 Brief description of finite element model

3.2.1 Vehicle model

Numerical model of HGV consists of 32405 nodes. This model was adapted from [7] and subjected to minor modifications in order to remove some noticed numerical instabilities. Two accelerometers were placed in the model: the first in the centre of mass and second on the dashboard of the vehicle. The placement of the second accelerometer corresponds to the placement during full-scale crash test. The car's numerical model with both accelerometers is shown in fig. 2.

Fig. 2. Heavy Goods Vehicle numerical model. Accelerometers are marked with green and red colour.

3.2.2 Segment of the concrete barrier model

An important aspect of numerical model preparation was to recreate the exact reinforcement placement as it is the key in steel - concrete interaction modelling. The finite element mesh for the concrete part consists of 4-node 3d solid elements based on tetrahedron shape. Steel coupling parts were meshed using hexahedron solid elements. Both solid element types show constant stress across the closed volume with one point integration rule. Steel reinforcing bars were modelled using Hughes-Liu beam elements with integration over cross section. The reinforcement used in barrier segment has diameter of 10 and 20 mm while the stirrups have diameter of 8 mm. Interaction between reinforcement and concrete was modelled by constraining beam elements in solid structure [5, 6]. The method used in modelling does not require reinforcement nodes to coincide with concrete mesh. In order to describe the complex behaviour of the concrete finite element mesh with high density is needed and additionally advanced material law should be used e.g. [8, 9]. Therefore, calculation time for concrete barriers is longer when compared to the steel barrier. However, the material law used in this research includes a feature, that reduces the mesh size sensitivity, [8, 9] based on the calculated fracture energy. Additional simulations were carried out to investigate the influence of the mesh size on the stress distribution. Thus, based on these results and in order to reduce the calculation time two different finite element mesh densities were applied. A more detailed mesh consists of 46004 nodes and 210074 finite solid elements. It is used in a region of a direct impact. More coarse mesh was constructed from 9099 nodes and 22027 tetrahedron solid elements and was used in remaining segments. Mesh details of the quarter of the segment can be found in fig. 3.

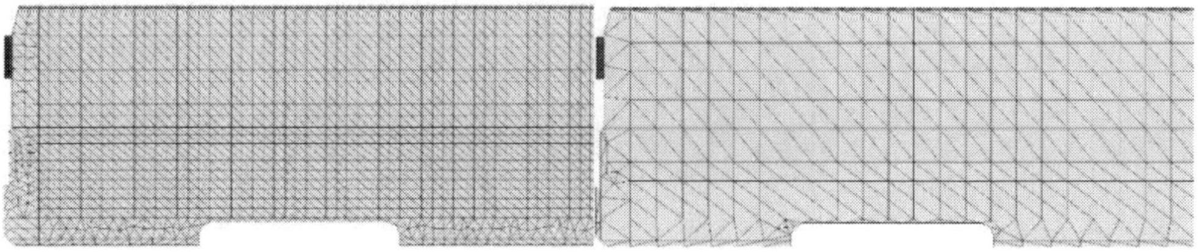

Fig. 3. Finite element mesh details for the quarter of the concrete barrier segment (segment 3 on the left and remaining segments on the right, see fig. 6).

3.2.3 Crash test assembly

Concrete road safety system consists of six segments. Segments are anchored to each other using coupling parts. There is a 5 mm gap between concrete segment faces. The segments are placed freely on the ground shell surface. Coulomb friction was specified using penalty based LS-DYNA's contact impact algorithm [5, 6]. Full length of the road restraining system is 46 m and the total weight is 24 tonnes. A vehicle weight of approx. 10 tonnes is placed at a varying distance from the fixed impact point. The angle between the velocity vector and the barrier axis is 8 degrees. Two views of the simulation model are shown on fig. 4. A detailed friction modelling was performed as it is a key factor for interaction between various types of materials.

Crash test simulation, described in the paper is based on full-scale crash test experiment made in Research Institute for Protective Systems (Inowrocław, Poland). It was performed as the part of the national programme aimed at the improvement of the safety of the roads in Poland [10]. Numerical simulations were carried out using finite element code with explicit time integration. This method is widely used in modelling vehicle collision against various types of road restrained systems [11-14]. Validation procedures have been described in European Standards [3,4], their direct use can be found in [15,16]. Validation of the concrete road safety barrier was performed showing that the obtained numerical response is correct. Therefore, the parametric analysis was conducted.

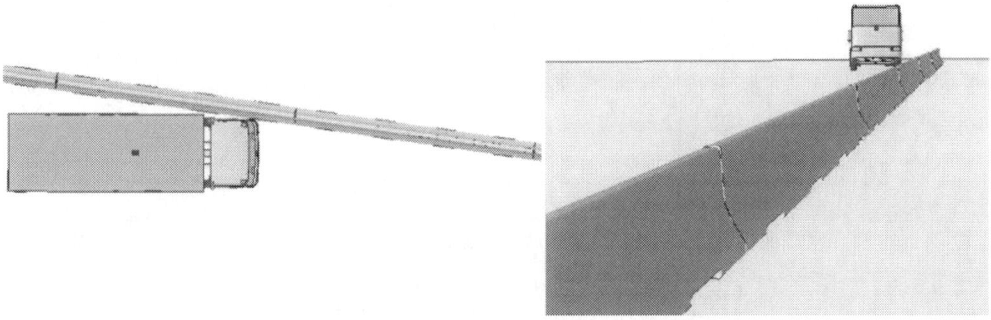

Fig. 4. Different views of the research ground with the vehicle and the barrier

4 Parametric analysis

4.1 Description of the analysis

Parametric analysis was carried out to assess the influence of the offset distance of the vehicle from the barrier on calculating impact severity indexes for numerical simulations. The general idea of the study was to confront normative requirements [1-2] for the full-scale crash tests with results obtained from numerical simulation. According to standard [1-2] to capture the information required in full-scale crash tests reports the duration time of the analysis should be at least approx. 2 seconds. The calculation time using 9 cores of the typical personal computer would take approx. 72 hours, doubling the number of cores to 18 yields a drop in calculation time to 48 hours. Taking into account the computation of concrete barrier 88 meters long the problem of available resources is significantly increasing. In order to decrease computation time it is necessary to exclude some elements of the model that may be assumed negligible. In [1] it is said that: *"Since the data will be filtered by recursive (Butterworth) filters, more data should be collected than is specifically required by the analysis. A recursive filter always produces "starting transients" at the beginning and end of the data, and requires time to "settle down". An additional 500 ms of data shall be collected at the beginning and end of the data; this extra data can then be discarded after filtering"* [1, pp.13]. According to this requirement crash test simulation should last additional second. Application of this requirement has adverse effect on the computation time. Therefore, it was found that the influence of the additional data collection time on computing impact severity indexes is worth investigating.

It was noticed that after setting initial velocity of the vehicle to the prescribed value, after some time, there is a drop of the initial velocity. Despite that there is no friction caused by air there are still energy dissipation due mutual interaction among parts of the model. Basing on the results obtained from simulations it was assumed that the impact of this factor is negligible. The variation of velocity when the actual impact occurs is less than 1 km/h. All the data can be seen in fig. 5. The graph presents curves for five different acceleration recording times before impact. Values at the intersection with the ordinate axis indicates velocities at the moment of impact. The curves were calculated with the accelerometer mounted in the middle of the vehicle's dashboard. All the data shown in fig. 5 was filtered with the Butterworth filter with the cut-off frequency 20 Hz build in Ls-PrePost software.

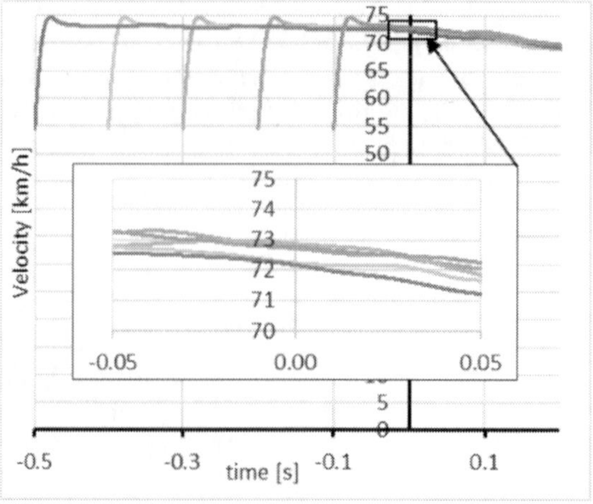

Fig. 5. Velocity in a function of time for different offset distances of the vehicle.

4.2 Results

The results were obtained for five different models. The only difference in the model input was the variable recording time of the acceleration data. The differences between crash tests results were found. They are probably due to complexity of the numerical modelling of the crash tests events. The explicit time integration method was used for time integration. It is known that this method causes errors to accumulate during the numerical integration in the time domain. Other deviations may occur due to the placement of the vehicle's centre of mass. This factor causes the wheels of the vehicle to turn over time. All the mentioned factors also contribute in the process of using the contact algorithm which may cause another differences in acquired results.

All of the five simulations show similar course, the differences are relatively small. Performance parameter values for concrete road safety barrier have been summed up in table 1. Depending on the offset distance the values slightly vary. Differences of the ASI and THIV parameters are more likely to be observed than in the working width (W). The difference probably is due to their calculation methods. ASI and THIV parameters are calculated directly from the data acquired from accelerometer. During simulation a whole vehicle vibrates, oscillates and this may lead to slight differences. On the other hand, working width is a resultant from all the model parts interaction, combined together. ASI parameter in most cases is equal to 0.4. Only the first case shows the ASI equal to 0.5. This difference may be caused by difference in the actual impact velocities. THIV was calculated, inter alia, from the angular velocity of the accelerometer. The higher values of the THIV (13-14 km/h) were observed when the wheels were turned left. Two cases, with the THIV parameter equal to the 12 km/h, had the wheels that remained straight.

The deformation of the system vary as much as +/- 4 cm. These variations may be attributed to the differences in used processors, approximations during time integration loop and others. All the information about the methods used in modelling and their limitations are contained in a book about theory used in LS-DYNA software [5]. Convention of the segments numbering for the concrete barrier H2/W5/B numerical model is shown in fig. 6. In the table 2 there is a summary of the static working width measures. The dynamic deflection values can be obtained by subtracting the width of the barrier from the static deformation values of certain positions. Fig. 7 presents the graph with static deformation's envelop on it. As it can be seen the maximum variation of the deformation is equal to 4 cm. Comparing this value to the length of the segment or it's width the ratio is equal to 1% and 7 % respectively.

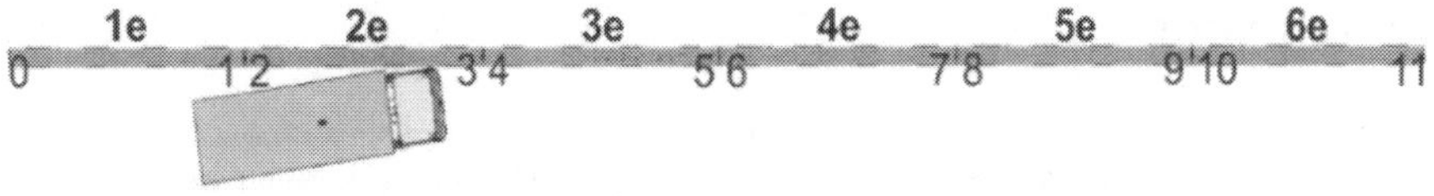

Fig. 6 Convention of the segments numbering

Table 1. Summary of functionality parameter values.

Collection time for additional data [s]	ASI [-]	THIV [km/h]	W [m]
0.1	0.45 (0.5)	13.7 (14)	0.91 (0.9)
0.2	0.40 (0.4)	12.0 (12)	0.93 (0.9)
0.3	0.44 (0.4)	13.1 (13)	0.93 (0.9)
0.4	0.4 (0.4)	13.5 (14)	0.89 (0.9)
0.5	0.38 (0.4)	11.9 (12)	0.92 (0.9)

Table 2. Static deformation for different collection times.

Collection time for additional data [s]	0.1	0.2	0.3	0.4	0.5	
Connection nr/ Element nr	Static deformation [cm]					Difference +/- [cm]
0	43	42	42	44	43	2
1E	56	55	54	54	55	2
1	68	67	66	64	67	4
2	68	67	66	64	67	4
2E	79	80	79	76	79	4
3	90	92	93	89	91	4
4	90	93	93	89	91	4
3E	88	91	91	89	88	3
5	83	86	86	85	83	3
6	83	86	86	85	83	3
4E	69	71	73	71	70	4
7	56	57	59	57	58	3
8	56	57	59	57	58	3
5E	56	55	55	56	56	1
9	55	53	51	55	54	4
10	55	53	51	55	54	4
6E	56	56	56	56	56	0
11	57	59	61	57	58	4

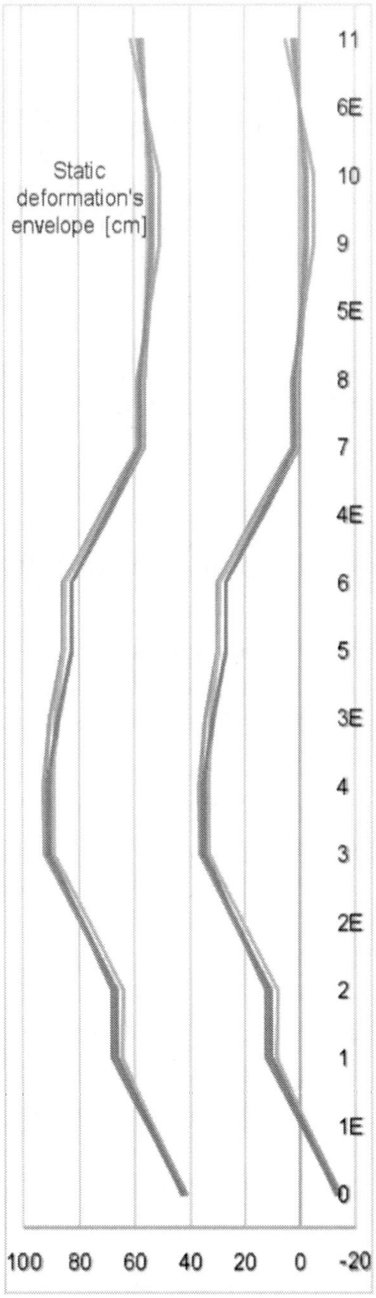

Fig. 7. Static deformation's envelope for different data collection times.

5 Conclusions

Based on the presented results the following conclusions may be drawn:

- The time necessary for the vehicle to approach the barrier has the influence on the actual impact velocity. This in turn causes the differences of the crash parameters values. However, basing on presented results, the influence of the additional acceleration data collection times on the calculation of impact severity indexes and other functionality parameters can be assumed negligible.
- Calculation time, as the key factor in explicit analysis, is strongly dependent on the mesh size. Therefore, the influence of the element size on the stress distribution must be investigated. In the analysed example case, based on full-scale test results, usage of segments element size approx.. 100 mm along with application of the fine meshed segment (el. size approx.. 38mm) in the place of direct impact provided sufficiently accurate results.

This work was supported by the National Centre for Research and Development (NCBiR) and General Directorate for National Roads and Motorways (GDDKiA), Poland. The research project name was "Life Cost analysis of Road Safety Elements" (contract number DZP/RID-I-64/12/NCBR/2016). Ls-Dyna calculations were carried out at the Academic Computer Centre in Gdańsk, Gdańsk University of Technology.

References

1. European Standard EN 1317-1:2010. Road Restraint Systems - Part 1 (2010)
2. European Standard EN 1317-2:2010. Road Restraint Systems - Part 2 (2010)
3. British Standard PD CEN/TR 16303-1:2012. Road Restraint Systems - Part 1 (2012)
4. British Standard PD CEN/TR 16303-1:2012. Road Restraint Systems - Part 4 (2012)
5. J.O. Halquist, *LS-DYNA Theory Manual*, USA (2006)
6. LSTC, *LS-DYNA Keyword User's Manual,* USA (2015)
7. NCAC, Crash Simulation Vehicle Models (accessed 10.03.2016)
8. Y. Murray, FHA, *User's Manual for LS-DYNA Concrete Material Model 159*, (2007)
9. Y. Murray, A. Abu-Odeh, R. Bligh, FHA, *Evaluation of LS-DYNA Concrete Material Model 159* (2007)
10. K. Jamroz, S. Burzyński, W. Witkowski, K. Wilde, Advances in Mechanics: Theoretical, Computational and Interdisciplinary Issues, 231-234 (2016)
11. W. Borkowski, Z. Hryciów, P. Rybak, J. Wysocki, JKONESPaT, **17**, 65-71 (2010)
12. M. Brovinsek, M. Vesenjak, M. Ulbin, Z. Ren, EFA, **14**, 1711-1718 (2007)
13. K. Wilde, K. Jamroz, D. Bruski, S. Burzyński, J. Chróścielewski, W. Witkowski, JCEEA, **XXXIII**, 455-467 (2016)
14. M. Klasztorny, D. Nycz, P. Szurgott, IJoC, **21**, 6, 644-659 (2016)
15. M. Klasztorny, K. Zielonka, D. Nycz, P. Posuniak, R. Romanowski, ACME, **18**, 339-355 (2018)
16. K. Wilde, K. Jamroz, D. Bruski, M. Budzyński, S. Burzyński, J. Chróścielewski, W. Witkowski, ACME, **63**, 187-199 (2017)

Use of DMDA method for production of heavyweight concrete

Mateusz Techman[1,*], and *Szymon* Skibicki[1]

[1] West Pomeranian University of Technology, Faculty of Civil Engineering and Architecture, Piastów 50, 70-311 Szczecin, Poland

Abstract. Waste materials are defined as all unwanted substances from manufacturing processes, including mining, industry, agriculture and others. One of the most commonly used waste product in concrete manufacturing is Fly Ash (FA). Studies have shown that the content of Fly Ash increases the workability of concrete, being useful in production of Self-Consolidating Concretes (SCC). One of the designing methods involving high amounts of fly ash is based on the Densified Mixture Design Algorithm (DMDA). This article attempts to work out HWSCC manufacturing method using the DMDA method with the use of barite and magnetite aggregates. The study integrates six concrete mixes. Concrete strength was tested on 10x10x10 cm samples after 3,7 and 28 days. The rheological properties were tested with Slump Test to find out if the mixes exhibit self-consolidating properties. Concretes with magnetite aggregate exhibit higher strength than concretes with barite aggregate. The study has shown that the DMDA method, designed to produce durable, low-cement concrete may not be applicable in the case of crushed aggregates. The study has shown that the method allows to decrease the cement amount and re-place it with pozzolan waste product to acquire concretes with similar properties to the ones acquired using different methods.

1 Introduction

Waste materials are defined as all unwanted substances from manufacturing processes, including mining, industry, agriculture and others. With regard to the production process the waste material can pose an environmental hazard, making it reusable or destined for disposal. The subject of the waste disposal and reuse has gained interest in the recent years. The overwhelming proof of human-caused global warming have been published in many journals. Scientists all over the world try to develop new technologies, upgrade or ban the harmful ones to reduce the overall carbon footprint left by people. One of the branches of industry that produces vast amounts of carbon dioxide is the cement industry. Studies have shown that production of cement constitutes to approximately 5% of the overall CO_2

* Corresponding author: mtechman@zut.edu.pl

production. Being the most used material in the world, reduction of Portland Cement production can significantly impact the global environment. Waste products can be used not only as an additive that strengthens the concrete properties, but also as an replacement for the binder [1,2].

One of the most commonly used waste product in concrete manufacturing is Fly Ash (FA). The FA have been used for many years, now proving to be one of the most useful waste products on the market. Fly Ash is acquired in the coal combustion processes. It can be used either as a partial replacement for the cement (up to 30%) or as an additive. The addition of the FA to the concrete increases its strength, durability and chemical resistance. Studies have shown that the content of Fly Ash increases the workability of concrete, being useful in production of Self-Consolidating Concretes (SCC) [3,4,5].

The SCCs for a group of concretes which, the name points it out to, have the ability to self-consolidate without any additional mechanical compaction. Self-Consolidating Concretes exhibit resistance to segregation, ability to flow past the rebar and self-removing of air bubbles. The use of Self-Consolidating Concretes requires less work, generates less noise and can be performed by unskilled workers [6].

There are many methods of designing the Self-Consolidating Concretes. Five major method groups may be distinguished including: empirical design methods, compressive strength methods, close aggregate packing methods, designing based on statistical factorial model, designing based on rheology of paste model. One of the groups, the close aggregate packing methods, focus on applying fine aggregate to fill the voids between the coarse aggregate. One of the methods that includes the least void parameter is based on the Densified Mixture Design Algorithm (DMDA) [7,8].

Hwang et al. proposed a method to employ the finer aggregates to fill the voids between greater particles. This iterative packing procedure involves Fly Ash as filler of the voids in the packed aggregate. The method was created based on maximum density theory and excess paste theory and tries to achieve the minimum usage of water and cement. This method tends to replace the cement by Fly Ash. The research shows that this method produces high flowable, low cost, durable concrete. However, the method depends strongly on the shape and space between the aggregate [7,8].

The method was proved successful in concrete manufacturing with normal, rounded aggregates. However, it was not proved anywhere if the method is feasible for concretes with crushed-stone aggregates with sharp edges. These types of aggregates reduce workability of concrete.

Heavyweight concrete has been studied for many years. Even now there are still studies regarding different properties of heavyweight concretes and mortars [9]. The literature on the topic of Heavyweight Self-Consolidating Concrete is however scarce [10,11]. Except of few attempts described in papers, the topic seems to be avoided. The main hardship in achieving this sort of concrete is the problem of segregation and aggregate type.

The article tries to work out manufacturing scheme for Heavyweight Self-Consolidating Concrete using the DMDA method with use of barite and magnetite aggregates.

2 Material and Method

The study uses the DMDA method to for concrete design. The method takes the form of a 7-step design procedure.

1. Selection of proper materials:

2. The first step requires selecting proper cement, fly ash or slag and superplasticizer. It is necessary to obtain information on content of the additives to exclude unwanted chemical reactions;
3. Achieving the maximum density by iterative aggregate packing:
4. The mix of fine aggregate and fly ash is densified in a 1l cylindrical container using vibration table to obtain the highest density of the mix;
5. Calculation of least void V_V:
6. This step is performed by calculating compaction coefficients of fly ash, sand and coarse aggregate mixes;
7. Determine the paste volume V_P:
8. This step is particularly complex, while determining the surface area of aggregate. The author of the method allows taking estimated values based on the experience of the scientist;
9. Calculation of aggregate volume V_{agg}:
10. Determination of SP content and water content:
11. The authors of the method recommend the water to cement ratio of 0.42, but water to binder ratio 0.28. This allows to save water during the production;
12. Iteration:
13. The last step requires repeating the previous steps to acquire necessary rheological and mechanical properties [7,8].

The study uses Fly Ash acquired from coal power plant Dolna Odra in Poland. The mixes were made using Ground Granulated Blast-furnace Slag Cement (GGBSC). The cement characterizes with better workability properties. The production of this cement is also more eco-friendly, considering lower CO_2 amount emitted during the production process. The superplasticizers used in the study were acquired from Sika Company. The Sike Viscocrete 3 is an polymer superplasticizer used for significant improvement of workability and reduction of water content in concrete. Sika 20 HE is a polycarboxylate superplasticizer for high reduction of water content.

The study uses two heavyweight types of aggregates and natural sand. Magnetite and barite aggregate were used to achieve high density of the mix. Natural sand was used as a fine aggregate to improve the workability of the mix. The round shape of the particles in the sand reduces the friction and improves the mix flow. On the contrary, crushed aggregates such as magnetite and barite show sharp edges that stop the flow of the mix [12,13].

The rheological properties were tested with Slump Test to see if the mixes are self-consolidating.

The study includes 6 concrete mixes. The concrete strength was tested on 10x10x10 cm samples after 3,7 and 28 days.

3 Results and Discussion

The first step of the method that includes the determination of properties was described in the second chapter of the article.

The results of compaction of the natural sand (NA) with fly ash (FA) are visible in the Table 1.

Table 1. Ratio (R1) of the natural sand (NS) to fly ash (FA).

Ratio NS:FA	1:0.05	1:0.1	**1:0.11**	1:0.15	1:0.19
Bulk density [kg/dm³]	1.84	1.86	**1.86**	1.83	1.82

As seen above, bulk density of the mix takes a curved shape, reaching its peak at ratio 1:0.1
and 1:0.11.

The ratio 1:0.11 was chosen for its higher FA content.

The next step requires calculating the ratio for different coarse aggregates. The results for magnetite and barite are visible in Tables 2 and 3.

Table 2. Ratio of the compacted fine aggregate (R1) and magnetite aggregate (MG).

Ratio MG:R1	1:0.28	**1:0.31**	1:0.33	1:0.46	1:0.66
Bulk density [kg/dm³]	3.05	**3.11**	3.06	3.00	2.83

Table 3. Ratio of the compacted fine aggregate (R1) and barite aggregate (BR).

Ratio BR:R1	1:0.29	1:0.38	1:0.54	**1:0.56**	1:0.59
Bulk density [kg/dm³]	2.60	2.62	2.66	**2.68**	2.67

The result of the compaction coefficient determination showed that the amount of voids in compacted barite aggregate is much higher than in magnetite aggregate.

In order to acquire the highest density of concrete the compacted aggregates with highest density were taken into consideration.

The final amount of cement, aggregate, water and superplasticizer were calculated using the algorithm and are not presented in this paper.

The first acquired mix recipe is visible in Table 4.

Table 4. The proportion of the prepared mixes.

	M1	**M2**	**M3**	**M4**	**M5**	**M6**
GGBSC 42.5	346	346	346	337	327	400
Fly Ash	71	71	71	89	110	-
Barite dust	-	-	-	-	-	160
Water	127	127	127	127	132	109
ViscoCrete 3	**6.92**	**8.30**	**17.29**	10,11	-	20.16
ViscoCrete 20 HE	-	-	-	-	8.16	-
Sand 0-2 mm	625	625	625	781	581	735
Magnetite 2-8 mm	2269	2269	2269	-	2264	-
Barite 2-8 mm	-	-	-	1558		1568

Approximately 70% of mixes volume is taken by aggregates which is recommended for SCC. The water cement ratio was set to ~0.36. All mixes were tested in the slump flow test.

Due to insufficient workability of the mix it could not be classified as SCC. The mix however had good viscosity and was resistant to segregation which lead to believing that the amount of the superplasticizer was insufficient.

Following two mixes were variation of the base mix with increased amount of superplasticizer. The first mix detected an increased amount of superplasticizer to 2.4% (maximum recommended value), the second the amount increased to 5.0%. None of the mixes reached required flow. This lead to believe that the proportions calculated according to algorithm were wrong. The re-calculations made for different fly ash content were made here. The acquired mix recipe is presented in Table 4 as M5. In order to avoid low flowability, the mix contains stronger superplasticizer dosed at its maximum recommended value. The calculations were also made for concrete on barite aggregate. The mixes were compared to a Heavyweight Self-Consolidating Concrete mix (M6) acquired with a different designing method in a previous study.

Table 5. Results of Slump Flow test.

	M1	M2	M3	M4	M5	M6
D_1 [cm]	39.5	41.0	41.0	38.0	59.5	66.6
D_2 [cm]	38.5	39.0	40.0	37.0	57.5	66.0
D_{mean} [cm]	39.0	40.0	40.5	37.5	58.5	66.3
t [s]	x	x	x	x	28.8	12.9

As shown in Table 5 none of the mixes on magnetite aggregate reached necessary slump flow. The mix on barite aggregate surpasses the recommended D500, however the flow time was extremely long.

Fig. 1. Left: Slump test of M4 mix, **Right:** Results of Slump Flow test HWSCC on barite.

As much as the method seems to not be able to produce self-consolidating concrete, the results gave us a workable, resistant to segregation concrete with heavyweight properties.

The results of density measurements and concrete strength are visible in Table 6.

Table 6. Density and compressive strength of the mixes.

	M1	M2	M3	M4	M5	M6
Density [kg/m^3]	3413.7	3430.4	3397.9	3370.4	2861.1	2964.4
Compressive strength 3-days [MPa]	47.1	51.8	38.2	47.7	39.5	38.5
Compressive strength 7-days [MPa]	66.8	63.5	54.6	65.0	52.9	57.7
Compressive strength 28-days [MPa]	86.3	85.2	76.0	85.9	70.5	68.2

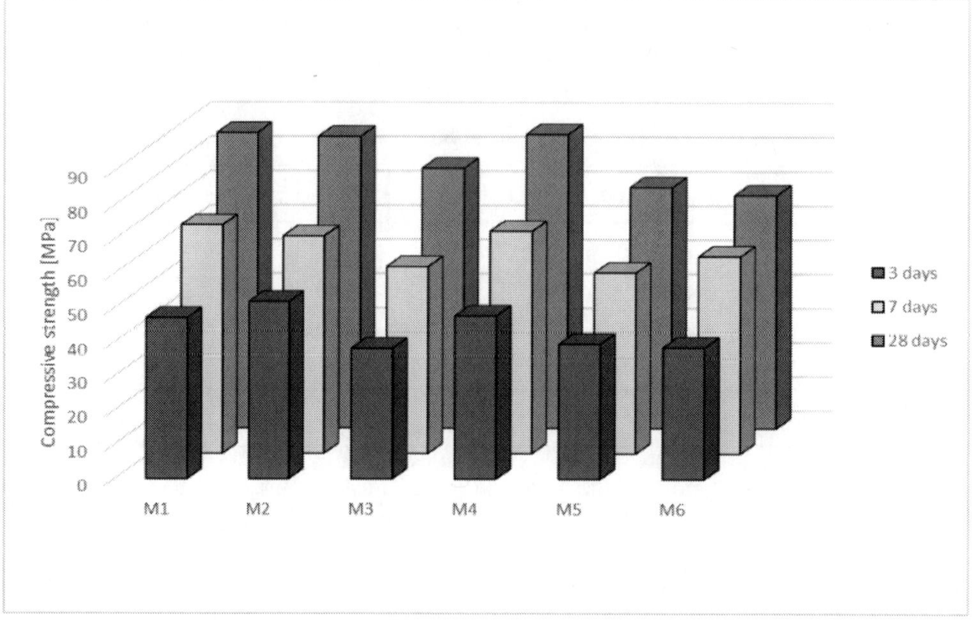

Fig. 2. Compressive strength of studied concrete after 3, 7 and 28 days.

According to the predictions, concretes with magnetite aggregate show higher strength than concretes with barite aggregate. The barite concrete produced using DMDA method showed similar strength and rheological properties to concrete produced with different method. However, the recipe for the DMDA concrete contained less cement, which is one of the most important factor considering eco-driven approach in designing of concrete.

4 Conclusions

The study has shown that the DMDA method, designed to produce durable, low-cement concrete can have issues if using crushed aggregates. The flow of such mixes is significantly reduced by the sharp edges of the aggregate. Increased amounts of fly ash or superplasticizer seems not to be able to overcome the friction between the particles. The solution is bound to reduce the overall amount of sharp edged aggregate and replace it with more round aggregates.

However, as the study tried to acquire heavyweight aggregate, this might be troublesome due to the availability of different aggregates. It is possible to use different type of heavyweight aggregate such as iron shots, but this is a topic for another consideration.

The study showed that the method allows to decrease the cement amount and replace it with pozzolan waste product to acquire concretes with similar properties to ones acquired using different methods. Reducing the amount of cement in the concrete mix is one of the most important aims in modern concrete technology.

References

1. R. Siddique, Utilization of Industrial By-products in Concrete, Procedia Engineering, **95**, 335-347 (2014)
2. G. Sua-iam, N. Makul, Incorporation of high-volume fly ash waste and high-volume recycled alumina waste in the production of self-consolidating concrete, Journal of Cleaner Production, **159**, 194-206 (2017)
3. J. M. Paris, J. G. Roessler, C. C. Ferraro, H. D. DeFord, T. G. Townsend, *A review of waste products utilized as supplements to Portland cement in concrete*, Journal of Cleaner Production, **121**, 1-18 (2016)
4. M. K. Dash, S. K. Patro, A. K. Rath, Sustainable use of industrial-waste as partial replacement of fine aggregate for preparation of concrete – A review, International Journal of Sustainable Build Environment, **5**, 484-516 (2016)
5. J. Borucka, *17th International Multidisciplinary Scientific GeoConference SGEM*, Imposed thermal stresses in high performance concrete modified with microsilica, (2017)
6. S. A. Miller, P. J. M. Monteiro, C. P. Ostertag, A. Horvath, Concrete mixture proportioning for desired strength and reduced global warming potential, Construction and Building Materials, **128**, 410-421 (2016)
7. Y. Y. Chen, B. L. A. Tuan, C. L. Hwang, Effect of paste amount on the properties of self - consolidating concrete containing fly ash and slag, Construction and Building Materials, **47**, 340-346 (2013)
8. C. L. Hwang, F. M. F. Sihotang, B. L. A. Tuan, C. T. Chen, *Green concrete for Sustainable Life-Cycle*, Trans Tech Publication, Switzerland (2011)
9. E. Horszczaruk, P. Sikora, K. Cendrowski, E. Mijowska, The effect of elevated temperaturę on the properties of cement mortars containing nanosilica and heavyweight aggregates, Construction and Building Materials, **137**, 420-431 (2017)
10. M. Kaszyńska, M. Techman, Conference *Dni Betonu, Właściwości reologiczne ciężkich betonów samozagęszczalnych*, Wisła (2016)
11. D. Revuelta, A. Barona, D. Navarro, Measurement of properties and of the resistance to segregation in heavyweight, self-compacting barite concrete, Materiales de Construcción, **59**, 31-44 (2009)
12. K. Saidani, L. Ajam, M. B. Ouezdou, Barite powder as sand substitution in concrete: Effect on some mechanical properties, Construction and Building Materials, **95**, 287-295 (2015)

13. K. Jankovic, S. Stankovic, D. Bojovic, M. Stojanovic, L. Antic, The influence of nano-silica and barite aggregate on properties of ultrahigh performance concrete, Construction and Building Materials, **126**, 147-156 (2016)

The SHM system as a data source for durability assessment. Case study of Rędziński Bridge stay cable system

Marco Teichgraeber[1,*], and *Paweł* Hawryszków[1]

[1]Wrocław University of Science and Technology, Faculty of Civil Engineering, Wybrzeże Wyspiańskiego 27, 50-370 Wrocław, Poland

Abstract. The paper is an introduction to a durability assessment algorithm based on the data collected by a Structural Health Monitoring (SHM) system of a cable-stayed concrete bridge. It contains basic information about the structure of the Rędziński Bridge and its monitoring system. The collected data from the first years are described and analysed. Later the algorithm to assess the durability of stay cables based on probabilistic method is presented.

1 Introduction

Important and innovative structures have been built over last decades. The usage of new materials and technologies allowed to create forms that have never been seen before, but working with innovative or improved materials brings some risk in the process of designing, constructing and maintenance, namely the assessment of its durability after years.

Nowadays a lot of older structures detect a number of some damages. The major example [1] is the cable-stayed bridge (1965) in Leverkusen in Germany where the traffic caused fatigue damages in the steel structure. Not only steel structures are sensitive to the effect of time. The Koror-Babeldaob Bridge (1978) collapsed in 1996 [2] because of project changes and the influence of rheology effects. Both examples show how a wrong assessment of loading after many years can affect the durability of a structure.

In the course of structural design the designer is given information about loadings and material properties taken from standards and laboratory experiments. The sources of information about real behaviour of the structure are long-term observations and unfortunately some failures or collapses. But in the last decades things have changed, namely Structural Health Monitoring (SHM) systems have become more common. At the beginning of that technology groups of sensors were installed in some old structures where visible damages had occurred [3]. After years also newly built objects have been equipped in that kind of systems, which have become more reliable and easy to use by the development of the internet. Nowadays SHM systems have become a convenient tool for engineers to supervise all kind of structures, not only bridges but also roof toppings

* Corresponding author: marco.teichgraeber@pwr.edu.pl

of stadiums or motorways. The system provides not only the information how the structure behaves at one moment, but it is also collecting the data and saving it, so the user may enter the SHM application to perform a long-term overview of a chosen structure element and values like stress, temperature or acceleration. At the turn of the 20th and 21st centuries a lot of major bridges were built in Poland, most of them have been connected to a SHM system, e.g. : the Solidarity Bridge over the Vistula River in Płock (2007) and the John Paul II arch bridge in Puławy also over the Vistula River (2008). After few years operation SHM systems brought data sets fir the researches to work with [4].

The paper presents the SHM system of the Rędziński Bridge. Furthermore an overview of the collected data will be presented and also the first attempt to use the data for durability assessment will be introduced.

2 SHM system of the Rędziński Bridge

2.1 The bridge structure

The Rędziński Bridge was built in 2011 and it is the main bridge on the Wrocław ring road motorway. The structure is a concrete cable-stay bridge with spans 49.00 m + + 256.00 m + 256.00 m + 49.00 m. The H-shaped pylon is 122.00 m high. The characteristic feature of the bridge are two separate box girder concrete decks under each road suspended to one pylon with 160 stay cables [5].

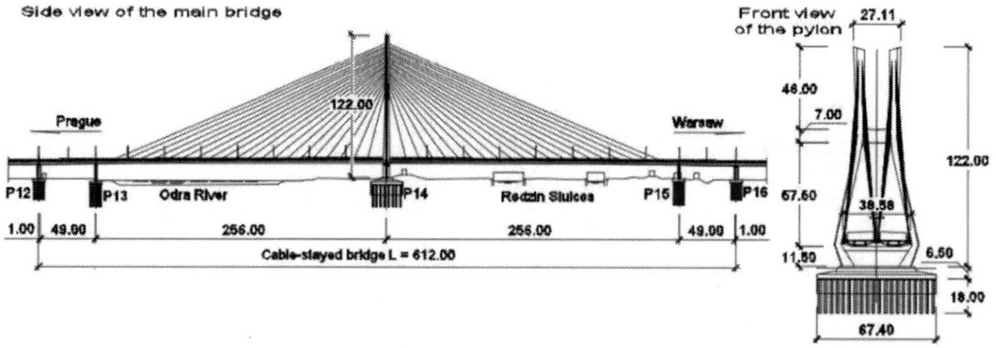

Fig. 1. General view of the Rędziński Bridge [5].

2.2 The bridge SHM system

Because of its size and outstanding structure the Rędziński Bridge was equipped with 222 sensors. Figure 2 shows the localization of sensors.

The connection with the system is possible by an internet browser application [6] that provides an overview of each sensor. It allows to see alerts and notification if some sensors are not working properly. Furthermore the application makes it possible to create diagrams of the measured values and to export them as .csv files compatible with calculation software.

The data in the system is measured with the frequency of 1 Hz, but only the maximum value form a 5 minutes period is saved on the server. For a long term analysis, a huge database was required, therefore first calculations based on the SHM system have been done after 5 years of use. In this time each sensor has saved on the server over 500000 values of the measured data – forces in cables, accelerations, stress values and temperature values.

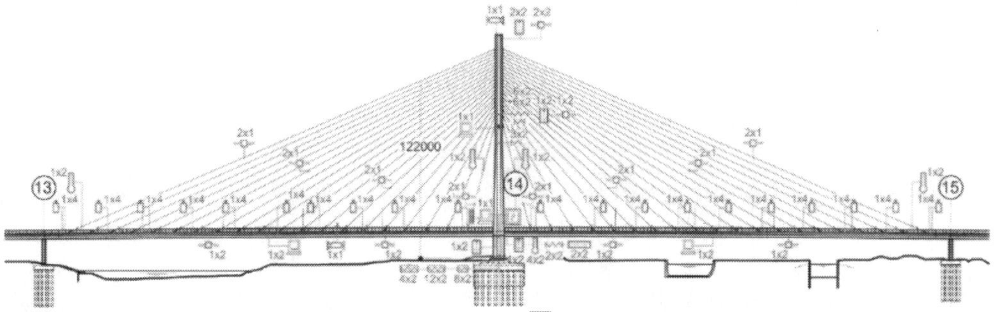

- F, force sensors in cable strands (80 pcs.),
- Tt, angle rotation sensors (10 pcs.),
- Tr, linear displacement sensors (4 pcs.),
- W, wind speed and direction sensors (2 pcs.),
- Te, temperature sensors (16 pcs.),
- Sc, sensors measuring linear displacements on concrete surface (16 pcs.)
- A, acceleration sensors (30 pcs.),
- CGD, global data server (1 pc.),
- EN, power connection (1 pc.),
- Sc1, sensors measuring linear displacements inside the concrete structure (24 pcs.),
- Sc2, sensors measuring stresses inside the concrete structure (8 pcs.),
- Si, sensors measuring linear displacements on steel surface (18 pcs.),
- Si1, sensors measuring stresses in the reinforcing rods (16 pcs.),
- SAD, Local Data Saver (5 pcs.).

Fig. 2. SHM system scheme and sensors description.

3 Data analysis

In the years 2016 and 2017 a first complex report about the SHM system was made [7]. The data collected by each sensor was precisely analysed and a comparison with the border values was carried out. The bridge behaves after 6 years unreservedly, but some characteristic phenomena were observed. As anexample, Figure 3 shows how the cable force changes during a month. There is a visible influence of temperature changes on the structure. Figure 4 describes how the global force in cable has changed for the last 5 years. It may have been caused by rheology processes in the decks and pylons. A more particular analysis of force changes is presented in the part 4.

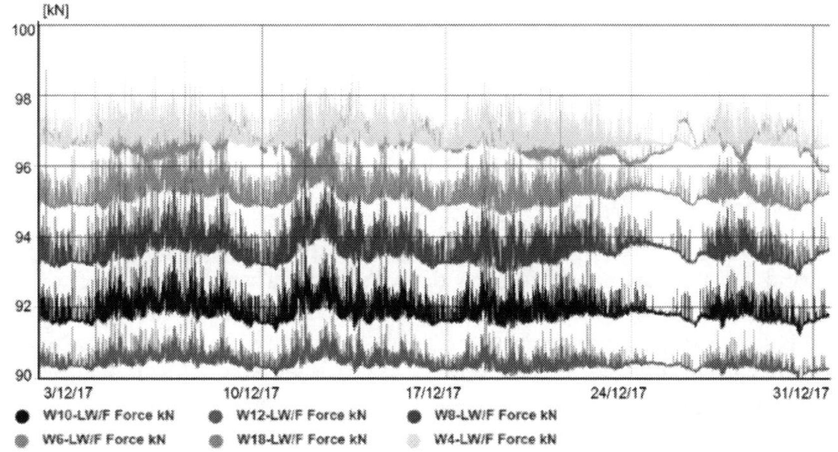

Fig. 3. Force changes in cables during December 2017. A diagram taken straight from the SHM application.

Another important observation was made about the temperature differences in elements of the bridge. Figure 5 shows the temperature course in the deck, pylon and a random cable.

After 6 years the extreme values have been noticed and compared to each other, what is shown in Table 1.

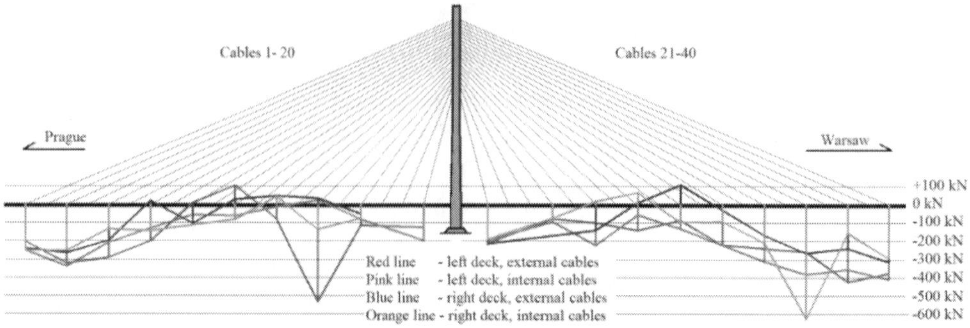

Fig. 4. Change of the monthly average force in cables, in each row by the end of 2015.

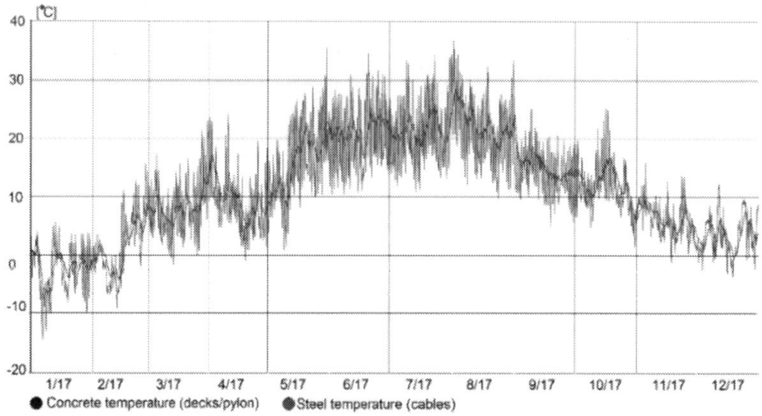

Fig. 5. The temperature changes in concrete and steel elements in 2017.

Tab. 1. The extreme temperature values.

Element	Minimum temperature [°C] (February 2015)	Maximum temperature [°C] (August 2015)
Cables	-20.89	44.37
Pylon	-9.05	36.03
Deck	-12.72	33.97

4 Durability assessment methods

Bridges are structures that are permanently subjected to moving and changing loads like traffic, wind and temperature. This sorts of loads cause a variable stress in each element, with an influence on the fatigue durability of an entire structure. SHM systems gives the opportunity to create a statistic database for a long-term durability assessment [8]. In a cable-stayed bridge the cables are the most sensitive part. Recently [9] a vast laboratory fatigue tests of cables have been carried out. The work was done under constant force changes and known imperfections in cables, but the diagrams in Figure 3. show how the force in bridge cable is randomly changing. In order to provide a reliable durability assessment not only the value of force is crucial but mostly the force amplitudes. Applying the collected data for a stay cable a diagram of measured stress values was created (Fig. 6). In this case it is one of the shortest cable of the Rędziński Bridge (cable number LZ-W1, which is the first external cable on the left deck counted from the pylon axis toward Prague). It is visible that the stress range is between 480 [MPa] and 550 [MPa]. After a rain-flow analysis of the force signal a histogram of particular amplitudes was made (Fig. 7).

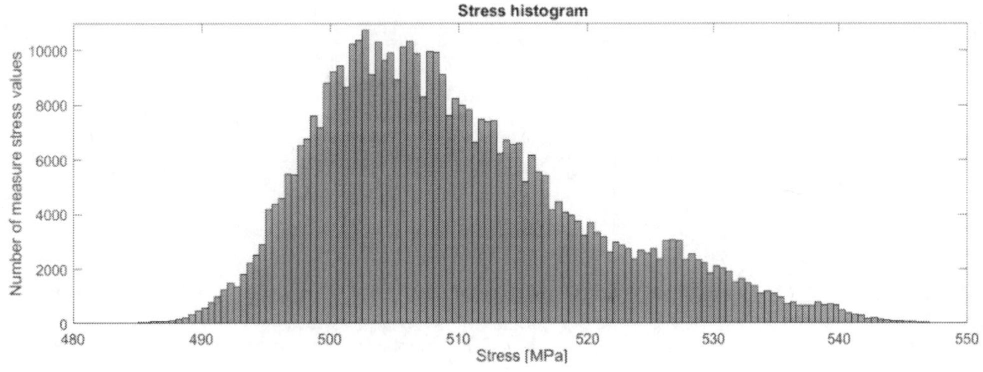

Fig. 6. Stress histogram.

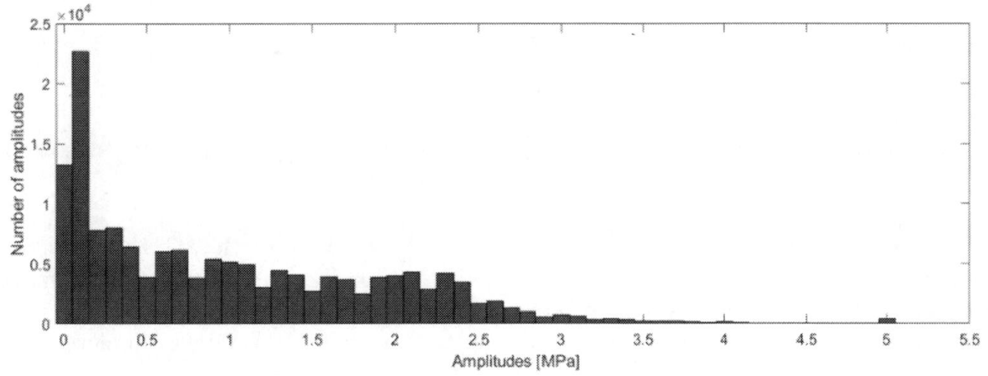

Fig. 7. Number of occurring amplitudes in the shortest stay cable in years 2011-2016.

According to the Palmgren-Miner rule [10] each cycle n_i (two amplitudes are a whole cycle) causes a small damage in the steel structure *n/N*. *N* is the amount of particular destructive cycles due to the Wöhler curve and *n* stands for the measured cycles. If *D* describes the sum of all damages caused by different cycles a simple equation can be formulated, namely:

$$D = \sum_{i=1}^{q} \frac{n_i}{N_i} \tag{1}$$

This equation is used in discreet calculations. The diagram in Figure 7. shows obviously the exact amount of calculated cycles, but a more convenient way is to create an amplitude probability density chart $p(\sigma_a)$ based on the calculated cycles. The probability of each amplitude is described by the equation (2), where N_c stands for the destroying amount of the cycles that have appeared in the particular cable so far and *n* is now the function that describes the amount of cycles.

$$P(\sigma_a) = \frac{n}{N_c} \quad \text{or} \quad N_c P(\sigma_a) = n \tag{2}$$

In this particular case the probability density equation is described as the exponential one:

$$\frac{dP(\sigma_a)}{d(\sigma_a)} = p(\sigma_a) = \frac{1}{\overline{\sigma_a}}\exp(-\frac{\sigma_a}{\overline{\sigma_a}}) \tag{3}$$

Where $\overline{\sigma_a}$ is the variable and s is the standard deviation of the counted amplitudes. Using a function to describe the course of amplitude values the Palmgren-Miner rule should be described by an integral

$$D = \int \frac{dn}{N} \tag{4}$$

Differentiating the equation (2) by $d\sigma_a$ a suitable formula is obtained

$$N_c p(\sigma_a) = \frac{dn}{d\sigma_a} \quad \text{or} \quad N_c p(\sigma_a) d\sigma_a = dn \tag{5}$$

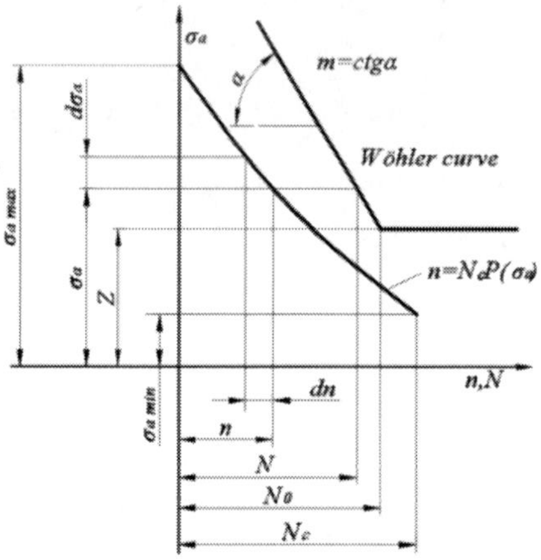

Fig. 8. Comparison of function *n* with the Wöhler curve [10].

The Wöhler curve is a function that stands for the laboratory assessment of the fatigue endurance for the particular kind of steel. Here Z is the amplitude value reached by N_c destroying cycles, after which the durability of steel is endless. While the Wöhler curve is described in the logarithmic system the m parameter appears in the equations.

$$\left(\frac{\sigma_a}{Z}\right)^m = \frac{N_0}{N} \quad \text{or} \quad \frac{1}{N} = \left(\frac{\sigma_a}{Z}\right)^m \cdot \frac{1}{N_0} \tag{6}$$

The merging action of the equations (4), (5) and (6) leads to the formula to compute the number of destroying cycles N_c. Figure 7 shows the function n with the comparison to the Wöhler curve.

$$D = \frac{N_c}{N_0} \int \left(\frac{\sigma_a}{Z}\right)^m p(\sigma_a) d\sigma_a \tag{7}$$

The last equation is a general formula to assess the durability by N_c cycles that a construction element can stand under current loading conditions. In order to assess the value some assumptions for D and the integration limits should be made. Table 2 shows the chosen values for different durability rules.

A construction element can keep its durability as long as the n curve stays below the Wöhler curve. Table 3 shows the example calulation for the durabilty assessment of the cable LZ-W1. All calcuations were done after each month according to the shown algorithm. The table presents only the months detecting an amplitude increase. The last column of the table presents the durabilty assessment as a percent value of caculated cycles number in the first shown month (December 2011).

Table 2. Parameters for the durability assessment depending on chosen rule [10].

Rule	**Palmgren-Miner**	**Serensen**	**Haibach**
D	*1*	*a ~ (amax, Z)*	*1*
integration limits	$\int_{Z}^{\sigma_{amax}} \left(\frac{\sigma_a}{Z}\right)^{m} p(\sigma_a)d\sigma_a$	$\int_{Z}^{\sigma_{amax}} \left(\frac{\sigma_a}{Z}\right)^{m} p(\sigma_a)d\sigma_a$	$\int_{\sigma_{amin}}^{Z} \left(\frac{\sigma_a}{Z}\right)^{m'} p(\sigma_a)d\sigma_a + \int_{Z}^{\sigma_{amax}} \left(\frac{\sigma_a}{Z}\right)^{m} p(\sigma_a)d\sigma_a$

Table 3. Parameters for the durability assessment according to Haibach modification.

Year	Month	Average amplitude	Standard deviation	Max. ampltude	Min. Ampltude	Average stress	Max. stress	Min. stress	N_c	
		$\overline{\sigma_a}$	a_sigma	$\sigma_{a,max}$	$\sigma_{a,min}$	s_sr	s_max	s_min	Number of cycles and its percentage coefficient	
		[Mpa]	[Mpa]	[Mpa]	[Mpa]	[Mpa]	[Mpa]	[Mpa]		
2011	December	0,9915	0,9244	25,6814	0	531,9398	575,1542	517,5746	7,29E+14	100%
2012	January	0,9618	0,9301	86,5916	0	531,0057	575,1542	365,9283	7,03E+14	96%
2012	April	0,959	0,9444	147,6867	0	529,4223	661,3018	365,9283	6,42E+14	88%
2012	May	0,9835	1,0424	248,0442	0	529,6494	862,0167	365,9283	3,56E+14	49%
2012	June	1,0094	1,1108	313,0618	0	529,5133	1152,7	365,9283	2,43E+14	33%
2013	June	0,9933	1,0567	313,0618	0	521,7808	1152,7	365,9283	3,28E+14	45%
2014	June	0,9961	1,0124	313,0618	0	514,6963	1152,7	365,9283	4,24E+14	58%
2015	June	1,0042	0,9885	313,0618	0	510,5295	1152,7	365,9283	4,89E+14	67%
2016	June	1,0178	0,98	313,0618	0	509,7471	1152,7	365,9283	5,09E+14	70%
2017	February	1,035	1,0326	313,0618	0	509,6253	1152,7	365,9283	3,76E+14	52%
2017	March	1,0374	1,035	488,9305	0	509,5813	1152,7	365,9283	3,71E+14	51%
2017	April	1,0385	1,037	488,9305	0	509,5982	1152,7	365,9283	3,67E+14	50%

5 Conclusions

The first assessment completed according to the algorithm shown for the cables LZ-W1, LZ-W14, LZ-W20 (Fig. 9) proves, that after more than 6 years of servise the durability of cables is preserved. The challenge in this case is to predict the durability after next years and decades today. The future work, based on the presentation above, will contain a specific description of the upcoming forces in the structure elements based on statistical data collected by the SHM system.

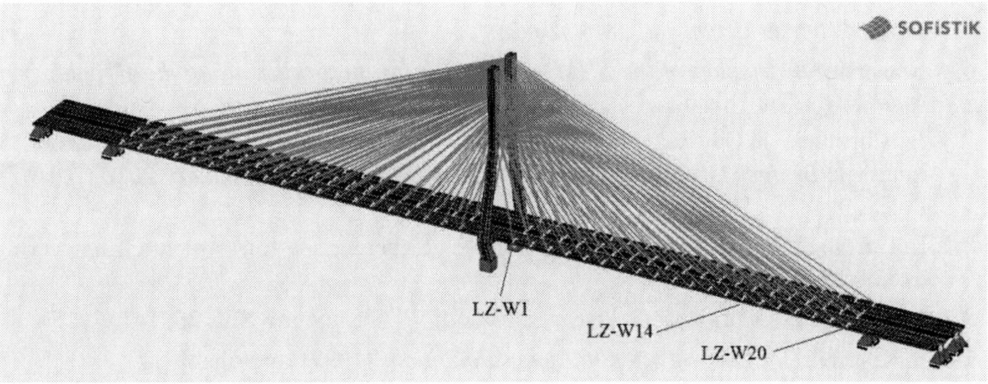

Fig. 9. The Rędziński Bridge FEM model.

The algorithm shown above is a simple and fast way to assess the durability of stay-cables based on the values given by the SHM system. The data used is only the total axial force. For an exact and long term assessment a FEM model should be created as a supporting tool. Creating an influence line for the axial force (Fig. 10) in the cables makes it possible to consider the model with the action of the loads estimated on the basis of the SHM system data. Regarding a vehicle traffic forecast for the motorway an assessment for next years can be made. Additionally not only axial force can be estimated, but also the bending moments in the cable anchorages, the most sensitive point of a cable stay. Furthermore using a FEM model the influence of temperature and wind can be evaluated separately.

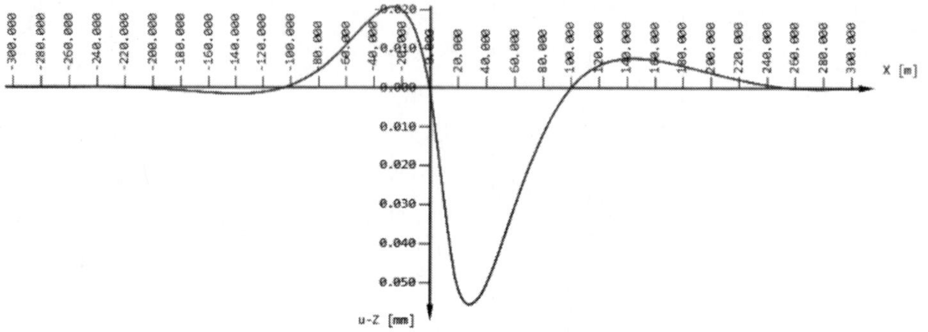

Fig. 10. Influence line for the axial force in the shortest stay cable WZ-L1.

The aim of the future work is to create a complete algorithm based on the SHM system data and a FEM model allowing to assess the durability of structural elements, not only the cables, under realistic traffic and weather conditions. Furthermore, the algorithm should be suitable for other bridges equipped in SHM systems.

References

1. G. Marzahn, P. Langenberg, G. Groten, M. Paschen, *25. Dresdner Brückenbausymposium*, 163 (2015)
2. J. Sheer, *Failed Bridges. Case Studies, Causes and Consequences* (2010)
3. H. Wenzel, *Health monitoring of bridges* (2009)
4. H. Onysyk, *Safety assessment of bridge in operation based on Structural Health Monitoring system data, PhD thesis* (2014) [in Polish]
5. P. Hawryszków, M. Hildebrand, *18th IABSE Congress "Innovative Infrastructures – Toward Human Urbanism"*, 811 (2012)
6. Neostrain, *Instruction of the SHM Application for the bridge along the Wrocław ring road* (2011) [in Polish]
7. J. Biliszczuk, J. Onysyk, M. Teichgraeber, *Data analysis of the SHM System of the Rędziński Bridge. 1st Rapport from the period August 2011 – January 2016* (2016) [in Polish]
8. L. Hui, L. Shunlong, O. Jinping, L. Hongwei, Structure and Infrastructure Engineering, **8**, 829 (2012)
9. J. Winkler, C. Georgakis, G. Fisher, S. Wood, W. Ghannoum, SEI, **25**, 141 (2015)
10. S. Kocańda, J. Szala, *Basics of fatigue calculations* (1997) [in Polish]

Influence of selected additives and admixtures on underwater concrete and the environment

Aleksandra Mariak[1], *Lucyna* Grabarczyk[1*], *Barbara* Wojtasik[2,3], and *Małgorzata* Zbawicka[4]

[1]Gdańsk University of Technology, Faculty of Civil and Environmental Engineering, ul. Narutowicza 11/12, 80-233 Gdańsk, Poland
[2]HydroBiolLab, Research Company and Hydrobiological Laboratory, ul. Żeliwna 23a, 81-159 Gdynia, Poland
[3]University of Gdańsk, Faculty of Biology, Department of Genetics and Biosystematics, ul. Wita Stwosza 59, 80-308 Gdańsk, Poland
[4]Institute of Oceanology PAS, Sopot, Department of Genetics and Marine Biotechnology, ul. Powstańców Warszawy 55, 81-712 Sopot, Poland

Abstract. The intensive civilization development implies searching for new possibilities connected with extension of city agglomerations, both the areas of flat building and the industrial areas. One of the most interesting solution is to use water reservoirs, rivers and sea areas. The extension of buildings affects the building materials, especially hydrotechnical concrete. Water structures are usually the objects of huge dimensions exposed to extreme conditions. In most cases these buildings are specific and complicated. They require an individual approach to concrete mix design and should be monitored by measurement technologies supported by automated numerical examinations. Nowadays health monitoring systems are applied to large variety of engineering structures. In order to get the required quality of hydrotechnical concrete, additives and admixtures are necessary. The material properties of underwater concrete can be also improved using polymer and steel fibers of various content and length. The majority of areas are built of concrete, that is the reason to be aware of admixture influence on the environment and living organisms. The attempt to evaluate the description of different kinds of hydrotechnical concrete as potential habitat of water organisms, was indicated in the paper.

1 Introduction

The majority of massive hydrotechnical structures require the usage of underwater concrete technology (UWC), where major issue refers to avoiding washout of binding agent from the concrete mixture. All underwater concrete techniques focus on minimization of mixture's contact with water during its supply for placement. Works organisation and uninterrupted concrete placement is a key factor here. Usually, concrete of class above C 20/25 are used. In case of underwater concreting, the requirements on technological process are more important than the concrete strength properties. In order to ensure a high liquidity of the mixture and to avoid binding agent losses, above 375 kg/m^3 of cement and additives per 1 m^3 of concrete it is assumed. Irrespectively of UWC method, special additives AWA (anti-washout admixture) and admixtures - HRWR (high-range water-reducing admixture) shall be used [1-3]. Application of new generation admixtures and additives, like AWA, HRWR is more and more popular. Economic reasons related to proper organisation of works [4] and variation-prone rheological and physical and mechanical properties of concrete make us believe that the structures operated in water are most popular here.

In most cases, pebble or crushed-stone aggregate is used in underwater concrete. Mixture placed under water should be resistant to segregation and water dilution, it must spread out readily and be self-compacting [5-6]. Mineral additives used mainly in UWC include silica fumes added in the quantity of up to 8% of cement weight and fly ashes, their content approximates to 20% of cement weight. Significant improvement to the underwater concrete technology was noted in previous years, to open an opportunity for concrete as the material of the future. Many academic centres carry out tests on mixtures of cement with the addition of fly ashes, calumite or fluidized bed slag, zeolite, metakaolin and microsilica [7]. Advantageous influence of additives on concrete properties has already been proved, therefore additives are commonly used in concrete production. In the case of underwater structures it is possible to use dispersed reinforcement in a form of steel and polymer fibres [8]. The possibility to modify mechanical properties of concrete makes the structure more resistant to abrasion, erosion and corrosion [9]. Additionally, surface protection by impregnation of precast elements immersed in water makes it possible to build structures in highly aggressive environment [10].

Sometimes, in order to avoid difficult process of casting, precast concrete elements are produced. They include foundations of wind power plant or lightweight concrete slabs as foundations of floating houses [11]. Another aspect is monitoring of concrete elements, both, during the concrete placement and exploitation of the structure. Advanced measurement techniques in combination with numerical simulations allow prediction and monitoring of the changes of thermal and strength parameters as well as condition of underwater concrete over time [12-14].

Technology of concrete placement in a structure is important, this is the key factor of quality and durability of the object. Any errors during underwater concrete casting are usually disastrous for the structure and their repair is very expensive. They are, however, not all the factors to be taken into account. Also, concrete components used and concreting technology are of significant importance for the water organisms. The aim of the paper is to determine the effect of underwater concrete components on water parameters (pH and electrolytic conductivity) and intensity of settlement by living organisms.

2 Selected admixtures and additives to underwater concrete

A characteristic admixture for underwater concrete is a superplasticizer, i.e. HRWR. At the moment, new generation admixtures are added to UWC, which are produced using the latest technology.

2.1 Admixtures based on hybrid polymers

Hybrid polymers-based admixture, due to the application of knowledge about synthesis of molecules, enables a high range of efficient dosing, strong reduction of batch water content, long-lasting consistency as well as homogeneity and cohesion of concrete mixture. It features very strong dispersing activity in relation to fine fractions facilitating liquid consistency of concrete mixture to be achieved. At the same time, it enables desired liquidity to be kept for a very long time. It contains 20% of sodium nitrate, fish toxicity of which is classified in Category 4 (Practically non-toxic and not an irritant, CL50 > 1 000 mg/l/96h). The product is not classified as environmentally hazardous.

2.2 Admixtures based on stabilised polycarboxylates

The produced admixture based on stabilised polycarboxylates was developed especially for concrete mixtures featuring high flowability and ability to fill up the moulds. It assures long-lasting, stable consistency, ensuring high immunity to any changes of water content at the same time. Additionally, concrete with polycarboxylate admixture features fast strength increase at early setting stages, also when CEM II and CEM III are used, or fly ash is added, which may be of high importance for placing concrete underwater. The product is not classified as environmentally hazardous.

2.3 Admixture based on modified polycarboxylates and phosphonates

Another admixture is an admixture based on modified polycarboxylates and phosphonates. Due to its special molecular structure it enables production of cohesive mixture of low viscosity and long-lasting consistency. It is recommended for mixtures of very high workability and low water-cement ratio. It ensures long-lasting properties of mixture without retardation of setting time. In the case of this admixture no aquatic toxicity information is available.

2.4 Anti-washout admixtures

Powder admixture preventing washout in water during placement enables achieving liquid consistency mixture at washout loss below 3%. It makes the mixture resistant to washout and separation, significantly facilitating underwater integration. AWA admixture contains 10% of calcium formate (fish toxicity of Category 4 - Practically non-toxic and not irritant, CL50 > 1000 mg/l/96h). Similarly to HRWR it is not classified as environmentally hazardous. In line with EN 934-2 standard [15], AWA belongs to a group of complex admixtures, while HRWR to plasticizing admixtures (superplasticizers).

2.5 Fly ash

Fly ash is a fine-grained fraction of ashes produced during fuel combustion. Addition of fly ash ensures better workability of concrete mixture, improves concrete durability and reduces hydration temperature, preventing the risk of shrinkage in fresh concrete [16]. Application of ashes (with some limitations) is also economically justified as it allows replacement of some portion of cement with much cheaper material [17].

While considering the impact of ash content in concrete on aquatic environment, it is important to mention ashes filtrate featuring high salt content which is easily and quickly washed out, without posing, however, any environmental hazard. Water extracts from ashes obtained in laboratory research feature increased alkalinity, increased pH and increased content of SO_4^{2-} ions. At the same time, the waters are free of organic contaminations, biogenic substances as well as other substances adversely influencing the environment [18].

Combined application of AWA, HRWR admixtures and fly ashes (Table 1) proved to be very efficient in production and placement of concrete mixture under water. Interaction of the above components enables washout of binding agents from concrete mixture and bleeding to be reduced to minimum. This provides significantly better visibility during underwater concreting operations. At the same time, the mixture maintains pumping consistency and self-compacting properties. Finally, it allows to obtain UWC of high strength and quality [19-21].

Table 1. Chemical composition of fly ash.

Component	[%]	Component	[%]
SiO_2	54,78	Na_2O	1,25
Al_2O_3	24,17	P_2O_5	0,53
Fe_2O_3	6,36	SO_3	0,57
CaO	4,28	TiO_2	1,35
MgO	3,06	Ig-loss	2,73
K_2O	3,37		

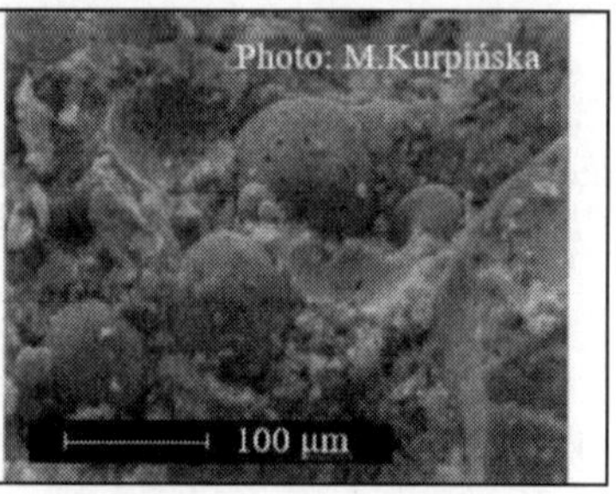

3 Influence of underwater concrete on aquatic organisms

The preliminary results of the authors' research [22] were published at the 8th International Scientific Conference, Toxic Substances in the Environment Toxe 2017. The aim of the research was to state if there is any influence of hydrotechnical concrete components on survivability of organisms inhabiting concrete elements.

The research was carried out on two concrete types: regular UWC (with AWA and HRWR added) and concrete with artificial aggregate modified with additives. Zebra mussel (*Dreissena polymorpha*), commonly present in freshwater reservoirs of low salt content, was used as a bioindicator [23]. Underwater concrete was ascertained to have lethal influence on *D. polymorpha*. Parameters of electrolytic conductivity and pH did not exceed the values present in lakes inhabited by that types of organisms. Therefore, the possibility of presence of compounds toxic for Zebra mussel emerging as a result of reaction in aquatic environment (lake) [22-25] shall be taken into account.

Up till now the aquatic organisms growing on concrete surface were a common phenomenon, considered the situation of adverse influence on concrete rather than the impact of living organisms. Natural components used in concrete production did not hinder mussels' settlement on concrete's surface, which were often growing in very large colonies. With regard to new concrete components which have appeared on the construction market, it was decided to carry out an experiment to check the influence of selected admixtures and additives on the settlement of living organisms on the surface of modified UWC.

4 Materials and methods

The experiments were conducted for three different compositions of UWC (no. 1, 2, 3). The compositions of concrete mixtures were shown in Table 2. Two types of lake water were selected, one originating from reservoir of good ecologic conditions: Wdzydze lake (*W*) and one from ecologically degraded lake: Klasztorne Małe lake in Kartuzy (*K*) (Fig. 1a).

a) b)

Fig. 1. a) Experimental site in natural environment, **b)** Method of samples seasoning.

Table 2. The compositions of concrete mixtures.

Component/ Properties	Concrete number		
	1	2	3
	[kg/m³]		
CEM I 52,5 N	420	-	-
CEM II/A-V 42,5 R	-	380	-
CEMIV/B(V) 32,5 R	-	-	260
Water	145	165	156
Silica sand 0/2mm	642	677	821
Gravel 2/8 mm	-	512	515
Gravel 8/16 mm	-	512	573
Granite grit 2/8	495	-	-
Granite grit 8/16	697	-	-
HRWR admixture based on hybrid polymers	-	-	1.6
HRWR admixture based on stabilised polycarboxylates	2.50	-	-
HRWR admixture based on modified polycarboxylates and phosphonates	-	3.50	-
AWA type admixture	-	1,50	-
water / cement ratio	0,35	0,43	0,60
f_{c2} [N/mm²]	59,2	26,8	9,5
f_{c28} [N/mm²]	87,7	44,9	26,2
f_{c56} [N/mm²]	92,6	50,2	33,2

The two-fold nature of the experiment was assumed: tests in natural environment and simultaneously in laboratory conditions (Fig. 1b). Before the experiments started, basic physical and chemical parameters of lake water had been measured: electrolytic conductivity (C) and pH (Table 3). Water temperature for experiment carried out in natural environment was adjusted to annual variability of temperature curve for a selected reservoir (*W* and *K*). In laboratory conditions, tests were carried out in the temperature of approximately 20°C (summer conditions in natural environment for coastal, littoral zone) with natural lighting and approx. 4-5°C without lighting (simulation of profundal zone).

Table 3. Water parameters before the experiment .

Water parameter	*W0*	*K0*
C [μS/cm]	251	385
pH	7,667	8,225

In natural environment experiment (Fig. 1a) 100 mm cubic concrete specimens were placed at the lakes bottoms in littoral zone. In laboratory part, 12 control reservoirs were established (2 water types, 3 concrete types, 2 storage environments) with 50 mm cubic concrete samples put inside (Fig. 1b).

Control measurements were planned 1, 3 and 6 months after experiments had started. The degree of growing of small aquatic organisms (microorganisms, algae, small invertebrates) on the concrete surface was evaluated using innovative dyeing method [26], in which the achieved images allow for evaluation of intensity and time of growing in different experimental conditions. Figure 2 shows pictures of concrete specimen without dyeing (Fig. 2a) as well as during dyeing (Fig. 2b). Based on images obtained as a result of dyeing with Rose Bengal a diverse degree of concrete inhabiting by small aquatic organisms was archived (Fig. 2c, 2d).

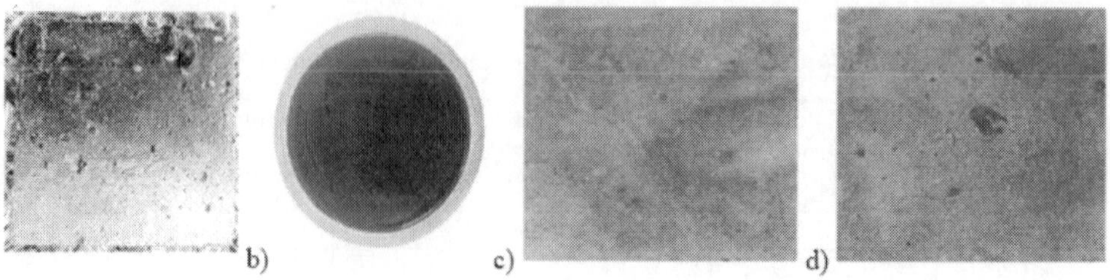

Fig. 2. a) Sample of concrete number 3 without dyeing, directly after taking it out from a lake, **b)** Dyeing process, **c)** Dyeing result - low degree of concrete inhabitation, **d)** Dyeing result - total coverage of concrete wall by small aquatic organisms; observations were carried out after 1 month of experiment time (photo: L. Grabarczyk).

4 Results and discussion

A month since the experiment had started, preliminary measurements of basic water parameters were taken and the process of dyeing of control samples was started. The results are presented in Table 4 and 5 (*W*0, *K*0 – initial parameters of water sampled from natural environment in *W* and *K* lakes).

Table 4. Change of water parameters (C and pH) in laboratory reservoirs at 20°C.

	***W*0**	***W*1-1**	***W*2-1**	***W*3-1**
C [μS/cm]	251	516	542	548
pH	7,667	8,645	8,721	8,814
	***K*0**	***K*1-1**	***K*2-1**	***K*3–1**
C [μS/cm]	385	697	745	771
pH	8,225	9,123	9,219	9,336

Table 5. Change of water parameters (C and pH) in laboratory reservoirs at 5°C.

	***W*0**	***W*1-1**	***W*2-1**	***W*3-1**
C [μS/cm]	251	364	397	401
pH	7,667	9,045	9,124	9,218
	***K*0**	***K*1-1**	***K*2-1**	***K*3–1**
C [μS/cm]	385	489	512	539
pH	8,225	9,538	9,689	9,831

The dyeing colour shade was evaluated on samples dried at 105 °C. At the time of publication drying still continues. Preliminary evaluation of freshly dyed samples allows to conclude that in the first 30 days of experiment, living organisms have grown on concretes in a quite large number, irrespectively of their type and environment. Detailed evaluation will allow for a more precise determination the concrete type and environment to create the most friendly conditions for microorganisms and algae settlement.

Regarding samples taken from natural environment single pieces of small invertebrates were observed, mainly on concrete no. 3 (Fig. 2a) - already such preliminary results allow for conclusion that not all underwater concretes (UWC) are lethally toxic for living organisms. The obtained images of concrete dyeing indicate diverse degree of inhabiting the concrete by small organisms at 20 °C, from insignificant degree up to the total coverage of the concrete's surface (Fig. 2c, 2d). Both the values of electrolytic conductivity and pH increased, whereas the value of electrolytic conductivity (C) has almost doubled. The results of C and pH values for *W* and *K* series samples are compared in Fig. 3 at 20°C and in Fig. 4 at 5°C. The conduced experiment proved the following:

- diverse degree of inhabiting concrete surface by aquatic organisms,

- influence of concrete on water parameters, like electrolytic conductivity and pH, while the increase of electrolytic conductivity is significant as compared to initial value,
- increase of electrolytic conductivity depends on the concrete sample used, the highest is for K3 sample incubated in ecologically degraded environment,
- differences in values of both, pH and conductivity, depend on experiment temperature,
- differences of concrete composition influence on diversification of water parameters, which may translate into various degree of environment attractiveness for living organisms,
- the largest changes of both pH and electrolytic conductivity were observed in ecologically degraded environment in concrete 3.

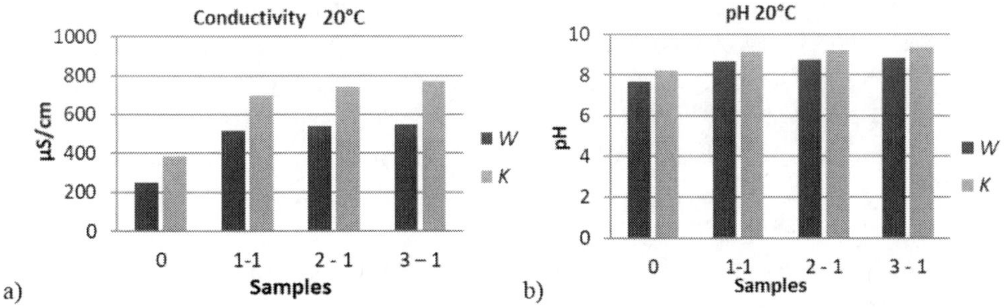

Fig. 3. Laboratory reservoirs at 20°C: **a)** changes of electrolytic conductivity, **b)** changes of water pH.

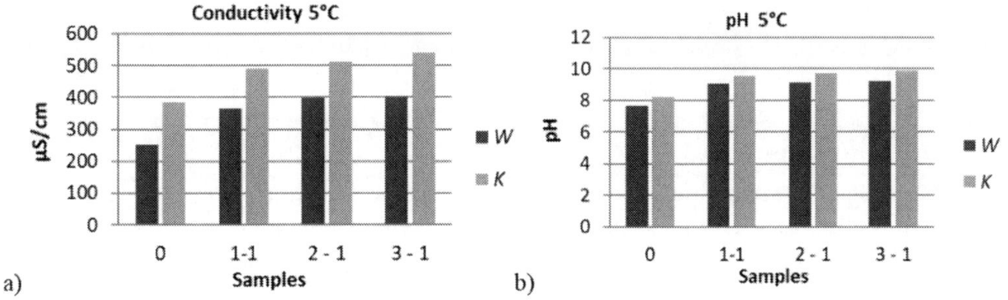

Fig. 4. Laboratory reservoirs at 5°C: **a)** changes of electrolytic conductivity, **b)** changes of water pH

5 Summary

Based on the results of the first stage of the studies, after 30 days of exposure of different concretes in water environments, it can be concluded that the highest impact on changes in water parameters concerns concrete number 3 with an HRWR admixture based on hybrid polymers. The tests showed an approximately two-fold increase in electrolytic conductivity. *W* reservoir the change ranges from 251 µS/cm to 548 µS/cm.

In the case of assessment of the degree of growing small organisms on concrete surface was carried out by an innovative dyeing method, at the same time as the water tests. After the first month of the experiment, the surface of the concrete with an HRWR admixture based on hybrid polymers was most often settled. On the other hand, in concrete with an AWA admixture and HRWR admixture based on modified polycarboxylates and phosphonates, a weak dyeing effect indicating a negligible amount of organisms has been noticed. The next research step will be carried out after 3 and 6 months of storage of concrete in the aquatic environment.

References

1. K. H. Khayat, M. Sonebi, ACI Materials Journal, **98**, 4, 289-295, (2001)
2. K. H. Khayat, J. Assaad, ACI Materials Journal, **100**, 3, 185-193, (2003)
3. V. S. Ramachandran, *Concrete Admixtures Handbook: Properties, Science and Technology*, (1995)
4. B. Grzyl, E. Miszewska-Urbańska, M. Apollo, *E3S Web of Conferences*, EDP Sciences, **17**, (2017)
5. M. Sonebi, Cement and Concrete Research, **31**, 11, 1553-1560, (2001)
6. M. Sonebi, K. H. Khayat, ACI Materials Journal, **98**, 3, 233-239, (2001)
7. M. Hara, S. Sogo, *Marine Concrete International Conference on Concrete in Marine Environment*, Concrete Society, London, 440-448, (1986)
8. A. Mariak, M. Kurpińska, *II Baltic Conference for Students and Young Researchers BalCon2018*, (to be published)
9. M. Kurpińska, T. Ferenc, *Shell Structures: Theory and Applications*, **4**, 549-552, (2018)
10. M. Kurpińska, Drogi i Mosty, **1-2**, (2011)
11. M. Kurpińska, T. Ferenc, *II International Conference of Computational Methods in Engineering Science*, (2017)
12. M. Miśkiewicz, Ł. Pyrzowski, K. Wilde, O. Mitrosz, Polish Maritime Research, **24**, 149-155, (2017)
13. A. Mariak, J. Chróścielewski, K. Wilde, *Shell Structures: Theory and Applications*, **4**, 557-560, (2018)
14. A. Mariak, M. Kurpińska, K. Wilde, *64 Scientific Conference Krynica Zdrój*, Poland, (2018), (to be published)
15. EN 934-2, Admixtures for concrete, mortar and grout. Concrete admixtures. Definitions, requirements, conformity, marking and labelling
16. K. H. Khayat, M. E. Gattioui, C. Nmai, *Fifth International CANMET/ACI Conference*, SP-173, ACI, Mich, (1997)
17. P. Freidenberg, E. Freidenberg, Przegląd budowlany, **10**, 32-35, (2007)
18. B. Broś, Aura Ochrona Środowiska, **10**, 13-16, (1992)
19. E. Horszczaruk, P. Brzozowski, Constr. Build. Mater., **72,** 167-173, (2014)
20. E. Horszczaruk, P. Brzozowski, Procedia Engineering, **196,** 97-104, (2017)
21. H. Falkner, V. Henke, Cement and Concrete Composites, **20**, 5, 377-385, (1998)
22. B. Wojtasik, M. Zbawicka, L. Grabarczyk, M. Kurpińska, *Toxic Substances in the Environemnt,* (2017)
23. P. Elliott, D. C. Aldridge, G. C. Moggridge, Water Res. **42**, 1664-1674, (2008)
24. B. R. Smith, D. R. Edds, Transactions of the Kansas Academy of Science, **117** (3 & 4), 159-166, (2014).
25. B. W. Kilgour, G. L. Mackie, *Zebra mussels: biology, impacts, and control*, Lewis Publishers, Boca Raton, Florida, 167-173, (1993)
26. B. Wojtasik, A method of assessing the biological corrosion of porous structures, including concrete, in particular hydrotechnical concrete. Patent application No. P.421947 Polish Patent Office (2017)

Use of radiography images and gray level co-occurrence matrix to investigate gravitational granular flow

Selam Waktola[1,*], *Krzysztof* Grudzień[1], *Maciej* Niedostatkiewicz[2], and *Laurent* Babout[1]

[1] Lodz University of Technology, Institute of Applied Computer Science, ul. Żeromskiego 116, 90-924 Lodz, Poland
[2] Gdańsk University of Technology, Faculty of Civil and Environmental Engineering, ul. Narutowcza 11/12, 80-233 Gdansk, Poland

Abstract. The paper presents analysis of granular gravitational flow based on radiography images processing. The investigations were conducted for silo model geometry with concentric/eccentric discharging modes. The continuous X-ray radiography scans of granular material distribution, acquired during flow, were obtained by means of an especially designed model silo with rectangular bin and different settings of hopper angles. Image processing involved texture analysis methods, the Gray-Level Co-Occurrence Matrix (GLCM). The texture analysis of radiography images provides information about changes of granular porosity in different silo zones during silo discharging process. This technique allows to divide the silo space into a number homogenous regions with similar porosity level. The proposed methodology was applied to analyse the flow in silo model with various hopper angles.

1 Introduction

A high level of automation of technological processes forces the necessity of storing and collecting materials for their further processing. Most of solid state industrial products appear, at various stages of production, in form of powders or grains [1]. Approximately 60% of industrial solid materials are stored, transported, or processed as bulk solids [2, 3]. Bulk solid materials are stored in containers; the largest groups are silos and bunkers, very often with reinforced concrete walls. Proper conveying and handling of bulk solids acts significantly on industry and society, where accurate designed silos and process flow control is crucial to efficiency and safety of industrial process and natural environment as well [4-6]. The control of granular flow phenomena in industrial installations has been investigated for many years, but it still remains a challenging task both in practical and theoretical point of view [7, 8]. Granular materials are ubiquitous by nature and for a scientific point of view, the study of their flow is seen among key chemical-physical phenomena without answers to many research questions, because of the peculiar interaction between particles during movement [1-3]. The analysis of bulk solid structure behaviour

* Corresponding author: kgrudzi@iis.p.lodz.pl

during flow allows to better understand the phenomena occurring in industrial processes. The silo discharging process is frequent in various branches of industry to introduce material in form of powders or grain to production line, after their storing in container. The main tasks of these containers are protection and storing solid materials and also as an important link between transporting lines in many industrial processes [1, 3, 5]. Silos are often used to heating or cooling, aeration or fumigation of powders. In industrial plants silos are applied for product mixing and blending.

Granular material behaviour is influenced by internal and external factors, such as granular properties, container structure, temperate or humidity, initial solid density, wall roughness, mean grain diameter of solid, pressure level, and flow velocity. There is a number of problems with silo utilization from uneven flow to blockage to silo collapse. These eventual disasters caused by uneven flow thus leads to major environmental and economic impacts due to the waste created [1]. This may be caused by the un- even material distribution, resulting in unsymmetrical stresses of walls and the bottom during loading and discharging of bulk materials. The main reasons influencing on stability of the silo is non-symmetrical flow of the bulk solid and non-controlled volume changes in the hopper [9, 10]. Both of them are connected with the shape and the dimension of the core of flow, created from the beginning of silo emptying process. The knowledge about the flow profile and mechanism of concentration changes, also at the core of flow allows to protect silo from the damage. This problem is crucial in the case of silos with funnel flow.

In literature numerous papers can be found describing different measurement techniques as well as numerical simulations dedicated to flow processes analysis [11-13]. Both approaches, based on experiment and computing calculations, provide significant results, enriching expert knowledge about flow dynamics. In order to conduct a complex flow analysis, high precision and accurate measurement systems are necessary. Many systems provide possibilities of flow monitoring using images [14]. The sets of 2D or 3D images of granular distribution during flow allow to conduct very deeply analysis of process behaviour in a spatio-temporal sense [15]. For such investigations process tomography systems provide valuable information about process, especially during in-situ experiments [18, 19]. X-ray tomography, one of the non-destructive techniques provide 3D information of the interior of the media, stands out as a unique tool to characterize processes that involve concentration changes, as in the case of gravitational flow [17-19].

The knowledge of density distribution within the bulk solid is of a major importance to understand and describe the flow behaviour of granular materials in silos while evaluating silo wall pressures, and to determine the flow type [6, 12, 20-22]. A comprehensive insight upon flow phenomena allows to develop more robust control system, save energy and achieve full quality product. This kind of collaboration between laboratory and industrial environment is based on rules, that wider knowledge about process can be used in control unit during real-time measurement data processing and improve the process.

The presented methodology aims at better understanding the flowing process of granular material in confined structure, thanks to use of X-ray imaging, development of dedicated image processing and analysis methods.

2 X-ray laboratory setup

In order to conduct a complex flow analysis, high precision and accurate measurement systems are necessary. Many systems provide possibilities of flow monitoring using images. The sets of 2D or 3D images of granular distribution during flow allow to conduct very deeply analysis of process behaviour in a spatio-temporal sense. For such an investigation process tomography systems provide valuable information about process, especially during in-situ experiments [19].

Visualizations of granular concentration distribution, based on the measurement of X-ray attenuation of sample located between X-ray source and detector, are very useful for flow diagnosis (fig. 1), [19]. The X-ray measurement can be presented in form of 2D radiography or 3D tomography images. In term of 2D radiography image, the integration of the X-ray attenuation is directly connected to the granular concentration located on the linear path between the X-ray source and pixel localization on detector. During the experiments granular material showcases different properties, rice, sorghum and glass particles.

Fig. 1. Experimental set-up a) picture of silo model in X-ray tomography hutch, b) silo model with rice and sorghum mixture.

In order to carry out in situ granular flow experiments, an especially designed model silo with rectangular bin was used. The bin part was 10 cm wide, 5 cm deep and 20 cm high. The left and right hopper angles can be independently set to generate different types of flows, i.e. mass and funnel flows, with concentric/eccentric discharging modes (see fig1b).

The silo material is polycarbonate with 5 mm thick. The outlet width is manually set, what allows to control the discharge velocity. The time-lapse studies were performed at INSA- Lyon (France) using a GE Phoenix v|tome|x device (see fig. 1). The device is equipped with a high- energy X-ray microfocus source (up to 160 kV) with a 4 μm spot size. The detector is equipped with a 1500x1920 pixels (with pixel size of 127X127 μm2), which allows to scan objects of height 20cm. During measurement the silo model was not rotated, as for 3D imaging, so position of silo was the same during all experiments. In figure 2 is presented example of raw radiography image.

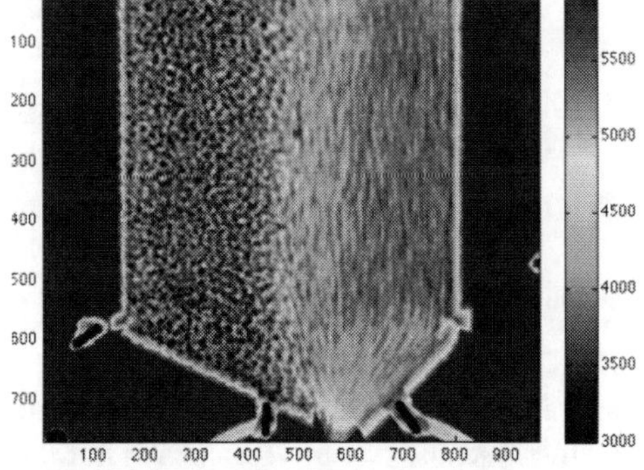

Fig. 2. Radiography image for eccentric flow - left hopper angle 20° and right 40°.

Before calculating the Gray-Level Co-Occurrence Matrix (GLCM) [23-25], the normalization procedure for raw radiography image was applied in order to improve the image contrast [19]. Such approach expedites the distinction between the stagnant and flowing zones inside granular material. However the most difficult problem is to segment the flowing zone into at least 2 different regions. Three different zones, creating characteristic spatial patterns, are visible in the presented image (fig. 2). For such type of images, texture analyzed can be applied to determine the silo area characterized by individual spatial distribution of gray pixels values distributions.

3 GLCM matrix analysis

The spatial relationship, texture analysis, of pixels in radiography images can be examined by applied the gray-level co-occurrence matrix (GLCM). The characterization of the texture of an image is conducted by calculating how often pairs of pixel in a specified spatial relationship occur in an image. This information is collected in GLCM matrix and after can be extracted by used statistical measures of this matrix [25].

In the first step of presented method the GLCM matrix is calculated and after its features are determined, which allow to analyze the texture patterns in images. The statistical parameters of GLCM are extracted based on the normalized, symmetric GLCM matrix. The matrix values calculation can be described as:

$$\boldsymbol{GLCM}_{\Delta i,\Delta j}(v,w) = \sum_{v,w=1}^{K} \begin{cases} 1, if \;\; I(i,j) = v \; and \; I(i+\Delta i, j+\Delta j) = w \\ 0, \;\; otherwise \end{cases} \quad (1)$$

where:

v and w	pixel values,
i and j	spatial positions of the pixel in the input image $\boldsymbol{I}$,
$(\Delta i, \Delta j)$	offsets – displacement vector, which specify direction and size of neighbors of the pixel of interest (see fig. 3),
$I(i, j)$	pixel value at position (i, j).

In the presented equation the size of GLCM matrix is equal to KxK, where K indicates number of grey levels in input image, so before calculation of GLCM matrix the grey level should decreased to reasonable value.

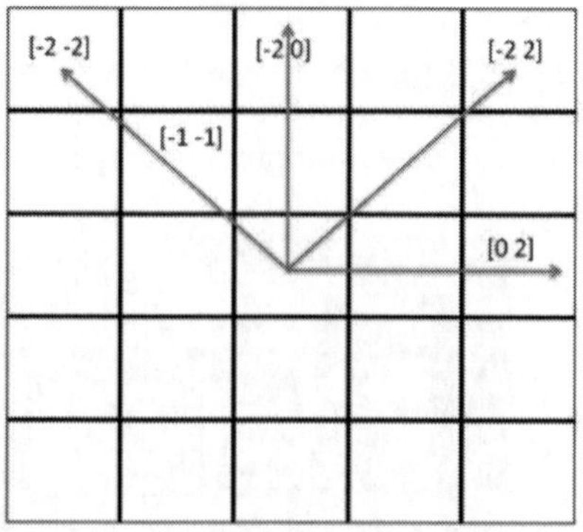

Fig. 3. Definition of size and orientation of neighbourhood for GLCM matrix calculation, example for displacement vector (2, 2).

The various features, which measure the gray-level distribution, can be calculated based on GLCM matrix values. In presented paper were chosen homogeneity and entropy:

$$homogeneity = \sum_{v,w=1}^{K} \frac{1}{1+|v-w|} p(v,w) \quad (2)$$

$$entropy = -\sum_{v,w=1}^{K} p(v,w)^2 * log(p(v,w)) \quad (3)$$

where:

$p(v,w)$ depicts GLCM's values as probabilities after normalization,

$\sigma_x, \sigma_y, \mu_x, \mu_y$ standard deviations and means of $p_x(v) = \sum_{j=1}^{K} p(v,w)$ and $p_y(j) = \sum_{i=1}^{K} p(v,w)$ respectively.

All features were calculated locally, for predefined displacement vector and number of grey levels *K*. The presented results were obtained for number of discretized grey level *K*=8 (in input radiography image) and for offsets [-1 0] and [1 0] (fig. 3). The size of neighbourhood for calculating local GLCM was 15x15. The calculations were conducted in *Matlab*® numerical computing environment.

4. Results and discussion

The radiography images were preprocessed before calculation of GLCM matrix. The applied normalization procedure of radiography image was described in [19]. The normalized radiography image allows to visualize the distribution of bulk solid concentration in comparison to initial state of silo discharging process (before open the silo outlet). Additionally the procedure eliminates influence of the silo wall on measurement results. In figure 4 are presented two radiography images, before normalization (fig. 4a), and after normalization (fig. 4b). The normalization calibrates data into range <0, 1>. The value 0 means empty space, radiography was obtained for empty silo and value 1 means values of material concentration obtained at initial stage of silo discharging process, radiography was obtained for full silo. The presented results ware taken for initially dense packing and for eccentric flow.

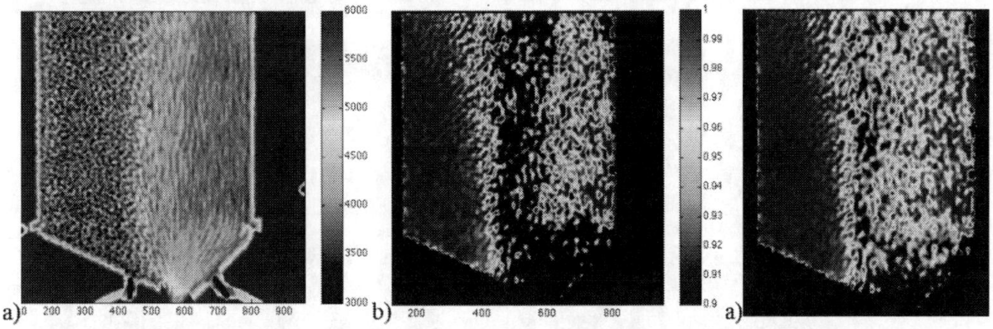

Fig. 4. Radiography images, a) scaled raw data, b) normalized data, c) filtering of normalized data.

The second step in pre-processing procedure was applied image filtering (average filter with size 7x7). In figure 4c is presented the example of normalized radiography image after filtering. The averaging filter allowed to visualize the granular material distribution in easier way to determine homogeneities of material structures. The examples presented in Figure 5 were obtained for middle and final step of silo discharging process. The homogeneity images (see fig. 5b) can be interpreted as level of similarity in granular material distribution; a higher value depicts more homogeneous area of pixels values, what is important for better interpretation of images. Entropy images provide similar information, however the scale values are much different than for homogeneity images.

The analysis of the obtained results allows to state, that it is possible to divide a silo space into part, characterized by similar flow pattern. Analysing of the homogenous image presented for other time point (fig. 5b) shows that high level of homogeneous is presented in two segments of silo (left and right part). It is also possible to distinguish, from the rest of silo, main funnel flow area with lower homogeneity value (~0.925). The combine of the normalized radiography image (fig. 6b) and GLCM homogenous image (fig. 6a) should allow to easier distinguish 3 zones of flow. In the future work this approach will be developed as well applied other GLCM features to prepared set of data for classification procedure using k-means algorithm.

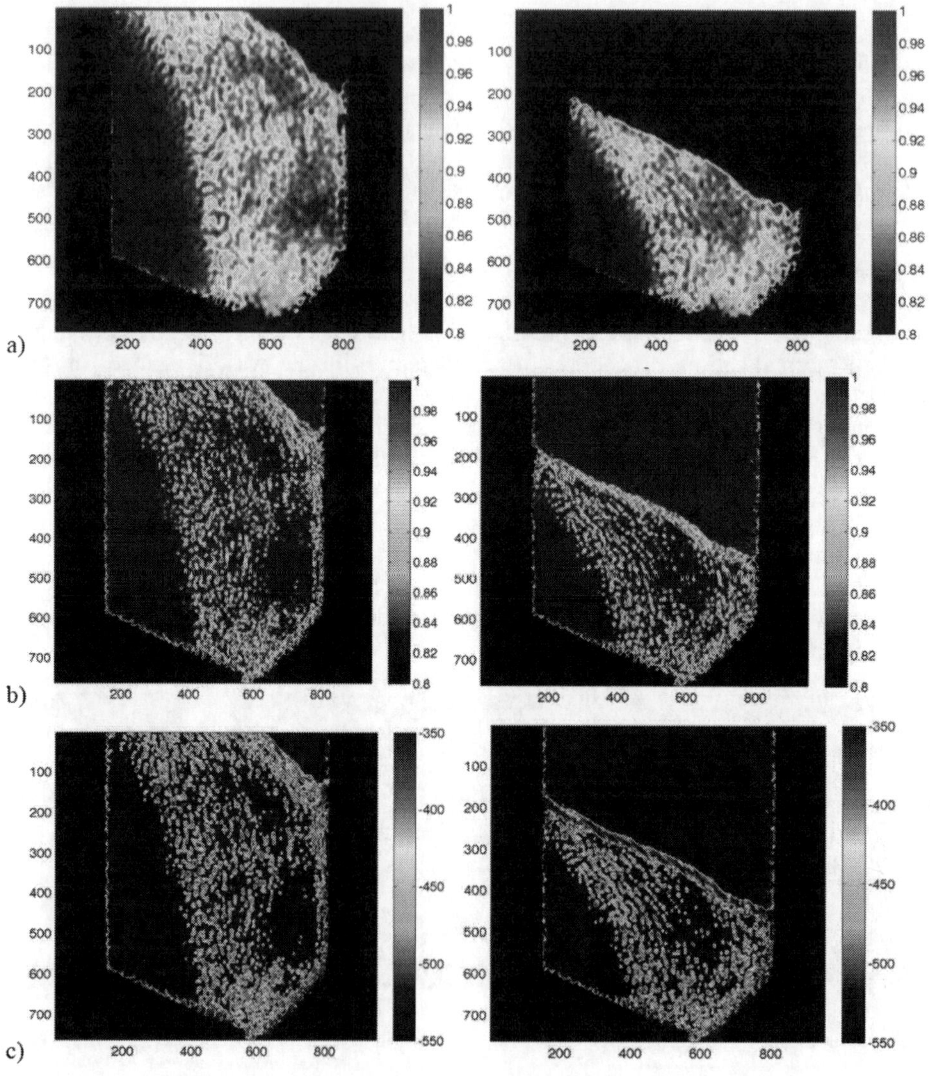

Fig. 5. Images of GLCM statistical features, a) normalized radiography after filtering, b) homogeneity, c) entropy.

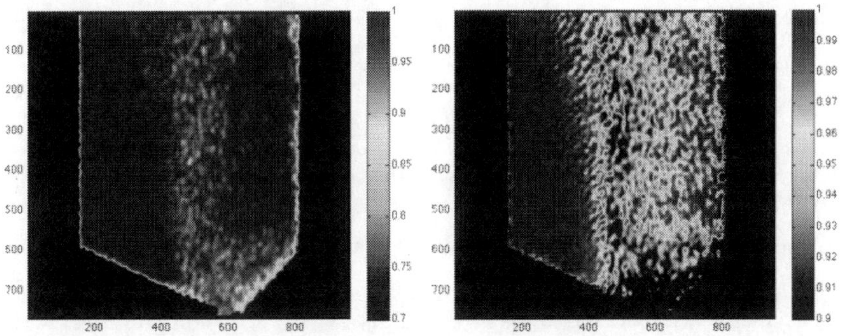

Fig. 6. Data after averaging filtering (7x7), a) the normalized radiography image, b) image of homogeneity.

Conclusions

The paper presents preliminary results of developing computer system for analysing of silo discharging process. The aim of instigation is a prepared image processing algorithm which allows to divide the silo area into homogenous zones in context of similarity in distribution of bulk solid concentration. The knowledge about the level of concentration changes during silo discharging in separate zones provides new point of view for investigated gravitational flow. Silo flow segmentation into homogeneous structures, in terms of granular material concentration changes, will improve understanding of flow phenomena for concentric and eccentric flow with various properties granular particles.

Similarity in pattern of material distribution occurring in radiography images can be analysed with the aid of texture analysis methods. The proposed concept bases on analysis of GLCM matrix features. The obtained GLCM features show that is possible to distinguish silo parts, much easier than for raw or even normalized radiography image. However in order to achieve the expected results necessary is to conduct more sophisticated analysis of GLCM matrix. Especially, the combination of different features and radiography images should provide vector of data, which can be used for conducting the image segmentation.

This work is supported by the National Science Centre, Poland (NCN OPUS10 2015/19/B/ST8/02773)

References

1. D. Schulze, Powders and Bulk Solids (Springer, 2008)
2. J.P.K.Seville, U. Tuzun, R. Clift, Processing of Particulate Solids, (Blackie Academic, London, 1997)
3. C. J. Brown, J. Nielsen, Silos: fundamentals of theory, behavior, and design, (E & FN Spon, 1998)
4. Buick J.M., J. Chavez-Sagarnaga, Z. Zhing, J.Y. Ooi, D.M. Pankaj, D.M. Cambell, C.A. Greated, Journal of Engineering Mechanics ASCE, **131**, 299 (2005)
5. K. Wilde, J. Tejchman, M. Rucka, N. Niedostatkiewicz, Powder Technology, **198**, 38 (2010)
6. M. Wojcik, J. Tejchman, Powder Technology, **192**, 298 (2009)
7. V. Mosorov, Flow Pattern Tracing for Mass Flow Rate Measurement in Pneumatic Conveying Using Twin Plane Electrical Capacitance Tomography, Particle & Particle Systems Characterization, **25**, 259, (2008)

8. A. Romanowski, K. Grudzien, R.G. Aykroyd, R.A. Williams, Part. Part. Syst. Charact., **23**, 289 (2006)
9. M.L. Dhoriyani, K.K. Jonnalagadda, R.K. Kandikatla, K.K. Rao, Powder Technology, **167**, 55 (2006)
10. R. Maiti, G. Das, P. K. Das, Powder Technology, **301**, 1054 (2016)
11. R. Balevičius, R. Kačianauskas, Z. Mróz, Irena Sielamowicz, Powder Technology, **187**, 159 (2008)
12. J. Tejchman J., Confined Granular Flow in Silos, Experimental and Numerical Investigations (Springer, 2013)
13. T. Rymarczyk, P. Tchórzewski, P. Adamkiewicz, K. Duda, J. Szumowski, J. Sikora, IEEE Sensors Journal, **17**, 8166 (2017)
14. J. Kryszyn, D. M. Wanta and W. T. Smolik, IEEE Sensors Journal, **17**, 8107 (2017)
15. R. Banasiak, R. Wajman, T. Jaworski, P. Fiderek, H. Fidos, J. Nowakowski, Dominik Sankowski, International Journal of Multiphase Flow, **58**, 1 (2014)
16. M. Niedostatkiewicz, J. Tejchman, K. Grudzień, Z. Chaniecki, Chemical Engineering Research and Design, **88**, 1037, (2010)
17. R.L. Michalowski, Powder Technology, **39**, 29 (1984)
18. C. Chen, P. W. Woźniak, A. Romanowski, M. Obaid, T. Jaworski, J. Kucharski, K. Grudzień, S. Zhao, and M. Fjeld, ACM Trans. Intell. Syst. Technol., **7**, 25, (2016)
19. K. Grudzień K., M. Niedostatkiewicz, J. Adrien, E. Maire, and L. Babout, Powder Technology, **224**, 196, (2012)
20. H. Ahn, Z. Başaranoğlu, M. Yılmaz, A. Buğutekin, M. Zafer Gül, Powder Technology, **186**, 65 (2008)
21. C. González-Montellano, Á. Ramírez, E. Gallego, F. Ayuga, Chemical Engineering Science, **66**, 5116 (2011)
22. C. González-Montellano, F. Ayuga, J. Y. Ooi, Granular Matter, **13**, 149 (2011)
23. R. Shenbagavalli, K. Ramar, Bonfring International Journal of Advances in Image Processing, **1**, 15, (2011)
24. R. M. Haralick, K. Shanmugam, I. Dinstein, IEEE Transactions on Systems, Man, and Cybernetics, **3**, 610, (1973)
25. Q. Chen, E. Agu, International Conference on Healthcare Informatics, 453 (Dallas, TX, 2015)

Verification of selected calculation methods regarding shear strength in beams without web reinforcement

Marta Hirsz[1,*], and *Krystyna* Nagrodzka-Godycka[1]

[1] Gdańsk University of Technology, Faculty of Civil and Environmental Engineering, ul. Narutowicza 11/12, 80-233 Gdańsk, Poland

Abstract. The purpose of the article was to compare selected calculation methods regarding shear strength in reinforced concrete beams without web reinforcement. Several calculation methods were tested. This included codes: PN-EN 1992-1-1:2008, ACI 318-14 and *fib* Model Code for Concrete Structures 2010. The analysis also consists of authorial methods published in technical literature. Calculations of shear strengths were made based on experimental works found in literature. The shear strength ratios V_{test}/V_{calc} were chosen to be the yardstick of comparison, where V_{test} is the experimental shear strength and V_{calc} is the calculated shear strength.. A wide range of variables including shear span/depth ratio, compressive strength of concrete, longitudinal steel percentage helped to verify the applicability of calculation methods. Although most of authorial techniques proved to be unstable, they succeeded to show that codes' formulas for shear strength may still be improved. The presented article is a part of Authors' long term research in the matter and a new chapter of their study now concerning beams without web reinforcement.

1 Introduction

Over the years much research and many debates have taken place all around the world to better understand the shear mechanism in beams both with and without web reinforcement [1, 2]. Although many experiments and analysis have been carried out, the provisions regarding calculating shear strength provide results that often differ from experimental data [3, 4, 5]. As it was prior proved by the Authors in the context of reinforced and prestressed concrete beams with shear reinforcement [6], the understanding of stress distribution in the support zone is still to be improved. Although the reliability of most calculation methods was undeniable, economics was lacking. However is this observation also valid for beams without shear reinforcement?

1.1 Primary assumptions

In order to ensure the coherence of the analysis, initial assumptions were taken:

- there was no shear reinforcement in beams,

* Corresponding author: marta.wisniowska@pg.edu.pl

- longitudinal reinforcement was either steel or GFRP bars,
- all cross sections were either rectangular or T-beams,
- all beams failed in shear,
- all beams were single – span,
- no limitation on material properties was imposed,
- beams with fiber - reinforced concrete were excluded from the analysis,
- in order to compare analytical results with experimental data, all units were taken without reduction factors.

1.2 Nomenclature

In order to ensure the clear - cut of the analysis, main parameters are named below:

a	shear span
b_w	web width
d	effective beam depth
f_c	compressive stress in concrete
f_y	yield stress of steel
ρ_l	longitudinal reinforcement ratio
KHH	method of Kuo, Hsu, Hwang
RFF	method of Rebeiz, Fente, Frabizzio
AFR	method of Ahmad, Fareed, Rafeeqi

2 Calculation methods

The fundamental approach to the shear problem is the truss analogy model presented by Mörsch [7]. The Strut-and-Tie model was based on two basic types of elements: compression struts and tension ties. Being modified many times, mostly on the value of the inclination angle, it became a leading model in most European Codes. In the following analysis PN-EN 1992-1-1:2008 [8] (eq. (1)) was chosen to be its representative. Experiments held by members of the American Concrete Institute proved that the truss analogy model does not cover all the variables that were affecting shear strength during laboratory tests. Therefore a semi - empirical formula was developed and adopted in ACI 318 - 14 [9] (eq. (4)). The last standard that was taken into consideration in the following analysis is *fib* Model Code 2010 [10] with its levels of approximation that each base on different shear model. In the presented analysis the first (eq. (5)) and the second (eq. (8)) level of approximation were analyzed.

In addition to the standards, four authorial methods published in technical literature were taken into consideration. The first two pose an attempt to improve ACI 318 provisions. Frosch Method [11] (eq. (10)) modified Standard's provisions in two ways. First it replaced the effective beam depth (*d*) with a cracked transformed section neutral axis depth (*c*). Secondly it eliminated longitudinal reinforcement ratio (ρ_l) as it is already taken into account in computing the value of c. Another attempt to modify ACI 318 provisions was carried by Korean scientists Kuo, Hsu and Hwang [12] (eq.(13)). The KHH method adopted a function of a shear span to effective depth radio $((a/d)^{-0,7})$ introduced at Huston University [12] that reflects the arch action in the beam. Following Frosch's expression it uses the depth of uncracked compression zone. A different method, introduced by Rebeiz, Fente, Frabizzio [13] (eq.(14)) was developed using the techniques of dimensional analysis and multiple regression analysis. The RFF method also takes into account the differences between short and long beams in terms of shear behavior. The last method by Ahmad, Fareed, Rafeeqi [14] (eq.(16)) for normal and light weight reinforced concrete slender beams without web reinforcement was based on the analysis of predictive accuracy of empirical equations used in different codes and also authorial methods.

2.1 Empirical equations for shear capacity in concrete

$$V_{PN\text{-}EN} = max\{[C_{rd,c}\, k\, (100\, \rho_l)^{1/3} + k_1\, \sigma_{cp}]\, b_w\, d;\ (v_{min} + k_1\, \sigma_{cp})\, b_w\, d\} \quad (1)$$

where:

$$C_{rd,c} = 0.18/\gamma_c \quad (2)$$

(note: following primary assumptions all units were taken without reduction factors)

$$k = min\{1+(200/d)^{1/2};\ 2\} \quad (3)$$

$$V_{ACI} = 0.17\, f_c^{1/2} b_w\, d \quad (4)$$

$$V_{MC,I} = k_v\, f_c^{1/2} b_w\, z \quad (5)$$

where:

$$k_v = 180/(1000+1{,}25z) \quad (6)$$

$$z = 0{,}9d \quad (7)$$

$$V_{MC,II} = k_v\, f_c^{1/2} b_w\, z \quad (8)$$

where:

$$k_v = (0.4/(1+1500\varepsilon_x))/(1300/(1000+k_{dg}\, z)) \quad (9)$$

$$V_{Frosch} = 0.42\, f_c^{1/2} b_w\, c \quad (10)$$

where:

$$c = (2\, \rho_l\, n + (\rho_l\, n)^2)^{1/2} - \rho_l\, n \quad (11)$$

$$n = E_S/E_C \quad (12)$$

$$V_{KHH} = min\{1.17(a/d)^{-0,7} f_c^{1/2} b_w\, c;\ 0.83\, f_c^{1/2} b_w\, c\} \quad (13)$$

$$V_{RFF} = (0{,}4 + (f_c\, \rho_l (d/a))^{1/2} (10 - 3A_d))\, b_w\, d \quad (14)$$

where:

$$A_d = \{a/d \text{ for } 1.0 < a/d < 2.5;\ 2.5 \text{ for } a/d \geq 2.5\} \quad (15)$$

$$V_{AFR} = \xi\, 0{,}35(\, a/d\, f_c\, c)^{0.33}\, \rho_l^{0.1} \quad (16)$$

where:

$$\xi = \{1 \text{ for } d < 300 \text{ mm};\ 17.32/(d)^{1/2} \text{ for } d \geq 300 \text{ mm}\} \quad (17)$$

3 Beams database

The assessment of the effectiveness of the selected calculation methods was based on shear test results of beams published in technical literature. A numbet of 76 beams without shear reinforcement was considered. Table 1. presents properties of the analysed beams that were used while computing shear strength.

Table 1. Properties of specimens.

Author	No.	f_c[MPa]	P_l [%]	b_w [cm]	a/d
Sokołowski [15]	3	55.0	4.49	10	1.5 - 3.5
Thamrin et. al [16]	14	13.0 - 33.5	0.6-2.5	12.5 - 13	2.3 - 3.8
Palaskas et. al[17]	3	32.0 - 32.8	0.5 - 0.7	19	3.9 - 4.1
Hanoon et. al[18]	4	19.9 - 60.1	1.1 - 3.5	18	1.9 - 2.1
Słowik [19]	25	35.0	0.9 - 1.8	12	1.8 - 4.1
Bakhmari, Ahmad [20]	27	44.8 - 52.7	0.6 - 1.1	15	2.0 - 6.0

4 Shear strength analysis

Fig. 1 presents distributions of shear strength ratios V_{test}/V_{calc} where V_{test} is the experimental shear strength and V_{calc} is the calculated shear strength by Codes' design procedures. Table 2 presents statistic data of the results. It can be seen that statistically ACI 318 - 14 [9] was the most accurate, however as the only method it happened to underestimate shear strengths. Underestimation means that the calculated value of shear strength was higher than the actual shear strength that was observed in laboratory tests. Therefore looking at the presented data it would be most reasonable to calculate shear strengths based on PN-EN 1992-1-1:2008 [8] or second level of approximation of *fib* Model Code 2010 [10]. The first level of approximation of *fib* Model Code 2010 [10] highly overestimated (calculated shear strength much lower than tested shear strength) most of the results.

Table 2. Statistic data for Codes.

method	PN-EN	ACI	MC I	MC II
average	2.03	1.34	3.04	2.29
standard diviation	0.78	0.52	1.14	0.86
variation coefficient [%]	38.59	38.50	37.72	37.62

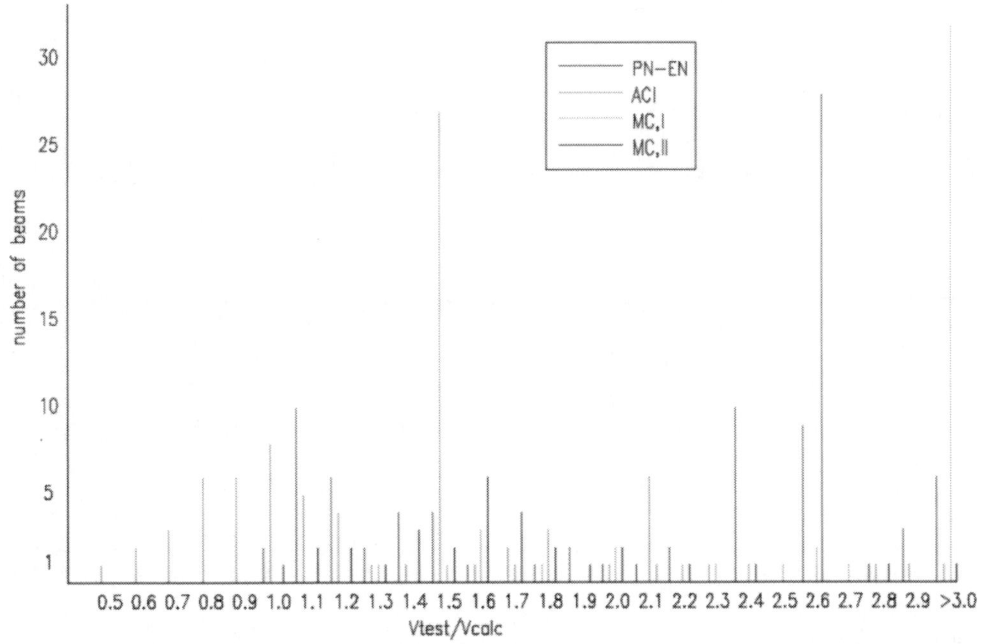

Fig. 1. Distributions of shear strength ratios for Codes.

Fig. 2 presents distributions of shear strength ratios V_{test}/V_{calc} where V_{test} is the experimental shear strength and V_{calc} is the calculated shear strength by Authorial design procedures. Table 3 presents statistic data of the results. The average ratios V_{test}/V_{calc} were not so much improved in comparison with Codes' results. Moreover the variation coefficients were higher comparing with Codes. Frosch method [11] and KHH method [12] led to a high number of uneconomical overestimations ($V_{test}/V_{calc} > 3.0$). In the Authors' opinion the only Authorial method that seems promising was the RFF method [13].

Table 3. Statistic data for Authorial Methods.

method	Frosch	RFF	KHH	AFR
average	3.64	1.79	2.92	1.25
standard diviation	1.37	0.72	1.18	0.56
variation coefficient [%]	37.62	40.38	40.44	44.79

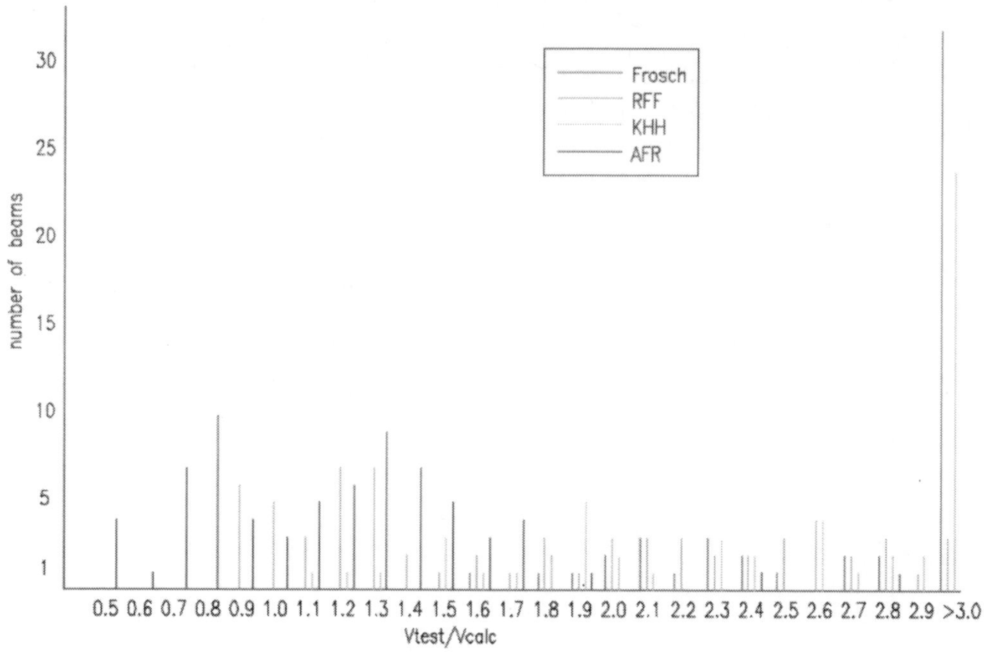

Fig. 2. Distributions of shear strength ratios for authorial methods.

5 Conclusions

The aim of the article was to compare selected calculation methods regarding shear strength in beams without web reinforcement. The analysis was limited due to the number of beams (76) and variety of properties affecting shear strength. Nevertheless it succeeded to prove that the provisions regarding shear strength could still be improved to give more accurate and reliable results. It should be noted that the aim of the article was not to decide which method is the most accurate, but to check if the available methods provide us with reliable results.

References

1. T. Godycki - Ćwirko, *Ścinanie w Żelbecie*, Arkady, Warszawa, (1968)
2. G. N. J. Kani, ACI Journal, **61**, 4, 441-468 (1964)
3. M. P. Collins, D. Mitchell, P. Adebar, F. J. Vecchio, ACI Structural Journal, **93**, 1, 93-S5 (1996)
4. E. L. Labib., Y. L. Mo, T. T. C. Hsu, International Journal of Concrete Structures and Materials, **7**, 1, 71-78, (2013)
5. T. L. Resende, L. C. D. Shehata, I. A. E. M. Shehata, Ibracon Structures and Materials Journal, **10**, 4, 886 - 894, (2017)
6. M. Wiśniowska, K. Nagrodzka - Godycka, Procedia Engineering, **193**, 136-143, (2017)
7. E. Mörsch, Der Eisenbetonbau, seine Theorie und Anwendung, Bd I, Verlag K. Wittwer, Stuttgart, (1929)
8. PN-EN 1992-1-1 Eurokod 2: Projektowanie konstrukcji z betonu. Część 1-1: Reguły ogólne i reguły dla budynków. PKN, Warszawa (2008)

9. ACI Committee Building Code Requirements for Structural Concrete (ACI 318-14) and Commentary. American Concrete Institute, Farmington Hills, MI (2014)
10. *fib* Model Code for Concrete Structures 2010, Ernst und Sohn, Berlin, 2013; polish version: Ajdukiewicz A.: Pre - Norma Konstrukcji Betonowych *fib* Model Code 2010, tom 2, Stowarzyszenie Producentów Cementu, Polska Grupa Narodowa fib (2014)
11. E. J. Tompos, R. J. Frosch, *ACI Structural Journal,* **99**, 5, 559 - 567, (2002)
12. W. W. Kuo, T. T. C. Hsu, S. J. Hwang, ACI Structural Journal, **97**, 4, 809-818, (2004)
13. K. S. Rebeiz, J. Fente, M. Frabizzio, 8th ASCE Speciality Conference on Probabilistic Mechanics and Structural Reliability,
14. S. H. Ahmad, S. Fareed, S. F. A. Rafeeqi, Civil Engineering and Architecture, **2**, 1, 33-41 (2014)
15. J. Sokołowski, Badania eksperymentalne strefy przypodporowej żelbetowych belek częściowo sprężonych, Praca doktorska, Politechnika Gdańska, (2010)
16. R. Thamrin, J. Tanjung, Journal of Engineering Science and Technology, **5**, 5 (2016)
17. M. N. Palaskas, E. K. Attiobge, D. Darwin, ACI Structural Journal, **78**, 6, 447-455 (1981)
18. V. Hanoon, M. S. Jaafar, H. J. Abed, Constructii, **2**, 15-23, (2014)
19. M. Słowik, *Nośność na ścinanie zginanych elementów żelbetowych bez zbrojenia poprzecznego*, Politechnika Lubelska, (2016)
20. I. A. Bukhari, S. Ahmad, The Arabian Journal for Science and Engineering, **33**, 2B, 321-336 (2008)

Multi-criteria analysis of the construction technologies in the aspect of sustainable development

Magdalena Gicala[1,*], and *Anna* Sobotka[1]

[1]AGH University of Science and Technology, Faculty of Mining and Geoengineering, al. Mickiewicza 30, 30-059 Cracow, Poland

Abstract. The complexity of sustainability complicates the design work and requires the implementation of Integrated Life Cycle Design. Due to the need of balance between environmental, social and economic aspects and the multitude of analysis indicators, the assessment of existing buildings is a multi-criteria problem too. The aim of the study is to compare three construction technologies – two traditional solutions and an innovative technology – in the aspect of sustainable development. The assessment was limited to the building envelope materials and included in entirety of 13 environmental, social and economic indicators, characteristic for the first stage of building life cycle. The absence of explicit dominance of one technology over another in terms of these factors was a starting point for the multi-criteria analysis. Based on the comparison of given technologies by WSM and TOPSIS methods, the multi-criteria analysis was carried out and the most advantageous technology was indicated. The results allow for an indirect selection of building materials to fulfil the sustainability requirements for the building.

1 Introduction

Since the United Nations Conference on Environment and Development (UNCED) in Rio de Janeiro, called the Earth Summit (1992), attention has been paid to the issue of sustainable development and to the implementation of its elements in various sectors of the economy. It is aimed at meeting the needs of the present without compromising the ability of future generations by harmonizing environmental, social and economic activities. These assumptions are reflected in construction industry too. The research undertaken in this area have different scopes and cover specific aspects of sustainability [1], selected life cycle stages [2], impacts or parts of a building, especially building materials [3]. However, the dynamic socio-economic development and changes in the construction industry require the update of research, correction of the assessment factors and the analysis of new technologies. The results obtained in this way form a basis to determine the development potential of selected technologies or indicating their shortcomings in the face of dynamic changes in construction industry.

* Corresponding author: mgicala@agh.edu.pl

2 Research methodology

The purpose of the study was to rank selected construction technologies (T1-T3) from the one that best satisfies sustainability requirements to the worst one. The assessment was limited to the building envelope materials and included in entirety of 13 environmental, social and economic indicators, characteristic for the first stage of building life cycle. The information contained in Environmental Product Declarations, catalogues and price lists of construction product manufacturers, and the data obtained from calculations were used in the analysis. The studies were carried out according to the currently applicable standards concerning the issue of sustainable building [4,5]. Due to the complexity of the sustainability issue, the multi-criteria approach was proposed in the second part, as it would allow considering simultaneous influence of various factors on the solution priority. The technologies were analysed using the WSM (Weighted Sum Method) and TOPSIS method (The Technique for Order of Preference by Similarity to Ideal Solution). The most advantageous technology was indicated on the basis of the obtained results.

2.1 The subject of the analysis

The subject of the analysis was a single-family one-storey residential building, located in Poland, made with the technology of self-supporting building development systems based on foamed polystyrene (T1) with a usable area of 110.75 m^2. For the purpose of the study, alternative conventional solutions such as: technology of timber frame building (T2) and technology using ceramic construction materials (T3) were considered. Comparison was made on the basis of building material impacts per 1 m^2.

2.2 Environmental aspect analysis

Environmental analysis covered 11 indicators selected from those proposed by the standard EN 15643. The values of the indicators were determined on the basis of Environmental Product Declarations (EPD) and the results of material consumption analysis for building envelope. The calculation was carried out using the following formula (1):

$$P_{te} = \sum_{j=1}^{m} k_{ej} m_j \tag{1}$$

where:

k_{ej} environmental impact indicator for unit volume of the jth material in e^{th} environmental impact category,

m_j amount of the jth material in $1m^2$ of partition area,

P_{te} environmental impact of materials in $1m^2$ of the partition made using the tth technology in e^{th} category.

The values of the indicators were referred to the unit of building usable area, and the S_{te} values were obtained. The Ecopoint unit [6,7] determined on the basis of relation (2) was used to rank the analysed technologies:

$$E_p = \sum_{e=1}^{n_e} \left(\frac{S_{te}}{R_e} \times w_e \right) \tag{2}$$

where:

e number of environmental impact category, $e = 1,2,...,n_e$, $n_e = 11$,

w_e weight of eth environmental impact category,

R_e reference value for e[th] category of environmental impacts (annual impacts of 1 European citizen),

S_{te} characterized indicator describing e[th] category of environmental impacts for t[th] technology (per 1 m^2).

The reference values and weights of the environmental indicators were taken according to [6, 8-10] and shown in Table 1.

Table 1. Reference values (R_e) per person per year and weights (W_e) of environmental indicators.

Indicator (Abbreviation of indicator)	Units	R_e	W_e [%]
Global warming potential (GWP)	kgCO_2 eq	12,300	24.1
Net use of fresh water (FW)	m^3	377	15.2
Depletion potential of the stratospheric ozone layer (ODP)	kgCFC11 eq.	0.22	13.5
Acidification potential of land and water (AP)	kgSO_2 eq.	71.2	8.4
Eutrophication potential (EP)	kg$(PO_4)^{3-}$ eq.	32.5	8.2
Radioactive waste disposed (RWD)	kg	3.91	7.0
Abiotic Resources Depletion Potential for elements (ADPE)	kg Sb eq.	39.1	6.6
Photochemical ozone creation potential (POCP)	kg Ethene eq.	21.5	5.8
Hazardous waste disposed (HWD)	kg	187.4	5.0
Abiotic Resources Depletion Potential of fossil fuels (ADPF)	MJ	273,000	4.0
Non-hazardous waste disposed (NHWD)	kg	3,750	2.1

2.3 Social aspect analysis

The social aspect analysis was based on one of the categories indicated by the EN 15643-3: *health and comfort*, taking into account thermal characteristics. This criterion was represented by the heat transfer coefficient of the external wall, its value was calculated on the basis of the relation:

$$U = \sum_{j=1}^{m} (d_j / \lambda_j) \tag{3}$$

where:

U heat transfer coefficient of external wall, $W / m^2 K$,

d_j thickness of jth material of external wall layer, m ,

λ_j thermal conductivity of jth material of external wall layer, W / mK .

2.4 Economic aspect analysis

According to the convention and scope of the research, the prices of building envelopes materials per usable area unit were assumed representative for the assessment and comparison of the technologies. The prices were calculated on the basis of data contained in the catalogues of material manufacturers. Moreover, using the formula (4), relative increase in prices for the technologies was determined, as compared to the lowest price:

$$\delta_c = [(c_t - c_{\min}) / c_{\min}] \times 100\% \tag{4}$$

where:

c_t price of materials per 1 m^2 of usable area for tth technology [PLN],

c_{min} the lowest estimated price [PLN].

2.5 Multi-criteria analysis

According to the general principles of building sustainability [5], all 3 aspects of sustainable development are necessary elements in the systemic approach. Therefore, it is required to consider simultaneous influence of environmental, social and economic impacts. A multi-criteria analysis based on WSM and TOPSIS methods is applicable in this issue, in order to minimize the subjectivism and assurance of result reliability. Environmental, social and economic aspects (analysis criteria) were represented by Ecopoint (Ep), heat transfer coefficient (U) and price of material (Pm), respectively.

WSM: The Weighted Sum Method is a mathematical method to optimize solutions and to support the decision-making process while required to consider various factors simultaneously. The values of synthetic indicators proper for each technology were determined using the formula [11]:

$$J_t = \sum_{i=1}^{n} (x_{ti}' / x_{imax}') \times W_i \tag{5}$$

where:

J_t value of synthetic indicator for tth technology,

W_i weight of ith sustainability aspect (criteria),

i sustainability aspect, $i = \{1,2,3\}$,

x_{ti} the original values of Ep, U and Pm for each technology,

$x_{ti}' = 1 / x_{ti}$

$x_{i\,max}'$ maximum value of the ith criterion's measures after maximization.

TOPSIS: The essence of this method is to find among the analysed solutions the one, closest to the ideal solution to satisfy the required criteria as much as possible, at the same time most distant from the anti-ideal solution. This method is currently widely used in relation to the sustainable development and sustainable building as well. The search for the most advantageous solution is carried out in several stages, described in detail e.g. in [13]. Relative closeness to an ideal solution has been determined using the formula [14]:

$$d_t = d_t^- / (d_t^+ + d_t^-) \tag{6}$$

where:

d_t^+ distance of the tth technology from the ideal solution,

d_t^- distance of the tth technology from the anti-ideal solution.

The standardisation procedure for ith analysis criterion (sustainability aspect) measure of the tth technology was carried out according to the formula:

$$z_{ti} = x_{ti} / \sqrt{\sum_{t=1}^{T} x_{ti}^2} \tag{7}$$

where:

x_{ti} - original i[th] criterion measure of the tth technology.

In both – WSM and TOPSIS method - the equivalence of the environmental, social and economic aspects, according to the idea of sustainable development, was assumed. Thus the weights W_i for sustainability aspects, being analysis criteria, are equal (W_i=1/3).

3 Research results

3.1 Environmental aspect analysis

The environmental indicators chosen in the analysis are dis-stimulants. Therefore, there is an attempt to find a technology, which would reach lowest Ecopoints value and thus would have the most advantageous environmental profile. Fig. 1 shows research results in form of a cumulative diagrams of environmental indicators and Ecopoint values for each technology.

The most advantageous result (E_p=0.108) is achieved by the timber frame technology. In a large extent this is caused by the impact of T2 technology on global warming. Timber is a renewable resource, forest vegetation contributes to CO_2 reduction in the atmosphere. Other indicators (apart from FW) remain also at a relatively low level. The worst result is achieved by the T1 technology (E_p=0.457). The reasons are considerable impact on global warming and high photochemical ozone creation potential.

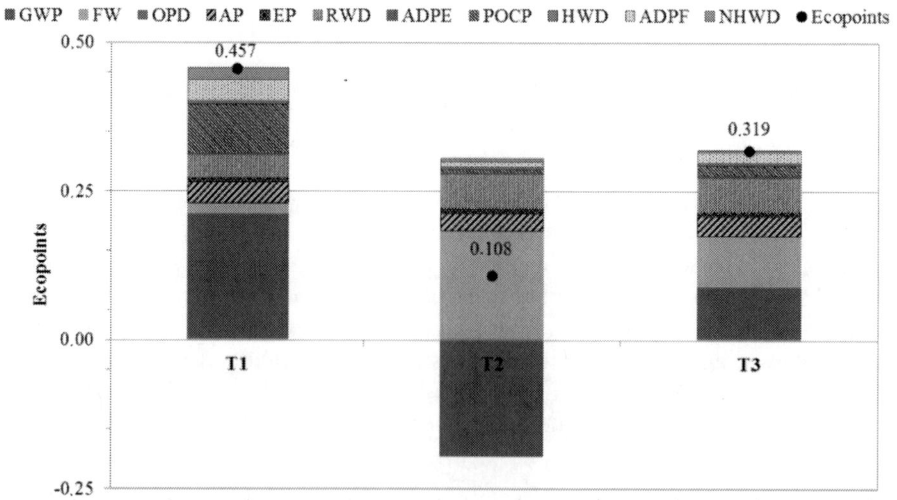

Fig.1. Ecopoints value for analysed technologies (T1, T2, T3).

3.2 Social aspect analysis

The tables 2-4 illustrate arrangement of layers, heat transfer coefficient values and thicknesses of external walls for T1-T3 technologies.

Table 2. External wall layers and heat transfer coefficient for technology T1.

Material of layer	λ (W/mK)	Thickness (m)	U (W/m²K)
Plasterboard	0.25	0.013	0.11
Adhesive mass with reinforced lattice	0.8	0.005	
Self-supporting polystyrene profiles	0.034	0.02	
Cement-lime plaster	0.083	0.02	
Wall thickness (m)		0.338	

Table 3. External wall layers and heat transfer coefficient for technology T2.

Material of layer	λ (W/mK)	Thickness (m)	U (W/m²K)
Plasterboard	0.25	0.013	0.21
Non ventilated air gap	-	0.025	
Vapour- insulation foil	0.18	0.002	
Mineral wool	0.034	0.16	
OSB	0.13	0.015	
Wind-insulation foil	0.18	0.002	
Wooden facade	-	-	
Wall thickness (m)		0.217	

Table 4. External wall layers and heat transfer coefficient for technology T3

Material of layer	λ (W/mK)	Thickness (m)	U (W/m²K)
Cement-lime plaster	0.83	0.01	0.23
Hollow brick	0.39	0.29	
Polystyrene foam	0.034	0.12	
Cement-lime plaster	0.83	0.015	
Wall thickness (m)		0.435	

Note: λ= thermal conductivity of material; U= heat transfer coefficient.

The coefficient U in technology T2 is close to the value for technology T3 (wall being twice thinner). The lowest value of coefficient U is observed for technology T1, which results from considerable thickness of the main material (30-74 cm) and low λ value of this material.

3.3 Economic aspect analysis

In order to assess and compare the technologies, the prices of building envelopes materials were calculated per 1 m² of usable area and the results are shown in Table 5.

Table 5. Prices of building materials of external walls and roofs in analyzed technologies, PLN.

Price of materials[a]	Technology T1	Technology T2	Technology T3
Envelope building materials, PLN	41,699.18	56,655.23	53,595.76
Per 1m² of usable area (Pm), PLN/m²	376.52	511.56	483.93
Relative increase of prices, %	-	35.9	28.5

[a] *Based on catalogs of materials manufacturers.*

The technology T1 detects the lowest price of materials per usable area unit. Thus, the price of materials used in technology T3 is almost 29% higher. The highest price of materials is observed for technology T2. Erection of a building in this technology requires that properly selected, assorted and processed timber is used, what generates high material costs. The price is almost 36% higher than the lowest price, assessed and shown in Table 5.

3.4 Multi-criteria analysis

WSM: The original and normalized measures of criteria (sustainable aspects) and their weights are shown in Table 6.

Table 6. The values of criteria measures (original - x_{ti}, normalized - z_{ti}) and synthetic indicator.

Technology	Ep		U		Pm		J_t
	x_{ti}	z_{ti}	x_{ti}	z_{ti}	x_{ti}	z_{ti}	
T1	0.457	0.236	0.11	1.000	376.52	1.000	0.745
T2	0.108	1.000	0.21	0.547	511.56	0.736	0.761
T3	0.319	0.338	0.23	0.483	483.93	0.778	0.533
Weight	33.3%		33.3%		33.3%		

The most advantageous technology shows the highest value of synthetic indicator (J_t). It is observed that technology T2 is the prevailing one due to the significant predominance of this solution with reference to the environmental aspect (Table 6). Technology T1 is the second in ranking as a result of low environmental efficiency characterised by the Ecopoint rate value and high social and economic performance

TOPSIS: Comparisons shown in Table 7 were obtained as a result of applying the TOPSIS method to solve the problem of selecting the technology that would best satisfy all of the main criteria.

Table 7. Results of TOPSIS method– distances of technologies from ideal and anti-ideal solutions.

Technology	Results of TOPSIS method		
	d_t^+	d_t^-	d_t
T1	0.2049	0.1339	0.395
T2	0.1096	0.2067	0.653
T3	0.1793	0.0817	0.313

Therefore there is an attempt to find a technology which would reach highest value d_t indicating relative closeness to the ideal solution. Technology T2 is the most distant from anti-ideal solution (highest value of d_t^-) and at the same time closest to the ideal solution (lowest value of d_t^+). As regards the other two technologies T1 is closer to the ideal. In case of the TOPSIS method, the position of a given technology is determined by the values of the criteria measures for other technologies.

4 Conclusions

In the study the authors have proved usefulness of the multi-criteria analysis in selecting a technology, which would most effectively fulfil the sustainable building assumptions. Indirectly this analysis simplifies the materials selection in the process of integrated design and evaluation of innovative technologies due to the sustainable development principles. Following the entries in the standard EN 15643 and studies of other authors it was proposed to apply 13 assessment indicators from the environmental, social and economic groups to the three building construction technologies. The ranking process was carried out using two methods of multi-criteria analysis - WSM and TOPSIS. The completed research proves that the timber frames technology is the most advantageous among all considered in the analysis, what is largely the result of the environmental efficiency. The technology of self-supporting building development systems based on foamed polystyrene is in the second place, due to the most advantageous social and economic profile and simultaneously very low environmental performance. It should be emphasised that the currently promoted issue of sustainable development in construction industry is still not adequately popularised among users of single-family residential buildings, hence it should be expected that the economic and social aspects will be more important for the decision makers (users, inhabitants) than the environmental ones. Considering this assumption and the obtained research results, it is found that new and only recently being developed technologies are close to optimal in given conditions. It is necessary to carry out further studies, taking into account other stages of the building life cycle and adding to the analyses other important criteria, particularly social and economic. Relations between research methods and results are also found, different rankings are anticipated, while other methods will be used.

References

1. A. Karatas, K. El-Rayes , Optimal trade-offs between social quality of life and Life-Cycle Cost in housing units, J. Constr. Eng. Manage. **140/12**, 04014058 (2014)
2. Y. Kang et al., Comparison of preproject planning for green and conventional buildings, J. Constr. Eng. Manage., **139/11**, 1-9 (2013)
3. O. Tatari, M. Kucukvar, Eco-efficiency of construction materials: Data envelopment analysis, J. Constr. Eng. Manage., **138/6**, 733-741 (2012)
4. CEN (European Committee for Standardization), *Sustainability of Construction Works - Sustainability Assessment of Buildings Parts 1-4: General Framework*, EN 15643-1 (2010-2012)
5. ISO (International Organization for Standardization), *Sustainability in building construction- General principles*, ISO 15392 (2008)
6. J. Mundy, *The Green Guide Explained. BRE Centre for Sustainable Products*, <http://www.bre.co.uk/filelibrary/greenguide/PDF/The-Green-Guide-Explained_March2015.pdf>
7. V. Aymard, V. Botta-Genoulaz, Normalisation in life-cycle assessment: consequences of new European factors on decision-making, *Proc., 6th International Conference on Information Systems, Logistics and Supply-chain ILS Conference 2016*, Bordeaux, France (2017)
8. BRE (Building Research Establishment), *Green Guide to Specification BRE Materials Industry Briefing Note 3b: Normalisation,* <http://www.bre.co.uk/greenguide/files/NormalisationBriefingDocumentFinal.pdf>
9. OECD (Nuclear Energy Agency Organisation for Economic Co-Operation and Development), *Radioactive Waste in Perspective*, <https://www.oecd-nea.org/ndd/pubs/2010/6350-waste-perspective.pdf>
10. O. Abbe, L. Hamilton L, BRE Global Environmental Weighting for Construction Products using Selected Parameters from EN 15804, <https://www.bre.co.uk/filelibrary/Materials/Environmental-weightings-15804_final.pdf>
11. I.Y. Kim, O.L. de Weck, Adaptive weighted sum method for multiobjective optimization: a new method for Pareto front generation, Struct. Multidiscipl. Optim., **31/2**, 105-116 (2006)
12. J. Szwabowski, J. Deszcz, Elementy procesu decyzyjnego. In: *Metody wielokryterialnej analizy porównawczej. Podstawy teoretyczne i przykłady zastosowań w budownictwie,* 9-24 (Wydawnictwo Politechniki Śląskiej, Gliwice, 2001) [in Polish]
13. E.K. Zavadskas, Assessment of individual residential buildings external walls alternatives applying TOPSIS method, *Proc., 9th International Conference on Modern Building Materials, Structures and Techniques*, Vilnius, Lithuania, **1-3**, 442-446 (2008)
14. D. Mierzyńska, Wielowymiarowa analiza dobrobytu społeczno-ekonomicznego w Polsce [Multidimensional analysis of socio-economic well-being in Poland], *Prace i Materiały Wydziału Zarządzania Uniwersytetu Gdańskiego*, **4**, 8, 421-434 (2011) [in Polish]

The collusion detection in public procurements – selected methods applied for the road construction industry in Poland

Andrzej Foremny[1], and *Hubert* Anysz[1,*]

[1]Warsaw University of Technology, Faculty of Civil Engineering, Institute of Building Engineering, Division of Production Engineering and Construction Management, Al. Armii Ludowej 16, 00-637 Warsaw, Poland

Abstract. Based on successful international examples of collusion detection, four indicators of the collusion have been chosen and analysed in this paper. The research was done for the majority of tender procedures in the road construction industry, in Poland that was finished in 2016 with the selection of a general contractor. Based on [1, 2, 3] the following four indicators were chosen: the number of offers placed in the tender procedure, the range of bid prices being offered, exactly the same set of offerors in a few tender procedures, repeating ordering the works execution to the same contractor in a given geographical region.

1 Introduction

The reason for introducing the system of public procurements was to provide a real market competition for the orders placed by the municipal or national offices, to ensure that public orders can achieve efficiency as high as the private ones [4]. The specific character of the construction industry (extremely high value of orders, very limited number of companies capable to handle the road construction contracts, specialization of contractors) creates favourable conditions for non-concurrent behaviours of the participants of the tender procedure [1, 5]. Sometimes a forbidden practice of discussing bid prices between offerors before placing the offers takes place. These factors of non-market origin [6] influence the contractors' decisions about participation in a tender procedure. As a result, the client receive offers where all bid prices are over the market level. Next – even choosing the lowest one – the efficiency of spending public money decreases [5] and market is abused.

It is rather difficult to prove that collusion took place. As it is a forbidden practice, the police and other secret service forces have been involved to discover the matter of the secret agreements between offerors. It is easier – based on the information collected from the offers already placed in the past – to evaluate if collusion took place in a given tender procedure. The client, suspecting the collusion, can revoke the tender procedure not choosing any offer as the best one [4]. If collusion takes place and is detected before a contract execution, a client can save money, but it significantly delays the start of work

* Corresponding author: h.anysz@il.pw.edu.pl

because the new invitation to tender has to be announced. It is crucial then, that the suspicion of collusion should be detected as soon as possible backed by a strong base. For the national economy and the future road users, significant delay of road construction can be near as much inefficient as building it on planned time paying the over-the-market price. It can be calculated based on [7].

2 Input data description

The authors of the paper prepared a database of tenders gathering information from web pages of institutions that were public ordering parties. The database consists of 98 public tenders for road construction or reconstruction whose results were published in 2016 in Poland, of a winning tender offer value 10 million PLN or more. Each record consists of the following data: organizer name, localization of construction works, road type (highway, expressway, national way, voivodship road, county road), type of works (construction, reconstruction), contract scope (design, design and build), list of all offerors not excluded, list of all price offers submitted.

3 Criteria applied for the collusion detecting

Numerous studies into bid collusion at auctions have been conducted in the past. The empirical literature is focused on the detecting of collusion using past examples. Porter and Zona [5], Zona [4] propose the tests for collusion detection in the context of highway construction data and school milk markets. In the case of the Ohio school milk case [5], the economic analysis has shown how prices were lower when (transportation) costs were higher which is inconsistent with competition but quite consistent with a particular collusive scheme [8].

The collusion agreements can be complicated to detect as they are typically negotiated secretly. In industries where collusion is common, however, suppliers and purchasers may be aware of longstanding bid-rigging conspiracies [9]. The authors based on official publications of the organization OECD and UOKiK selected and applied four methods of searching for collusion behaviour in this paper [3, 9].

4 The number of offers placed in a tender procedure

A small number of offers in a particular tender in comparison to other similar tender procedures may signalize a higher possibility of collusion. The authors has taken into consideration only the offers that were accepted and published on an official tender results sheet. Thus the list does not show the offers that were not valid or that were withdrawn by the offerors before the results publication. The most frequent number of the offers placed was 5 (in 18 procedures). The minimal number of offerors was 3, the maximal number of offerors was 20, the average number of offerors was 7,8. The average number of offerors in one tender calculated in this research is very similar to 8,2 that was calculated in the previous study in 2014 [1]. A histogram was created (see fig. 1). In the analyzed database there are 6 tenders with only 3 offers placed. It can be a sign of collusion in comparison to the average number of offers in a single tender which was 7,8.

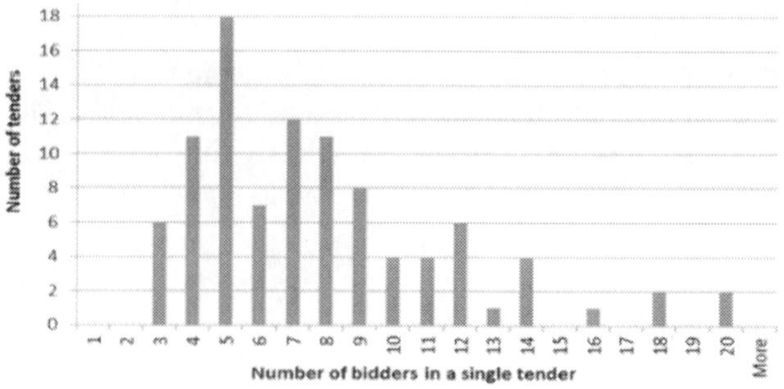

Fig. 1. Histogram of the number of tenders with the number of bidders given on the horizontal axis [own elaboration].

5 The range of bid prices being offered

The decision to bid and to calculate a bid price taken by a general contracting company depends on many factors: technical, economic, geographical and others [6, 10, 11, 12, 13]. There are many available methods of multicriteria assessment [14]. A small range of bid prices in comparison to other, similar tender procedures may signalize higher possibility of collusion. The range R_i, showing the percentage by which the lowest (winning) bid is lower than an average bid in a procedure i, was calculated using the formula:

$$R_i = \frac{V_{av\,i} - V_{\min i}}{V_{av\,1}} * 100\ \% \tag{1}$$

where:

R_i	the range (defined above) for procedure i
$V_{min\,i}$	the lowest value of an offer in procedure i
$V_{av\,i}$	the mean average value of offers in procedure i

In the case of calculation in this chapter, 14 records from the database had to be eliminated because they missed bidding price data. The results are presented in fig. 2.

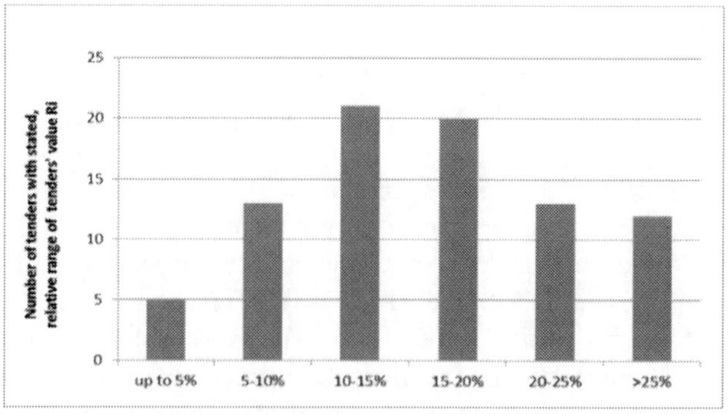

Fig. 2. Histogram of the number of tenders with the value of range R_i within specified percentage intervals [own elaboration].

The average range R for analyzed tenders was 16,9 %. There were 5 tenders that had an extremely narrow range and can be suspected of collusion behavior between offerors. The minimal range observed in a single tender was lower than 1 %, this was a very valuable tender – the winning offer showed the price higher than 400 million PLN. The ranges observed in this research are similar to the recorded in a previous study in 2014 [1].

6 The geographical criterion of the collusion detection

On an abused market a geographic allocation of winning tenders may occur. Within the analysed year 2016 a list of 98 tenders were checked. There were 756 offers placed by 138 companies or consortia. Collusion can be done between companies that repeatedly place offers and – from time to time – win. The tab. 1 shows how many procedures were won by a certain number of companies.

Table 1. Numbers of wins and number of companies [own elaboration].

	Number of wins			
	0	1	2	3 and more
Number of companies that won given number of times	99	27	9	3

It was assumed that the company involved in a collusion has to win one or more bidding procedures. The more price bids are placed by a single company, the more orders are won thus higher probability of participation in a collusion. It is not a kind of a proof that collusion took place. It is rather excluding from the collusion by searching the areas where entities have not won any contract.

Table 2. Attempts and wins of the key players [own elaboration].

	Company A	Company B	Company C
The number of tenders attempted	90	56	72
The number of tenders won	17	23	13

There are the three leading players in the road construction industry in Poland (taking into consideration analysed database of tenders in 2016). Each of these companies won 3 or more tender procedures for building or designing and building a road, where the contract values were higher than 10 million PLN. The data concerning their attempts are summarized in tab. 2.

Fig. 3. The procedures won by A, B or C in voivodships (the total number of procedures is shown after slash) [own elaboration], with the use the graphics from [15].

The leading three players have won 54% (53 out of 98) tender procedures, signing the contracts for ca. 5.537 million PLN (out of ca. 11.803 million PLN). The fig. 3 shows voivodships where the companies A, B and C have won the procedures.

Based on fig. 3 the following findings can be emphasised:

- the company A was the only one (from A, B, and C) that has won in 3 voivodships,

- the company B was the only one (from A, B, and C) that has won in 3 voivodships,
- the company C hasn't won any tender procedure in Southern Poland,
- there are 6 voivodships where A, B and C have won proportionally much more procedures than average; they are shown in tab. 4.

Table 4. Percentage of procedures won by A, B or C in selected voivodships [own elaboration].

Voivodship by its capital	Procedures won by A or B, or C
Olsztyn	100 %
Lublin	83 %
Poznań	81 %
Kielce	80 %
Wrocław	75 %

The irregularities found, based on the geographical criterion are essential, but part of them can arise from natural reasons. One of them is the lack of the company C among winners in Southern Poland. This company has won 13 tender procedures, but there are 17 voivodships in Poland (4 more that procedures won by C). Special attention should be paid to the tenders listed in the tab. 4. In these voivodships, only 5 out of 30 procedures were not won by the three key players (83% of orders there has gone to A, B or C company, compared to 54% totally).

7 The coincidence of participants and the winners as a criterion of collusion detection

The same set of offerors in a few tender procedures may signalize higher possibility of collusion. One of the best tools for searching coincidences is an association analysis [16]. The collected and prepared data is searched in order to find the set of factors that is repeatedly accompanied by the examined phenomenon (the phenomenon appears repeatedly too). These factors are often called the $bodies$ (b), and the phenomenon is called the $head$ (h) [17]. Measurements of the frequency of appearing the rule $b \rightarrow h$ (if the $body$ has appeared, then the $head$ appears too) is executed by calculating two proportions called $support$ (sup) and $confidence$ $(conf)$. They are defined as:

$$\sup(b \rightarrow h) = \frac{n(b \cap h)}{N} \tag{2}$$

and

$$conf(b \rightarrow h) = \frac{n(b \cap h)}{n(b)} \tag{3}$$

where:

$n(b \cap h)$	the number of cases when b and h appear simultaneously,
N	the total number of examined cases,
$n(b)$	the number of cases when b appears (regardless the $head$).

The usefulness of association analysis is in the fact that the $body$ can be represented by a coincidence of many factors accompanying the examined phenomenon [18]. Then:

$$b = (b_1 \cap b_2 \cap b_3 \cap \ldots \cap b_m) \tag{4}$$

where:

m — the number of factors being checked every time together with the phenomenon appearance.

The head can be described as a simultaneous appearance of more than a single phenomenon (e.g. joint appearance of three sub-phenomena creating the examined phenomenon). For the analysis of tender procedures, the *STATISTICA* 12 computer program was used with its function *association analysis*. There were 7, following *bodies*:

- the order scope (0 for ordering building works; 1 for the order "design and build");
- the type of a road (H for highways, E for express roads, K for national roads, W for voivodship roads, P for county roads);
- placing an offer by the company A (1 for an offer placed, 0 for not placing an offer);
- placing an offer by the company B (1 for an offer placed, 0 for not placing an offer);
- placing an offer by the company C (1 for an offer placed, 0 for not placing an offer);
- number of participants in a tender procedure (0 for 5 or less, 1 for more than 5);
- value of an order meant as the lowest (winning) bid price (0 for the contracts with value lower than 100 million PLN, 1 for the contracts of higher value);

The *head* was defined as:

$$h = (h_A \cup h_B \cup h_C) \tag{5}$$

so the value of the head could be

$$h = \begin{cases} 1 & \textit{when the winner was A or B or C} \\ 0 & \textit{when the winner was other} \end{cases} \tag{6}$$

In a total number of98 analyzed procedures, there were only 4 where no offer has been placed, by any company from the three leading players. The thresholds for the value of a contract and a number of participants in a tender procedure had to be set manually. The software – after setting the minimum values separately for the confidence and for the support – search for all association rules fulfilling the limits of $conf$ and sup [19]. As a collusion happen very rarely (due to the authors' belief) the support was set to 5%. The lowest limit of confidence was set to 50%. The number of the rules found was 39. The results with the highest confidence are shown in tab. 5.

Table 5. The rules found with the highest confidence [own elaboration].

Label	If body	than	Head	sup	Conf
R1	WAY = W, B = 1, VALUE = 0	==>	1	21,4	75,0
R2	WAY = W, A = 1, B = 1, C = 1	==>	1	20,4	76,9
R3	WAY = W, B = 1, C = 1	==>	1	21,4	77,8
R4	WAY = W, C= 1, B = 1, VALUE = 0	==>	1	17,3	81,0
R5	WAY = W, A = 1, B = 1, C= 1, VALUE = 0	==>	1	17,3	81,0

Within all 34 rules found, if the type of the road was mentioned, it was always a voivodship road. The most interesting rules are R2 and R5. If all leading players (the companies labelled as A, B, C) have placed an offer in a given tender, then the probability of winning the contract was 76,9% (for the analyzed database from the year 2016). If the condition about the contract value was considered (VALUE=0 when the contract value is below 100 million PLN), the confidence increases to 81% (rule R5).

8 Conclusions

In the cases of collusion suspected the bidding procedures were listed separately for each criterion in tab. 6.

Table 6. The numbers of bidding procedures fulfilling each criterion of the collusion detection [own elaboration].

	The low number of bids (criterion No. 1)	The small range of the price bids (criterion No. 2)	The geographical criterion (criterion No. 3)	Repeating joint participation (criterion No. 4)
Numbers of procedures fulfilling the criteria	6	5	25	20

The thresholds in each criterion were set by the authors, so they are subjective and relative. The thresholds of criteria 1 and 2 were based on [1, 3, 9]. Next, the number of the suspected procedures is extended. It was assumed that the more criteria is fulfilled by a certain procedure, the higher is the probability of a collusion appearance. The result of the check is shown in tab. 7.

Table 7. The numbers of bidding procedures fulfilling the criteria of the collusion detection [own elaboration].

	1 criterion	2 criteria	3 criteria	4 criteria
Procedures fulfilling the criteria	28	12	0	1

It is necessary to emphasize, that the collusion is found if there is a court confirming sentence. Otherwise – like in this paper – only the risk of appearing of a collusion can be discussed. For the time being, there is no confirmation of any collusion in the analyzed tenders. Thus, there is no possibility to discuss more, how many criteria should be fulfilled to predict a high risk of collusion as high. It can be stated that the procedure fulfilling the four criteria (formulated by the international organization, and some national institutions set up to protect the market from non-concurrent behaviors) is much more suspected for a collusion, than the procedure that fulfill only one.

If it is confirmed, that fulfilling only the one criterion of a collusion is a strong indicator of a fraud, it will mean that in 42 % procedures for the road construction in Poland in 2016 (41 out of 98 analyzed) achieved price was not-concurrent. If 3 or 4 criteria are required, the road construction market will look much better. Only one procedure would be suspected as non-concurrent. Stating the collusion requires a huge effort, and it is a time-consuming process. It is much easier to extend requirements of the scope of information provided by offerors in their tender documents [20]. A higher level of details would allow the clients for easier inconsistencies finding in price bids and for easier excluding inconsistent, overpriced bids. The current scope of offers, as well as ordering construction works together with a design does not allow for successful, quick exclusion of inconsistent, overpriced bids from tender procedures.

References

1. H. Anysz, A. Foremny, J. Kulejewski, *Estimating potential losses of the client in public procurement in case of collusion utilizing a MLP neural networks*, Technical Transactions **111/1–B**, 105–118, DOI: 10.13140/2.1.2409.6321 (2014)

2. Y.C.A. Capobianco, *Ex Officio Cartel Investigations And The Use Of Screens to Detect Cartels*, OECD (2013).
3. Urząd Ochrony Konkurencji i Konsumentów, *Zmowy przetargowe*" (2017)
4. J.D. Zona, *Bid-rigging and the Competitive Bidding Process: Theory and Evidence*, Ph.D. Dissertation, State University (1986)
5. R.H. Porter, J.D. Zona, *Detection of Bid Rigging in Procurement Auctions*, Journal of Political Economy **101/ 3**, 518–538 (1993)
6. A. Leśniak, *Classification of the Bid/No Bid Criteria – Factor Analysis*, Archives of Civil Engineering **61/4**, 79-88, DOI: 10.1515/ace-2015-0037 (2015)
7. Instytut Badawczy Dróg i Mostów, *Instrukcja oceny efektywności ekonomicznej przedsięwzięć drogowych i mostowych dla dróg wojewódzkich*, (2008)
8. J. Harrington, *Behavioral screening and the detection of cartels*, European Competition Law Annual (2006)
9. OECD, *OECD recommendation of the council on fighting bid rigging in public procurement* (2012)
10. A.S. Bageis, C. Fortune, *Factors affecting the bid/no bid decision in the Saudi Arabian construction contractors*, Construction Management and Economics **27/1**, 53–71 (2009)
11. M.S. El-Mashaleh, *Decision to bid or not to bid: a data envelopment analysis approach*, Canadian Journal of Civil Engineering (2010)
12. A. Enshassi, S. Mohamed, A. El Karriri, *Factors affecting the bid/no bid decision in the Palestinian construction industry*, Journal of Financial Management of Property and Construction **15/2**, 118–142 (2010)
13. A. Leśniak, E. Plebankiewicz, *Modeling the Decision-Making Process Concerning Participation in Construction Bidding*, Journal of Management in Engineering **31/2**, 4014032, DOI: 10.1061/(ASCE)ME.1943-5479.0000237 (2015)
14. S. Biruk, P. Jaśkowski, A. Czarnigowska, *Modeling Contractor's Bidding Decisions*, Procedia Engineering **182**, 91–98, DOI: 10.1016/j.proeng.2017.03.125 (2017)
15. Centrum edukacji i nauczania, [Online:] http://www.nowowydane.pl/cen [Access: 15-Jan-2018]
16. D.T. Larose, C.D. Larose, *Discovering Knowledge in Data*, Willey IEEE, ISBN978-81-265-5834-6 (2016)
17. T. Morzy, *Eksploracja danych. Metody i algorytmy*, PWN (2013)
18. M. Pęczkowski, M. Lasek, *Analiza asocjacji i reguły asocjacyjne w badaniu wyborów zajęć dydaktycznych dokonywanych przez studentów. Zastosowanie algorytmu Apriori*, Ekonomia. Rynek. Gospodarka. Społeczeństwo (2013)
19. Internetowy podręcznik statystyki, [Online:] https://www.statsoft.pl/textbook/stathome.html. [Access: 10-Jan-2018]
20. H. Anysz, *Założenia systemu rozliczeń TKW + M (techniczny koszt wytworzenia plus marża)*, Przegląd Budowlany **1** (2018)

The Conjugated Triangle Method CTM of the detection of inconsistent bids in the construction industry

Hubert Anysz[1,*], and *Krzysztof* Kaczorek[1]

[1]Warsaw University of Technology, Faculty of Civil Engineering, Institute of Building Engineering, Division of Production Engineering and Construction Management, Al. Armii Ludowej 16, 00-637 Warsaw, Poland

Abstract. The existing system of bids evaluation in the construction industry does not bring the clients the certainty if all placed bids comprise real, achievable prices. Especially in the road construction sector, where the earthworks create significant part of the works, comparing e.g. the price of 1 km of roads often gives misleading results. The paper comprises the complete proposal to modify the scope of information provided by offerors in the public procurements (in the construction industry) allowing the clients to determine easily if the contractors have placed the consistent offers i.e. if the bid prices are not too high and if the lowest bid price is not a dumping price. The Conjugated Triangle Method (CTM) of inconsistent bid detection in construction industry is based on the three inter-related parameters: means of production, time of works, on the value of works. Regardless of the offered price it should match the other two parameters.

1 Introduction

The system of public procurement provides public investors with the tool allowing for the choice of the optimum bid for building works execution. Both design documentation and technical specification describe the planned object in detail. The client sets essential assumptions, terms of a tender procedure giving (among others) required time for works execution, required building materials, the technology of selected works [1, 2]. All the aforementioned means make the bid price the most important criterion for the contractor choice. Even if it is not the only criterion, the bid price remains the decisive one for the client to choose the offer. Sometimes it happens that one of the bid prices is exceptionally low. The choice of that one is often a source of many problems appearing during the construction process [3]. In the case of the collusion – which is a forbidden practise – the participants discuss the bid values before placing offers. It leads to the situation when all bids take a relatively high value [4]. Both cases make the client's investment much less financially effective that it was planned.

* Corresponding author: h.anysz@il.pw.edu.pl

2 Existing system of the bids evaluation

Currently public investors set the time of work execution and ask the offerors for price offers and proposed time schedules. For the planned building objects of higher value, some information about access to the critical building machinery and experienced personnel is required too. According to the limited information about production resources [3] and according to the fact the price offers and the time schedules are being prepared on different levels of detail [5], it is really difficult to evaluate if the offers are not too low or not too high. No problem occurs to recognize which bids are overpriced, not to be considered. None of them is usually chosen as the best one. As the bid price is the most influencing factor on the client's decision, they are naturally prone to choose the bid with the lowest value or near the lowest value. A key problem arises if the lowest bid is under-priced or not. When the under-priced offer is chosen as the best one, the contractor is not able to execute all works within the offered price. Thus the search for artificial problems during construction work execution emerges to force the client to increase the contract value [6]. On the other hand, if all bids fulfil the client's requirements and there are no formal mistakes in preparing offers, the client is forced to choose the cheapest one. The non-financial criteria rarely make a different choice, unless the client proves that the lowest bid represents unrealistic, under-the-market value of building works. The Polish legal system provides the contractors with legal tools to claim this kind decision of the client. This makes the tender procedure much longer than planned. The evidence of under-pricing a given bid can be made using the three following checks:

- comparing the value of bids in a given tender procedure,
- comparing the index prices of similar, completed already objects,
- comparing the unit prices for the specific types of planned construction works to the unit prices achieved on already completed projects.

All the checks mentioned above do totally assure if a given offer is under-priced. The evaluation based on the index prices of similar, already completed objects may not give the high-level certainty in assessing the bid as an under-priced. The massive and varying amount of earthworks, the different number of overpasses etc. has made the prices of 1 km of express roads in Poland varying from several to a few dozen of million PLN (2009-2013) [7]. Relying on the unit prices of building works is doubtful too, as it was proved the unit prices shown in tenders comprise indirect costs and the profit of the contractor [8, 9, 10]. While there is no obligation to assign them to each building work in the same proportion, it is hard to prove if the price is an under-market price for a given type of building work. Applying all of the above mentioned procedures to check if a given bid is on the market level or below, the answer is formulated with a given confidence. Finally, after long-lasting analysis and sessions, the decision is made in Poland by a dedicated chamber for public procurements KIO (*pl. Krajowa Izba Odwoławcza*). The time-consuming process of excluding the under-priced bids and their high negative impact on the clients financial effectiveness has pushed the authors to invent the method of evaluating the bids in construction sector, allowing for quick working out the decision about exceptionally low bid in a given procedure.

3 The Conjugated Triangle Method (CTM) of the bids evaluation

The Conjugated Triangle Method (CTM) of the detection of inconsistent bids in the construction industry is based on the following three inter-related parameters [9, 11]:

- the means of production (M),
- the time of works (T),
- the value of works (C).

The liaisons (conjunctions) between pairs of these parameters exist every time [9, 12, 13] and arise from their nature (not from any subjective evaluation).

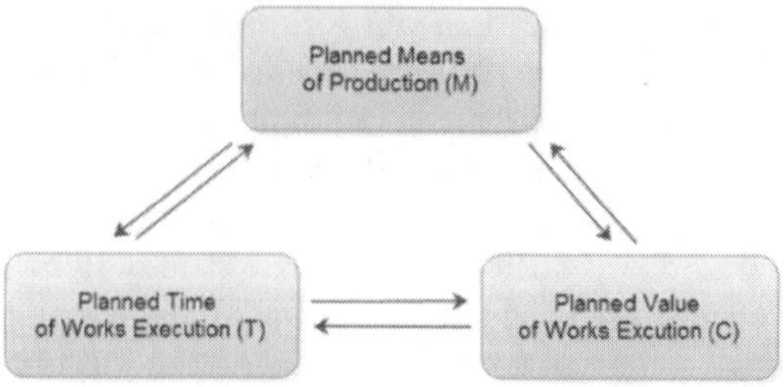

Fig. 1. The Conjugated Triangle and its elements [own elaboration].

If there is proper information provided in the offers, the clients – based on a number and types of means of production – can estimate the time of the work execution and compare it to the planned time [14, 15]. The next check is to compare estimated cost of the production means, to the bid prices. This double check can provide precise information which of the low price bids are inconsistent and under-priced.

3.1 The means of production and time of works execution (M-T)

The conjunction (M-T) between the means of production assigned by the contractor to perform a given type of building work and the time necessary to execute them is evident. A pair of diggers can make the cut twice quicker than the one only [16]. This (M-T) conjunction is discrete and proportional, but only in case of a low number of machines. When too many diggers are working at the same time on the limited area, their total efficiency is lower than the efficiency of a single machine multiplied by the number of machines [17]. This finding shows a low practical meaning as the contractors are not equipped with an unlimited number of machines. It can be assumed that every machine added to the already working, increase their total efficiency proportionally.

3.2 The means of production and the cost of works execution (M-C)

This (M-C) conjunction shows a linear character too. Every means of production has its price (not considered a purchase price but an amount of money to pay for utilizing a given resource in a unit of time). The more resources we hire, the more we have to pay for them [18]. The differences in the prices of resources hired by different contractors exist, but there is no market reason that would make these differences high. The other problem that can be recognized are indirect costs. In fact, comparing the tenders for erecting the same building object it can be assumed that indirect costs influence the total cost of every contractor on near the same level.

3.3 The time of works execution and the cost of it (T-C)

The time-cost dependence is not linear in construction industry. There is optimum time t_{opt} of work execution a given type (and number) of works for a chosen technology (see fig. 4) providing the minimum cost c_{min} of work execution.

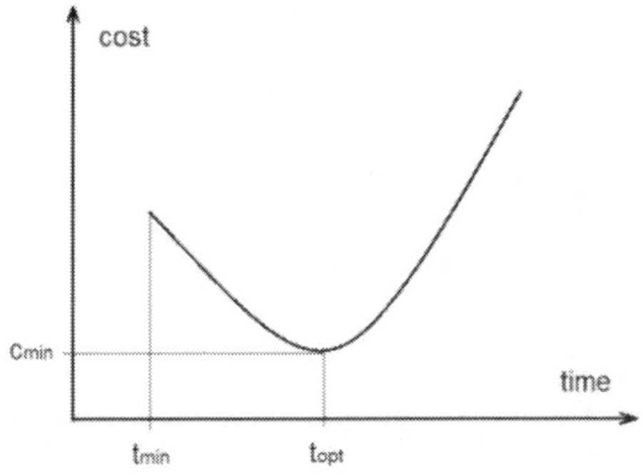

Fig. 2. The time-cost relationship for the construction works [19].

The technology is a lower limit of shortening the time of work execution. Applying the means of production very inefficiently we can extend the time of the work execution to infinity. The evaluation of the cost of works based on the time predicted for their execution is almost impossible (or the error will be high) according to following facts:

- there is no proof that the assumed time is an optimum one,
- there is no information what is the real level of work intensity assumed by a particular contractor.

The attempt to estimate the time for work execution based on the cost, results in an approximate answer too. As shown in fig. 3, for the cost c, it is possible to estimate the minimum and maximum time of the work execution.

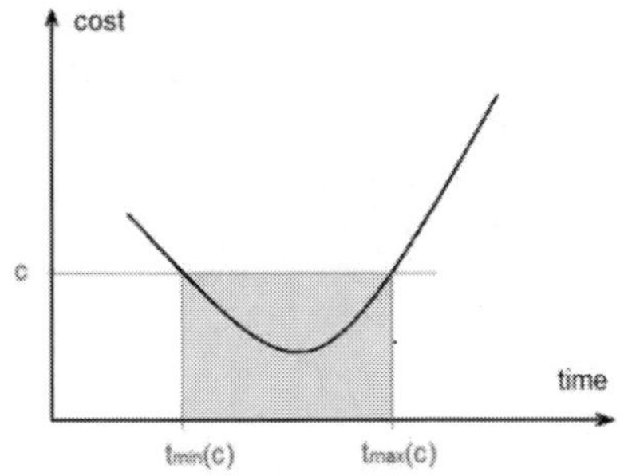

Fig. 3. The low precision of the evaluation of the time of work execution basing on its cost [own elaboration].

The conjunction T-C between time and cost of building work execution is strong as remaining two (M-C, M-T), but according to its non-linear character, it is the most difficult to evaluate. In fact it is the only conjunction that can be currently applied by the clients (without applying CTM), as the information about means of production is strictly limited in tender documents.

3.4 The requirements for CTM application

In order to overcome the current situation, when the wide range of possible bid price can be accepted – even these under-priced – it is necessary to extend the scope of information provided by offerors in the tender documents. This will allow for CTM application for evaluation of the bid prices in a given tender procedure.

The tender documents should comprise:

- the price offer,
- the time schedule based on Critical Path Method [19],

- the schedules of employment of the core building machinery,
- the schedule of the workers' employment
- the list of core building materials with their volume (or number, or weight).

The critical point is that all schedules mentioned above, should detect an identical level of details as the price offer, i.e. every element from the price offer (having there its number, description and the price) should be represented in time schedules.

3.5 The detection of the inconsistent offers with CTM

The abovementioned conjunctions (T-C, M-C, M-T) are not equally strong, meaning that the evaluation simplicity of one conjugated value by the other is not identical.

Table 1. The level of difficulty in estimating C, T and M basing on their conjugated parameters [own elaboration].

Conjugated parameter as the estimation basis	**Level of difficulty in estimating the value of a conjugated parameter**		
	Cost	Time	Means of production
Cost	n/a	High	Moderate
Time	High	n/a	Low
Means of production	Moderate	Low	n/a

Table 1 summarizes previous findings and gives an advise which of the checks of the offers placed should be done as a first one – the easiest one – M-T. While analysing a single task e.g making a cut, given:

- the number of machines (from machinery time schedule) with their key parameters,
- preparation time of the cut (from the time schedule),
- typical efficiencies for the set of assumed machines from [20],

it is easy to estimate if the assumed set of machines to make a certain cut in the assumed time [14]. Assuming one shift per day, the following formula can be used:

$$t_i = \frac{V_i}{e_i} \tag{1}$$

where:

t_i calculated time of $i-th$ element of building works in days,
V_i volume of $i-th$ element of building works,
e_i efficiency of the set of machines assigned to $i-th$ element of building works per shift

The six exemplary offers for preparing 30.000 m^3 of a cut in 20 days are shown in tab. 2. The types and the number of diggers of each type can be read from machinery employment schedule, the time of cut preparing from the time schedule. The unit efficiency of each type of a digger is based on [20]. It can be noticed that the offer A should not be considered as it proposes an insufficiently low number of low-efficiency diggers [15]. It is practically impossible to make 30.000 m^3 cut in 20 days with the machinery proposed in offer A. Keeping the scheduled 20 days based on the offer D is doubtful too. In order to compare the prices of the bid, the exemplary hiring prices of the machines (per day) were set here. Next, based on tab. 2. and tab. 3 the average cost for hiring the sets of machines – separately for each bid – can be calculated and compared to the exemplary bid values – see tab. 4. The margin g can be calculated too, for each offer as:

$$g = p - e \quad (2)$$

where:

p the bid price,

e the estimated cost of the machinery

Table 2. Analysing the M-T conjunction for a cut **making** (planned 30.000 m³ in 20 days) for exemplary offers labelled from A to F [own elaboration] .

Bid label	**Offered set of machines**	**Hoe volume [m³]**	**Number of machines**	**Unit efficiency [m³/shift] [20]**	**Total efficiency [m³/shift]**	**Calculated time of work execution [days]**	**Scheduled time of work execution days**	**Calculating possible volume of a cut [m³]**	**Planned volume of the cut [m³]**
A	Digger type X1	0,6	2	330	1080	27,8	20,0	21.600	30.000
	Digger type X2	1,0	1	420					
B	Digger type X1	0,6	1	330	2010	14,9		40.200	
	Digger type X2	1,0	1	420					
	Digger type X5	2,5	1	1260					
C	Digger type X3	1,2	2	470	1590	18,9		31.800	
	Digger type X4	1,6	1	650					
D	Digger type X2	1,0	3	420	1290	23,8		25.200	
E	Digger type X1	0,6	2	330	2430	12,3		48.600	
	Digger type X3	1,2	1	470					
	Digger type X4	1,6	2	650					
F	Digger type X4	1,6	2	650	2560	11,7		51.200	
	Digger type X5	2,5	1	1260					

Table 3. The exemplary average prices of hiring the machine per day [own elaboration].

Digger type	**X1**	**X2**	**X3**	**X4**	**X5**
Average hiring rate per day	400	500	600	700	900

Table 4. The estimated costs and bid prices comparison [own elaboration].

Bid label	**Price per day**	**Cost estimation for 20 days**	**Bid price**	**Margin**
A	1300	26 000	37 000	11 000
B	2000	40 000	45 000	5 000
C	1900	38 000	40 000	2 000
D	1500	30 000	39 000	9 000
E	2800	56 000	55 000	-1 000
F	2300	46 000	48 000	2 000

The margin is bound to be positive. It comprises all overheads and the necessary profit. The real cases of tender procedures, data collected from the actual bids can help the clients to evaluate which level of margin is usual, acceptable and typical for a given branch of the construction industry. Analysing only the bid prices from tab. 4 no one would say the price of the offer *A* is under-estimated (see fig. 4).

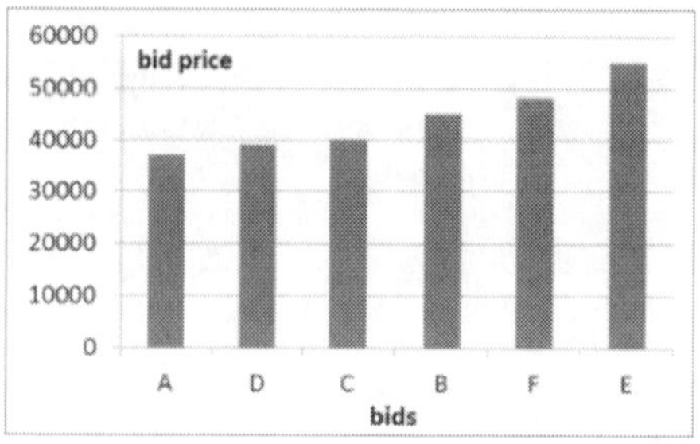

Fig. 4. The bid prices of the exemplary tender procedure [own elaboration].

The application of CTM allows for a double confirmation, that the lowest *A* bid is inconsistent and has to be excluded (they are not technically prepared to make the 30.000 m^3 cut in 20 days, and the margin is extremely high).

4 Conclusions

The Conjugated Triangle Method of the detection of inconsistent bids can be helpful especially in searching the under-priced offers – the source of many problems arising during construction work execution. It is based on the analysis of mutual conjunctions of three basic parameters: the cost, the time and means of production necessary to execute every building work. The relationships between the three parameters are common and natural. Application of the CTM does not require to disclose any information that currently is kept by the contractors as a secret one. Estimating the time of work execution (based on the proposed – in the offer – set of means of production), then estimating the contractor's price and comparing to the bid price can provide a precise answer if a given offer is under-priced. For the first part of analysis, the CTM can be applied for a single bid – no comparison to the other bids is necessary (it is not available without the CTM application). The two disadvantages of the CTM were found: the higher effort of the tender participants put in the bid preparing and the lack of possibility of CTM application in "design+build" projects. Nevertheless, where only the building works are ordered, a significant advantage of CTM, i.e. the precise information which offers are inconsistent, protect the clients from the choice of the offeror, whose falsely attractive bid price would be the reason for the client's loss.

References

1. B. Grzyl, M. Apollo, *Umowa o roboty budowlane w aspekcie podziału ryzyka stron.* Inżynieria Morska i Geotechnika **6**, 838-843 (2015)
2. B. Grzyl, E. Maliszewska-Urbańska, A. Siemaszko, *Safety of investment process parties in the aspect of construction work contract.* Journal of Science of the Gen. Tadeusz Kosciuszko Military Academy of Land Forces **49**/4, 208-215 (2017)
3. B. Dao, S. Kermanshachi, J. Shane, S. Anderson, E. Hare, *Exploring and assessing project complexity.* Journal of Construction Engineering and Management, 10.1061/(ASCE)CO.1943-7862.0001275, 04016126 (2017)
4. H. Anysz, A. Foremny, J. Kulejewski, *Estimating potential losses of the client in public procurement in case of collusion utilizing a MLP neural networks*, Technical Transactions **111/1–B**, 105–118, DOI: 10.13140/2.1.2409.6321 (2014)
5. S. Biruk, P. Jaśkowski, A. Czarnigowska, *Modeling Contractor's Bidding Decisions*, Procedia Engineering **182**, 91-98, DOI: 10.1016/j.proeng.2017.03.125 (2017)

6. S. Dubas, J. Pasławski, *The concept of improving communication in BIM during transfer to operation phase on the Polish market*, Procedia Engineering **208**, 14-19, DOI: 10.1016/j.proeng.2017.11.015 (2017)
7. H. Anysz, *Wykorzystanie sztucznych sieci neuronowych do oceny możliwości wystąpienia opóźnień w realizacji kontraktów budowlanych. Rozprawa doktorska*, Wydział Inżynierii Lądowej PW, Oficyna Wydawnicza Politechniki Warszawskiej, DOI: 10.13140/RG.2.2.14129.68960 (2017)
8. H. Anysz, *The profit as in-company evaluation of the construction site effectiveness*. MATEC Web of Conferences **117**, 1-6, DOI: 10.1051/matecconf/201711700009 (2017)
9. H. Anysz, *Założenia systemu rozliczeń TKW + M (techniczny koszt wytworzenia plus marża)*, Przegląd Budowlany **1** (2018)
10. A. Dziadosz, O. Kapliński, A. Tomczyk, M. Rejment, *Analiza i ocena ryzyka finansowego w przedsięwzięciu budowlanym*, Materiały Budowlanej **8**, 112-113, DOI: 10.15199/33.2016.08.34 (2016)
11. J. Kulejewski, *Zarządzanie projektem według PMBoK (cz. 1)*. Przegląd Budowlany **80/4**, 47-53 (2009)
12. M. Banach, R. Marcinkowski, *Harmonogramowanie robót budowlanych z wykorzystaniem przedmiarów dodatkowych w kosztorysowaniu*, Materiały Budowlane **6**, 6-7, DOI: 10.15199/33.2016.06.02 (2016)
13. M. Rogalska, Z. Hejducki, *Time coupling methods construction scheduling and time/cost optimization*, Wrocław: Oficyna Wydawnicza Politechniki Wrocławskiej (2011)
14. M. Krzemiński, *The scheduling of construction work under the assumption of brigade multitasking*, Procedia Engineering **208**, 63-68, DOI: 10.1016/j.proeng.2017.11.021 (2017)
15. E. Plebankiewicz, J. Malara, *The algorithm for the evaluation of construction workers' labour productivity*, Technical Transactions, Y. **114/6**, 153-161, DOI: 10.4467/2353737XCT.17.095.6571 (2017)
16. E. Radziszewska-Zielina, A. Sobotka, E. Plebankiewicz, K. Zima, *Wstępna identyfikacja i ocena parametrów wpływających na wydajność układu operator-maszyna do robót ziemnych*, Budownictwo i Architektura **12(1)**, 53-60 (2013)
17. N. Ibadov, *Selection of Construction Project Taking into Account Technological and Organizational Risk*, Acta Physica Polonica A **132**, 974-977, DOI: 10.12693/APhysPolA.132.974 (2017)
18. M. Juszczyk, A. Leśniak, K. Zima, *Selected issues of bills of quantities in construction works in Poland*, Journal of Science of the Gen. Tadeusz Kosciuszko Military Academy of Land Forces **49**, 1(183), 189-200 (2017)
19. K. Jaworski, *Metodologia Projektowania Realizacji Budowy*, Wydawnictwo Naukowe PWN, Warsaw (2009)
20. *Katalog Nakładów Rzeczowych 2-02*, tab. 0218, Wacetob (1997)

Predicting the impact of traffic–induced vibrations on buildings using artificial neural networks

Anna Jakubczyk-Gałczyńska[1,*]

[1]Gdańsk University of Technology, Faculty of Civil and Environmental Engineering, Narutowicza 11/12, 80-233 Gdańsk, Poland

Abstract. Traffic–induced vibrations may constitute a considerable load to a building, cause cracking of plaster, cracks in load–bearing elements or even a global structural collapse of the whole structure [1-4]. Vibrations measurements of real structures are costly and laborious, not justified in all cases. The aim of the paper is to create an original algorithm, to predict the negative dynamic impact on the examined residential building with a high probability. The model to forecast the impact of vibrations on buildings is based on artificial neural networks [5]. The author's own field studies carried out according to the Polish standard [6] and literature examples [7-10] have been used to create the algorithms. The results of the conducted analysis show that an artificial neural network can be considered a good tool to predict the impact of traffic–induced vibrations on residential buildings, with a sufficiently high reliability.

1 Introduction

The problem of traffic-induced vibrations is important because more and more areas are urbanized. New roads are being built, both in cities, on suburbs. In rural areas more buildings are being built near roads. Areas that are well communicated are readily built and inhabited. As a result, vibrations in buildings caused by vehicles concern both existing as well designed houses.

Taking into account the impact of traffic-induced vibrations caused on buildings, they can be unnoticeable or even causing a structural failure. Vibrations may cause cracking of plaster, cracks in load-bearing elements or even collapse of the structures [11, 12]. The impact of vibrations of the building is affected by many factors related to the road on which vehicles pass. The dynamic parameters of the building are also important. Measurements of vibrations on real constructions are labor-intensive and costly tests, and importantly, not justified in every case. It may be economically unprofitable to carry out such tests for all buildings located near the road. Modern, still growing technology brings a lot of possibilities to solve such as problems (for example see [13-15]). Therefore, there is a need to develop effective methods for forecasting the impact of traffic-induced vibrations on residential buildings. Thus the aim is to create an expert system, which it is possible to

* Corresponding author: annjakub@pg.gda.pl

predict with a certain probability the threat of negative dynamic impact on the tested building, without performing labor-intensive measurements. There are many methods for forecasting various types of phenomena. An example here are Artificial Neural Networks (ANN).

2 Traffic-induced vibrations

2.1 Definition and factors affecting the size of interaction

Vibrations are a phenomenon of the movement of building structure molecules most frequently caused by waves propagating in the subsoil and reaching to the foundations [16]. Vibrations transmitted through the ground can be divided into: seismic (independent of human activity, caused by earthquakes, see [17-23]) and paraseismic caused by human activity (see, for example [1-4]). Traffic-induced vibrations are a subset of paraseismic effects and are caused by the passage of wheeled and rail vehicles (also underground). The subject of the research in the paper are vibrations caused by road traffic.

The size of vibration depends on many factors, are associated with the following [16]:

- distance of the object from the source of vibrations (the range of dynamic effect is usually about 25 m from the egde of the road),
- soil: ground water, partitions, wells, absorption of soil,
- vehicles: shape, weight, technical condition, speed, number of vehicles passing at once, starting, stopping,
- surface: type, technical condition, unevenness (damage, protruding sumps causing a change in the trajectory of the vehicle, vertical and horizontal hit of wheels, which increases the emission of waves propagating in the ground),
- building: type, construction, foundation, technical condition, damping.

These factors depend on the intensity of the transmitted vibrations to the building. The more vibrations are transmitted to the building, the greater damage can occur.

2.2 Measurement methodology

In Poland, each building, should be carried out a detailed standard analysis according to [6]. Acceleration values recorded during the measurements are analyzed in one-third octave frequency bands. As a result, the amplitude values are obtained for all the centre frequencies in the range of 1-100 Hz. Then, peak values are selected and plotted on Dynamic Influence Scale (DIS) (see Fig. 1). The standard [6] concerns residential buildings, which have been divided into two groups:

- densely-shaped brick buildings, vertical projection of which should not exceed 15 meters, th building should not exceed two floors and its height must not exceed the measurement of the projection (assessment by DIS I),
- buildings of maximum 5 floors, their height being lower than doubled width (assessment by DIS II).

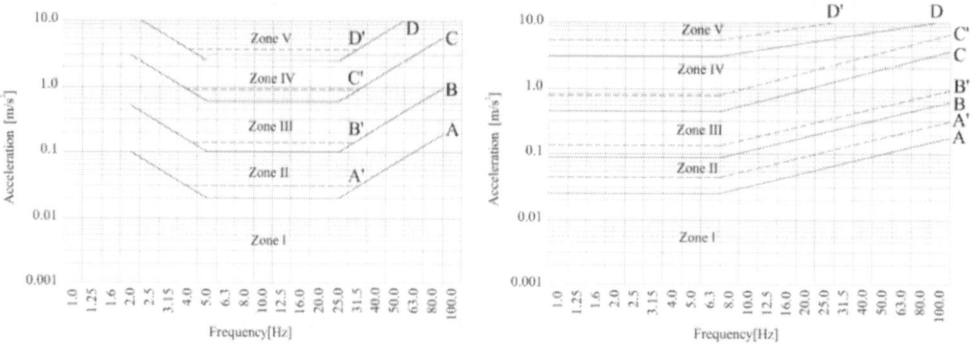

Fig. 1. Dynamic Influence Scale (DIS I on the left, DIS II on the right) [6].

The size of the impact and possible effects are described in [6]:
- Zone I: no impact of vibrations on building;
- Zone II: vibrations are noticeable but do not pose a threat to the structure;
- Zone III: the overall load bearing capacity of building may be weakened;
- Zone IV: vibrations have major influence on building, the amplitudes are high enough to cause various objects in the apartments to tremble, there is a risk for health of inhabitants;
- Zone V: The load bearing capacity is dysfunctional as a result of large amplitudes of vibrations what may lead to a major malfunction or even total collapse of the structure.

3 Experimental research

The measurements were made using specialist equipment (see Fig. 2). Piezoelectric acceleration sensors were connected to the measuring apparatus in their assigned sockets, which were previously configured (accordingly). During the tests 6 sensors were used mounted on the foundation wall of a building parallel to the street, just above ground level according to [6]. After connecting the sensors and setting the apparatus, calibration was carried out. On the computer connected to the measuring apparatus, the time histories shall be performed during passing of various types of vehicles (two-axle buses, vans and trucks with a total weight up to 10 t and for buses and trucks weighing over 10 t and have more than 2 axles.

Fig. 2. Specialized measuring equipment and piezoelectric accelerometers mounted on the wall of the tested building.

The measurements for 11 buildings were analyzed, filtering each measurement in 21 third bands (in each measurement independent values for several sensors were obtained, but ultimately the sensor that showed the highest amplitudes was selected), receiving 693 time history. An extreme acceleration was determined for each of the mid-frequencies one-third octave bands and the results were plotted on DIS graphs. They were data for artificial neural networks.

4 Artificial neural networks

4.1 Definition of artificial neural networks

Artificial neural network (ANN) is a machine learning algorithm based on the principle of the human brain's prosperity [5]. The human brain consists of about 10^{10} nerve cells, called neurons, and performs even 10^{18} operations per second. The neuron consists of a body called somme and spikes. The core element of the soma is the nucleus, in which all processes essential for the functioning of the neuron occur [24]. The spikes are divided into dendrites (there can be thousands of them) introducing information to the neuron and axon - one projection leading information from the soma. In order to create an algorithm it is necessary to build an artificial neuron (Fig. 3). The information is given, processed, saved and even corrected by the system. The process of building network consists of three basic steps: learning, validation and testing.

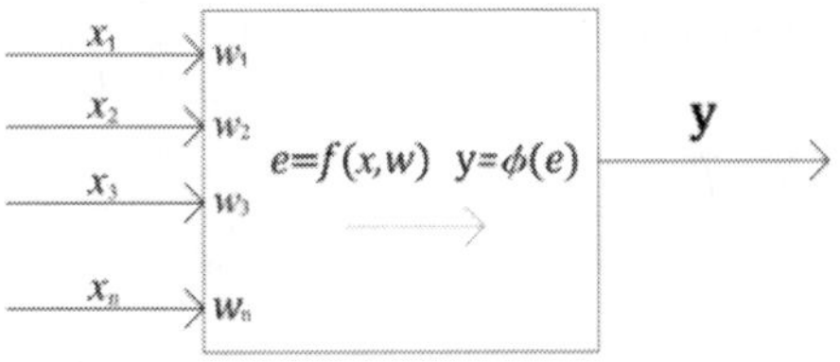

Fig. 3. Model of artificial neuron (perceptron): x_i– input signals; y – output vector (signal); w_i– weight; e – internal processing function; ϕ – activation function [5].

In classification problems, the most common types of ANN are Multilayer Perceptrons. These are basic neural networks, based on the principle: input – internal processing and activation functions - output [25]. This is a global approximation. Another tool is Radial Basis Functions that allow local type approximation. In the hidden layer, the neuron is subjected to functions that change radially around the selected center c.

4.2 Created algorithm of ANN

This subchapter presents the classification problem defined in such a way that the ANN algorithm (on the basis of several parameters given by the user) will be able to make a predict of the impact of traffic-induced vibrations on the building with a satisfactory probability. The network was built on the basis of principles described in the literature [5, 24-26]. The first step was to create a database necessary to start the builiding of the algorithm, i.e. to perform vibration measurements on the buildings. 63 samples were collected, including 33 samples of input data, performing their own measurements and 30 samples based on measurements of other researchers [7-10].

The input neurons were assumed to be factors (independent variables) determined during measurements based on standard [6] and book [16]:

- B_{TC} – technical condition of the building;
- D_{BR} – distance of the building from the edge of the road;
- S_A – absorption of soil;
- R_S – type of road surface;
- R_{STC} – technical condition of the road surface;

- V – vehicle type.

Input vectors for subsequent network adopted at random, in the form of:

- ANN no. I: $\mathbf{x}_{(6x1)} = \{B_{TC}, D_{BR}, S_A, R_S, R_{STC}, V\}$;
- ANN no. II: $\mathbf{x}_{(5x1)} = \{B_{TC}, D_{BR}, S_A, R_{STC}, V\}$;
- ANN no. III: $\mathbf{x}_{(4x1)} = \{B_{TC}, D_{BR}, S_A, R_{STC}\}$;
- ANN no. IV: $\mathbf{x}_{(4x1)} = \{B_{TC}, D_{BR}, R_S, V\}$;
- ANN no. V: $\mathbf{x}_{(4x1)} = \{D_{BR}, S_A, R_{STC}, V\}$.

As an output signal, a neuron with possible values was determined:

- 0 – no impact on building – zone I according to standard [6];
- 1 – possible impact on building – zone II or above according to standard [6].

All samples were randomly divided into 3 sets, i.e. learning, verification and testing set. Two variants of division were made:

- 1st variant: 44 samples were allocated randomly to the learning set (69.84% of all samples), 10 cases assigned to the verifying collection (15.87% of all cases) and 9 samples for the testing set (14.29% of the total);
- 2nd variant: 30 samples were allocated randomly to the learning set (47.62% of all samples), 16 cases were assigned to the verification set (25.40% of all cases) and 17 samples for the testing set (26.98% of the total).

Depending on the type of ANN, the BFGS learning algorithm was used, a strong second-order algorithm with fast convergence, local extremes of functions were found [26]. In the case of radial networks, Radial Basic Functions were used in the learning process. In the created algorithms, the activation functions given in [24].

4.3 Results

Five different network characteristics have been created. A summary of the results of the reliability of all networks is presented in Table 1. The general reliability of the ANN was calculated as a weighted average, taking into account the size of the learning, verification and testing sets.

Table 1. Summary of reliability of the analyzed artificial neural networks.

NETWORK No.	VARIANT OF DIVISION	RELIABILITY [%]			
		LEARNING SAMPLES	VERIFICATION SAMPLES	TESTING SAMPLES	GENERAL
ANN I	1	81.82	90.00	77.78	82.54
	2	76.67	81.25	88.24	80.96
ANN II	1	88.64	100.00	88.89	90.48
	2	96.67	87.50	82.35	90.48
ANN III	1	86.36	60.00	66.67	79.36
	2	76.67	68.75	88.24	77.78
ANN IV	1	88.64	70.00	66.67	82.54
	2	83.33	81.25	88.24	84.13
ANN V	1	86.36	90.00	88.89	87.30
	2	83.33	87.50	88.24	85.71

Taking into account all the combinations of the factors in different variants, it can be concluded that the best prediction, with a general reliability of 90.48%, was achieved by network No. II, in which the input vector has the form: $\mathbf{x}_{(5x1)} = \{B_{TC}, D_{BR}, S_A, R_{STC}, V\}$. This means that the best combination of input factors for the analyzed cases of vibration prediction is: technical condition of the building, distance of the building from the edge of the road, absorption of soil, technical condition of the road surface and vehicle type.

Network no. II, both in the first and in the second variant of the division into samples, reached very good results. In the first case-division variant, the reliability for the learning sample was 88.64%, the verification samples reached 100.00% correctness, while for the testing cases, the algorithm received 88.89% correct results. On the other hand, in the second variant the distribution of the cases, the credibility of the learning sample was 96.67%, for the verification sample obtained 87.50% accuracy, for the testing cases for algorithm obtained 82.35% correct results. Obtaining the best prediction results for the network no. II means that the type of surface is the least important for the analyzed cases. After performing sensitivity analysis for the testing samples (i.e., cases not used to build the algorithm or to validate it), it can be determined that the condition of the road surface is the most important issue.

5 Conclusions

The conclusion of results is that ANNs is suggested to be a relevant tool to predict the impact of traffic-induced vibrations on residential buildings with a relatively high probability. In addition, the presented algorithm can be applied both to existing and designed buildings, while the source of vibration is already present or may appear in the future.

The author's method may be helpful when user must decide about the location of the proposed investment and, in the case of existing buildings provides the basis for specialized vibration measurements to confirm or exclude risks.

References

1. A. Jakubczyk–Gałczyńska, R. Jankowski, Traffic–induced vibrations. The impact on buildings and people, Environmental Engineering, *Proceedings of the International Conference on Environmental Engineering*, Vilnius, Lithuania, 1–8 (2014)
2. A. Jakubczyk–Gałczyńska, R. Jankowski, A. Kristowski, The application of neural networks in forecasting the influence of traffic–induced vibrations on residential buildings, *Technical Transactions*, **9(3–B)**, 75–82 (2016)
3. A. Jakubczyk–Gałczyńska, A. Kristowski, R. Jankowski, Comparing the effectiveness of ANNs and SVMs in forecasting the impact of traffic–induced vibrations on building, Geodetic Congress (BGC Geomatics), Gdańsk, Poland, 121–125 (2017)
4. A. Jakubczyk–Gałczyńska, A. Kristowski, R. Jankowski, Application of support vector machine for determination of impact of traffic–induced vibrations on buildings, *Advances in Intelligent Systems and Computing*, **637**, 161–167 (2018)
5. R. Tadeusiewicz, *Sieci neuronowe*, Akademicka Oficyna Wydawnicza, Warszawa (1993)
6. PN-B-02170:2016-12 (in Polish): Ocena szkodliwości drgań przekazywanych przez podłoże na budynki (Evaluation of the harmfulness of building vibrations due to ground motion) (2016)
7. T. Chyży, K.R. Czech, M. Malesza, C. Miedziałowski, Badania i ocena wpływów drgań drogowych w zespole budynków zabytkowych, *Wiadomości Konserwatorskie*, **26**, 499–509 (2009)
8. R. Czech, C. Miedziałowski, T. Chyży, Wpływ poprawy stanu drogi na redukcję drgań w zabytkowym kompleksie budynków, *Materiały Budowlane*, **6**, 105–106 (2015)

9. J. Kawecki, K. Stypuła, Diagnozy a posteriori wpływów drgań drogowych na budynki, XXIV Konferencja Naukowo–Techniczna Awarie Budowlane, Szczecin–Międzyzdroje, 539–546 (2009)
10. K. Stypuła, Wpływ drgań na budynki i ludzi w budynkach – rola badań in situ, [on-line], [access 17 December 2017], http://www.map.piib.org.pl/materialy-szkoleniowe (2015)
11. R. Ciesielski, E. Maciąg, *Drgania drogowe i ich wpływ na budynki*. Wydawnictwa Komunikacji i Łączności, Warszawa (1990)
12. J. Kawecki, Kryteria oceny wpływu drgań komunikacyjnych na budynki zabytkowe i ludzi w budynkach w ujęciu normowym, *Przegląd Budowlany*, **86(11)**, 43–50 (2015)
13. A. Siemaszko, M.W. Kembłowski, Ocena efektywności monitoringu obiektów inżynierskich za pomocą sieci Bayesa, *Materiały Budowlane,* **6**, 139-140 (2016)
14. M. Apollo, B. Grzyl, E. Miszewska-Urbańska, Application of BN in Risk Diagnostics Arising from the Degree of Urban Regeneration Area Degradation, Geodetic Congress (BGC Geomatics), Gdańsk, Poland, (2017)
15. M.W. Kembłowski, B. Grzyl, A. Kristowski, A. Siemaszko, Risk Modelling with Bayesian Networks-Case Study: Construction of Tunnel under the Dead Vistula River in Gdansk. *Procedia Engineering,* **196**, 585-591 (2017)
16. J. Dulińska, J. Kawecki, K. Kozioł, K. Stypuła, T. Tatara, *Oddziaływania Parasejsmiczne Przekazywane na Obiekty Budowlane*, Wydawnictwo Politechniki Krakowskiej, Kraków (2014)
17. H. Elwardany, A. Seleemah, R. Jankowski, Seismic pounding behavior of multi-story buildings in series considering the effect of infill panels, *Engineering Structures*, **144**, 139-150 (2017)
18. T. Falborski, R. Jankowski, A. Kwiecień, Experimental study on polymer mass used to repair damaged structures, *Key Engineering Materials*, **488-489**, 347-350 (2012)
19. R. Jankowski, Pounding between superstructure segments in multi-supported elevated bridge with three-span continuous deck under 3D non-uniform earthquake excitation, *Journal of Earthquake and Tsunami*, **9**, Paper no. 1550012 (2015)
20. R. Jankowski, S. Mahmoud, Earthquake-Induced Structural Pounding, Springer, Cham, Switzerland (2015)
21. R. Jankowski, S. Mahmoud, Linking of adjacent three-storey buildings for mitigation of structural pounding during earthquakes, *Bulletin of Earthquake Engineering*, **14**, 3075-3097 (2016)
22. H. Naderpour, R. C. Barros, S.M. Khatami, R. Jankowski, Numerical study on pounding between two adjacent buildings under earthquake excitation, *Shock and Vibration*, article ID 1504783 (2016)
23. M. Szczepanski, W. Migda, R. Jankowski, Modal analysis of real timber frame houses with different insulation materials. *Advances in Science and Technology Research Journal*, vol. **10**, 215-221 (2016)
24. S. Haykin, *Neural Networks and Machine Learning*, Pearson Prentice Hall, New Jersey, USA (2009)
25. S. Osowski, *Sieci Neuronowe do Przetwarzania Informacji*, Oficyna Wydawnicza Politechniki Warszawskiej, Warszawa (2000)
26. L. Rutkowski, *Metody i Techniki Sztucznej Inteligencji*, Wydawnictwo Naukowe PWN, Warszawa (2009)

Multicriteria assessment of selected types of the single-layer exterior walls

Piotr Plebankiewicz[1], and *Agnieszka* Leśniak[2*]

[1]Cracow University of Technology, Department of Civil Engineering, Warszawska 24, 31-155 Kraków, Poland

Abstract. The choice of technology for the construction of external walls and materials is made at the building design stage. This is one of the key decisions, any change after the beginning of construction is bound to cause serious consequences. It is therefore worth analysing various design solutions beforehand, taking into account the cost, quality, execution time, as well as the partition parameters obtained. The paper presents three variants of the solution for single-layer external walls that can be used in single-family buildings. Costs were calculated and selected technical parameters were listed. The evaluation of the proposed variants was made using the multi-criteria analysis method.

1 Introduction

Walls are structural elements of buildings to perform the function of load transfer (mainly vertical load components). They constitute a barrier separating the interior of the building from the external space, as well as protect it from the influence of the external environment (external walls), or separate rooms from each other (internal walls) [8]. Static, technical and economic considerations mainly determine the choice of structure and thickness of the walls. From static viewpoint the most important factors are: the method of fixing the wall resulting from the structural layout of the building and the size of the loads acting on it. From the economic and technical point of view, the choice of wall construction is mainly determined by thermal and acoustic insulation.

The properties of the wall are determined primarily by its constituting the materials from which it was made. Their selection depends on the purpose and function of the wall in the building. Depending on the location in the building, the walls can be divided into external and internal ones. Within this categorization, a fundamental difference in the requirements for both types of walls is encountered. External walls should be characterized mostly, by a suitable load-bearing capacity (strength), as well as thermal-humidity and acoustic insulation. It is also important to ensure their required fire resistance, long-term service life and high accumulation and thermal stability. Vital properties also include resistance to the impact of the atmosphere; i.e, frost resistance, biological resistance and corrosion resistance. Further requirements are: low humidity in the air-dry state, the ability to quickly dry out, rather low absorbability and small capillary rise. External walls should

* Corresponding author: alesniak@L3.pk.edu.pl

trigger a positive effect on the health conditions in the rooms. An important feature is also the aesthetic appearance and the possible low cost of materials and implementation [9]. Nowadays, when choosing building materials, attention is paid not only to the cost of purchase but also to the cost of the building's life cycle [1, 5, 10, 13].

Producers of materials for building walls for years have ensured that single-layer walls are an ideal solution for everyone who values the speed of bricklaying work, as well as the accuracy of works implementation. Due to the lack of additional insulation layers, the duration of construction is much shorter than that of other types of walls. An important advantage is also a lower probability of execution errors, often occurring during the construction of thermal insulation layers in double-layer partitions.

The aim of the paper is to present selected three solutions for single-layer walls used in single-family housing. The proposed variants were compared in terms of the necessary costs of their construction and their selected technical parameters. To evaluate the variants the multi-criteria analysis method was used.

2 External walls – division and requirements

Considering the structural function and the type of loads carried, the following walls are differentiated [8, 11]: load-bearing, self-supporting, shielding and partition ones. A division according to the technology of implementation involves [8, 11]: brick, prefabricated and monolithic ones. Due to the technology of constructing vertical brick partitions, they can be divided into [8, 11]:

- Three-layer walls which consist of two vertical, parallel layers of the wall. They are connected with each other by anchors and the space that lies between them forms a gap filled, unfilled or partially filled with non-structural material [11]. The arrangement of the layers in the diaphragm wall is beneficial from the perspective of the building physics. The shielding wall can be made of bricks and fittings, it is durable and allows the attachment of heavy decorations, yet it generates additional costs.
- Two-layer walls are partitions in which the wall is a construction element and insulation forms a barrier to the escape of heat from the building. Instead of the shielding layer, the partition has a plaster that protects the insulation from damage. It is the insulation of the partition that determines its thermal efficiency. It prevents the formation of thermal bridges and any additional loss of energy. Due to the external plaster on the mesh, the partition has a low resistance to mechanical loads.
- Single-layer walls are made of only one layer – structural one with plaster on its both sides. This type of walls is mainly characterized by the speed of bricklaying work and the accuracy of execution which is easier to attain. To make single-layer walls, the warmest varieties of hollow clay bricks, expanded clay concrete or aerated concrete are used. The choice of materials for building walls is diverse, and their use brings both advantages and disadvantages.

The increasing awareness of the value of the share of energy necessary for heating and cooling of residential buildings in the overall assessment of energy consumption made the requirements for thermal insulation of external walls consistently heightened. In Poland in accordance with the Regulation of the Minister of Transport, Construction and Maritime Economy of July 5, 2013 from January 1, 2017, amending the regulation on technical conditions which should be met by buildings and their location (Journal of Laws of 2013, item 926), heat transfer coefficient UC (max) for external walls, at the temperature of the heated room $\geq 16^{o}C$, was lowered and amounts to $0.23 W/m^2K$. What is more, from 2021 this value will be only $0.2 W/m^2K$.

In the case of walls, acoustic insulation is also important, which in Poland is defined in the Polish Standard PN-B-02151-3: 2015-10 Building acoustics – Protection against noise in buildings – Part 3: Requirements for sound insulation of partitions in buildings and building elements. RA2 acoustic insulation determined for external walls depends on the type of the wall (with windows / no windows) and on the sound level A in dB of external noise in the surrounding of the building.

The technical conditions that should be met by the partition in the event of a fire have also been developed. The fire resistance of individual building elements is characterized by two or three criteria whose definitions and the required values can be found in the Notice of the Minister of Infrastructure and Development of July 17, 2015, regarding the technical conditions that should be met by buildings and their location (Journal of Laws 2015, 1422). Section VI Fire safety, Chapter 2. Fire resistance of buildings.

3 Variantization of solutions for the external walls of single-layer housing buildings

The choice of variants of the implementation of single-layer external walls is dictated by the choice of such solutions that currently meet the heat transfer coefficient of 0.2W/m2K, required from 2021.

3.1 Variant I – single-layer wall made of ceramics filled with mineral wool

Ceramic products are characterized by good compressive strength, frost resistance and fire resistance and their important advantage is the favorable microclimate that such products create in the room [8]. Ceramic materials for building walls include bricks and hollow blocks. Considering the low thermal insulation of ceramic bricks, they are usually avoided in non-insulated single-layer walls. Porous ceramics deserves attention, since due to the process of achieving air-filled micropores with an enlarged surface in the ceramic material and a special drill system and shape, the heat path that penetrates the wall is significantly longer. As a result, the heat conduction coefficient is more than twice smaller than in ceramic hollow blocks [8]. One of the newer solutions on the market is a ceramic hollow block filled with mineral wool Porotherm 44 T Dryfix [2]. The heat transfer coefficient is 0.17 [W/m^{2}K]. The wool inside is hydrophobic. It also has the ability to pass water vapor, so there is no risk of mold growth in the block. The technology of constructing walls in this system consists in bricking dry ceramic blocks. The producer [2] also offers a thin-layer mortar based on polyurethane, which is completely water-free.

3.2 Variant II – single-layer wall made of cellular concrete blocks

Cellular concrete is a building material produced from aerated cement-lime, lime or cement mortar. As a result of the introduction of gas under appropriate pressure into the mixture, homogeneous pores, called cells, are formed [11]. Different methods of producing pores in concrete result in the following types [12]: autoclaved aerated concrete, foamed concrete and foamed-aerated concrete. An important advantage of cellular concrete is the ease of processing and bricklaying, as well as its lightness. In the considered variant, a block of cellular concrete of YTONG ENERGO [14] was adopted, for which the heat transfer coefficient is 0.19 [W/m^{2}K]. In order to maintain high thermal insulation properties of the material on an entire partition, it is extremely important to guarantee a sufficiently high quality of execution. Special mortars should be used in this case [14].

3.3 Variant III – single-layer wall from light expanded clay aggregate concrete blocks with an insulating expanded polystyrene insert

While adding expanded clay to concrete (mineral aggregate produced as a result of foaming and sintering of clay raw materials in rotary kilns) light expanded clay aggregate (leca) concrete is formed. The leca concrete elements are observed as the most popular products made of lightweight aggregate concretes. Depending on the destination, products of various sizes can be found on the market. For constructing single-layer walls hollow blocks 36.5 and 38 cm wide are used most often. They are characterized by a heat transfer coefficient U of about 0.33W/m^{2}K [3]. Regarding thermal insulation, blocks filled with expanded polystyrene are a more advantageous solution. In the third variant of the single-layer wall, the HOTBLOK system was proposed – leca concrete blocks with a core in the form of thermal insulation inserts for which the heat transfer coefficient is 0.17 [W/m^{2}K] [3].

Figure 1 shows selected hollow bricks for the construction of the proposed variants of the single-layer external wall.

a) b) c)

Fig. 1. Example construction materials for laying single-layered external walls: a) Hollow brick Porotherm 44 T Dryfix [2]; b) YTONG ENERGO block 48cm thick [13]; c) HOTBLOK leca concrete block with a polystyrene insert [3].

3 Costs and selected technical parameters of the proposed single-layer wall variants

The cost calculation includes 1m^2 of the full partition. The cost of lintels, window openings and plastering was not included. The cost of selected variants were estimated according to catalogues: KNR K-49 (for variant I), KNR K-28 (for variant II) and KNRW 2-02 used by analogy (for variant III). The calculation included only direct costs: labour, material and equipment.

Table 1. The unit price and labour intensity for the selected variants of a single-layer wall.

Variant	Unit price [PLN/m^2]	Labour intensity [w-h/m^2]
Variant I A wall made of ceramic hollow blocks filled with mineral wool	328.43	1.32
Variant II Wall made of cellular concrete blocks	182.38	1.02
Variant III Wall made of leca concrete blocks with an insulating expanded polystyrene insert	317.28	1.97

Financial basis used for calculations include:
- labour cost -14.80 PLN/w-h.
- materials and equipment prices were taken from the market or the price publications.

Costs of purchases of materials are included in the prices of materials. Table 1 presents the calculation results.

Due to the assumptions made, the most expensive solution is manufacturing 1 m^2 of a ceramic hollow wall filled with mineral wool. The cheapest is the wall built from a cellular concrete block. The greatest labour intensity (based on the data from catalogues KNR K-49, KNRK-28, KNR 2-02) is involved in building the leca concrete block wall, while the lowest – the cellular concrete block.

Table 2 illustrates the selected technical parameters of the materials used in the variants under consideration.

Table 2. Selected technical parameters for the materials used in the analyzed variants.

Parameter	Porotherm 44 T Dryfix	YTONG ENERGO	HOTBLOK
Heat transfer coefficient U [W/m^{2}K]	0.17	0.19	0.15
Fire classification	REI 90	REI 240	REI 120
Sound insulation [dB]	48	42	46
Compression endurance [MPa]	7.5	2	1.5
Wall thickness [cm]	44	48	42
Hollow block mass [kg]	19	20	22-23

Source: analysis based on the producers' data [2, 3, 14].

Based on the information in the table, in terms of technical parameters, Porotherm 44 T Dryfix block shows the most favorable values.

4 A multi-criteria analysis for selected walls variants

The applications of multi-criteria analysis supporting decision making in the construction industry were presented, e.g. in [4, 6, 7]. The methods of a multi-criteria comparative analysis include a class of algorithms that create a scalar whose numerical value becomes a synthetic evaluation index – the so-called "mathematical methods" [15]. The application of the synthetic evaluation indices requires division of the evaluation criteria into stimulants which are positively correlated with the dependent variable, and negatively correlated destimulants. The most frequently used discrete multi-criteria method is the corrected summation index [16].

In order to provide the comparison and evaluation of the variants, the chosen evaluations of the criteria should be given dimensionless quantities, due to which it is possible to compare not only the quantitative but also the qualitative features. The procedure involves three steps [15]:

1. determining the corrected values of criteria evaluations for individual variants (stimulant, destimulant),
2. determining the coded values of criteria evaluations for individual variants,
3. determining the corrected summation index *Ji* according to formula (1).

$$J_i = \sum_{j=1}^{m} (z_{ij} \cdot v_j) \tag{1}$$

where:

J_i the synthetic evaluation of the *i-th* variant,
M the number of criteria,
z_{ij} the coded measure of the *i-th* variant in relation to *j-th* criterion,
v_j the weight of the *j-th* criterion.

To perform the multi-criteria analysis the authors proposed four criteria. The criteria and their values are presented in Table 3.

Table 3. Proposed criteria and their measures (analysis based on the producers' data [2, 3, 14]).

Criterion	The nature of the criterion	Weight of the criterion	Measure of the criterion for the variant		
			V_1	V_2	V_3
			Ceramics with mineral wool	Cellular concrete	Leca concrete with styrofoam insert
K_1 coefficient U [W/m^{2}K]	Destimulant	0.20	0.17	0.19	0.15
K_2 execution cost [zł/m2]	Destimulant	0.60	328.43	182.38	317.28
K_3 labour intensity [w-h]	Destimulant	0.10	1.32	1.02	1.97
K_4 sound insulation [dB]	Stimulant	0.10	48	42	46

In order to compare the variants, one has to use a method of data coding because it is impossible to equate such specific factors as price and time. Many different methods of data coding can be found in literature [4, 6, 7]. To encode the data, the Neumann-Morgenstern method was used to maximize the objectivity of the result. It is a one of the most popular method of data coding used in construction decision problems [6, 15]. A detailed description of the method can be found in [15]. Coding was made using the formulas (2) i (3):

- for stimulants:

$$\frac{x_{ij}-x_{i\,min}}{x_{jmax}-x_{jmin}} \tag{2}$$

- for destimulants:

$$z_{ij} = z_{ij} = \frac{x_{jmax} - x_{ij}}{x_{jmax} - x_{jmin}} \tag{3}$$

where:

x_{jmin} the minimum value of the j-th criterion measure,
x_{jmax} the maximum value of the j-th criterion measure,
x_{ij} partial measure according to the criterion K_j for the variant W_i.

Table 4 illustrates the encoded measures of criteria values for the proposed variants and the summation index J_i is established according to formula (1).

Table 4. Encoded measures of criteria values for the proposed variants.

Criterion	Weight	Encoded measure of the criterion for the variant		
		V_1	V_2	V_3
		Ceramics with mineral wool	Cellular concrete	Leca concrete with styrofoam insert
K_1 coefficient U	0.20	1.00	0.00	0.92
K_2 execution cost	0.60	0.00	1.00	0.08
K_3 labour intensity	0.10	0.68	1.00	0.00
K_4 sound insulation	0.10	1.00	0.00	0.67
Summary indicators J_i		0.37	0.70	0.30
Evaluation		2	1	3

The values of the summation index for the considered variants are presented in Figure 2.

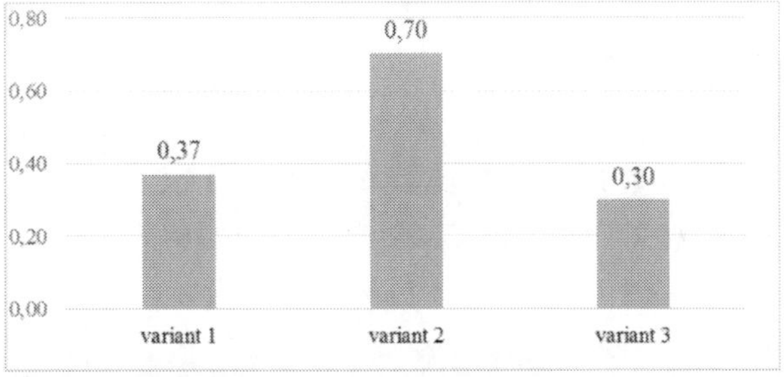

Fig. 2. The values of the summation index for the individual variants.

The multicriteria analysis performed for the adopted criteria and the weights assigned to them revealed that the most advantageous solution is the variant of the wall made of cellular concrete blocks. This result was probably influenced by the decision to accept the price criterion as the criterion with the highest weight. The labour intensity of erecting the partition made of this material is also the smallest. Blocks made of light material can significantly speed up the construction process. When comparing the U coefficient, this material, in comparison to the other two (Table 2), will take the highest value; nevertheless, it still meets the standards currently in force in Poland (0.23 [W/m2K]) as well as those that will apply after 2021 (0.2 [W/m2K]).

The difference between the values of the summation index for the other two variants is very small, although it should be noted that with the assumptions made, the second option was the variant involving the ceramic brick filled with mineral wool, in this case the cost of making 1 m^2 of the wall of this material was the highest.

5 Summary

The paper includes a comparative analysis of three variants of single-layer walls. In order to calculate the costs and determine the labour intensity the authors directly used the following catalogues: KNR K-49 (for variant I), KNR K-28 (for variant II) and KNRW 2-02 that was used by analogy (for variant III).

In order to select the best among the tree variants proposed, the multi-criteria analysis was conducted. Four criteria were adopted to assess the variants discussed: the U coefficient, cost per $1m^2$, labour intensity of constructing $1m^2$ of the wall, and sound insulation.

The results obtained revealed that ultimately the best solution is a single-layer wall made of cellular concrete blocks. It should be mentioned that this variant was also the cheapest, even though the weight of the criterion: costs per $1m^2$ was the highest. It should be emphasized that the weights of the criteria were accepted by the authors subjectively. Also the method of data coding (the Neumann-Morgenstern method) was accepted a priori. In the case of a change in weight or the data coding method, the evaluation of individual variants may be different.

References

1. B. Grzyl, E. Miszewska-Urbańska, M. Apollo, E3S Web of Conf., **17**, (EDP Sciences 2017)
2. http://wienerberger.pl/informacje/porotherm-dryfix-innowacja-w-murowaniu/
3. http://www.hotblok.pl/index.php/pl/oferta/system-budowy-scian/
4. M. Kaftanowicz, M. Krzeminski, Proc. Eng. **111,** doi:10.1016/j.proeng.2015.07.102 (2016)
5. M. Kozlovska, M. Spisakova, D. Mackova, WMCAUS 2017, **245,** UNSP 072022, (2017)
6. A. Leśniak, J. Balicki, EBER, **4**, 2 doi: 10.15678/EBER.2016.040206 (2016)
7. A. Leśniak, K. Zima, SGEM 2015 Conf. Proc., **3**, 5 (2015)
8. P. Markiewicz, *Budownictwo ogólne dla architektów*, (Wyd. Archi-Plus, Kraków 2011)
9. E. Moj, M. Śliwiński, *Podstawy budownictwa. Część I. Skrypt dla studentów wyższych szkół technicznych do przedmiotu: budownictwo ogóle*, (Wyd. PK, Kraków 1998)
10. B. Nowogońska, Engineering Mechanics. 20th international conference, May 12-15, 2014, Svratka, Czech Republic (2014)
11. J. Panas (edit), *Budownictwo ogólne. Elementy budynków, podstawy projektowania. Tom 3,* (Wyd. Arkady, Warszawa 2008)
12. J. Panas (edit), *Budownictwo ogólne. Konstrukcje budynków, Tom 4,* (Wyd. Arkady, Warszawa, 2010)
13. E. Plebankiewicz, K. Zima, D. Wieczorek, Arch. Civ. Eng. **62**, 2 (2016)
14. *Poradnik murowania Ytong,* (Xella Polska, Warszawa, 2017)
15. J. Szwabowski, J. Deszcz, *Metody wielokryterialnej analizy porównawczej*, (Wyd. PŚl, Gliwice, 2001)
16. T. Trzaskalik, *Wielokryterialne wspomaganie decyzji. przegląd metod i zastosowań*, Zesz. Nauk. Org. Zarz. P. Śl. **74** (2014)

The criteria for evaluation and selection the best tender applied by the authorities in Poland and selected EU countries

Beata Grzyl[1,*], *Magdalena* Apollo[1], *Emilia* Miszewska-Urbańska[1] , and *Adam* Kristowski[1]

[1]Gdańsk University of Technology, Faculty of Civil and Environmental Engineering, Narutowicza 11/12, 80-233 Gdansk, Poland

Abstract. The aim of the regulation in EU countries concerning public procurement is to ensure efficient spending of funds by evaluating and choosing the most advantageous offer. The aim of the conducted research is to determine the extent the Polish contracting entities benefit from the possibility of broad selection of various criteria for tender evaluation. On the basis of the analysis of 500 notices on work procedures conducted in the third quarter of 2016 - the third quarter of 2017 in open and restricted tenders in Poland and 500 tenders in open and restricted proceedings in five selected EU countries, the authors compared and classified of applied the practical criteria to evaluate and select the best offer. In order to collect the information presented in the article, the authors applied the documentary method using domestic and foreign literature regarding the subject and official publications.

1 Introduction

One of the EU's key priorities under the Europe 2020 strategy is sustainable development, including to support a competitive, resource-efficient and environmentally-friendly economy. An important objective of EU public procurement legislation is to provide contractors with the ability to compete effectively for public contracts in other Member Countries and to increase the efficiency of public spending. The purpose of the Directives [1, 2] is the simplification of procurement procedures, improving the procedure implementation, to enhance the sustainable development and the economy innovativeness and to increase involvement of small and medium-sized entities in the public procurement market [3, 4]. The changes in EU legislation aim to achieve a right relationship between the three complementary elements: the environment, the society and the economy [5, 6]. The measures carried out in this area are related to the monitoring of the mineral resource usage, carbon dioxide emission, waste management and economic phenomena (e.g. supporting the entities by means of renewable energy sources) and social (social security and economic stability of the population) [7]. The coordination of activities in the above areas is aimed at developing a compromise between social needs, technology development and rational use of environmental resources [8, 9]. One of the methods of working towards a sustainable

* Corresponding author: beata.grzyl@pg.edu.pl

development strategy is to implement the rules governing the above relationships, including the award of public procurement and the application of additional criteria to select the best offer.

2 The types of award criteria

The contracting authority is not restricted in setting the criteria for tender evaluation, however, each criterion must refer to the subject of the contract.

Exemplary criteria for evaluating tenders and awarding for work contracts ("build and design" and "build" options) may include [10]:
1. The cost criteria, which include: the offer price, the amount of the discount granted to the contracting authority, the gross work rate, the life cycle costs of the facility [11, 12, 13], including operating costs, terms of payment (e.g. payment schedule, payment terms),
2. The objective criteria, which may include: deadline for realization of the contract subject, the quality determined by the percentage share of the contractor's self-realization of the contract and the technical parameters of materials applied for the execution of works [14, 15], the methodology, functionality, warranty terms, the organization, the social and environmental aspects [16, 17, 18].

3 The analysis of selection criteria for the best offer applied in selected EU countries

3.1 The analysis of award criteria applied by awarding entities in Poland

The obligation to implement the Directive by Poland [1] imposed an obligation to amend the Public Procurement Law from 22 June 2016 [19]. The Directive introduced a number of changes in terms of cost and non-cost award criteria.

In order to determine the type and number of award criteria applied by Polish contracting authorities and the selection of the most favourable tender for construction works, the authors analyzed 500 proceedings of the value above the EU thresholds, conducted through open and restricted tendering in the third quarter of 2016 - the third quarter of 2017. The information has based on the notices contained in the Public Procurement Information Bulletin of the Public Procurement Office [20] regarding the construction works, implemented in the "build" and "design and build" system.

Table 1. The criteria for evaluation and selection of the best offer applied by the awarding entities in Poland. Source: own study based on [20].

The applied criterion of evaluation and selection of the best offer	The mean weight of a given criterion [%]	The percentage of proceedings involving a given criterion [%]
Price	58,6	100
Implementation deadline	45,5	24,6
The period of guarantee/warranty	43,3	48,8
Quality, functionality, technical parameters	32,4	25,3
Knowledge, experience, etc.	26,8	20,6
Terms and conditions of payment	23,1	20,2

The main objective of the amendment of the Public Procurement Law [19] is to strengthen the non-price criteria for tender evaluation. The introduced changes, may result in seen broader (compared to the previous years) application of non-price criteria by awarding entities such as the deadline and guarantee period (Table 1). Attention is paid to the importance of other non-price criteria and considering them important enough by the authorities. In every fourth procedure, the criteria of awarding the contract is the quality of execution of the works and the functionality of the proposed solutions, in every fifth - the knowledge and experience of persons directed to the performance of the contract and the terms of payment. The results of the analysis of selected investigations also indicate that the criteria applied to evaluate and select the most advantageous offer do not include criteria dealing with the social, environmental or innovation aspects [21, 22].

In Poland and in other EU Member Countries, the same Directive [1] is in force, however, in assessing and selecting the best offer, each country may choose slightly different criteria. In order to determine the nature and number of award criteria applied by awarding entities from five selected EU countries, the authors analyzed 500 selected open and restricted construction procedures from the third quarter 2016 – to the third quarter 2017, whose values rise above the EU thresholds.

3.2 The analysis of award criteria applied by awarding entities in Sweden

The results of the analysis indicate that in Sweden the contracting authorities take into account the price criterion (in each analyzed case), but also the qualifications and experience of the persons participating in the performance of the contract, the way of planning the implementation of the investment and the environmental aspects (Table 2). In 83.1% of cases conducted in Sweden one of the decisive selection criteria is the experience of persons performing the order, and the mean weight of this criterion is 34.2% (the maximum weight in the analyzed proceedings was 60%). The attention is also drawn to the fact that almost half of the analyzed cases use the environmental criterion. The average weight of this criterion is 15.3%. This means that in most cases this is a complementary criterion. Similarly looks the criteria of "completion date of the order subject". It occurs in 33.5% of the analyzed cases, its overage weight is 13.1%. The criterion "social aspects" is applied in average in every 5 cases.

Table 2. The criteria for evaluation and selection of the best offer applied by the contracting authorities in Sweden. Source: own study based on [23].

The applied criterion of evaluation and selection of the best offer	The mean weight of a given criterion [%]	The percentage of proceedings involving a given criterion [%]
Price	51,5	100
Professional qualifications, experience	34,2	83,1
Organization, planning, consistency of tasks	31,6	21,5
Social aspects	20,4	20,1
Environmental aspects	15,3	45,2
Implementation deadline	13,1	33,5
Architectural approach (e.g. using BIM)	11,9	15,4

3.3 The analysis of award criteria applied by contracting authorities in Spain

While choosing the best offer, the Spanish contracting authorities mainly use the criteria of price, technical capacity, financial and professional capacity of the entity (average weight: 34.9%) and warranty period (average weight: 11.0%). In Spain, the price criterion is normally used in each of the analyzed proceedings. Its average weight is 58.8%. There are a number of criteria applied in Spanish procurement (Table 3). The most frequently used criteria for choosing the best offer are: time of realization, environmental and social aspects, quality, functionality, efficiency of solutions, quality of applied materials. In every 4 proceedings the best offer is chosen based on the criteria "completion date", "quality of applied solutions and materials" and "organization, planning and coherence of tasks, including application of BIM technology in planning and execution of investment". In Spanish proceedings social and environmental criteria also appear. Their average weight is about 12%, which in most cases does not determine the choice.

Table 3. The criteria for evaluation and selection of the best offer applied by the contracting authorities in Spain. Source: own study based on [23].

The applied criterion of evaluation and selection of the best offer	The mean weight of a given criterion [%]	The percentage of proceedings involving a given criterion [%]
Price	58,8	100
Financial and professional capacity	34,9	46,2
Implementation deadline	24,2	23,1
Social aspects	13,6	32,8
Organization, planning (the use of BIM)	13,4	24,8
Quality, functionality, technical parameters	12,3	39,4
Environmental aspects	11,3	33,3
The period of guarantee	11,0	42,4
Quality of applied solutions and materials	9,7	24,9

3.4 The analysis of award criteria applied by contracting authorities in France

Table 4. The criteria for evaluating and selecting the best offer applied by the contracting authorities in France. Source: own study based on [23].

The applied criterion of evaluation and selection of the best offer	The mean weight of a given criterion [%]	The percentage of proceedings involving a given criterion [%]
Price	52,3	100
Financial and professional capacity	41,6	81,3
The method of execution	21,4	5,4
Implementation deadline	18,2	14,6
The cost analysis	17,8	5,9
Environmental aspects	15,7	21,3
Quality, functionality, technical parameters	15,3	4,8
Organization, planning, consistency of tasks	14,7	12,6
Professional qualifications, experience	13,1	11,8
Quality of the applied materials	12,2	7,3
The architectural approach (e.g. using BIM)	12,1	1,8

The price criterion occurs in each of the analyzed French proceedings. The average weight is 52.3%, which means that it does not determine in any case the result of the selection. An important, frequently applied criterion deals with technical, financial and professional capacity of the contractor. It reports on the financial capacity of the company, takes into account technical and human resources involved in the implementation of the contract subject, describes the principles of safety and effectiveness of activities undertaken during the investment implementation. Under this criterion, the technical solutions applied for the execution of the order, as well as environmental requirements, may be taken into account. This criterion is found in 81.3% of analyzed orders with an average weight of 41.6%. Both implementation method criterion with an average weight of 21.4% and a "cost analysis" criterion with an average weight of 18.8% are also worth mentioning. In 21.3% of the investigated procedures, the contracting entities take into consideration the environmental aspects while assessing and selecting the most advantageous tender. The average weight of this criterion is not too high - 15.7%. The results of the analysis indicate that the selection of the most advantageous offer also applies to other criteria of smaller importance (Table 4) e.g. BIM [24].

3.5 The analysis of award criteria applied by awarding entities in Germany

The most advantageous offer in Germany is chosen primarily on the basis of price and the quality criteria. The price criterion is a standard in every German case with an average weight of 54.8%. The quality criterion is applied in every second procedure in Germany. Another important criterion for selection is organization and task planning (average weight: 36.2%). The weight of the following criteria "completion date" and "environmental aspects" (24.8%, 23.5%), indicate that in some cases they may be decisive in choosing the best offer. In almost every case, risk management criterion occurs (Table 5). The innovation criterion occurs in individual proceedings only.

Table 5. Evaluation and selection criteria of the best offer applied by the contracting authorities in Germany. Source: own study based on [23].

The applied criterion of evaluation and selection of the best offer	The mean weight of a given criterion [%]	The percentage of proceedings involving a given criterion [%]
Price	54,8	100
Quality, technical parameters, use of BIM	41,2	48,3
Organization, task planning	36,2	26,9
Implementation deadline	24,8	14,2
Environmental aspects	23,5	28,5
Risk management	21,7	19,4
Guarantee period	19,0	10,2
Experience	17,9	17,3
Innovation	7,5	3,9

3.6 The analysis of award criteria applied by awarding entities in Czech Republic

Table 6. The criteria for evaluation and selection of the best offer applied by the contracting authorities in the Czech Republic. Source: own study based on [23].

The applied criterion of evaluation and selection of the best offer	The mean weight of a given criterion [%]	The percentage of proceedings involving a given criterion [%]
Price	75,3	100
Terms and conditions of warranty	27,4	38,9
Implementation deadline	25,2	65,4
Experience	18,1	26,3
Quality, functionality, technical parameters	16,9	7,6
Environmental aspects	15,3	5,7
Organization, task planning	9,8	3,5

In Czech Republic, when choosing the most advantageous offer, three criteria are most important: price, guarantee and the date of realization of the order subject. The average weight of the price criterion is as much as 75.3%. Other criteria ("experience", "quality, functionality", "technical specifications") are complementary only, in most cases they do not prejudge the choice of the offer. The criteria for evaluation and selection of the most favorable environmental aspects and organization and task planning are applied in individual cases (Table 6).

4 Conclusions

The analysis of selected procedures concerning the execution of construction works in the "build" and "design and build "system in the aspect of evaluation criteria and selection of the best offer applied in Poland and selected EU countries justifies the formulation of the following conclusions and statements.

1. The Directive [1, 2] applicable in EU countries, promotes the application of numerous non-price criteria for tender evaluation. The results of the analysis indicate that Polish and Czech contracting authorities do not apply this possibility. In tendering procedures conducted in these countries, one or two non-price criteria (term of performance, guarantee period) are most frequent, but due to their low weight, in most cases they do not act strongly on the final evaluation and selection of the best offer. The price plays a decisive role. The results show that contracting authorities from Sweden, Spain, France and Germany are more likely to use different criteria for evaluating and choosing the most advantageous tender - they generally use 4 to 5 criteria. In many cases, the criterion of decisive importance, besides price is "experience", "innovation" and "environmental aspects". Comparing the most commonly applied criteria in Poland and France, many differences can be seen (Fig.1).
In French orders, the practical criteria such as guarantee or social aspects do not are observed. Technical, financial and professional ability, a method and time of execution of the contract play an important role. Contrary to Polish, in the French public procurement, the cost of life cycle analysis is taken into account while choosing the offer. In the case of French and Swedish proceedings, the criterion of the extension of the warranty period does not apply, although in Poland it occurs in every second procedure. In Sweden, contracting authorities highly point out the qualifications and experience of staff, organization and task planning, and social and environmental aspects. In the Swedish and French cases, the criteria for job planning, work organization are applied. In Poland such criteria are practically not used (Fig.2).

2. The Directive [1] grants Member Countries the freedom to regulate these matters under national legislation. This means that the Member Countries themselves decide whether the price criterion in the contracts awarded in these countries is limited. Despite the significant reduction of the price criterion for construction works in Poland (up to 60%), criteria of a qualitative, environmental, social, innovative nature have not been widely applied yet.

3. In many cases where the evaluation and selection of the best offer is based on two criteria only, the weight of the first criterion is so large that the second criterion is insignificant. In the case of proceedings to take a greater number of criteria into account in tender evaluation a greater competition occurs. Thus the contractor selects "the most economically advantageous tender", to bring about a high work quality and additionally takes into the consideration the prosocial and environmental aspects of the investment. On the other hand, the contractor, who fulfills all additional non-price requirements of the ordering party may win the proceeding despite offering higher prices than the competitors.

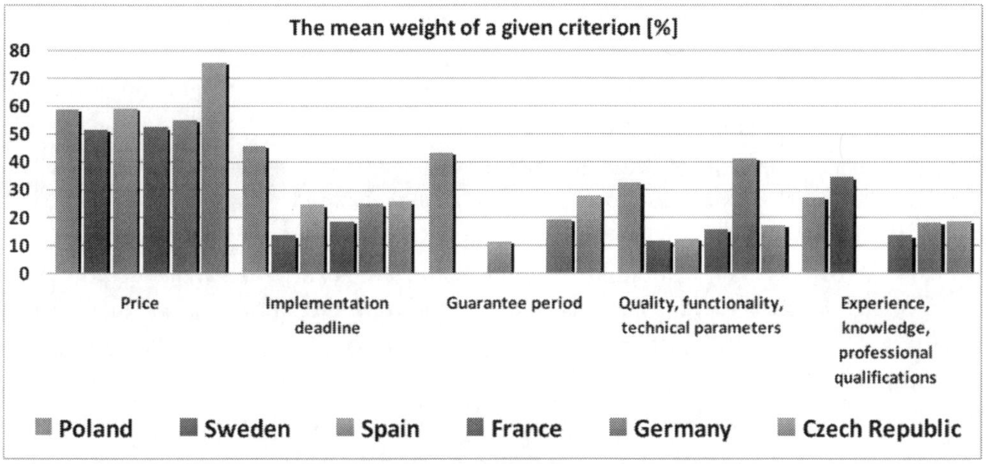

Fig. 1. Used criterion. Source: own study based on [23].

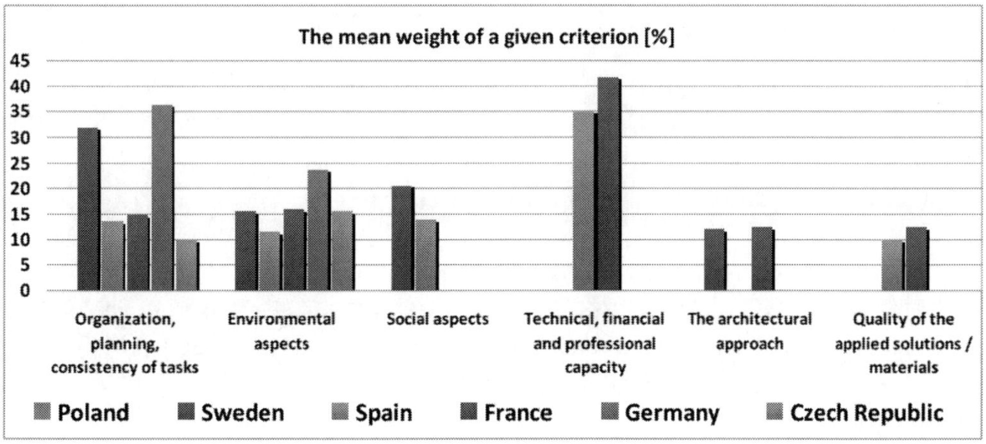

Fig. 2. Unused criterion. Source: own study based on [23].

4. The application of the price as the decisive criterion for the selection of the offer has its advantages (e.g. speed, simple application, safety of the contracting authority, low cost of preparation and conducting of the proceedings) but also disadvantages (e.g. no possibility to award specific features of the order, low operating costs, modern technological solutions). The application of price and other criteria of significant importance gives numerous benefits, e.g. it contributes to increase the efficiency of spending public funds like the ability to precisely match the selection of the offer to the specific needs of the contracting entity, a better preparation of the description of the subject of the order and a precise definition of the essential features of the product, the promotion of its desirable qualities e.g. innovation, quality, limited environmental impact. In practice, however, it raises a lot of difficulties for the procuring entity, e.g. increasing the cost and effort of preparing and carrying out the procedure, resulting in a more complex and longer procedure for evaluating submitted tenders, sometimes involving experts to evaluate proposed solutions, improves the possibility of incorrect definition of evaluation criteria and questioning them by tenderers.

References

1. Dyrektywa Parlamentu Europejskiego i Rady 2014/24/UE z dnia 26 lutego 2014 r. w sprawie zamówień publicznych.
2. Dyrektywa Parlamentu Europejskiego i Rady 2014/25/UE z dnia 26 lutego 2014 r. w sprawie udzielania zamówień przez podmioty działające w sektorach gospodarki wodnej, energetyki, transportu i usług pocztowych.
3. D. Łapacz, International Business and Global Economy, *Zamówienia publiczne w UE – analiza porównawcza procedur w wybranych państwach członkowskich*, **34**, 67–77, (2015)
4. B. Wieczerzyńska, Rocznik Administracji Publicznej, *Efektywność zamówień publicznych w wymiarze prawnym, ekonomicznym i społecznym*, **2**, 558-570, (2016)
5. J. Szymonek, *Zrównoważone zamówienia publiczne w Polsce: możliwości, bariery, strategie*, (2017)
6. E. Dziawgo, Prace Naukowe Uniwersytetu Ekonomicznego we Wrocławiu, *Realizacja strategii „Europa 2020" w krajach Unii Europejskiej*, **478**, (2017)
7. M. Wojcińska, D. Wojciński, Kontrola Państwowa, Klauzule społeczne w zamówieniach publicznych–szanse na upowszechnienie ważnych rozwiązań, **62** (2(373)), 113-126 (2017)
8. B. Grzyl, E. Miszewska-Urbańska, M. Apollo, E3S Web Conf., **17**, (2017)
9. M. Apollo, E. Miszewska-Urbańska, Adv. Sci. Technol. Res. J., **9(28)**, 68–74, (2015)
10. F. Łapecki, G. Wicik, 2016. Zamówienia publiczne po nowemu – poradnik dla małych i średnich przedsiębiorców, PARP, Warszawa (2016)
11. E. Plebankiewicz, K. Zima, D. Wieczorek, Technical Transactions, Quantification of the risk addition in life cycle cost of a building object, **5**, 35-45, (2017)
12. M. Juszczyk, A. Leśniak, K. Zima, Z. Nauk./WSOWL im. gen. T. Kościuszki, Selected issues of bills of quantities in construction works in Poland, **1**, 189-200, (2017)
13. Leśniak, K. Zima, Sustainability, **10**, 5 (2018)
14. M. Kurpińska, T. Ferenc, ITM Web Conf., **15**, (2017)
15. M. Pszczoła, J. Judycki, D. Ryś, Transportation Research Procedia, **14**, 738-747, (2016)
16. M. Łojko, Studia Ekonomiczne, Klauzule społeczne jako instrument wsparcia rozwoju ekonomii społecznej w procesie zamówień publicznych, **326**, 108-121, (2017)
17. M. Łysakowski, Magazyn Autostrady, Pozacenowe kryteria oceny ofert w przetargach na budowę dróg, **5,** 142-146, (2018)
18. M. Maksimiuk, Magazyn Autostrady, Przetargi na oświetlenie drogowe–nowe kryteria oceny ofert, **1-2**, 36-38, (2017)
19. Tekst jednolity ustawy Prawo zamówień publicznych, poz. 1579 (2017)
20. Biuletyn Informacji Publicznej UZP BIP, 2017, https://www.bip.gov.pl
21. Z. Walczak, A. Szymczak-Graczyk, Przegl. Bud., BIM jako narzędzie przyszłości w projektowaniu i rewitalizacji obiektów budowlanych, **88**, 20-26, (2017).
22. K. Zima Procedia Engineering, **208**, 203-210 (2017)
23. The supplement to the Official Journal of the EU., 2017, http://www.ted.europa.eu
24. B. Grzyl, E. Miszewska-Urbańska, M. Apollo M., Procedia Engineering, **196,** 1026-1033, **(**2017)

Simulation Method for Scheduling Linear Construction Projects Using the Learning–Forgetting Effect

Łukasz Rzepecki[1,*], and *Sławomir* Biruk[1]

[1]Lublin University of Technology, Faculty of Civil Engineering and Architecture, Nadbystrzycka 38D, 20-618 Lublin, Poland

Abstract. Linear projects such as highways, multi-storey buildings or pipelines are implemented by repeating the sequences of processes on units. Repetitive operations affect acquiring more and more experience, which shortens the duration of subsequent activities. On the other hand, the possible breaks in work make this effect disappear. Breaks at work are most often caused by the lack of crew work continuity. While delaying the subsequent sequences of processes properly, it is possible to reduce interruptions and limit the forgetting effect. A simulation method for scheduling linear construction projects has been developed, taking into account the learning and forgetting effect. It allows to increase the probability of the crew work continuity while meeting the project deadline with a specific probability. As a result, the estimation of the project duration is more accurate, which may provide an advantage at the stage of submitting offers in tenders for construction works.

1 Introduction

Construction projects encompass within their scope frequently repeated works conducted on identical or similar work units. Such projects are exemplified by multi-storey buildings, highways and pipelines. Identically composed crews carry out construction processes on successive work units in the same cyclical and parallel order, with each subsequent process beginning only after work on a given unit has been completed by the previous crew.

The Critical Path Method (CPM) is not effective for scheduling repetitive processes. Methods based on the Line of Balance (LOB) concept are usually applied to plan linear projects. These methods focus on scheduling tasks for carefully selected work crews by coordinating activities in time in such a way as to eliminate unnecessary idle periods in the operation of resources. The most well-known methods deriving from LOB include Linear Scheduling Method (LSM) [1], Productivity Scheduling Method (PSM) [2–3], Vertical Production Method (VPM) [4] and Repetitive Scheduling Method (RSM) [5]. They apply graphic modelling, optimization methods, elements of network planning and, most importantly, take into account constraints between activities executed on the same floor and a different floor [6]. All these planning techniques fall under the umbrella term of Linear

* Corresponding author: l.rzepecki@pollub.pl

Scheduling Method (LSM). The analysis of interconnections between processes executed on successive work units allows establishing the controlling sequence of processes, whose practical importance is the same as that of the critical path for CPM [5].

The majority of existing methods for scheduling construction projects is applicable under deterministic conditions. While schedules developed this way greatly facilitate modelling complex projects, they are prone to fall out of date under real-life conditions. This is true particularly of the building industry which differs largely from manufacturing where production takes place under stable technical and organizational conditions. Construction is more exposed to risk factors and uncertainty than other branches of national economy [7]. The effect of external factors on the implementation of a building project is magnified by a large number of participants and long implementation cycles. Complexity of an undertaking and the period of its implementation increase in step with the force of risk factors, as it becomes more difficult to evaluate their probability and predict their possible effects [8]. A schedule should be developed for an acceptable risk level. A commonly used way to make schedules more robust against disruptions is to allocate time buffers (reserves). Time buffers are used to ensure the timely completion of individual processes and the project as a whole. The main problem is to size the time buffers adequately, as this may affect the duration of the works.

2 The learning-forgetting effect

As the same work processes are repeated in successive units, work crews acquire experience which helps to shorten lead times and reduce labour costs. Developing a schedule requires determining work efficiency to allow tracking the time and cost of the undertaking. In the case of projects consisting of repetitive processes, the scheduling stage should allow for past experience contributing to reducing the work time, in order to gain an edge at the bidding stage.

The learning theory aims to mathematically describe the relations between work repetition and efficiency. This is true where workers carry out standard productive operations, whereby they improve their skills and possibly also learn how to use their workplace in a more ergonomic way.

The oldest basic learning model is Wright's exponential model [9] from 1936:

$$t_n = t_1 \cdot n^{-l}, \tag{1}$$

where:

- t_n lead time of process performed on work unit n,
- t_1 lead time of process performed on the first work unit,
- n number of process repetitions,
- l reduction parameter reflecting the slope of logarithmic curve; it is assumed that $l = -\frac{\log s}{\log 2}$, and
- s known as learning rate – is defined as a time reduction percentage that results in doubling the production output.

Work interruptions mean that crews need longer times to complete operations on successive units. Short interruptions lead to a partial loss of experience, while in the case of sufficiently long ones, the loss may be complete [10]. Forgetting as a result of work interruptions is a reverse process to acquiring experience [11, 12] and contributes to longer lead times. Globerson and Levin [11] proposed the so-called S-shaped Forgetting Function, which is easily applicable to the simulation method for the reason that the forgetting effect depends directly on the length of work interruption Δt:

$$t_b = t_1 - [t_1 - t_w] \cdot (a \cdot \Delta t + 1) \cdot e^{-a \cdot t}, \quad (2)$$

where:

- t_w duration of subsequent process if no interruption occurred,
- a forgetting coefficient,
- e base of natural logarithm.

The effect of experience on work efficiency was also surveyed in the construction industry, for example, in works related to reinforced concrete floors [13], roof coating [14], reinforced concrete piles [15] or installing rebars [16]. The learning curve was also used for planning motorway construction [17] as well as the installation of underwater caissons [18]. It is usually assumed that the rate of experience for most processes within the construction industry ranges from 70% to 95%.

The learning-forgetting theory was also used for scheduling construction projects. Lam et al. [10] used Wright's model [9] and S-shaped Forgetting Function [11] to plan undertakings by the LOB method under deterministic conditions. Lutz et al. [19] took into account the learning effect only in a repetitive process simulation. They assumed that the duration times of construction processes, might be shortened by up to 25% due to the experience curve.

3 The method of determining time buffers for linear construction project planning

A sequence of processes i ($i = 1, 2, \ldots, n$) is to be performed on each unit j ($j = 1, 2, \ldots, m$) in unchanged order. The time for completion of works by crew i on unit j equals t_{ij}. The starting date of the first process on the first unit is $t_{11}^s = 0$ (Fig. 1).

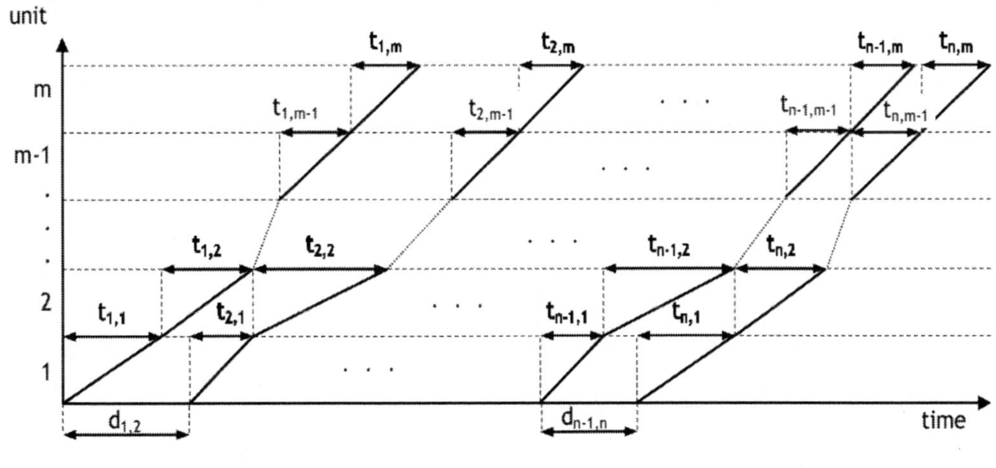

Fig. 1. Line of Balance diagram.

The shortest project cycle with crews operating continuously can be achieved at critical (maximum) proximity of the neighbouring broken lines (sections on the Line of Balance diagram) reflecting the progress of individual processes. The process i+1 should start relative to i with a delay of $d_{i,}(i$+1). This delay can be calculated under deterministic conditions using the formula below:

$$d_{i,(i+1)} = \max \begin{cases} t_{i1} \\ t_{i1} + t_{i2} - t_{(i+1)1} \\ \dots \\ \sum_{j=1}^{m} t_{ij} - \sum_{j=1}^{m-1} t_{(i+1)j} \end{cases}, \quad (3)$$

The following formula defines finish dates for processes within the first series string on successive work units j ($j = 1, 2, \dots, m$) and start dates for subsequent processes ($j + 1$):

$$t_{1j}^{f} = \sum_{k=1}^{j} t_{1k}, \quad (4)$$

while the start date for the subsequent process ($i \geq 2$) can be defined using another formula, namely:

$$t_{ij}^{s} = \sum_{s=1}^{i} d_{(s-1),s} + \sum_{k=1}^{j-1} t_{ik}. \quad (5)$$

Under random conditions, lead times t_{ij} for processes i ($i = 1, 2$) on successive units j ($j = 1, 2, \dots, m$) are random variables T_{ij} with expected (average) value $E(T_{ij})$ and variance $D^2(T_{ij})$. Random variable parameters can be established by analogy to PERT for the following variants: t_{ij}^{a} – optimistic time estimate, t_{ij}^{m} – most likely time estimate, and t_{ij}^{b} – pessimistic time estimate. Random variables exhibit the effect that the process finish and start dates scheduled according to formulas (4) – (5) fall out of date, leading to a possible discontinuity in crew operations and extending the lead time for the entire project.

The following procedure is proposed as a way for repetitive-process project schedules to increase the probability of work continuity, while also ensuring that the crews can meet the baseline finish date with a specific probability.

1. Define parameters types for the time distribution of the duration of construction processes. The formula below should be used to define the distribution function of project duration by the Monte Carlo method, without taking into account the learning effect but considering that the processes begin at the earliest possible dates (directly after the completion of a previous process on the same unit and the completion of works on the prior unit):

$$t_{ij}^{s} = \max \begin{cases} t_{(i-1),j}^{s} + t_{(i-1),j} \\ t_{i,(j-1)}^{s} + t_{i,(j-1)} \end{cases}, \quad (6)$$

where:

t_{ij} crew work time i ($i = 1, 2, \dots, n$) on unit j ($j = 1, 2, \dots, m$) and $t_{11}^{s} = 0$.

2. Agree with investor on baseline finish date T_d and assess probability R of meeting that time-limit based on a previously calculated distribution function.

3. Perform model simulation with regard to the learning-forgetting effect and assuming that processes start at earliest possible dates. The learning rate and forgetting coefficient should be based on the contractor's real-life expertise or literature describing relevant tests performed under comparable conditions. Calculate total idle time for all crews. The baseline finish date T_d will be met with probability R due to process duration shortened

as a result of learning. This allows for a time reserve that can be used to increase the work continuity without altering the probability of timely completion on baseline finish date.

4. Delay start date of works (process strings performed by the same crew) and improve solution – introduce Buffer Times $d_{i,(i+1)}$. Delaying the start date of works contributes to reducing idle times, whereby forgetting effect is eliminated. Minimalizing of forgetting effect causes, as a result, shortens process duration times. Buffer Times may not affect the project's baseline finish date. This stage analyses project variants for different values of $d_{i,(i+1)}$, where process start date ij is the greater of the values calculated on the basis of (5) and (6). The acceptable variant is a solution which meets the baseline finish date. The total crew idle time is calculated for each variant. This test aims to find the minimum idle time variant. This paper seeks to find a suboptimal solution through a heuristic algorithm – simple hill climb algorithm. In the next step, an optimization algorithm is developed and implemented in simulation software.

4 The learning-forgetting theory applied to schedule crew works related to the implementation of repetitive construction processes under random conditions

This section presents the proposed method used as an example to schedule interior finish works in a multi-storey building. It is assumed that each floor requires the same amount of work, while crew efficiency depends on equipment efficiency (impossible to modify continuously duration time of processes), leading to variation in lead times for successive construction processes conducted on the same work unit. Each floor requires five construction processes in succession. The following triangular time distributions for work duration have been adopted here (Table 1).

Table 1. Durations of construction processes conducted on each floor.

Process	Optimistic time estimate t_a [days]	Most likely time estimate t_m [days]	Pessimistic time estimate t_b [days]
Partition walls	30	31	35
Plasters	13	15	19
Backings	36	38	43
Painting	17	19	22
Floors	14	15	17

The simulation model implements Wright's exponential learning model – formula (1) and Globerson and Levin's forgetting model – formula (2).

It is assumed that the learning rate and forgetting coefficient are the same for each process, standing at $s = 0.95$ and $a = 0.15$, respectively. Independent random number generators (in the range from 0 to 1) were used to produce random variables representing the duration of construction work. The simulator was developed in GPSS World language from Minuteman Software.

The linear project simulation tests were performed in line with the proposed method but disregarding the learning-forgetting effect, with the earliest possible times adopted as process start dates. Based on the results, a baseline contract period of $T_d = 668$ days was agreed upon, with 60% probability of timely completion. The overall idle time for all work crews on all floors was 834 days. The tests were repeated with the learning-forgetting effect factored in. The probability of $T_d = 594$ days was 60%, with idle time equal to 624 days. Fig. 2 presents distribution functions, while fig. 3 contains total idle times for selected solutions adjusted for time buffers. The suboptimal solution features the following buffer lengths: $d_{12} = 87$, $d_{23} = 15$, $d_{34} = 271$, $d_{45} = 67$ days. The idle time was reduced to 136 days in all, which means 9.07 idle days for each floor, or 1.81 average idle time for each crew.

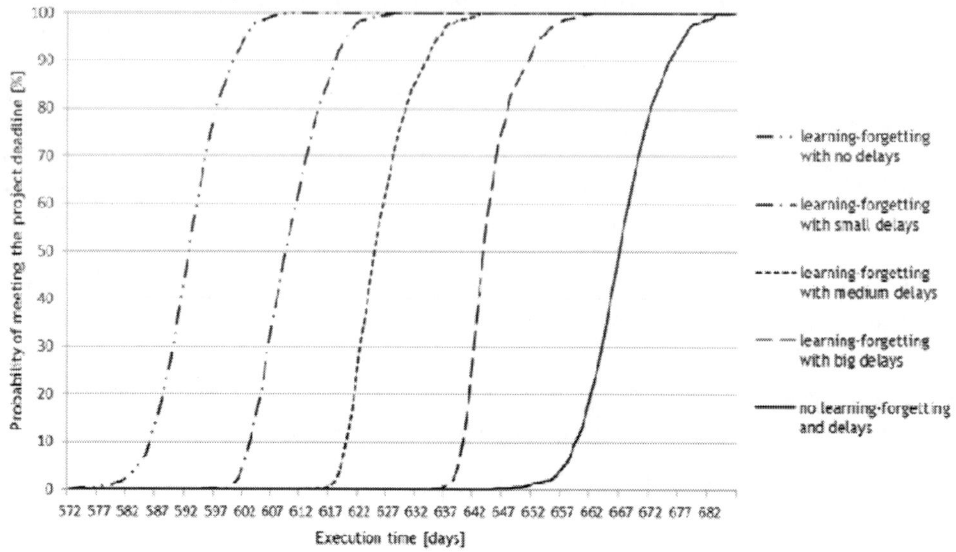

Fig. 2. Probability of meeting the project deadline.

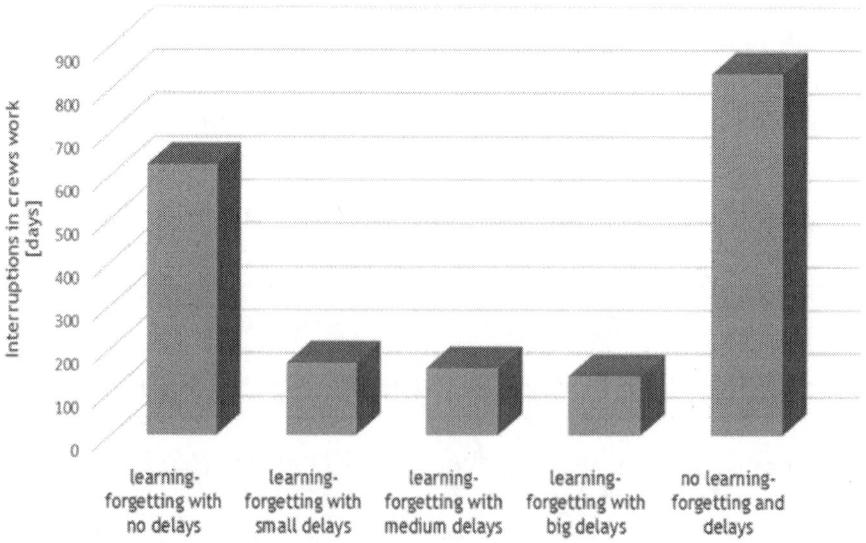

Fig. 3. Summed idle times in crews work.

5 Conclusions

Reference should be made to the learning-forgetting effect in scheduling linear projects such as high-rise buildings, roads and pipelines. In this type of projects it is possible to apply a division into units where the same of processes sequence is conducted. Learning effect allows the project lead time to be estimated with more precision as well as ensures a competitive edge at the bidding stage. Scheduling project work requires construction-specific risk factors to be considered in order to prevent the schedule from becoming outdated and, in particular, to anticipate finish date delays. Linear project schedules can be protected from disruptions allowing for delayed beginnings of successive processes (time buffers). It is also important to strive to eliminate idle periods which result when a preceding crew takes more time than planned to complete processes on a unit. Experience shows that delaying the work of crews carrying out repetitive processes contributes to reducing idle time as well as minimizing the forgetting effect, which leads to shorter lead times. At the same time, time buffers do not affect the baseline finish date. Research on the influence of experience and forgetting effect as well as time dispersion should be of interest to construction professionals and scholars alike, as it allows a more precise estimation of the duration of construction processes. Incorporating the Monte Carlo simulation, the proposed method allows linear construction project planning to take account of random conditions and learning–forgetting effect, while time buffers help to keep to the base line finish date at a given probability level, while also reducing idle time. The main problem in this kind of analyses is the need to have a high-quality inputs, eg data from completed projects.

The majority of construction projects, including linear projects, are affected by risk factors. The variety of processes, a large number of participants in the project as well as long implementation cycles make the construction project exposed to many external factors. If the project is very complicated and is carried out for a long time, the impact of risk factors increases due to problems related to the probability estimation of their occurrence and the assessment of their possible effects. The main limitations for the proposed method is the selection of a linear project with a few activities to minimize the impact of risk factors, as well as determining the optimal size of time buffers, which will not disturb the implementation cycles duration, thus shifting the project deadline.

Other of the primary limitations of the developed methodology in its present form is that it cannot handle intermittent activities. Also, the study focused on a project with unlimited access to resources. Nonetheless, the system can be adapted in the future for accommodating the intermittent activities, varying resource rates, limited resources, among other aspects.

This work was financially supported by Ministry of Science and Higher Education within the statutory research number S/63/2018.

References

1. D.W. Johnston, Linear scheduling method for highway construction, Journal of the Construction Division, **107(2)**, 247–261 (1981)
2. G. Lucko, Productivity scheduling method compared to linear and repetitive project scheduling methods, Journal of Construction Engineering and Management, **134(9)**, 711–720 (2008)
3. G. Lucko, Productivity Scheduling Method: Linear Schedule Analysis with Singularity Functions, Journal of Construction Engineering and Management, **135(4)**, 246–253 (2009)
4. J.J. O'Brien, VPM scheduling for high-rise buildings, Journal of the Construction Division, **101(4)**, 895–905 (1975)
5. R. Harris, P. Ioannou, Scheduling projects with repeating activities, Journal of Construction Engineering and Management, **124(4)**, 269–278, (1998)

6. W.Y. Thabet, Y.J. Beliveau, HVLS: Horizontal and vertical logic scheduling for multistory projects, Journal of Construction Engineering and Management, **120(4)**, 875–892 (1994)
7. R. Flanagan, G. Norman, *Risk management and construction*, Blackwell Publishing. Oxford, UK, 1993
8. P. Zou, G. Zhang, J. Wang, Understanding the key risks in construction projects in China, International Journal of Project Management, **25**, 601–614 (2007)
9. T.P. Wright, Factors affecting the cost of airplanes, Journal of the aeronautical sciences, **3(4)**, 122–128 (1936)
10. K.C. Lam, D. Lee, T. Hu, Understanding the effect of the learning-forgetting phenomenon to duration of projects construction, International Journal of Project Management, **19**, 411–420 (2001)
11. S. Globerson,, N. Levi, A. Shtub, Impact of breaks on forgetting when performing a repetitive task, IIE Transactions, **21(4)**, 376–381, (1989)
12. M.Y. Jaber, M. Bonney, Production breaks and the learning curve: the forgetting phenomenon, Applied Mathematical Modelling, **20(2)**, 162–169, (1996)
13. R. Pellegrino, N. Costantino, R. Pietrroforte, S. Sancilio, Construction of multi-storey concrete structures in Italy: patterns of productivity and learning curves. Construction Management and Economics, **30(2)**, 103–115 (2012)
14. M. Mályusz, A. Pém, Prediction of the learning curve in roof insulation, Automation in Construction, **36**, 191–195 (2013)
15. J. Hinze, S. Olbina, Empirical analysis of the learning curve principle in prestressed concrete piles, Journal of Construction Engineering and Management, **135(5)**, 425–31 (2009)
16. A.M. Jarkas, Critical investigation into the applicability of the learning curve theory to rebar fixing labor productivity, Journal of Construction Engineering and Management, **136(12)**, 1279–1288 (2010)
17. A. Hassanein, O. Moselhi, Planning and scheduling highway construction, Journal of Construction Engineering and Management, **130(5)**, 638–646, (2004)
18. A. Panas, J.P. Pantouvakis, Simulation-based and statistical analysis of the learning effect in floating caisson construction operations, Journal of Construction Engineering and Management, **140(1)**, 04013033 (2013)
19. J. Lutz, D. Halpin, J. Wilson, Simulation of learning development in repetitive construction, Journal of Construction Engineering and Management, **120(4)**, 753–73 (1994)

Application possibilities of LBN for civil engineering issues

Agata Siemaszko[1,*], and *Magdalena* Apollo[1]

[1]Gdańsk University of Technology, Faculty of Civil and Environmental Engineering, ul. Narutowicza 11/12, 80-233 Gdańsk, Poland

Abstract. Bayesian Networks (BN) are efficient to represent knowledge and for the reasoning in uncertainty. However the classic BN requires manual definition of the network structure by an expert, who also defines the values entered into the conditional probability tables. In practice, it can be time-consuming, hence the article proposes the use of Learning Bayesian Networks (LBN). The aim of the study is not only to present LBN, which can be helpful in civil engineering problems, but also to analyze and evaluate the potential of a selected software. Based on a real example the functional values of the Open Markov, Hugin and AgenaRisk applications were compared.

1 Introduction

A highly competitive and dynamic environment, for the participants of the investment process, forces efficient operation and decision-making [1]. Complex construction projects involving a number of interrelated and interacting variables result in the development of methods and tools supporting the decision-making process. Contemporary methods permanently rooted in solving civil engineering problems include e.g. Artificial Neural Networks (ANNs) [2, 3]. Bayesian networks, allowing not only to capture the issue of data uncertainty, but also to identify interrelations between variables, creating a network of interactions, are still much less common in practice.

Bayesian networks, also known as a causal network or Bayesian Belief Networks (BBN), are a powerful tool for knowledge representation and reasoning under uncertainty, which visually presents the relations between a set of probability variables. In addition, by defining conditional probability tables, the BBN take into account correlation relationships applied in the back propagation mechanism updating *a prori* probability and introducing this way new information in the form of expert knowledge to the network. Updating of the information introduced leads to the construction of the current risk scenario, thus making it possible to bring more reasonable decisions.

Owing the currently used software, the BBN presents probabilistic relationships between a set of variables in a simple and transparent manner [4]. Nevertheless, the classic BBN requires manual definition of the network structure (relations) by an expert, who also

* Corresponding author: agata.siemaszko@pg.edu.pl

defines the values entered into the conditional probability tables [5]. In practice such an action may be quite cumbersome and time-consuming, hence the article proposes the use of Learning Bayesian Networks (LBN), so far rarely used in the field of construction projects. The LBNs employing the available source data are not only able to automatically create a network structure, but also incorporate a formalized mechanism for learning the values of conditional probabilities [5]. The research indicates that this approach not only relieves experts but also eliminates widely commented and criticized subjectivism of Bayesian Networks, which can lead to distortions in the estimation process [6, 7].

The aim of the study is to present Learning Bayesian Networks, which can be helpful in civil engineering problems and also to analyze and evaluate the potential of a selected software. Based on a real example of a problem regarding the operation management of a building facility, the functional values of the Open Markov [8], Hugin [9] and AgenaRisk [10] applications were compared. The authors investigated symptoms related to the technical condition of a building and thus, with the help of LBN, assessed its technical condition.

2 Learning Bayesian Networks (LBN)

According to [11] the BBN consists of two parts:
- qualitative (structural) part: graphical representation of relationships between variables in the form of a graph,
- quantitative (parameter) part: definition of the quantitative dependence between variables in the form of conditional probability tables (calculation of the total probability resulting from relationships between elements).

The BBN is represented by an acyclic graph consisting of a set of vertices and arrows illustrating the nature of the relationship. The vertices represent probabilistic variables while arrows illustrate causal relationships between variables.

The LBNs consist of the same two parts as the traditional BBN: structural and parametric. An important difference, however, is the network building process itself. Eliciting Bayesian networks from experts can be a laborious and difficult procedure especially in the case of large networks. Thus researchers developed methods that could learn the network structure from available data. Furthermore, they formalized methods for learning the conditional probabilities also from input data [5].

The programs the analysis is focused on use three most popular algorithms that allow the automatic construction of the structural part and the parametric LBN. These algorithms include:

1) Expectation-Maximisation (EM):
   - algorithm defining the number of iterations, so the number of times the entire data file is used to learn the model parameters,
   - the allowed range of iterations is from 1 to 50, although it is possible that the EM algorithm run less iterations, if the convergence threshold is satisfied earlier (the EM convergence threshold is the difference between the expected log-likelihood of current and previous iterations),
   - the algorithm continues until it either converges or the maximum number of iterations is reached.
2) The K2 algorithm belonging to the group of heuristic algorithms, i.e. those that allow to find an approximate solution, and only in specific cases, accurate thanks to the Bayesian evaluation function. This function determines the degree of adjustment of the configuration of a given Xi node to the actual distribution. The K2 algorithm adds

new parents to each node until it improves the value of the matching function or when the maximum number of parents specified by the user is reached.

3) The PC algorithm relies on minimizing the amount of necessary d-separation tests, which means quick and effective execution of lower-order tests. In the first phase of the algorithm's operation, the structure of the network is determined, next some of the edges are oriented as far as possible. Thus, a partially oriented structure is generated representing the whole class of potential Bayes networks. Algorithms take two input parameters. First is the confidence level for the independence test. The lower the confidence level, the more attributes are considered independent and a lower number of edges detect the resulting network. The second parameter is the strategy of selecting pairs for the d-separation test. The d-separation test is performed for all attribute pairs in the sample. The complexity of the test for each pair depends on the number of edges eliminated in previous tests.

The algorithms used in the applications proposed in the article are described in more detail in the publications [8, 9, 10]. The subject of this study is simplicity of the problem solution to manage construction work operation.

2.1 Open Markov [8]

OpenMarkov is an open-source software tool for Learning Bayesian Networks (LBN) from data interactively, but also to editing and evaluating several types of several types of PGMs, such as Bayesian networks, influence diagram, as well as to cost-effectiveness analysis

The first method to build a LBN in OpenMarkov is to do it automatically, learning the structure of the network (the directed graph) and its parameters (the conditional probabilities) from a database. The second approach is interactive learning, have an algorithm proposes some modifications of the network, called edits (typically, the addition or the removal of a link), which can be accepted or rejected by the user based on their common sense, their expert knowledge or just their preferences; additionally, the user can modify the network at any moment using the graphical user interface and then resume the learning process with the edits suggested by the learning algorithm. It is also possible to use a model network as the departure point of any learning algorithm, or just to indicate the positions of the nodes in the network learned, or to impose some links.

2.2 Hugin [9]

Hugin software is based on Bayesian networks and influence diagram technology, an advanced artificial intelligence technique widely used for supporting decision-making under uncertainty. The Hugin supports two kinds of LBN: structure learning and parameter learning. Structure learning is the process where the system learns the dependencies between the variables that exist in the data. Parameter learning is the task where you fill in the parameters describing the strength of the dependencies in the learned (or built) structure.

Structure learning in Hugin is supported through the PC-algorithm. Parametric learning is adaptive learning and EM (Estimation-Maximization) learning.

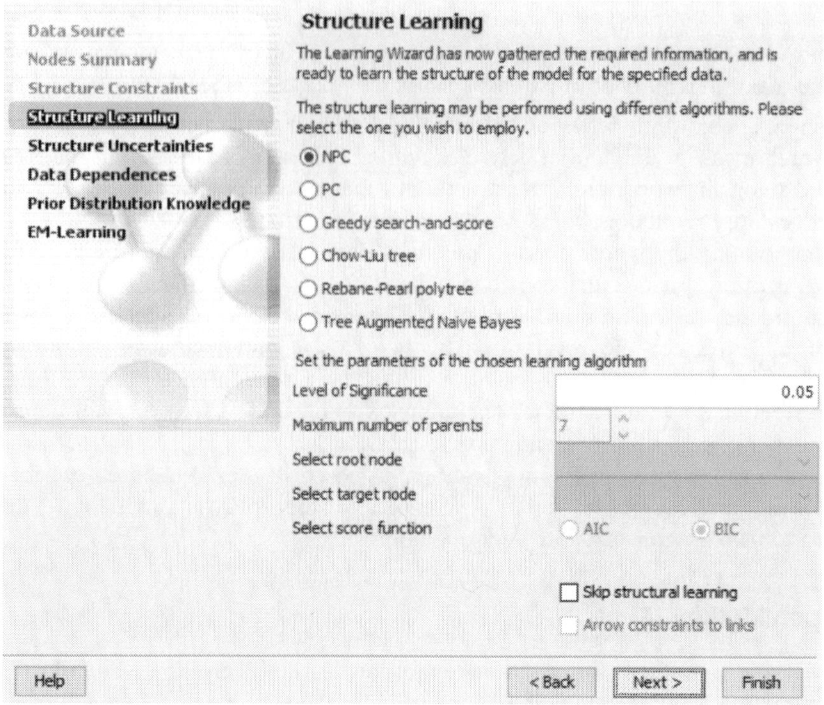

Fig. 1. Structure learning dialog box in Hugin [9].

2.3 AgenaRisk [10]

AgenaRisk enables to automatically learn conditional probability tables using three different ways of table learning:

- learning from data alone,
- learning from data with expert judgement,
- learning from data with expert judgement and custom settings.

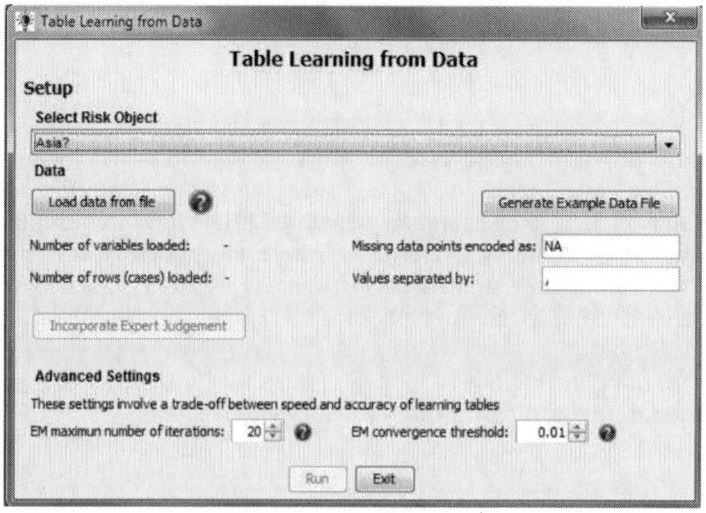

Fig. 2. Table learning from data dialog box in AgenaRisk [10].

In all the above mentioned cases learning can be performed even in the case of the so-called "missing data". Nevertheless, the time necessary to learn a NPT is more a function of the amount of missing data than the amount of data, so more processing time is essential for large data files with lots of missing values. The learning process is performed using an Expectation-Maximisation (EM) algorithm which supports table learning for both Boolean and Labelled types of nodes.

Before running the table learning process it is necessary to prepare and load the data from a file. The dataset should be in "csv" format and if we are not sure what is the format in which the data should be prepared, it is possible to use a button "Generate Example Data File" (Figure 2). This way it is possible to use the generated file as a template.

In the table learning process AgenaRisk also makes it possible to incorporate expert judgement. In this case it is possible to set the ratio of the confidence in knowledge (as represented by the existing node probability tables) compared to the data, from the loaded data file. By default, the application defines a situation where there is no confidence in the knowledge already encoded in the NPTs and a 100% data confidence holds. In this case the existing NPT values will be ignored and the new NPTs will be learnt solely from data.

In the last variant of table learning it is possible to incorporate expert judgement for each node individually.

3 Example problem

The subject of a simplified problem presented on Figure 1 is to determine the current technical condition of a building based on two available observations. The first one (qualitative) concerns the appearance of scratches and cracks on the building's structural walls. There are two states to choose from: YES - scratches or cracks are observed and NO - monitoring jumps into the lack of scratches. The second observation (quantitative) concerns the measurement of heat loss, e.g. by performing the air leak test: BLOWER DOOR TEST. Two states were also taken into account here: YES - the measurement indicates deviations from the assumed norms of air tightness of the facility and NO - no air leaks. Updating the information about the object state, i.e. entering observations into the network, the data is backed up and the probability of occurrence of a specific technical state of the object (GOOD or BAD) is updated.

The main task of the discussed model is to present the possibility of using self-learning Bayesian networks to the problems of managing the operation of buildings by describing the mechanism of learning the structure and parameters of the presented grid in selected programs.

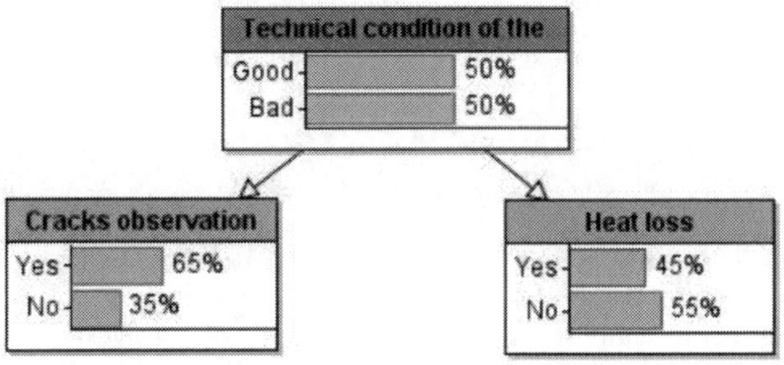

Fig. 3. Problem model: structure and parameters presented in the AgenaRisk program (own study).

3.1 Data for LBN

Bayesian networks can be defined by an expert and /or can learn from the data provided. These data are regarded a set of m vectors (x1 (i), ..., xn (i)), i = 1, ..., m, generated independently of each other from the probability distribution represented by a "real" Bayesian network (X, S, P), where S is a structure, and P - parameters. Based on the data loaded into the program, a Bayesian network (X, S ', P') is created, which in the assumption should bring as close as possible the output network (X, S, P). The data used in the network learning process in three alternative programs are presented in Table 1.

Table 1. Data for LBN saved as .csv file (own study).

Technical condition of the building	Cracks observation	Heat loss	Conditional probability
GOOD	YES	YES	40·30=1200
GOOD	YES	NO	40·70=2800
GOOD	NO	YES	60·30=1800
GOOD	NO	NO	60·70=4200
BAD	YES	YES	90·60=5400
BAD	YES	NO	90·40=3600
BAD	NO	YES	10·60=600
BAD	NO	NO	10·40=400

3.2 Problem analysis – Open Markov

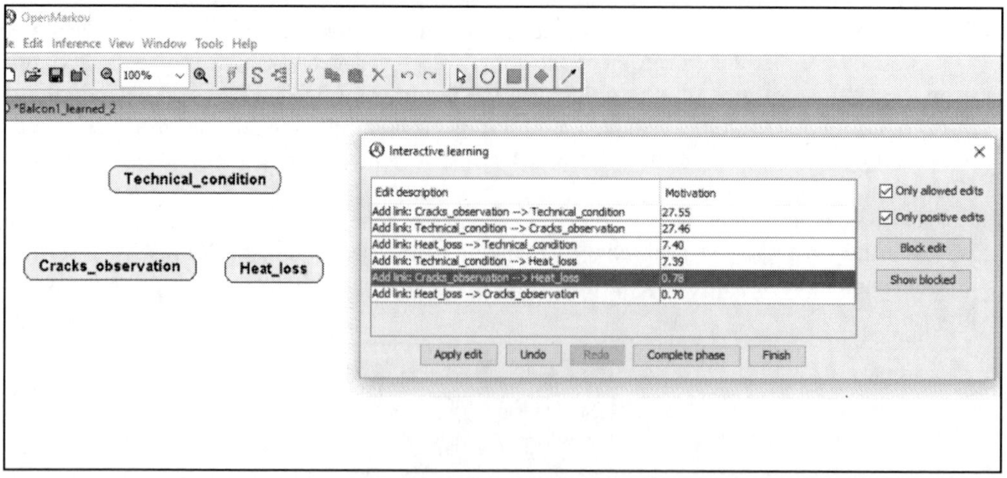

Fig. 4. Problem model in Open Markov: structure and parameters after loading data from .csv file.

The Open Markov program makes it possible to learn the network based on the data contained in the .csv file. Autmomatic learning does not bring the expected results, because in the constructed network there is a wrong connection of HEAT LOSS and TECHNICAL CONDIOTION nodes. The solution that gives the best results is interactive network learning. It is based on the fact that the algorithm proposes some network modifications, called changes (adding or deleting a link) that can be accepted or rejected by the user based on common sense, expert knowledge or only preferences. In addition, the user can modify the network at any time using the graphical user interface, and then resume the learning process in the edition proposed by the learning algorithm. An example of interactive learning is shown in Figure 4.

3.3 Problem analysis – Hugin

The Hugin program turns out to be the easiest to use. During the construction of the network, the user is able to modify all elements starting from the structure of the model, ending with the conditional probabilities tables. The model built from the data contained in the .csv file is identical to the original model and has been shown in the Figure 5. The visual side of the program is exemplary too. The user can easily read all the necessary information from the model.

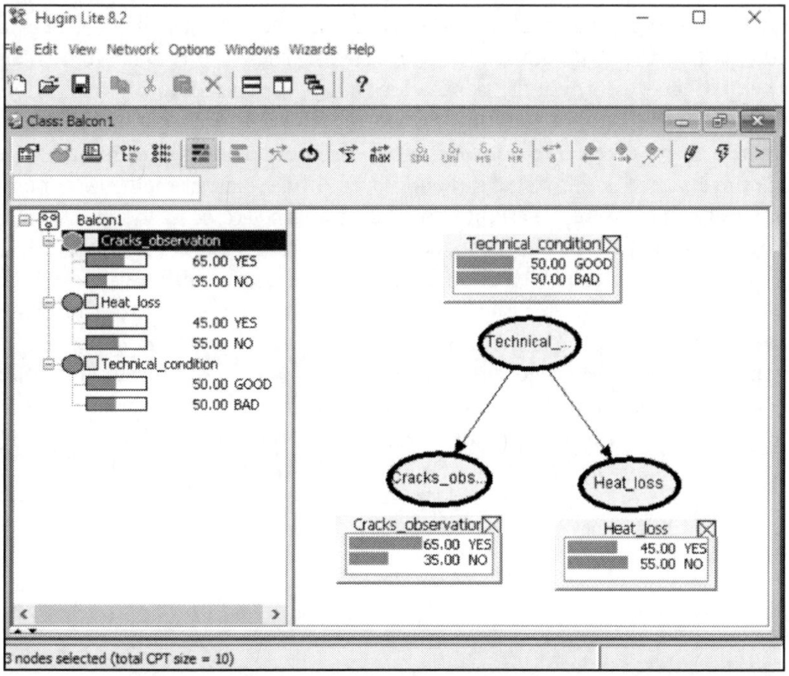

Fig. 5. Problem model in Hugin: structure and parameters after loading data from .csv file.

3.4 Problem analysis – AgenaRisk

According to the information presented in chapter 2.3, the module teaching Bayesian network in AgenaRisk program allows learning the value of conditional probability based on the data provided in the .csv file (the so-called: learning the parameter values from data). A serious limitation of the application possibilities of the program seems to be the inability to automatically build a network structure (the so-called learning the structure from data). What's more, the data exported by AgenaRisk to the .csv file does not include all the necessary combinations of responses, which in other applications are necessary to accurately define the structure. Using the initial model presented in Figure 3, it was possible to load the created .csv file (data presented in table 1). This allowed to verify the value of conditional probabilities, which proved to be in line with the responses of the Open Markov and Hugin programs.

4 Conclusions and recommendations

Based on the conducted analysis, it was shown that only two of the above programs allow automatic network structure creation - Open Markov and Hugin. Consequently, only these applications seem to provide full benefits of the Bayesian self-learning networks. However, in order to prepare the LBN, it is necessary to prepare the data properly. Definition of all available combinations of existing parameters in practice turns out to be very time-consuming, requires a vast theoretical background. Such an operation does not facilitate automatic construction of the network structure. It turns out that insufficient number of rows of data defined in the .csv file, or a possible mistake, may consequently result in improper network construction, and thus erroneous propagation of information. The AgenaRisk is dedicated to learning the values of conditional probabilities only neglecting structure building. In engineering practice, however, automatic parameters learning is much more valuable than the structure itself. The network structure is built intuitively/based on the expert experience. Learning the parameter values, especially in the case of missing data, as well as smooth application functioning are therefore, in the opinion of the authors, the key parameters to be taken into consideration while choosing a program for the LBN. An additional convenience is the visual side and ease of use, and at the same time building the relationship between different variables of the problem. Here, in the authors' opinion, the Hugin program seems to be the best, dialogue operations are very intuitive and all the functions of the program can be easily used without excessive training.

In addition, the article shows that it is possible and desirable to use the LBN to manage the operation of engineering structures. In many publications, it has been proved that Bayesian networks are an effective tool and have been used in various fields so far, including technical diagnostics and medical diagnostics. The success of Bayesian network shows positive results in construction industry, so there is a clear need to apply this methodology to manage the operation of engineering structures.

References

1. B. Grzyl, A. Kristowski, E. Miszewska-Urbańska, *IOP Conf. Ser. Mater. Sci. Eng.* **245**, 8 (2017)
2. S. Arangio, J. L.Beck, Struct. Control Health Monit. **19**, 1 (2012)
3. A. Jakubczyk-Gałczyńska, A. Kristowski, R. Jankowski, *Geodetic Congress (BGC Geomatics)* (2017)
4. M. Apollo, *Procedia Eng.* **196** (2017)
5. Neapolitan, R.E., *Learning Bayesian Networks* (Pearson, 2003)
6. S.S. Leu, P.V.H. Son, P.T.H. Nhung, J. Comput. Civ. Eng. **29**, 6 (2015)
7. P. C. Liao, H. Shi, Y. Su, X. Luo, J. Constr. Eng. Manag. **144**, 3 (2018)
8. OpenMarkov Tutorial - Version 0.2.0 (2017)
9. A. Hanson: Hugin Tutorial (2015)
10. AgenaRisk 7.0 User Manual, Agena Ltd, (2016)
11. M. W. Kembłowski, B. Grzyl, A. Kristowski, A. Siemaszko, *Procedia Eng.* **196** (2017)

Transparent façades – selection of construction materials with the use of modified multi-criteria spider's network analysis method

Oliwier Zapała[1], and *Damian* Wieczorek[1,*]

[1] Cracow University of Technology, Faculty of Civil Engineering, Institute of Construction Management, 31-155 Kraków, Warszawska street 24, Poland

Abstract. The article deals with transparent façades of public buildings which are the main and most impressive elevations of the building. It also focuses on some issues related to high requirements placed on façade systems in terms of insulation from environmental conditions and comfort of the future users of the building. Due to the fact that there are many materials that can be used for the main structural elements (columns, spandrel beams and ribs) of load-bearing façade structures, the authors of the article attempted to do a multi-criteria comparative analysis of three different materials (safety glass, glulam and rolled steel) of a 6-meter-long reference column. The set of criteria, consisting of twelve elements, was adopted for comparative analysis. A combination of two methods was selected for performing the multi-criteria analysis: a geometric method of spider's network and a mathematical method of the adjusted index of summation.

1 Introduction

At present transparent façades dynamically begin to mark their presence in construction. With increasing frequency, architects try to use glass as the basic façade material in their projects, and its application is primarily determined by aesthetic qualities. Due to the developing trend, new technologies and new possibilities for using façades of this type in construction are emerging [1].

According to the definition presented in [2], the façade is the main and the most impressive façade, with a particularly large decoration, which fulfills the representative functions of the entire building, thus standing out from other façades. Transparent façades can perform representative functions; however, due to the specificity of their construction (as in, e.g. frameless glass façades), they often require a merge of glass panes with the load-bearing façade structure, which can be constructed of many materials used for the main structural elements (columns, spandrel beams and ribs).

The main aim of the study is a multi-criteria comparative analysis of three different materials (safety glass, glued laminated timber - glulam and rolled steel) of a 6-meter-long reference column. The material selection process for the load-bearing structure is treated

* Corresponding author: dwieczorek@L3.pk.edu.pl

with regard to the decision making problem that one may face in the process of choosing the best solution based on the criteria that interest the decision maker [3-5].

2 Types and basic properties of building glass

Glass is a material that in construction is used not only for the production of windows, thermally insulating elements, hollow blocks or glass roof tiles, but also for panes, glazing units and main structural elements on the elevations of cubature buildings.

Glass is an amorphous inorganic material which is created in the process of rapid cooling of the liquid to a solid state without the crystallization step [6]. Glass has numerous physical and mechanical features [6, 7]. Glass is a substance without an ordered internal structure, because of which it lacks the constant melting temperature and is a poor conductor for electricity. Glass is an isotropic material, with high chemical resistance, with good acoustic insulation but poor thermal insulation (its thermal conductivity [λ] is about 0.80-1.00 W/mK). The density of construction glass is 2400-2600 kg/m^3 and is comparable to the density of reinforced concrete. The glass hardness is measured on the Mohs scale and ranges from 5 to 7. The bending strength of glass varies from 30 to 50 MPa and is greater than 32 MPa for top-class glulam (GL32h, GL32c). Its compressive strength is in the range from 800 to 1000 MPa, that is, up to 5 times higher than for very high strength concrete (about 180-200 MPa) and comparable to the ultra-high strength concrete, but very rarely used in construction. Young's modulus for glass is 70 GPa.

The most negative feature of glass is its brittleness. Glass, when reaching maximum strength, undergoes sudden cracks without giving any indication of loss of load capacity, as opposed to steel (which undergoes deformation), wood (which creaks before the loss of load capacity) or concrete (which is scratched in an emergency condition) [1].

In construction the following types of glass are used [1, 6-9]:

- Sodium-calcium-silicate glass and boron-silicate glass, which are characterized by high resistance to aggressive chemical compounds,
- Float glass, which is the basic type of glass obtained in the melting process of raw materials (quartz sand, additives in the form of sodium and calcium carbonate, fluxes, i.e. boron and lead oxides, and cullet),
- ESG, i.e. toughened glass, which arises in the process of heating Float glass at a temperature of approx. 700°C and (then) rapid cooling, which shapes such a system of internal stresses, which increases the strength parameters of glass several times and heightens the range of glass resistance to rapid changes in temperature; ESG glass breaks (after shattering) into small, blunt pieces, which is why it is classified as safety glass; ESG glass should not be machined,
- TVG, i.e. semi-hardened glass, one that is heat-strengthened in the heating process at a temperature of approx. 625°C, which is (then) cooled more slowly than in the case of the ESG glass, using water, which results in approx. twice the bending strength compared to Float glass; TVG glass is also a safety glass, breaking into slightly larger, but also blunt pieces as in the case of ESG glass; TVG glass should not be machined,
- VSG, i.e. glued (or laminated) glass, which is created by connecting several sheets of Float, ESG or TVG glass with polyvinyl butyral PVB film or acrylic resins cured chemically with UV radiation; glass joining technology via PVB requires the use of autoclaving which consists in heating the package to 115°C, at a pressure of about 10 bars; VSG glass is also a type of safety glass, because in the case of a crack, the laminated glass does not break down into small fragments but sticks to the PVB film.

3 Characteristics of transparent façades

3.1 Basic requirements for transparent façades

Transparent façades, as well as other outer partitions, need to meet requirements concerning insulation from environmental conditions and providing comfort to users of the building. These requirements can be reduced to the following issues related to: thermal protection, anti-noise protection, fire resistance and protection against biological and/or chemical corrosion [8]. Due to the dissimilarity of glass products to other, most common building materials (e.g. glulam, concrete, ceramic, or steel), the authors will describe more broadly only the requirements related to thermal protection and fire resistance. As it was already indicated in chapter 2 of the paper, building glass is a material that is perfectly resistant to chemical agents and makes a good acoustic insulation.

The requirements concerning heat protection are directly related to the need to save energy. In the Regulation of the Minister of Infrastructure of April 12, 2002 regarding technical conditions which should be met by buildings and their location (Journal of Laws of 2015, item 1422), which pursuant to the amending Regulation of the Minister of Infrastructure and Construction of November 14, 2017 (item 2285), correction were made in the entries related to, e.g. the limit values of heat transfer coefficient (U_{max}) for windows, balcony doors, external doors and transparent non-opening surfaces (including transparent façades), for all types of buildings, in such a way that the values (U_{max}) can not be greater than 1.1 W/m^{2}K if the temperature inside the heated space (t_i) is equal to or greater than 16°C. It is worth adding that before the legal changes mentioned above, the limit value (U_{max}) was 1.8 W/m^{2}K and was by as much as 63.6% higher than the current one.

In the case of requirements related to fire resistance, it should be noted that although glass products become increasingly more common in construction, are characterized by differences in issues related to fire protection. As the publication by [8] states, single-layer silicate glass cracks and falls out of the frame at a temperature of about 200°C. A much better fire resistance is the feature of glued glass (VSG) which, combined with an additional reinforcing mesh between the layers, gives the fire resistance class (fire integrity expressed in minutes) E30. Single-layer glass with embedded reinforcing mesh can achieve fire integrity at the E60 level (unless the glass panes are too large), while the modified boron-silicate glass, for reasons of much longer resistance to high temperatures, can achieve the fire resistance class E90, and even E120, if the panes are not too large.

3.2 Types of transparent façades

There are two basic types of glass façades [8, 9]: single, which are usually composed of two or three panes of glass between which there is a partition in the form of space separated from the surroundings and double, which consist of two glass surfaces forming a gap through which the air flows (the so-called ventilated façades).

In the case of single façades, it is not possible to determine the recommended thickness of the façade clearly, because individual systems differ considerably in terms of thermal insulation properties. The typical thickness range of such façades varies from a few to even 150 mm. The single façades are usually made of ordinary glass, glued VSG, hardened ESG or semi-hardened TVG. All glass types are characterized by low thermal insulation due to the high value of the thermal conductivity coefficient (λ) [9].

Double façades have been used for a short time and because of their better thermal insulation parameters are called climatic façades [10]. They consist of two glass surfaces that create a gap that allows the air to flow. The distance between these surfaces typically ranges from 100 to even 2000 mm. Double façades are characterized by acoustic protection, thermal insulation and high efficiency of heat recovery [10].

Due to the type of fixing, glass façades are divided into [7, 9]: frameless ones, the essence of which is point fixing of glass panels to an independent load-bearing structure and with framing, in which glass panes are typically fastened to steel or aluminium profiles.

Frameless façades are characterized by greater freedom in forming glazing in buildings. The most important issue in the design and implementation of these structures is the appropriate selection of connections [7]. Connections between glazing units should also be selected on the grounds of aesthetics and economics considerations of the solution applied.

In the case of the façade with framing, the most important issue is the appropriate selection of static schemes and the system used in relation to the entire facility. Depending on the chosen façade system, the designer should provide an additional supporting structure, which will carry permanent loads from the tare weight of the façade and the variables from the wind [7]. The adoption of the right concept and selection of appropriate construction schemes can significantly reduce the investment costs of a construction project, which is particularly noticeable in the case of high façades [1].

3.3 Examples of implementation of transparent façades

Transparent façades are increasingly used in building construction. One can identify many implementations of office buildings or public utilities (sports halls, indoor riding schools, swimming pools, or cultural facilities) whose façades were designed and made as glass façades with frame or as frameless structures [1]. As shown by numerous frameless façades, load-bearing structures can now be made not only from the most popular construction materials (glulam, steel or aluminium), but also from safety glass [11].

Figures 1 and 2 illustrate example implementations of transparent façades in public buildings related to culture. Figure 1 (Lesser Poland Garden of Art in Krakow) depicts a frame façade attached to steel elements. Figure 2 (The Chopin Museum in Żelazowa Wola) shows a frameless façade, in which an independent supporting structure for glazing elements creates a column and spandrel beams system made of safety glass VSG.

Fig. 1. Main façade of the Lesser Poland Garden of Art in Krakow; source: [12].

Fig. 2. Main façade of the The Chopin Museum in Żelazowa Wola; source: [13].

4 Multi-criteria comparative analysis of construction materials for load-bearing façade structures

As already mentioned in chapter 1, the specificity of frameless transparent façades requires the connection of glass panes with the load-bearing façade structure, which can be made of many materials used for the main structural elements (columns, spandrel beams and ribs). Due to this fact, the authors of the study attempted to perform a multi-criteria comparative analysis of three different materials on a 6-meter-long reference column: 1 - safety glass with cross-sectional dimensions 80×400 mm, 2 - glulam with cross-sectional dimensions 160×500 mm and 3 - steel rolled section IPE360.

In this study, the material selection process has been treated with regard to the decision making problem that one may face in the process of choosing the best solution based on the criteria that interest the decision maker [3-5]. In the Polish literature, there are many publications that discuss a multi-criteria choice of alternative solutions for different decision problems. In the case of construction, the following solution variants are multi-criteria evaluated and, typically, hierarchized (from the most to the least favourable): constructional material [14, 15], technological [16, 17] and installation [18]. One can also indicate examples of multi-criteria comparative analyses focused only on environmental aspects for material solutions [19, 20] or supporting the bidding decision [21].

In the case of selecting the construction material for the load-bearing façade structure, a combination of two methods was chosen for performing the multi-criteria analysis: a geometric method of spider's network (assuming the local evaluation of variants in specific groups of criteria) and a mathematical method for formulating synthetic evaluations of the adjusted index of summation (using partial measures encoded for variants according to Neuman-Morgenstern) [3, 4].

The comparative analysis was performed in two stages. In the first stage, the material variants were assessed against each of the twelve criteria (locally, three criteria in each of the four groups). The spider's network method was used as there was no need to determine the weight of each criterion [4]. In the second stage, material variants were compared globally based on the adjusted index of summation (AIS, where the maximum can be 1.00).

4.1 Typification and characteristics of the set of criteria

The set of criteria (C_i), consisting of twelve elements, was adopted for the comparative analysis, where: criterion (C_1) is Cost of material, (C_2) - Total initial cost, (C_3) - Life cycle cost, (C_4) - Degree of similarity to the main façade material, (C_5) - Maintaining purity of element, (C_6) - Resistance to mechanical damage, (C_7) - Content of primary energy, (C_8) - Specific heat, (C_9) - Thermal conductivity, (C_{10}) - Weight of component, (C_{11}) - Way of ensuring fire resistance and (C_{12}) - Degree of chemical and biological corrosion resistance.

All criteria were categorized into one of four (three-element) groups: I - economic (criteria C_1-C_3), II - aesthetic (C_4-C_6), III - energetic (C_7-C_9) and IV - others (C_{10}-C_{12}). For each criterion, the type (quantitative, qualitative) and character (stimulants which are positively correlated with the criterion, and negatively correlated - destimulants) were determined. Qualitative criteria have been described by applying a three-point scale. The value of the partial assessment of all qualitative criteria (for a given material variant) was determined by the experts. For each qualitative criterion, the scale has been selected: for (C_4) 3 means identical material (glass), 2 - natural material (glulam), 1 - different material (steel), for (C_5) 3 - easy, 2 - quite easy, 1 - quite difficult, for (C_6) 3 - good, 2 - average, 1 - poor, for (C_{11}) 3 - by cross section (no coating required), 2 - by cross-section and coating, 1 - by coating or element housing, and for (C_{12}) 3 - good, 2 - average, 1 - poorly resistant.

In the case of quantitative criteria (C_3), (C_7) and (C_8), the Life cycle cost methodology according to [22] was used accordingly, including the costs incurred in the course of the operation for the continuous assurance of elements cleanliness and renewal of impregnating coatings (every 6 years but only for glulam) and content values of primary energy and specific heat for material variants that have been taken from the publication [19, 20].

4.2 Locally evaluation of material variants in specific groups of criteria

In the first stage, the material variants were assessed locally against each of the twelve criteria, which were listed in chapter 4.1. Due to the fact that all criteria were categorized into one of four three-element groups, the authors decided to use the spider's network method since it is quite simple and generates objective results for the three assessment criteria (as mentioned above). Partial measures of material variants were encoded to the form of the so-called measure of *i-th* variant according to *j-th* criterion (z_{ij}), which is calculated for stimulants and destimulants according to the respective formulas [4]:

$$z_{ij} = \frac{x_{ij} - x_{j,min}}{x_{j,max} - x_{j,min}} \qquad z_{ij} = \frac{x_{j,max} - x_{ij}}{x_{j,max} - x_{j,min}} \tag{1}$$

where: x_{ij} - partial measure of assessed variant according to the *j-th* criterion, $x_{j,min}$ - minimum value of partial measure for *j-th* criterion and $x_{j,max}$ - maximum value of partial measure for *j-th* criterion.

In the next step, the material variants were individually assessed in each of the four groups of criteria. For this purpose, a three-dimensional polar system was built for each variant (with [*x, y, z*] axes) corresponding to three criteria in a given group (i.e. [C_1], [C_2] and [C_3]) on the scale adopted with values from 0 to 1 on each axis. On the axes points corresponding to the encoded measures of variants are placed, which became the vertices of the polygon representing the analysed solution variant. Finally, the surfaces of the created polygons (SA) whose sizes could be compared with each other were calculated.

Table 1 presents the results of the analysis obtained in the first stage.

Table 1. The results of the analysis obtained in the first stage.

Criterion		Type, char.	Unit	Partial and encoded* measures of variants			Variants assessment (SA_i - surface area)		
				Safety glass	**Glulam**	**Rolled Steel**	**Safety glass**	**Glulam**	**Rolled Steel**
I economic	C_1	*Qn, D*	PLN/pcs	3770.64 (0.00)	1137.60 (1.00)	2946.36 (0.31)	0.00	0.73	0.35
	C_2	*Qn, D*	PLN/pcs	5450.64 (0.00)	1972.38 (1.00)	3357.48 (0.60)			
	C_3	*Qn, D*	PLN/pcs	45743.60 (0.00)	19405.56 (0.75)	10401.88 (1.00)			
II aesthetic	C_4	*Ql, S*	*Points 1-3*	3 (1.00)	2 (0.50)	1 (0.00)	0.50	0.13	0.25
	C_5	*Ql, S*	*Points 1-3*	1 (0.00)	2 (0.50)	3 (1.00)			
	C_6	*Ql, S*	*Points 1-3*	3 (1.00)	1 (0.00)	2 (0.50)			
III energetic	C_7	*Qn, D*	MJ/pcs	9600.00 (0.14)	2481.60 (1.00)	10723.38 (0.00)	0.41	0.50	0.00
	C_8	*Qn, D*	J/kgK	840.00 (0.81)	2510.00 (0.00)	440.00 (1.00)			
	C_9	*Qn, D*	W/mK	0.80 (0.99)	0.16 (1.00)	58.00 (0.00)			
IV others	C_{10}	*Qn, D*	kg/pcs	480.00 (0.00)	264.00 (1.00)	324.60 (0.64)	0.50	0.61	0.00
	C_{11}	*Ql, S*	*Points 1-3*	3 (1.00)	3 (1.00)	1 (0.00)			
	C_{12}	*Ql, S*	*Points 1-3*	3 (1.00)	2 (0.50)	1 (0.00)			

where: * - encoded measure in brackets, char. - character of criterion,
Qn - quantitative criterion, *Ql* - qualitative criterion, *S* - stimulant, *D* - destimulant

4.3 Globally evaluation of material variants

In the second stage, material variants were compared globally based on the adjusted index of summation (AIS) which is calculated on the basis of the following formula [4]:

$$AIS_i = \sum_{j=1}^{m} (z_{ij} \cdot v_j) \tag{2}$$

where: i - variant's number, j - criterion's number, m - number of all criteria, z_{ij} - measure of *i-th* variant according to *j-th* criterion and v_j - weight of *j-th* criterion.

The results obtained after the first part of the analysis for the four groups of criteria were adopted "*a priori*" for further calculations as partial measures of material variants, which were then encoded according to Neuman-Morgenstern. The degrees of importance (weights) of the groups of criteria, were determined using the expert knowledge in the field of architectural design, construction design and management of building investment projects. In the last step the encoded measures for material variants were calculated including weights of criteria ([z_{ij}] multiplied by [v_j]), which were then added to the form of the adjusted index of summation (AIS) to compare and prioritize solutions.

Table 2 presents the results of the analysis obtained in the second stage.

Table 2. The results of the analysis obtained in the second stage.

Groups of criteria	Weight	Type, char.	Partial and encoded* measures of variants			Encoded measures with criteria weights		
			Safety glass	**Glulam**	**Rolled Steel**	**Safety glass**	**Glulam**	**Rolled Steel**
I economic	0.350	*Qn, S*	0.00 (0.00)	0.73 (1.00)	0.35 (0.48)	0.000	0.350	0.168
II aesthetic	0.225	*Qn, S*	0.50 (1.00)	0.13 (0.00)	0.25 (0.32)	0.225	0.000	0.073
III energetic	0.175	*Qn, S*	0.41 (0.82)	0.50 (1.00)	0.00 (0.00)	0.143	0.175	0.000
IV others	0.250	*Qn, S*	0.50 (0.82)	0.61 (1.00)	0.00 (0.00)	0.205	0.250	0.000
Adjusted index of summation (AIS_i)						**0.573**	**0.775**	**0.241**

where: * - encoded measure in brackets, char. - character of groups of criteria,
Qn - quantitative criterion, *S* - stimulant

5 Conclusions

The authors of the study paid particular attention to the following two facts: that today transparent façades clearly mark their presence in building construction and that the requirements placed on them in terms of thermal protection will continue to grow. Legal changes from November 2017 caused a lowering of the heat transfer coefficient values (U_{max}), for example, for transparent, non-opening surfaces (including transparent façades, regardless of the type of building) as much as 63.6% from 1.8 to 1.1 W/m^{2}K, when the temperature inside the heated rooms (t_i) is at least 16°C. Forecasts suggest that the next legal changes will cause that from the end of 2020 the value (U_{max}) for transparent non-opening surfaces will not be more than 0.9 W/m^{2}K. This is related to the European Union's assumptions about the Member States' pursuit of sustainable development, also by the constant reduction of energy consumption by buildings.

The multi-criteria comparative analysis shows that the best material solution for the 6-meter-long reference column is glulam - AIS = 0.775. The results for other materials were respectively: for safety glass - AIS = 0.573 and for rolled steel - AIS = 0.241. The superiority of safety glass over rolled steel was primarily determined by aesthetic and energetic groups of criteria. The comparative analysis also reveals that safety glass is the most attractive material in terms of aesthetics, since the synthetic assessment for safety glass in the first stage of the analysis was the highest in the group of the aesthetic criteria ($SA_1 = 0.50 > SA_3 = 0.25 > SA_2 = 0.13$).

References

1. A. Jóźwik, Cz. Tech. Arch. **104**(4-A), 219-224, (2007)
2. W. Szolginia, *Architektura* (Wyd. Sigma NOT, Warszawa, 1992)
3. M. Dytczak, *Wybrane metody rozwiązywania wielokryterialnych problemów decyzyjnych w budownictwie* (Politechnika Opolska, Opole, 2010)
4. J. Szwabowski, J. Deszcz, *Metody wielokryterialnej analizy porównawczej* (Politechnika Śląska, Gliwice, 2001)
5. T. Trzaskalik, Zesz. Nauk. PŚl, seria : Organizacja i Zarządzanie **74**, 239-263, (2014)
6. M. Meller, P. Siwiński, *Szkło budowlane* (Politechnika Koszalińska, Koszalin, 2002)
7. A. Piekarczuk, *Elementy konstrukcyjne ze szkła budowlanego* (ITB, Instrukcje, wytyczne, poradniki, Warszawa, 2013)
8. Z. Mielczarek, *Nowoczesne konstrukcje w budownictwie ogólnym* (Arkady, Warszawa, 2009)
9. E. Neufert, *Podręcznik projektowania architektoniczno–budowlanego* (Arkady, Warszawa, 2000)
10. A. Bugaj, Rynek Inst. **11**, 48-51, (2013)
11. B. Stelmach, *Poszukiwania struktur. Prace architektoniczne 1997 - 2011. Tom 2 - Zdjęcia z realizacji* (Wyd. PAN, oddział w Lublinie, 2012)
12. http://www.ots-ip.pl/wp-content/uploads/2014/03/Untitled-11.jpg (access date: 15. Jan. 2018)
13. http://www.polskiekrajobrazy.pl/images/stories/big/27418.._Zelazowa_Wola_i_okolice_.JPG (access date: 15. Jan. 2018)
14. M. Gicala, A. Sobotka, Prz. Nauk. Inż. i Środ. **26**(2), 159-170, (2017)
15. A. Leśniak, K. Zima, *International Multidisciplinary Scientific GeoConference Surveying Geology and Mining Ecology Management* (SGEM, Albena, Bulgaria) **3**(5), 157-164, (2015)
16. W. Drozd, Prz. Bud. **84**(4), 32-37, (2013)
17. A. Leśniak, J. Balicki, Entrepr. Bus. and Econ. Rev. **4**(2), (2016)
18. A. Radziejowska, K. Zima, *E3S Web of Conferences* **10**(00078), (EDP Sciences, 2016)
19. M. Golański, Bud. i Inż. Środ. **3**, 39-53, (2012)
20. J. Górzyński, *Podstawy analizy środowiskowej wyrobów i obiektów* (WNT, Warszawa, 2007)
21. A. Leśniak, A. Radziejowska, Proc. Eng. **208**, 76-81, (2017)
22. E. Plebankiewicz, K. Zima, D. Wieczorek, Arch. of Civ. Eng. **62**(2), 149-166, (2016)

Critical evaluation of modern concrete applications

Jarosław Malara[1,*], and *Justyna Szafraniec*[1]

[1]Cracow University of Technology, Faculty of Civil Engineering, Poland, ul. Warszawska 24, 31-155 Cracow, Poland

Abstract. In the article, the modern and innovative applications of concrete mix have been critically analyzed. The authors focused first of all on transparent concrete - LiTraCon. The article mainly analyzed its technical parameters. Another modern way of using the concrete mix, which has been included in the publication, is paving stone absorbed in daylight. This material description is primarily, focused on its serviceability. Another material presented is white concrete, its advantage is the ease of colour choice. The authors also described the possibilities of using properly shaped architectural concrete as well as its unusual use in the form of doors. The entire publication concludes with a summary.

1 Concrete as a building material

The history of concrete reaches 7000 years BC [1]. The traces of this material were found in the area of present-day Israel. Also in Europe, around 5600 BC, concrete was used in the area of present Serbia. In addition to the Mediterranean basin, traces of the use of this material were discovered in Mexico (around 1100 BC). The most recognizable traces of the widespread use of concrete are found in Italy and come from antiquity [2].

Increasing the possibility of using concrete and its freeforming have been achieved thanks to the use of steel and later in other components. In the first place, steel bar reinforcement was used to increase the concrete parameters. The next stage was the pre-stressing of concrete with cables and strings, nowadays various types of fibers are used very intensively, which form scattered reinforcement [3, 4, 5].

The modern meaning of this material, despite the passage of time from the beginning of its invention, has not lost its value. It is by far the most popular structural material commonly used in cubature, industrial, hydrotechnical or single-family construction. In recent years, the authors of the article have observed an increasing interest of the concrete designers considering concrete as a material both structural and architectural material. The further part of the publication presents, examples of concrete mix application to bring not only the strength parameters of manufactured building elements, but also an attempt to extract the previously unknown beauty of concrete [6].

* Corresponding author: jmalara@L3.pk.edu.pl

2 Contemporary examples of concrete use

2.1 LiTraCon

Aron of Losonczi, a Hungarian architect, invented light-transmitting concrete. However, this is not full transparency, but only semipermeability. Generally, this material should be considered a breakthrough and discussed in this article. The name of the concrete LiTraCon is an abbreviation for the words Light Transmitting Conrete [7]. An interesting fact is the possibility of applying sort of this concrete for the production of artistic handicrafts. Elements such as lamps, furniture or sculptures are made of this material.

LiTraCon includes approx 96% concrete mixture with fine grain size and 4% fiber optics. The grain size and diameter of the optical fibers have been selected to allow for correct light construction of the finished element structure. Optical fibers are arranged parallel to one another and perpendicular to the edge of a concrete block. This allows for partial light transmission. This means that applying LiTraCon you can see both the shapes and colors of objects on the other side of the material. An example of LiTraCon blocks is presented in Figure 1.

Fig. 1. Photo of LiTraCon's blocks [8].

Turning to the issue of technical parameters, LiTraCon detects the following features:
- density - 2100-2400kg / m3,
- compressive strength - 50MPa,
- bending strength - 7MPa,
- tensile strength - 7MPa,
- composition: about 96% concrete mix and 4% optical fibers,
- production form - prefabricated blocks, standard dimensions 600x300mm
- thickness of blocks - 25-500mm
- available colours - gray, black or white.

A small demand for LiTraCon in Poland is undoubtedly related to the price of this product. The cost of 1m2 of 30mm thick blocks is estimated at 1300 euros. It should be pointed out that as the thickness of the blocks increases, the price increases considerably. This is due to technological difficulties to provide light transmission for larger element thickness.

Turning to the critical assessment of the innovative solution, among the positives, first of all necessary the architectural value. LiTraCon is a kind of concrete to make the interior of the room more attractive. Technical parameters, especially compressive strength, are high and comparable with high quality concrete, which should also be considered an asset. Despite the theoretical possibility of forming blocks with a thickness of up to 500mm, without an incredibly large financial outlay, it is difficult to treat LiTraCon as a construction material. The prices of elements with a thickness of 30 mm are already high enough, and thus this element can only be treated as a finishing material or architectural curiosity, which significantly reduces the possibility of its use. In the climatic conditions prevailing in Poland, it cannot be used even as a filling of curtain walls, because in the case of a thermal insulation layer, it would not let light through. In summary, LiTraCon is a very expensive concrete material that can only serve as an architectural feature.

2.2 Paving stone absorbed in daylight

The Polish producer of concrete products Polbruk has developed a paving stone enriched with phosphors. Their use allows you to achieve the effect of lighting the sett with a blue light in the darkness. This is due to the ability to absorb solar energy through the phosphors and release it in the form of light at night. This phenomenon is called luminescence. The trade name of the product is Polbruk Lumia. Fig.2 displays the visual effect of phosphor application.

Fig. 2. The possibility of creating patterns with Polbruk Lumia [9].

The use of a shining cube at night SHOWS several major advantages. The first one is the possibility to emphasize horizontal surface markings. Thanks to Polbruk Lumia, pedestrian crossings can be made visible, parking spaces can be marked or lanes can be marked in places where pavement paving or concrete blocks are laid. Another issue is related to energy efficiency. The use of this material does not make it necessary to supply electricity and illuminate key areas for the investor. In addition, the aesthetic value of the concrete block should be emphasized. The characteristic blue light raises the visual dimension of the surface.

On the other hand it should be pointed out that currently the paving stone is produced in three color variants only the shades of gray. Its format is 10x20x6cm, which makes it difficult to connect it with other concrete cube formats. Afixed thickness of 6cm makes it impossible to use it in surfaces intended for heavy car traffic.

According to the authors, the presented solution is an innovative solution on the Polish market. It shows an undoubted advantage over defects, which promises well in the context of a wider application of this product. However, it is necessary to consider expanding the range both in terms of paving stone size and color. Certainly, increasing the universality of the solution will contribute to its widespread use.

2.3 White cement

The search for alternatives to the existing building materials increases investors' interest in concrete products based on white cement. In other to get an imitation of stone slabs with a very similar look, similar strength parameters, but a lower price, this may be the answer. The use of white cement can be found in such elements as:

- terrazzo,
- bricks production,
- plastering mortar,
- element of small architecture (fountains, barbecues, balcony balustrades, etc.),
- plaster,
- architectural concrete,
- glues,
- weld.

While describing white cement, it is necessary to mention its technical parameters. The manufacturing process of white cement is almost identical to the typical gray Portland cement. The difference lies primarily in the selection of appropriate raw materials. The clinker, which is a cement intermediate, is intended to show an appropriate chemical composition and content of individual minerals. The sulfate compounds are very important, in the case of white cement they serve as a regulator of the setting time. Due to a high degree of product fineness and excellent homogenization, the white cement properties are more advantageous than the classic gray cement parameters. The comparison is shown in Table 1.

Table 1. Comparison of exemplary gray and white cements - cement from the Górażdże factory.

Technical Parameters	Portland White Cement CEM I 42,5 N	Portland Gray Cement CEM I 42,5 N	Norm [10]
Compressive strength after 2 days [MPa]	42,0	28,1	$\geq 20{,}0$ PN-EN 197-1
Compressive strength after 28 days [MPa]	67,0	59,0	$\geq 42{,}5 \leq 62{,}5$ PN-EN 197-1
The beginning of binding time [min]	145	188	$\geq 60{,}0$ PN-EN 197-1
Sulphate content SO_3 [%]	2,9	2,55	≤ 5 PN-EN 197-1
Chloride content Cl^- [%]	0,01	0,065	$\leq 0{,}1$ PN-EN 197-1
Loss of roasting [%]	2,5	2,39	$\leq 5{,}0$ PN-EN 197-1
The residue is insoluble [%]	0,2	0,67	$\leq 5{,}0$ PN-EN 197-1

More favorable technical parameters of white cements in relation to their gray counterparts have been confirmed many times. The data presented in Table 1 also testify to this. However, due to a higher degree of complexity in the production of white cements as well as the need for even more precise selection of aggregates used for their production, combined with their availability, the price of white cements is about three times higher compared to gray cements. This difference acts upon in the value of the final concrete product.

According to the authors, white cement shows potential that should be conducive to increased use. High technical parameters convince to a positive opinion, but above all, the versatility of applications, combined with aesthetic values. Due to the fact that cement is only a component of a concrete mix, its use does not cause large differences in the price of the products obtained.

2.4 Architectural concrete

In recent years, architects have begun to appreciate the appearance of natural concrete in a significant way. The concept of architectural concrete has become a common circulation. In the most general terms, architectural concrete should be understood as any kind of surface for which a concrete mix is a finishing layer. Architectural concrete may serve its structural functions too.

Due to a vast definition of architectural concrete, it is difficult to assess this material in terms of chemical composition or the content of individual aggregates or cements. Depending on the desired final effect, the composition of the concrete mix can be individually selected to the needs of the investor. The texture of the concrete surface depends, first of all, on the grain size of the designed mixture. Plasticisers and other fluids are often used to ensure complete filling of molds and boards, depriving finished products of cancers or other imperfections.

The aim of excellent precision and ideal surface in building conditions is difficult to get. That is why architectural concrete products are frequently prefabricated. The prefabrication process allows for making concrete casts in molds, instead of system formworks, this issue significantly affects the visual effects of the obtained element.

A very interesting example of a concrete mix application is a door made of concrete offered by one of the companies. With the exception of the hardware and lock elements, the rest is a concrete mix. Extremely important for this element is the use of dispersed reinforcement, which increases the strength of the door and virtually eliminates the need to preserve them. The example mentioned by the authors is however somewhat abstract. The basic disadvantage in addition to weight is its price. According to catalogs, it ranges between 8,000 and 12,000 euros. It is therefore a unique, but at the same time very luxurious product.

Architectural concrete is an example of a solution that, according to the authors, will certainly be constantly developed in the construction industry. Its advantage is the ability to obtain both design and finishing features in one element. Architectural concrete shows clear trends in contemporary architecture and construction, which often use raw materials emphasizing their natural beauty.

3 Summary

The article presents modern examples of the use of concrete mix. According to the examples given by the authors, concrete is currently not only a structural material, but can be a decorative element. The development of construction chemistry and technology has allowed the creation of transparent concrete or shining in the dark absorbed light. Architectural trends caused that the natural beauty of concrete began to be used, without the need to cover it with plaster or other finishing layers.

A natural barrier to using innovative solutions from concrete is their price. In particular, mention should be made of doors made of concrete or light-transmitting LiTraCon blocks. Another issue is fashion in modern construction. Certainly, for some time the material and technological solutions described in this article will no longer be appreciated by designers. However, the objects made with the use of architectural concrete can remain and survive for dozens, or hundreds of years.

The aim of the article, achieved by the authors, was to indicate modern and unusual applications of concrete. Many different products have been analyzed up till now. The most interesting ones were presented here. Due to their small scale of application, all these products detected a negative feature of high price. The most important advantage is the aesthetic value. A novel and creative contribution to the development of science, presented in the article, was to find atypical solutions for the use of concrete mix. In addition, the authors made a critical evaluation of the presented technologies. Indicating such applications will contribute to popularizing the use of concrete for less common applications.

References

1. W. Raczkiewicz, Beton – materiał budowlany znany od wieków (Przegląd Budowlany, **10**, 2012)
2. W. Affelt, *Dziedzictwo w budownictwie* (Wydawnictwo Politechniki Gdańskiej, Gdańsk, 1999)
3. A.M. Brandt, *Cement Based Composites: Materials, Mechanical Properties and Performance* (Taylor and Francis, London and New York, 2009)
4. W. Sun, Y.M. Zhang, H.D. Yan, R. Mu, Damage and damage resistance of high strength concrete under the action of load and freeze-thaw cycles (Cement and Concrete Research, **29**, 1999)
5. T. Zych, Współczesny fibrobeton – możliwość kształtowania elementów konstrukcyjnych i form architektonicznych (Czasopismo Techniczne, Kraków, **18**, 371-386, 2010)
6. M. Głuchowski, Zastosowanie nowoczesnych rozwiązań technologii betonu w kształtowaniu architektury (Czasopismo Techniczne, Kraków, 2010)
7. Anonymous, Translucent concrete developed in Europe. (Civil Engineering : Magazine of the South African Institution of Civil Engineering. Yeoville, **13**, 10)
8. http://litracon.hu/en/products/litracon-blokk (visiting at: 02.02.2018)
9. http://www.polbruk.pl/pl/produkty/styl/polbruk-lumia (visiting at: 02.02.2018)
10. PN-EN 197-1:2012 Cement. Część 1: Skład, wymagania i kryteria zgodności dotyczące cementów powszechnego użytku

Multicriteria assessment of selected types of ceilings

Julia Nowak[1,*], *and Edyta* Plebankiewicz[1]

[1] Cracow University of Technology, Faculty of Civil Engineering, Warszawska 24, 31-155 Kraków, Poland

Abstract. The paper compares three systems of the rib-and-slab ceiling: Vector, Rector and Teriva. The following comparison criteria were selected: the tare weight of ceiling beams, execution time, slab faulting, sound insulation, thermal insulation and aesthetics of the lower surface. The results of the multicriteria comparative analysis clearly indicated the advantage of the Vector ceiling, assuming the same weights for all criteria taken into account.

1 Introduction

The basic function of the ceiling is to transfer the load to the walls, and subsequently to the foundations of the building. The ceilings bear their own weight, service loads and load of partition walls. They stiffen the walls of the building and separate storeys. Moreover, they function as acoustic and thermal barriers between storeys. In the event of fire, the ceiling prevents the fire from spreading between storeys.

One of the most widespread, especially in housing construction, are rib-and-slab ceilings. These are the ceilings indicating prefabricated beams as load-bearing elements with spacing between centres not exceeding 90 cm [1]. As a rule, beams are stretched towards a smaller span, leaning them on the walls of the building and buntlines. Supplementary elements located between the beams include, e.g.: ceramic blocks, light expanded clay blocks, gravel concrete bricks, light concrete bricks, as well as styrofoam or gypsum moldings. The upper part of the ceiling is a concrete overlay poured on the construction site.

The paper briefly describes and compares three popular rib-and-slab ceiling systems: Vector, Rector and Teriva. The analysis points out the most attractive solutions out of a total number of three.

* Corresponding author: j.nowak96@gmail.com

2 Ceiling systems

2.1 The Teriva ceiling

Rib-and-slab ceilings the Teriva are one of the longest and most popular ceilings in housing construction, especially single-family buildings. The Teriva involves ribs in the form of truss beams, its reinforcement is concreted and, typically, light expanded clay aggregate (LECA) block or autoclaved cellular concrete block. In the case of the ceiling span of over 4.5 m, partition gilts have to be constructed, laid transversally to the ceiling ribs. Several variants of Teriva are available on the market, variable due to e.g. load capacity, spacing of beams or spread.

The Teriva ceilings are slightly easier to construct than, for example, monolithic ceilings. The low weight of the ceiling elements (hollow blocks or beams with reinforcement) means that they can be arranged by hand. The popularity of Teriva acts positively on impact on its availability.

This kind of ceiling is practically impossible to construct with irregular, complex projects. They are also susceptible to slab faulting, their sound insulation is much worse than that of the majority of ceiling solutions [2].

2.2 The Vector ceiling

The Vector ceiling is a relatively new type of ceiling, as it appeared on the Polish market in 2015. The Vector is a composite beam-and-block ceiling. The Vector ceilings consist of panels 4 cm high and 60 cm wide. In addition a Vector ceiling is reinforced by a centrally embedded steel truss. Due to the applied truss, the installation of panels on the construction site is easier. Moreover, the trusses are covered with a layer of concrete overlay, thus strengthening the connection with the panels. An additional connection between the layers is strengthened with a bonding layer, which also performs a waterproof function.

The undeniable benefits of this ceiling system are: short assembly time, as well as high load capacity and sound insulation parameters. The additional advantages of this system include no need for an individual project for each room [3–6].

2.3 The Rector ceiling

The Rector ceilings have been available on the Polish market for over 10 years. They involve prefabricated prestressed RS beams in several variants, differing in: section height, number and location of the prestressing strands and stiffening with a space truss. They are made of concrete and prestressing steel of high strength parameters (C 50/60 concrete, steel grade 2060 MPa), thus ensuring high strength and quality of the ceilings designed. The main prestressing reinforcement are strands of high strength steel strings.

The Rector ceilings are distinguished by a combination of prefabrication and advantages resulting from the compression of concrete, which ensure: greater durability, high flexibility of the systems applied (from 16 to over 30 cm of ceiling thickness and ranging from 1m to 10m of ceiling span), the deflection reduced to a half, no faulting or scratch effect, neither is the usage of distributing ribs necessary [7, 8].

3 Criteria for comparison of ceilings

The implementation cost of the analysed ceilings is similar. In order to choose the best solution, it is worth taking into account the technical parameters specific for each of the ceiling systems.

In order to compare the ceiling solutions, the authors decided to choose six criteria, particularly important from the viewpoint of both the contractor and the future user.

The following criteria were selected:

1. The tare weight of the beams [kg/mb] – this parameter largely determines, among others, the necessity of using a crane for transporting ceiling elements.
2. Implementation time [r-g] – the assembly time was estimated for 100 m2 of ceiling area with the assumption that the team involves 4 people.
3. Slab faulting – associated with the stiffness of the ceiling - excessive deflection will result in squeaking of the floor and cracking of the plaster on the ceiling of the lower storey.
4. Sound insulation – the ability to suppress noise is essential from the point of view of the comfort of use. In residential buildings, ceilings should have adequate insulation from airborne sounds (conversation, music, etc.) and sound impact (footsteps, furniture being moved, etc.). This feature is particularly important in the case of ceilings parting residential storeys.
5. Thermal insulation – a ceiling is usually an internal partition that divides rooms with low differences in temperature. In the case of flat roofs and structures separating the basement or a non-use attic, the ceiling should meet appropriate thermal insulation requirements. The lower the heat transfer coefficient of the ceiling, the lower the costs associated with thermal insulation.
6. Aesthetics of the lower surface – smooth and even bottom surface of the ceiling minimizes the cost of plastering or general finishing of the lower surface of the ceiling.

4 Multicriteria comparative analysis

In order to find the most advantageous and attractive solution of the ceiling, a multicriterial comparative analysis was conducted. Wide applications of multi-criteria tools facilitating decision making in the construction industry were presented, e.g. in [9-13].

The methods of multi-criteria comparative analysis include a class of algorithms that create a scalar denoting a synthetic evaluation index – the so-called "mathematical methods" [11]. The corrected summation index is the most frequently used discrete multi-criteria method [13]. The application of the synthetic evaluation indices requires a division of the evaluation criteria into stimulants which are positively correlated with the dependent variable, and negatively correlated destimulants.

In order to perform the comparison and evaluation of the variants, the chosen evaluation variants of the criteria should be given dimensionless quantities, to perform both quantitative and qualitative comparison.

The procedure involves three steps:

- determining the corrected values of criteria evaluations for individual variants (stimulant, destimulant),
- determining the coded values of criteria evaluations for individual variants,
- determining the corrected a synthetic evaluation index.

Table 1 presents description of criteria applied in the analysis.

Table 1. Description of criteria using in the analysis.

Variants	Criteria					
	The tare weight of the beams [kg/mb]	Implementation time [r-g]	Thermal insulation [W/m2K]	Slab faulting [points 1-5]	Sound insulation [dB]	Aesthetics of the lower surface [points 1-5]
	K1	K2	K3	K4	K5	K6
	Destimulants			Stimulants		
Teriva	20	24	0,3	2	43,3	2
Vector	70	11	0,3	5	61,2	5
Rector	16	13,5	0,43	5	47,5	2

Once the criteria were tentatively defined, they were assigned specific values and to subsequently break down into classes. Subsequently, four methods of coding were performed – standardization; normalization; Neuman-Morgenstern method and Pattern method.

Standardization

The measure of the *i*-th variant according to the *j*-th criterion:

- for stimulants:

$$z_{ij} = \frac{x_{ij} - \overline{x_j}}{s_j} \tag{1}$$

where:

x_{ij}	the value of criteria evaluation
$\overline{x_j}$	arithmetic mean
s_j	standard deviation

- for destimulants:

$$z_{ij} = (-1) \cdot \frac{x_{ij} - \overline{x_j}}{s_j} \tag{2}$$

The standarization method coding result are present in table 2.

Table 2. Results of coding by the standardization method.

Variants	Criteria					
	Stimulants			Destimulants		
	Value z_{ij}			Value z_{ij}		
	K4	K5	K6	K1	K2	K3
1	-2,4495	-0,0836	-1,2247	0,0060	-1,6872	0,0919
2	1,2247	0,1196	2,4495	-0,0136	1,1128	0,0919
3	1,2247	-0,0360	-1,2247	0,0076	0,5744	-0,1838

Normalization

The measure of the *i*-th variant according to the *j*-th criterion:

- for stimulants:

$$z_{ij} = \frac{x_{ij}}{x_{j\,max}} \tag{3}$$

where:

$x_{j\,max}$	maximum value of the measure of weights according to j-th criterion

- for destimulants:

$$z_{ij} = \frac{x_{ij}'}{x_{j\,max}'} \tag{4}$$

where:

x_{ij}' reducing descending values to increasing ones (maximizing):

$$x_{ij}' = \frac{1}{x_{ij}} \tag{5}$$

$x_{j\,max}'$ maximum value of the measure of weights according to j-th criterion after maximizing.

The normalization method coding results are present in table 3.

Table 3. Results of coding by the normalization method.

Variants	**Criteria**					
	Stimulants			**Destimulants**		
	Value zij			**Value zij**		
	K4	**K5**	**K6**	**K1**	**K2**	**K3**
1	0,4000	0,7075	0,4000	0,8000	0,4583	1,0000
2	1,0000	1,0000	1,0000	0,2286	1,0000	1,0000
3	1,0000	0,7761	0,4000	1,0000	0,8148	0,6977

Neumana - Morgensterna

The measure of the *i*-th variant according to the *j*-th criterion:

- for stimulants:

$$z_{ij} = \frac{x_{ij} - x_{j\,min}}{x_{j\,max} - x_{j\,min}} \tag{6}$$

- for destimulants:

$$z_{ij} = \frac{x_{j\,max} - x_{ij}}{x_{j\,max} - x_{j\,min}} \tag{7}$$

Results of coding by the Neumana - Morgensterna method are present in table 4.

Table 4. Results of coding by the Neumana-Morgensterna method.

Variants	**Criteria**					
	Stimulants			**Destimulants**		
	Value zij			**Value zij**		
	K4	**K5**	**K6**	**K1**	**K2**	**K3**
1	0,0000	0,0000	0,0000	0,9259	0,0000	1,0000
2	1,0000	1,0000	1,0000	0,0000	1,0000	1,0000
3	1,0000	0,2346	0,0000	1,0000	0,8077	0,0000

The **Pattern coding** was also performed, consisting of replacing the partial measure by its quotient with the sum of measures of all variants according to the given criterion.

The measure of the *i*-th variant according to the *j*-th criterion:

- for stimulants:

$$z_{ij} = \frac{x_{ij}}{\sum_{i=1}^{n} x_{ij}} \tag{8}$$

- for destimulants:

$$z_{ij} = \frac{x_{ij'}}{\sum_{i=1}^{n} x_{ij'}} \tag{9}$$

The Pattern method coding results are present in table 5.

Table 5. Results of coding by the Pattern method.

Variants	Criteria					
	Stimulants			Destimulants		
	Value zij			Value zij		
	K4	K5	K6	K1	K2	K3
1	0,1667	0,2849	0,2222	0,3944	0,2016	0,3661
2	0,4167	0,4026	0,5556	0,1127	0,4399	0,3661
3	0,4167	0,3125	0,2222	0,4930	0,3585	0,2679

The next step was a synthetic evaluation of the variant solutions. Three different indicators were selected for a more accurate analysis:

- Multiplication indicator – which is the product of the ratings for a given variant of all criteria;
- Summation indicator – the sum of all evaluations;
- Additive indicator – the sum of the scores is divided by the number of criteria.
- Results of analysis are presented in Table 6, Table 7, Table 8 and Table 9.

Table 6. Results of analysis for standardization coding.

Variants	Summation indicator	Additive indicator	ORDER
1	-5,3471	-0,8912	III
2	4,9850	0,8308	I
3	0,3622	0,0604	II

Table 7. Results of analysis for normalization coding.

Variants	Multiplication indicator	Summation indicator	Additive indicator	ORDER
1	0,0415	3,7658	0,6276	III
2	0,2286	5,2286	0,8714	I
3	0,1765	4,6886	0,7814	II

Table 8. Results of analysis for N-M coding.

Variants	Summation indicator	Additive indicator	ORDER
1	1,9259	0,3210	III
2	5,0000	0,8333	I
3	3,0423	0,5071	II

Table 9. Results of analysis for normalization coding.

Variants	Multiplication indicator	Summation indicator	Additive indicator	ORDER
1	0,0003	1,6358	0,2726	III
2	0,0017	2,2935	0,3822	I
3	0,0014	2,0707	0,3451	II

5 Summary

The paper compares three popular systems of rib-and-slab ceilings. Taking into account the similar cost of the solutions, the technical parameters characterizing the selected floor systems were adopted for the analysis. It was also assumed that all the adopted parameters were of the same importance for the decision maker. With such assumptions, the methods used for the multi-criteria analysis indicated unambiguously the advantage of the Vector ceiling over the other solutions. Yet it should be remembered that the investor's preferences can change the ranking of the solutions obtained. For example, the need to use a crane may, in some cases, eliminate the choice of the Vector ceiling.

References

1. http://www.stropy.pl/ (access date: 10. Jan. 2018)
2. http://www.stropex.com.pl/ (access date: 10. Jan. 2018)
3. http://strop-vector.eu/strop-vector/ (access date: 10. Jan. 2018)
4. http://www.fabrykastropow.pl/zespolony-strop-gestozebrowy-vector (access date: 10. Jan. 2018)
5. http://stropvector.pl/ (access date: 10. Jan. 2018)
6. http://www.konbet.com.pl/ (access date: 10. Jan. 2018)
7. http://www.rector.pl/ (access date: 10. Jan. 2018)
8. https://rector.informatorbudownictwa.pl/ (access date: 10. Jan. 2018)
9. R. T. Marler, J. S. Arora, Struct Multidiscipl Optim, **26**(6), 369-395, (2004)
10. M. Dytczak, *Wybrane metody rozwiązywania wielokryterialnych problemów decyzyjnych w budownictwie* (Politechnika Opolska, Opole, 2010)
11. J. Szwabowski, J. Deszcz, *Metody wielokryterialnej analizy porównawczej* (Politechnika Śląska, Gliwice, 2001)
12. A. Leśniak, B. Szewczyk, Tech Trans, **28**, 4-B, 185-190, (2015)
13. T. Trzaskalik, Zesz. Nauk. PŚl, seria : Organizacja i Zarządzanie **74**, 239-263, (2014)

Overview of vibroacoustic isolators used in railway tracks

Cezary Kraśkiewicz[1,*], *Wojciech* Oleksiewicz [1], *Monika* Płudowska-Zagrajek[1] , and *Cezary* Lipko[2]

[1]Warsaw University of Technology, Faculty of Civil Engineering, Institute of Roads and Bridges, 16 Lecha Kaczyńskiego Ave., 00-637 Warsaw, Poland
[2] TINES Railway, TINES Capital Group SA, Ciepłownicza 23 Street, 31-574 Cracow, Poland

Abstract. The paper presents an overview of vibroacoustic isolators used in railway tracks to reduce negative effects from railway traffic. These include single components or their assemblies, with different material characteristics and location in the track structure due to track structure type (ballasted or ballastless track system). The reduction of negative effects relates mainly to the reduction of vibration and structure-borne noise. The practical solutions for railway structures include various components that perform different functions at the same time: vibroacoustic isolation and mechanical function to ensure safe and durable use of rail track (i.e. rail fastening systems, rail supporting structures). The main goals of this paper are systematization and description of basic material characteristics of vibroacoustic isolators applied sufficiently close to the place of vibration generation during the movement of rail vehicles.

1 Introduction

Rail track structure consists of many components, with different functions. One of them are vibroacoustic isolators used in railway tracks to reduce negative effects from railway traffic. These elements vary depending on material characteristics and location in the track structure due to track structure type (ballasted or ballastless track system).

In the classification according to [1] kinds of rail track superstructures it is appropriate to assume the main criterion linked with occurrence in a given structure of the ballast, as the main superstructure layer and its function of shaping the geometry of the track. When this feature is fulfilled, the track structure is qualified as ballasted track system (Fig. 1). Otherwise, it is qualified as ballastless track system (Fig. 2, 3). It is important for ballast layer to meet both of the above conditions, because ballast layer can occur in ballastless track system - in that case it serves different functions i.e.: drainage, thermal insulation or noise reduction.

* Corresponding author: c.kraskiewicz@il.pw.edu.pl

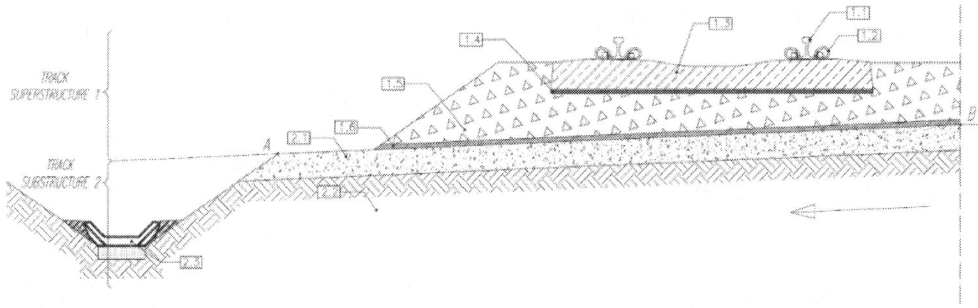

Fig. 1. Components of ballasted track system.
Key: Track superstructure (1): 1.1 – Vignole rail profile, 1.2 – rail fastening system (type SB or W14), 1.3 – rail sleeper, 1.4 – (option) under sleeper pad, 1.5 - ballast, 1.6 – (option) under-ballast mat. Track substructure (2): 2.1 – blanket layer, 2.2 – subgrade, 2.3 – surface drainage [1].

An example of ballastless track system on rail crossing in LC-L system is given in Fig. 2. The principle of rail fastening in ERS system, illustrated on Detail A, entitles of fixing rail in rail channel, where it is supported on continuous, resilient strip pad (polyurethane composite) and continuous layer of reis grout. After accurate horizontal and vertical positioning of the rail, resilient casting elastomer (adapted to the operating conditions and required vibration damping effectiveness) is applied.

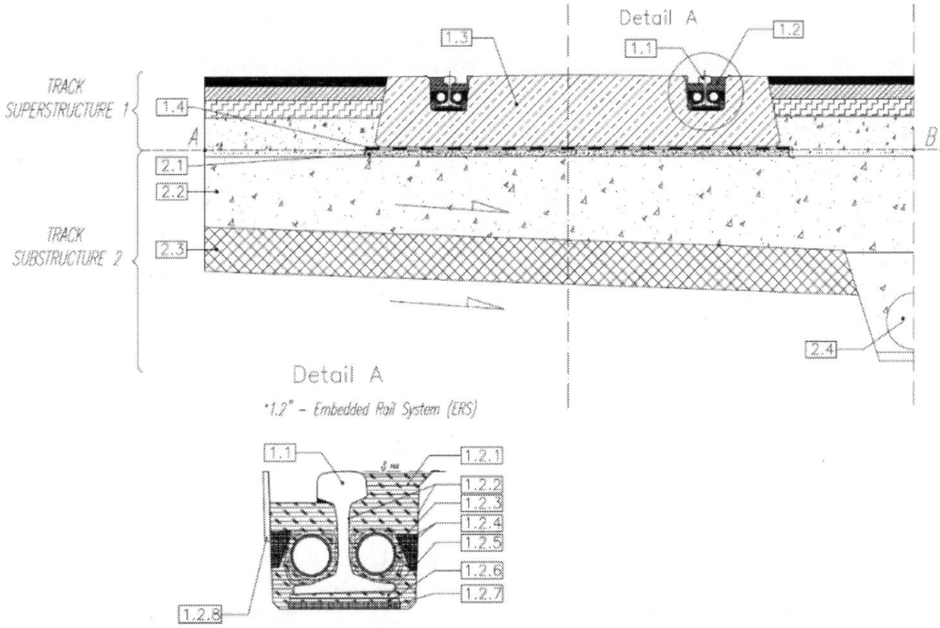

Fig. 2. Compounds of ballastless track system on railway crossing in Level Crossing-Length system (LC-L system) with integrated railway and road superstructure.
Key: Track structure (1): 1.1 – Vignole rail profile; 1.2 – rail fastening system – i.e. Embedded Rail System (ERS); 1.3 – concrete slab track; 1.4 – slab-track mat;
Track substructure (2): 2.1 – sand blinding layer; 2.2 – blanket layer; 2.3 – substructure - compacted and reinforced subsoil; 2.4 – drainage;
Detail A – 1.2 (Embedded Rail System) – key: 1.2.1 – resilient casting elastomer based on resins - element of rail fixing and support, 1.2.2 – primer, 1.2.3 – wedges as a tool for horizontal alignment of the rail, 1.2.4 – material saving item, 1.2.5 – polymer inclination shims for providing inclined position of the rail, 1.2.6 – polymer shims as a tool for vertical leveling of the rail, 1.2.7 – resilient strip mat, 1.2.8 - continuous, steel reinforcement of inner edge of the rail channel [1].

The following figure (Fig. 3) shows a use of a ballastless track system in a tunnel on example of the Warsaw Metro Line II.

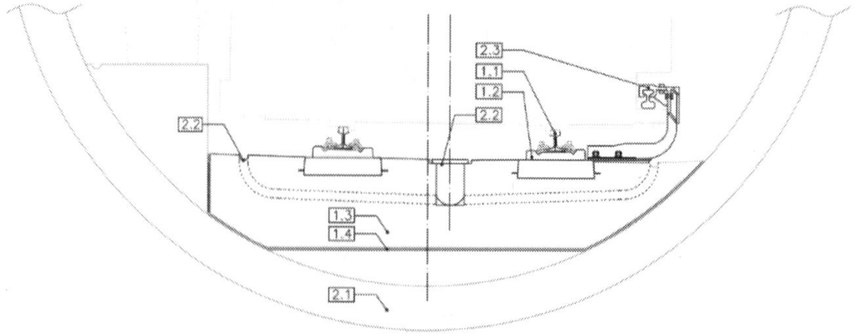

Fig. 3. Components of ballastless track system in a tunnel on example of the Warsaw Metro line II. Key: Track superstructure (1): 1.1 – Vignole rail profile type 49E1; 1.2 – rail fastening system – Embedded Blocks System (EBS); 1.3 – concrete track slab; 1.4 – slab-track mat. Track substructure (2): 2.1 – concrete inverted slab; 2.2 – surface drainage; 2.3 – conductor-rail [1]

2 Application areas of vibroacoustic isolators in railway tracks

The application areas of vibroacoustic isolators in railway tracks are given in Fig. 4.

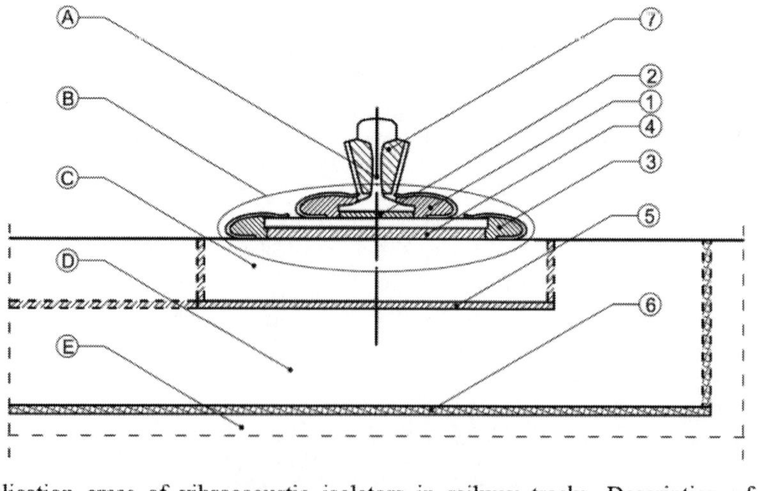

Fig. 4. Application areas of vibroacoustic isolators in railway tracks. Description of rail track components: A – Vignole rail; B – fastening system; C – rail supporting system (C.1. – rail sleeper in ballasted track system or C.2. – rail supporting structure in ballastless track system); D.1. – ballast layer in ballasted track system or D.2. – concrete base plate in ballastless track system; E – track substructure.

The application areas of vibroacoustic isolators in railway tracks may include the following elements (Fig. 4):

- resilient fastening (fixing) of rail foot (1): resilient clamps or grouting;
- resilient fastening (supporting) of rail foot (2): rail pad, continuous rail pad, discrete or continuous grouting;
- resilient fixing in indirect rail fastening system (3): tension clamps, anchor bolts with compression spring;

– resilient support of rail supporting structure in indirect rail fastening system (4): base plate pad, discrete grouting;
– resilient support or fixing of rail supporting structure (sleeper or block) (5): under sleeper pad, under block pad, grouting;
– resilient fixing and support of railway track (6): under-ballast mats or slab-track mats – which with rail track structure compose floating slab track system;
– filling gaps between fishplate and rail web (noise reduction) (7): grouting or rail web filler block.

2.1 Resilient rail fastening systems

In the entirety of the listed components of rail track structure rail fastening system demonstrates the greatest diversity of design solutions that constitute the basis for differing classifications [1]. This is particularly the case of a ballastless track structure, where resilient of load-bearing capability necessary for the proper functioning of the system must be provided by variously solved resilient elements of rail fastening system. The variety of possible design solutions is result of dynamic development of material engineering.

One of the basic classifications criterion of a rail fastening system is their continuity in the rail track. It is illustrated in Fig. 5, where two sorts of rail fastening systems are defined: discrete and continuous.

a) b)

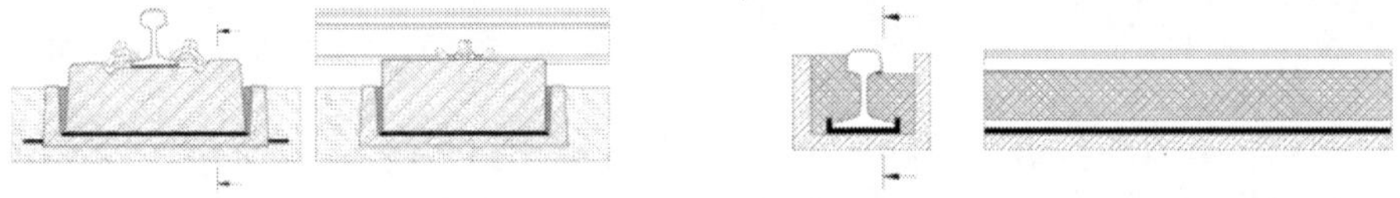

Fig. 5. Illustration principle of classification of rail fastening systems due to criterion of their continuity occurrence in rail track; a) discrete rail fastening (i.e. rail fastening in Embedded Blocks System - EBS); b) continuous rail fastening (i.e. Embedded Rail System - ERS) [1].

Another basic classification of rail fastening systems is classification due to criterion of the fixing rail with the rail supporting structure. The fastening system, where the rail is directly fixed to the supporting structure by anchors, bolts or screws is described as a direct rail fastening system (Fig. 6a, 6b). However, if the rail is fixed to the supporting structure with the usage of additional element (i.e. ribbed base plate), then these systems is described as indirect rail fastening system (Fig. 6c).

The term of rail fastening system used in standard [2] requires more detailed description of its two main functions specified in that standard as rail support and rail fixation.

The rail support system provides absorption of vertical loads and transfers them to a ballast layer (ballasted track system) or a substructure – a concrete base plate (ballastless track system). In the rail support the force vector (connected to the stiffness of rail supporting structure) acts downwards.

The elements of a supporting rail system (both continuous and discrete) are rail pads, resilient railfoot profiles, resilient elastomeric grout based on polyurethane reins and others, if included (i.e. ribbed base plate, elastic pad). The material features of a rail supporting system components act upon their functions: static and dynamic stiffness of support, protection against stray current and restraint against longitudinal rail displacements.

Rail fixing is the function of a rail fastening system to ensure:

– clamping force acting on a rail foot;

– restraint against longitudinal rail displacements (effect of clamping force and friction coefficient);
– torsional resistance against rail rollover (effect of lateral shear resistance of anchor elements);
– electrical insulation between the rail and the supporting structure (sleeper, block or concrete slab) providing protection against stray current.

Fig. 6. Illustration principle of classification of rail fastening systems due to criterion of fixing rail with rail supporting structure: a) direct rail fastening system (SB fastening system), b) direct rail fastening system (embedded blocks system with tension clamps type Skl 14), c) indirect fastening system (ribbed base plate) [1].

The presented classification variants of rail types fastening systems, apart from the classical division on direct and indirect, following systems can be specified due to design solution and interconnectedness of rail supporting and fixing:

– discrete rail fastening system - discrete both rail support and rail fixing, i.e. Embedded Blocks System (EBS);
– continuous rail fastening system - continuous both rail support and rail fixing, i.e. Embedded Rail System (ERS);
– mixed rail fastening system – continuous rail support and discrete rail fixing, i.e. direct rail fastening to concrete slab with continuous grout – fig. 7).

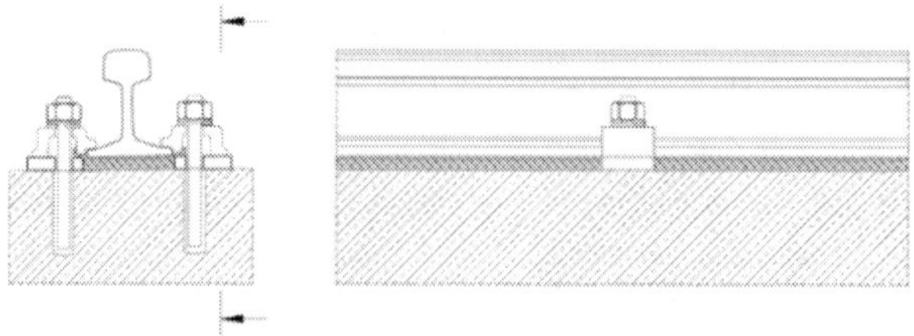

Fig. 7. Mixed rail fastening system (direct rail fastening to concrete slab) [1].

2.2 Resilient support and fixing of rail supporting structure

The Under Sleeper Pads (USP) are used in ballasted track systems (Fig.1) to reduce the dynamic effects from the rolling stock transferred through the rails, fastening system, rail supporting structure (sleepers or turnout bearers) to the ballast.

The Under Sleeper Pads are produced from elastomeric materials and showcase two variants due to the applied material:

– pads based on polyurethane in versions with closed or open pores;
– rubber pads (composites based on blends of natural rubber and/or synthetic rubber).

The Under Sleeper Pads are fixed to the bottom of the sleepers or turnout bearers and cover overall or partially (in under rail section) the bottom surface of the sleeper (Fig. 8).

Mounting of the Under Sleeper Pads takes place:

- during the production of the sleepers/turnout bearers by placing the pad on the bottom surface of the sleeper, in concrete mixture;
- on finished sleepers/turnout bearers by sticking the pad by means of using fast hardening adhesive (i.e. epoxy glue).

The testing procedures for Under Sleeper Pads are given in standard [3], but it does not specify the required parameter values.

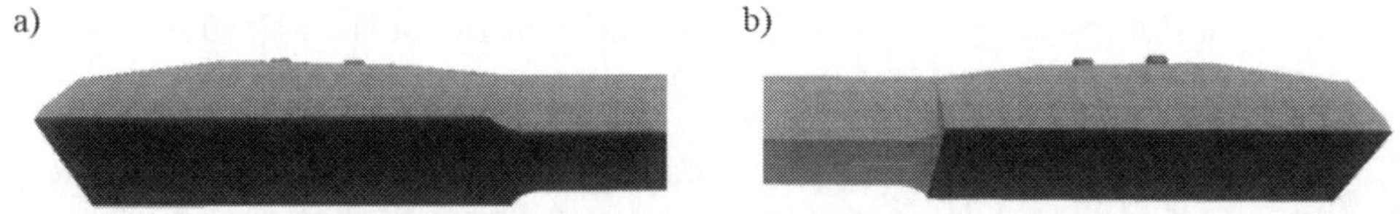

Fig. 8. Under Sleeper Pad: a) overall coverage the bottom surface of the sleeper; b) partial coverage the bottom surface of the sleeper (in the area of the main transmission of vertical pressure).

Details of the support block in EBS (Embedded Block System) and components of the rail fastening system are given in Fig. 9. The rail supporting structure in Embedded Block System is an example of floating slab track system, where a heavy support block is set at the resilient under-block pad. Significant vibration damping in EBS system is also attained due to elastic fixing of the support block in the socket by means of grouting, which serves as a support block coating.

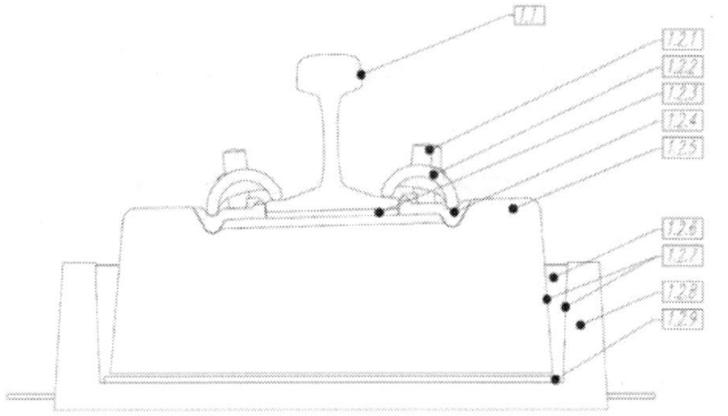

Fig. 9. Detail of support block in Embedded Block System (EBS) - Key: 1.1 – Vignole rail, 1.2.1 – sleeper screw; 1.2.2 – tension clamp type Skl-14; 1.2.3 – rail pad (resilient rail support), 1.2.4 – angled guide plate Wfp; 1.2.5 – reinforced concrete block; 1.2.6 – resilient compound; 1.2.7 – adhesive material; 1.2.8 – prefabricated concrete tray or concrete slab; 1.2.9 – resilient under block pad [1].

2.3 Resilient support and fixing of rail track structure

Resilient mats in railways are used for vibration isolation of railway tracks by means of reducing dynamic effects of rail traffic. The reduction relates mainly to vertical and transverse material vibration, but also to structure-borne noise [4, 5].

In a slab track system resilient mats are applied horizontally and vertically directly below and from the sides of the concrete track base plate (Fig. 11) to provide resilient support layer for railway track. The resulting floating slab track system is effective in mitigating vibrations, especially at low frequencies, and achieve a significant reduction in vibrations and structure-borne noise emissions at excitation frequencies above $\sqrt{2}$ times the tuning frequency [6, 7, 8, 9, 10].

Resilient mats can be grouped into two categories, because of the range of applications in various types of railways track construction:

– Under-Ballast Mats (UBM)/ Sub-Ballast Mats (SBM) – used in ballast railway track (Fig. 1); their two variants can be classified due to the main purpose of their use according to standard [11]:
  a. used primarily for isolation from vibration,
  b. used primarily for stress reduction in ballast [12, 13].
– Slab-Track Mats (STM) – used in ballastless railway track (applied under or from the sides of concrete track base plate – Fig. 11), with three variants of a slab support system according to standard [14]:
  c. discrete support (steel springs or elastomeric pads) – Fig. 10,
  d. linear support (strip mats) – Fig. 10,
  e. continuous support (elastomeric mats) – Fig. 11.

Due to the applied materials and production technologies, resilient mats can be grouped into mats made of various kinds of elastomers (elastomeric mats) or mineral wool mats. Elastomeric mats can be divided into two groups:
– mats based on polyurethane in versions with closed or open pores;
– rubber mats (composites based on blends of natural rubber and/or synthetic rubber), whose resiliency is ensured by shape (channels, grooves or protrusions of different shape) and cross-sectional structure of the mat (density and pore volume).

The given examples of floating slab track systems (Fig. 10–11) are individual, customized solutions, they differ in terms of basic dimensions of components, as well as kinds of applied resilient elements. Characteristics of resilient mats important for vibration isolation and their test methods were given in standard [11, 14].

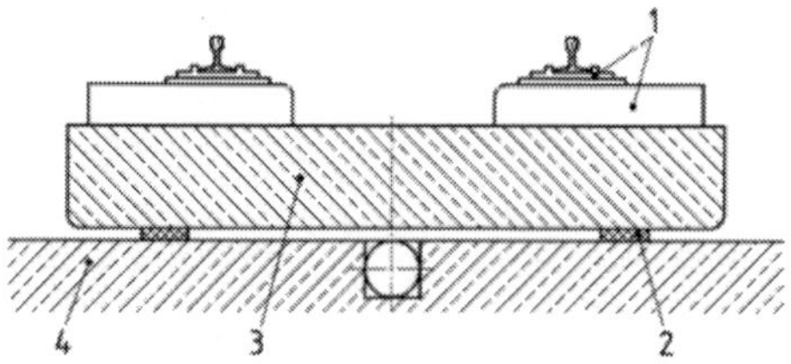

Fig. 10. Example of rail track structure in floating slab track system. Key: 1 – fastening system and support block, 2 – slab-track mat (discrete or linear support), 3 – concrete track base plate, 4 – tunnel invert slab with drainage pipe [15].

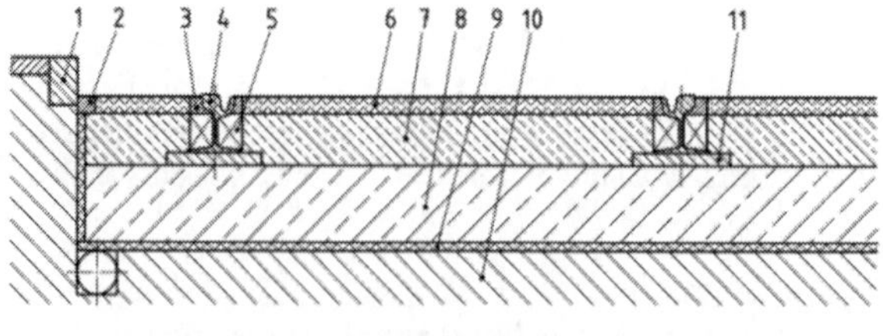

Fig. 11. Example of rail track structure in floating slab track system. Key: 1 – kerbstone, 2 – resilient joint, 3 – grouting, 4 – rail, 5 – rail web filler block, 6 – asphalt, 7 – concrete filling, 8 – reinforced concrete track base plate, 9 – resilient mat as continuous support, 10 – consolidated formation with drainage, 11 – resilient continuous strip [15].

3 Conclusions

Vibroacoustic isolators are components used to limit the negative effects (vibration and noise) generated by rail traffic on the track and are applied possibly close to the location of vibration generation during rail traffic (contact area of the wheel with the rail).

Along with further advances in engineering and material engineering, as well as, increasing requirements for protecting surroundings of transport routes from negative transport effects, it can be concluded that vibroacoustic isolators will find an increasingly widespread use in rail track structure.

References

1. C. Kraśkiewicz, Współczesne budownictwo w badaniach młodych naukowców, Politechnika Gdańska, Wydział Inżynierii Lądowej i Środowiska, 126–140 (2017)
2. *PN-EN 13481-5+A1:2017-04 Railway applications - Track - Performance requirements for fastening systems Part 5: Fastening systems for slab track with rail on the surface or rail embedded in a channel*
3. *PN-EN 16730:2016-08 Railway applications - Track - Concrete sleepers and bearers with under sleeper pads*
4. M. Butorina, N. Minina, P. Ivanov, A. Petryaev, Procedia Engineering **189**, 352–359 (2017)
5. P.Carels, Joint Baltic-Nordic Acoustic Meeting 2012, 1-16 (2012)
6. A. Zbiciak, C. Kraśkiewicz, C. Lipko, W. Oleksiewicz, IPICSE-2016, MATEC Web of Conferences **86**, 01015 (2016)
7. J. Sołkowski, Materiały Konferencyjne, Stowarzyszenie Inżynierów i Techników Komunikacji Rzeczpospolitej Polskiej. Oddział w Krakowie 2(106), 159–175 (2015)
8. A. Zbiciak, C. Kraśkiewicz, W. Oleksiewicz, M. Płudowska-Zagrajek, C. Lipko, XXVI R-S-P Seminar 2017, Theoretical Foundation of Civil Engineering, MATEC Web of Conferences **117**, 00090 (2017)
9. R. G. Wettschureck, M. Heim, S. Mühlbachler, Proceedings Inter-noise **97**, 577–580 (1997)
10. R. G. Wettschureck, F. Breuer, M. Tecklenburg, H. Widmann, Rail Engineering International 4, 12-16 (1999)
11. *DIN 45673-5:2010-08 Mechanical vibration. Resilient elements used in railway tracks. Part 5: Laboratory test procedures for under-ballast mats*
12. S. K. Navaratnarajah, B. Indraratna, Journal of Geotechnical and Geoenvironmental Engineering, **143**, 6 (2017)
13. S. K. Navaratnarajah, B. Indraratna, S. Nimbalkar, Procedia Engineering **143**, 1108–1119 (2016)
14. *DIN 45673-7:2010-08 Mechanical vibration - Resilient elements used in railway tracks - Part 7: Laboratory test procedures for resilient elements of floating slab track systems*
15. *DIN 45673-1:2010-08 Mechanical vibration. Resilient elements used in railway tracks. Part 1: Terms and definitions, classification, test procedures*

The influence of selected static and dynamic parameters of resilient mats on vibration reduction of railway tracks

Cezary Kraśkiewicz[1,*], *Artur* Zbiciak[1], *Wojciech* Oleksiewicz[1], and *Andrzej* Piotrowski[2]

[1]Warsaw University of Technology, Faculty of Civil Engineering, Institute of Roads and Bridges, Al. Lecha Kaczyńskiego 16, 00-637 Warsaw, Poland
[2]Warsaw University of Technology, Faculty of Civil Engineering, Institute of Building Engineering, Al. Lecha Kaczyńskiego 16, 00-637 Warsaw, Poland

Abstract. The paper presents laboratory methods of testing the prototype resilient mats, used in rail track systems in order to reduce vibrations transmitted to the vicinity of the railroad and to improve the resistance of the railroad structure. The laboratory testing program was carried out for elastomeric materials (polyurethane based). The key parameters being investigated in the Warsaw University of Technology laboratory unit were used in order to determine the insertion loss vibration level by applying analytical method. The paper presents the influence of selected parameters i.e. static and dynamic moduli on the reduction of vibration and structure-borne sound level – i.e. the value of insertion loss parameters.

1 Introduction

Resilient mats can be grouped into two categories, because of the range of applications in various types of railways track construction:

a) Under-Ballast Mats (UBM)/Sub-Ballast Mats (SBM) – used in ballast railway track; among which two varieties can be classified due to the main purpose of their use according to standard [1]:

- used primarily for isolation from vibration [2, 3],
- used primarily for stress reduction in ballast [4, 5].

b) Slab-Track Mats (STM) – used in ballastless railway track (applied under or from the sides of concrete track base plate, including three varieties of slab support system according to standard [6]:

- discrete support (steel springs or elastomeric pads),
- linear support (strip mats),
- continuous support (elastomeric mats).

* Corresponding author: c.kraskiewicz@il.pw.edu.pl

2 Parameters characterizing static and dynamic properties of resilient mats

As there are no European standards to define the requirements and methodology of research for resilient mats in railway tracks, it is justifiable to use German Standards: DIN 45673-1 [7], DIN 45673-5 [1] and DIN 45673-7 [6]. All the substantial characteristics of resilient mats and procedures for their research are presented in study [8].

Static and dynamic bedding moduli determine effectiveness of dumping vibrations transmission to the environment. It should be noted that the mats possessing very low value of static bedding modulus give in greater vertical deflection of rail and railway track. The value of the static and dynamic bedding modulus is defined for resilient mats for linear and continuous support. It includes a wide range of values from ~ 0.003 to ~ 0.33 N/mm^3, depending on the material type and structure, thickness of the mat, the load value in which bedding modulus is defined and the load frequency for dynamic modulus.

2.1 Laboratory tests

In the research laboratory unit at the Warsaw University of Technology (Fig. 1), UBM mats were tested in accordance with the requirements of the DIN 45673-5 [1] Standard and STM mats in accordance with DIN 45673-7 [6]. The polyurethane-based mats with a thickness of 15 mm and of diversified densities were tested: G180 (180 kg/m^3), G160 (160 kg/m^3) and G150 (150 kg/m^3). Dimensions of test samples: 300 mm × 300 mm. Test temperature: (23 ± 3)°C. The method of load application: flat loading plate – UBM/STM – flat loading plate (Fig. 1). The loading plate is made of steel with a sanding disc (K120 grit on a rigid linen backing cloth) inserted between the loading plate and the test object. The sample loading was applied by the INSTRON 8802 machine and the deflections were measured by four displacement sensors.

Fig. 1. View of tested sample STM G150.

2.2 Static bedding modulus

Static bedding modulus characterizes conditions of rail deflection under the pressure of non-moving rolling stock. It also affects the vertical deflection of railway track in floating slab track system. The relationship between static bedding modulus and the applied force is non-linear. In such cases, the static bedding modulus is determined within different ranges of loads, depending on the application purpose of resilient mat. Based on displacements s_1 and s_2 the static bedding modulus can be evaluated in N/mm^3 as a secant modulus with respect to the range of stresses to be assessed σ_1 and σ_2 as expressed via Eq. (1).

$$C_{stat} = \frac{\sigma_2 - \sigma_1}{s_2 - s_1} \quad [\text{N/mm}^3] \tag{1}$$

Static bedding modulus diagrams are shown in Fig. 2 and its numerical values within different evaluation ranges of loads in Tab. 1.

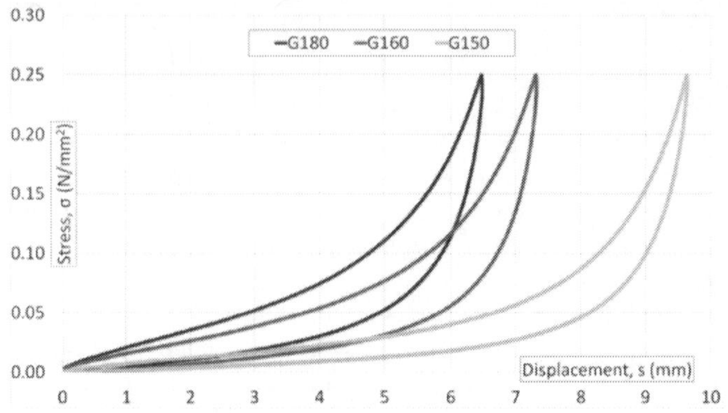

Fig. 2. Static bedding modulus of STM G180/160/150 according to DIN 45673-5 [1].

Table 1. Static bedding modulus of resilient mats.

Evaluation range of loads $[\mathrm{N/mm^2}]$	**Static bedding modulus** C_{stat} $[\mathrm{N/mm^3}]$		
	G180	G160	G150
0.02 – 0.05	0.016	0.013	0.010
0.02 – 0.07	0.017	0.015	0.013
0.02 – 0.10	0.021	0.019	0.017
0.02 – 0.20	0.035	0.032	0.031

2.3 Dynamic bedding modulus and dynamic stiffening ratio

Dynamic bedding modulus characterizes mat work under the pressure of moving rolling stock, therefore, it determines the limitation of vibration transmission. The value of dynamic bedding modulus is related not only to loading force, as in static bedding modulus, but also to frequencies. Therefore, it should be determined under the standard pressure and frequency ranges.

The DIN 45673-5 Standard [1] divided a test procedure into two groups. The first one is applied for dynamic bedding modulus within lower frequencies $C_{dyn1}(f)$ (5, 10, 20 and 30 Hz), the second one within higher frequencies $C_{dyn2}(f)$ (10 - 160 Hz). The parameter $C_{dyn1}(f)$ can be used to determine the lower-frequency bending deformation of the rail under the influence of the rolling wheel, as a result of the interplay of the bending elasticity of railway track structure. The parameter $C_{dyn2}(f)$ refers to under-ballast mats and can be used to determine the level of reduction of structure-borne noise. It influences the natural frequency f_0 of the elastically supported track as an oscillatory system and thus the insertion loss.

Dynamic stiffening ratio $r(f)$ is a proportion between dynamic and static bedding modulus – Eq. (2). It defines the possibility of achieving effective vibration damping for resilient mats. Because of the known property of elastomeric materials being rate-dependent, both bedding modulus as well as stiffness modulus are determined for various frequencies of excitation. The closer to 1 the value of dynamic stiffening ratio is (for $r(f) > 1$), the better vibration isolation properties has the test piece (e.g. resilient mat).

$$r(f) = \frac{C_{dyn(f)}}{C_{stat}} \tag{2}$$

Dynamic bedding modulus (5 Hz) diagrams for UBM are shown in Fig. 3 and numerical values of dynamic bedding modulus/dynamic stiffening ratio for evaluation range of loads 0.02 – 0.07 N/mm^2 according to DIN 45673-5 [1] in Tab. 2.

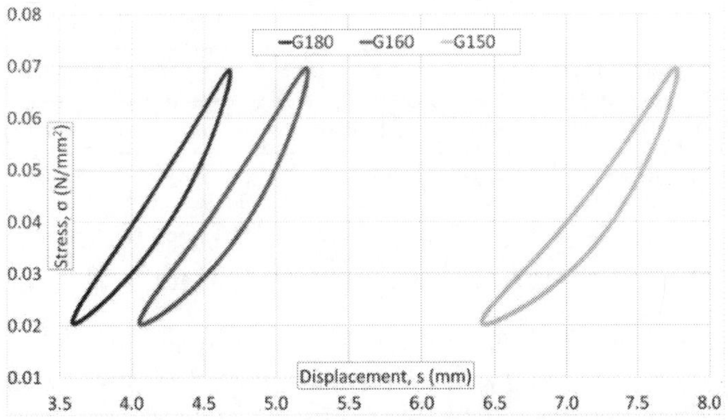

Fig. 3. Dynamic bedding modulus (5 Hz) of UBM G180/160/150 according to DIN 45673-5 [1].

Table 2. Dynamic bedding modulus/dynamic stiffening ratio of resilient mats UBM for evaluation range of loads 0.02 – 0.07 N/mm^2 according to DIN 45673-5 [1].

Parameter	**G180**	**G160**	**G150**
dynamic bedding modulus $C_{dyn1}(5\,\text{Hz})$ $[\text{N/mm}^3]$	0.046	0.043	0.037
dynamic stiffening ratio $r(5\,\text{Hz})$ [1]	2.68	2.85	2.84
dynamic bedding modulus $C_{dyn1}(10\,\text{Hz})$ $[\text{N/mm}^3]$	0.059	0.049	0.041
dynamic stiffening ratio $r(10\,\text{Hz})$ [1]	2.92	3.28	3.15

Dynamic bedding modulus (20 Hz) diagrams for STM are shown in Fig. 4 and numerical values of dynamic bedding modulus/dynamic stiffening ratio mats STM for harmonic excitation with a particle velocity amplitude of 7 mm/s (corresponding to a particle velocity level L_V=100 dB relative to the standard reference particle velocity of 5×10^{-8} m/s) and preloading $s_V = 0.04$ N/mm^2 according to DIN 45673-7 [6] in Tab. 3.

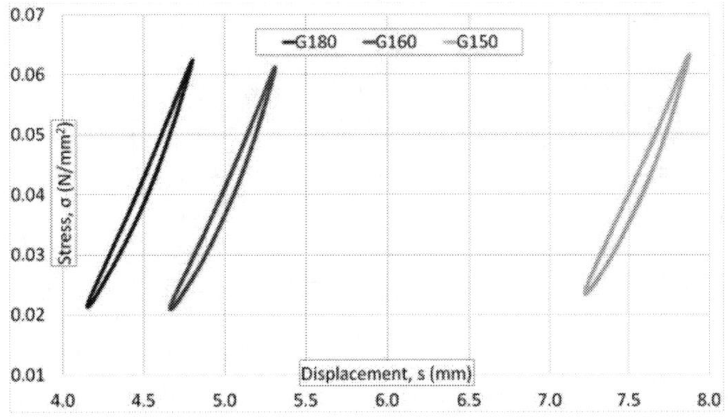

Fig. 4. Dynamic bedding modulus (20 Hz) of STM G180/160/150 according to DIN 45673-7 [6].

Table 3. Dynamic bedding modulus /dynamic stiffening ratio of resilient mats STM for harmonic excitation with a particle velocity amplitude of 7 mm/s and preloading $s_v = 0.04$ N/mm^2 according to DIN 45673-7 [6].

Parameter	G180	G160	G150
dynamic bedding modulus $C_{dyn1}(5\,\text{Hz})$ $[\text{N/mm}^3]$	0.028	0.028	0.030
dynamic stiffening ratio $r(5\,\text{Hz})$ [1]	1.63	1.86	2.34
dynamic bedding modulus $C_{dyn1}(10\,\text{Hz})$ $[\text{N/mm}^3]$	0.039	0.039	0.039
dynamic stiffening ratio $r(10\,\text{Hz})$ [1]	2.29	2.58	2.97
dynamic bedding modulus $C_{dyn1}(20\,\text{Hz})$ $[\text{N/mm}^3]$	0.058	0.057	0.056
dynamic stiffening ratio $r(20\,\text{Hz})$ [1]	3.40	3.78	4.31

4 Mechanical model of ballast railway track

Discrete mechanical model of the ballast railway track is shown in Fig. 5. Such a model was partially analysed in [9] and is also presented in Standard [10] for calculation of insertion loss. The state-space formulation of differential equations representing this model was given in [11] taking into account the algorithm of eigenvalue problem solution described in [12]. In this section we demonstrate dynamic behaviour of the system retrofitted with viscoelastic fractional model of Under-Ballast Mats (UBM) (Fig. 5b). Selected features of the system with UBM will be visualized comparing the reference system without UBM (see Fig. 5a).

The reference system is composed of two layers. The parameters $k_o = 220$ MN/m and $d_o = 0.265$ MNs/m represent stiffness and damping of upper layer of the system (rail, fastening, sleeper and ballast) while $k_u = 1500$ MN/m and $d_u = 1.2$ MNs/m define lower layer (track-bed). Effective mass of both reference as well as retrofitted (isolated) systems are $m_{eff} = 2600$ kg.

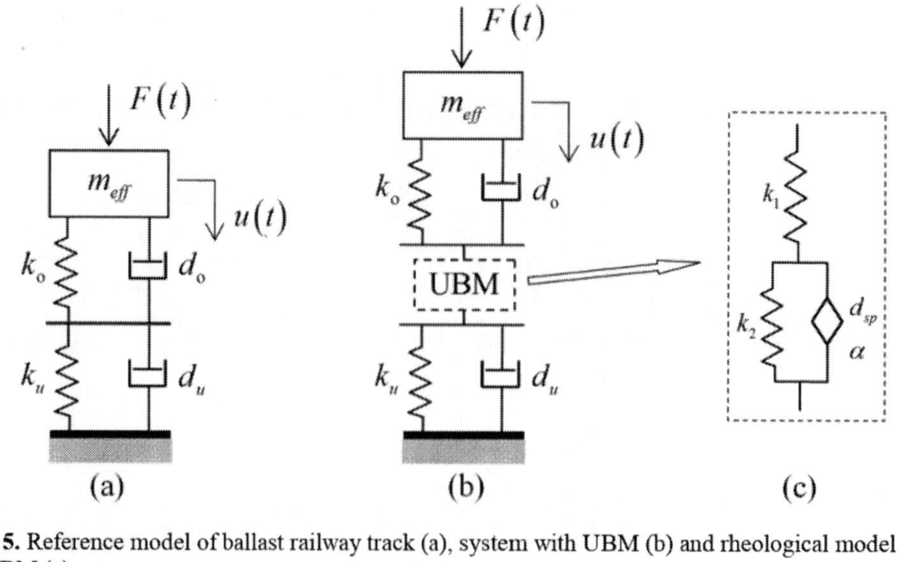

Fig. 5. Reference model of ballast railway track (a), system with UBM (b) and rheological model of UBM (c).

In the case of an isolated system, an additional layer is considered representing Under-Ballast Mat. Rheological viscoelastic model of UBM is shown in Fig. 5c. The model is composed of two springs k_1 and k_2 and of a fractional element (spring-pot) defined by two parameters d_{sp} and α. Constitutive relationships of the spring-pot element, relating the force $f(t)$ and the displacement $u_{sp}(t)$, can be expressed in a differential form (see [13])

$$f(t) := d_{sp}\, \mathrm{D}^{\alpha} u_{sp}(t); \quad \alpha \in (0,1) \tag{3}$$

where $\mathrm{D}^{\alpha} \equiv \frac{\mathrm{d}^{\alpha}}{\mathrm{d}t^{\alpha}}$ denotes fractional derivative operator:

$$\mathrm{D}^{\alpha} u_{sp}(t) := \frac{u_{sp}(0)}{\Gamma(1-\alpha)\, t^{\alpha}} + \frac{1}{\Gamma(1-\alpha)} \int_0^t \frac{\dot{u}_{sp}(\tau)}{(t-\tau)^{\alpha}} \mathrm{d}\tau \tag{4a}$$

$$\Gamma(1-\alpha) := \int_0^{\infty} t^{-\alpha} e^{-t} \mathrm{d}t \tag{4b}$$

The complex stiffness of the UBM model k^*_{UBM} and of entire isolated system k^*_{vib} can be defined as follows [14, 15]

$$k^*_{UBM}(i\omega) = \left(\frac{1}{k_1} + \frac{1}{k_2 + d_{sp}(i\omega)^{\alpha}} \right)^{-1} \tag{5a}$$

$$k^*_{vib}(i\omega) = \left[\frac{1}{k^*_{ref}(i\omega)} + \frac{1}{k^*_{UBM}(i\omega)} \right]^{-1} \tag{5b}$$

Another formulas defining complex transfer functions, insertion loss and complex compliance moduli were presented in [9].

Based on laboratory results presented in previous section, rheological parameters of three types of UBM were evaluated. The results of a curve fitting procedure are presented in Table 4 and depicted in Fig. 6. The curve fitting procedure is based on evaluation of dynamic stiffness moduli being the absolute value of the complex stiffness k^*_{UBM}.

Table 4. Curve fitting results of rheological models of UBMs.

UBM	k_1 [MN/m]	k_2 [MN/m]	d_{sp} [MNs$^{\alpha}$/m]	α [-]
G180	6.079	2.016	0.234	0.891
G160	4.764	1.850	0.259	0.898
G150	3.866	1.674	0.275	0.899

It should be emphasized that the values of bedding moduli evaluated based on the curve fitting procedure (Fig. 6) were divided by the area of the sample (300×300 mm) in order to obtain stiffness moduli shown in Table 4 (see [8]).

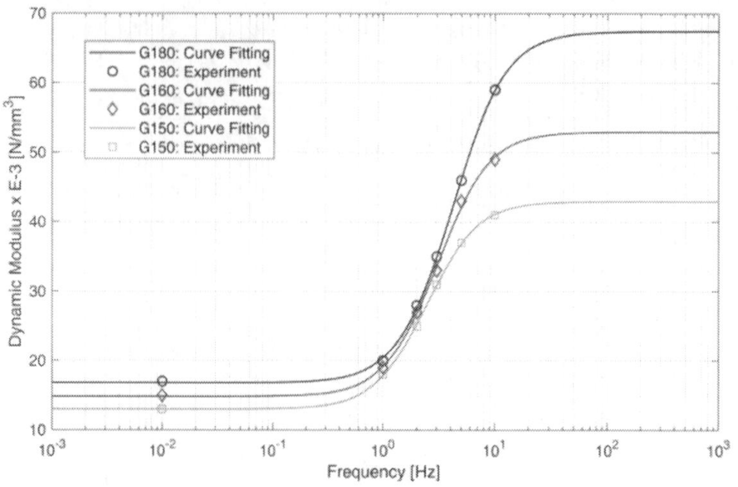

Fig. 6. Curve fitting results for three analysed UBM models.

The calculation results for both reference system and vibro-isolated system with UBM G180 are visualized in Fig. 7. Horizontal axis used in Fig. 7 represents centre frequencies of 1/3 octave bands. The plots depicted in Fig. 7 indicate a region of vibration-isolation effectiveness for approx. $\text{freq} > 9.5\ \text{Hz}$ where $IL > 0\ \text{dB}$ (see vertical dashed line). It was proved that in case of G160, the verge of vibration-isolation effectiveness equals $8.8\ \text{Hz}$ while for G150 it starts from $8.1\ \text{Hz}$.

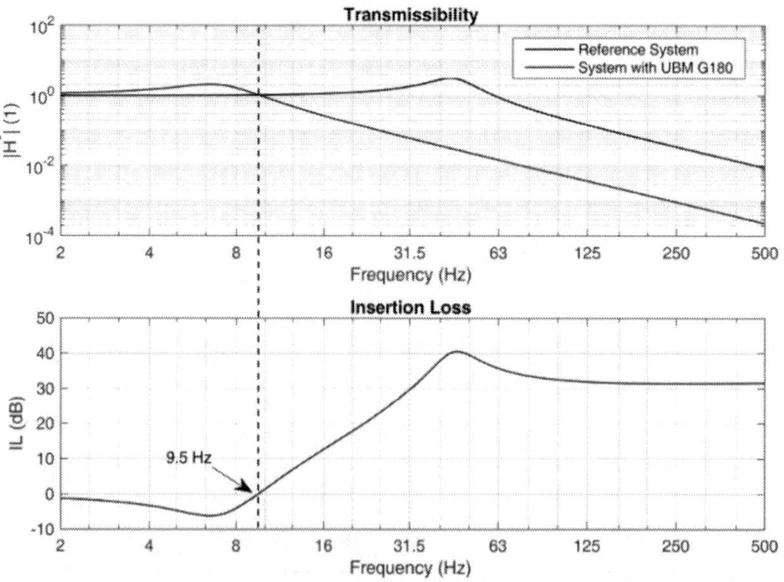

Fig. 7. Transmissibility and Insertion Loss: reference system and system with Under-Ballast Mat (UBM G180).

Such diagrams are presented in numerous papers [16, 17, 18] for various mechanical models representing railway track systems with vibroacoustic isolators.

In further studies, the results of vibration reduction forecasts will be compared with the results of measurements performed on test sites in accordance with the requirements of the standards [19, 20].

4 Conclusions

The key parameters (static bedding modulus, dynamic bedding modulus and dynamic stiffening ratio) being investigated at the Warsaw University of Technology laboratory unit were used in order to determine the insertion loss vibration level by applying analytical method. The paper confirmed the influence of selected parameters i.e. static and dynamic bedding moduli on the reduction of vibration and structure-borne sound level – i.e. the value of insertion loss parameters.

References

1. DIN 45673-5:2010-08 Mechanical vibration. Resilient elements used in railway tracks. Part 5: Laboratory test procedures for under-ballast mats
2. M. Butorina, N. Minina, P. Ivanov, A. Petryaev, Procedia Engineering **189**, 352–359 (2017)
3. P.Carels, Joint Baltic-Nordic Acoustic Meeting 2012, 1-16 (2012)
4. S. K. Navaratnarajah, B. Indraratna, Journal of Geotechnical and Geoenvironmental Engineering, **143**, 6 (2017)
5. S. K. Navaratnarajah, B. Indraratna, S. Nimbalkar, Procedia Engineering **143**, 1108–1119 (2016)
6. DIN 45673-7:2010-08 Mechanical vibration - Resilient elements used in railway tracks - Part 7: Laboratory test procedures for resilient elements of floating slab track systems
7. DIN 45673-1:2010-08 Mechanical vibration. Resilient elements used in railway tracks. Part 1: Terms and definitions, classification, test procedures
8. C. Kraśkiewicz, C. Lipko, M. Płudowska, W. Oleksiewicz, A. Zbiciak, Procedia Engineering **153**, 317–324 (2016)
9. A. Zbiciak, C. Kraśkiewicz, C. Lipko, W. Oleksiewicz, IPICSE-2016, MATEC Web of Conferences **86**, 01015 (2016)
10. DIN 45673-4:2008-07 Mechanical vibration. Resilient elements used in railway tracks. Part 4: Analytical evaluation of insertion loss of mounted track systems
11. A. Zbiciak, C. Kraśkiewicz, W. Oleksiewicz, M. Płudowska-Zagrajek, C. Lipko, *XXVI R-S-P Seminar 2017, TFoCE*, MATEC Web of Conferences **117**, 00090 (2017)
12. V. G. S. Simionatto, H. H. Miyasato, F. Melo, M. Junior, *Proceedings 21st Int. Congr. Mech. Engin.* (2011)
13. W. Grzesikiewicz, A. Wakulicz, A. Zbiciak, IJMS **70**, 90–98 (2013)
14. R. W. Clough, J. Penzien, *Dynamics of Structures* (McGraw-Hill, 1975)
15. J. Sołkowski, Proceeding, Research and Technical Papers of Polish Association for Transportation Engineers in Cracow **2(106)**, 159–175 (2015)
16. R. G. Wettschureck, U. J. Kurze, ACUSTICA **58**, 177–182 (1985)
17. R. G. Wettschureck, DAGA **87**, 217–220 (1987)
18. R. G. Wettschureck, M. Heim, S. Mühlbachler, Proceedings Inter-noise **97**, 577–580 (1997)
19. DIN 45672-1:2018-02 Vibration measurement associated with railway traffic systems - Part 1: Measuring method for vibration
20. DIN 45672-2:1995-07 Vibration measurement associated with railway traffic systems - Part 2: Evaluation method

Determination of p-y curves for offshore piles based on in-situ soil investigations

Kamila Mię̨dlarz[1,*], and *Lech* Bałachowski[1]

[1]Gdańsk University of Technology, Faculty of Civil and Environmental Engineering, ul. Narutowicza 11/12, 80-233 Gdańsk, Poland

Abstract. Offshore piles are subjected to complex loads with considerable lateral component. The pile-soil response to lateral loads can be described with the p-y method. For a given depth the load–deflection relationship is built to simulate the surrounding soil stiffness. This state-of-art paper presents a brief discussion of determination methods for the p-y curves using a standard approach based on the soil parameters derived from laboratory and in-situ tests or directly from field tests. The basic relationships for both cohesive and cohesionless soils are discussed. The advantage of direct design methods to describe the p-y curve relies in the reduction of necessary laboratory tests.

1 Introduction

Nowadays, most of offshore wind farms in Europe are located on waters with depths up to 20 m, and the most popular type of foundation structures are monopiles. These structures are appropriate for water depths up to 35 m. They are made of steel, cylindrical pipes with diameters up to 9 m and wall thickness up to 150 mm [1, 2].

This article focuses on free-headed piles subjected to horizontal loads only (no bending moments). The piles under lateral loads can be divided into three categories:

- flexible,
- rigid,
- of intermediate stiffness.

The lateral resistance for piles under horizontal loading can be calculated by constructing non-linear load-deflection (p-y) curves. This method is most common and recommended in e.g. the Offshore Standards DNV-OS-J101 [3] or in the API Recommended Practice 2A-WSD [4]. The API p-y formulation is proper for long flexible piles with diameters of up to 1m and the L/D ratio around 30. A pile is considered to be long flexible if it meets the condition [5, 6]:

$$L > 3 \cdot l_0 \ with \ l_0 = \left(\frac{4 \cdot E_p \cdot I}{K} \right)^{1/4} \tag{1}$$

* Corresponding author: kamila.miedlarz@pg.edu.pl

where:
l_0 transfer length,
E_p modulus of the pile material,
I pile moment of inertia,
K soil stiffness.

The monopiles are relatively short and rigid of 10 times greater diameters and length/diameter (L/D) ratio of values lower than 30. They can be considered short rigid piles if their length satisfies the relationship [5]:

$$L < l_0 \tag{2}$$

The cases between the relations from eq. (1) and (2) are intermediate and can be considered either flexible or rigid. The lateral response of rigid piles is still often modelled with the use of p-y curves, as for flexible piles, which may not accurately describe their behavior. For this reason, many studies and field testing have been carried out to develop a method more suitable for monopiles [6]. The p-y curves are used to model the soil stiffness as non-linear springs applied between beam-column elements - representing pile in foundation analysis. A division of methods used to determine p-y curves is presented in this article.

In the design of offshore support structures it is vital to perform geotechnical tests: both in-situ (CPT/less often DMT test) and laboratory tests (on undisturbed or evenly disturbed samples) [7].

The aim of the work is to collect and discuss the available methods applied to create p-y curves during the offshore pile modelling. The article is the introduction to further research, applying the presented calculation methods and formulas as an input to create an offshore pile model. The presented review of direct and indirect methods for piles subjected to lateral loads considers mainly flexible piles. The next step of the research intends to modify the given approach and adjust it to the behaviour of rigid monopile.

The determination of soil stiffness using CPT and DMT is a starting point for the study of monopile foundation. However, the behaviour mechanism of such a foundation is different from the flexible pile case. Lateral deformations of rigid foundations are more uniform, interaction occurs between the base and shaft mechanism, as the monopiles resembles a block foundation. Additional model tests and numerical analysis are necessary to better analyse this behaviour.

2 Lateral stiffness of piles up to 2,5 m of diameter

2.1 Indirect approach

The non-linear p-y curves presented in API RP 2A-WSD [4] and DNV CN 30.4 [7] are used to design laterally loaded pile structures with diameters up to 2,5 m. They can be also applied in case of not ideally flexible piles.

Construction of p-y curves depends on the soil type and differs for cohesive and cohesionless soils. In the case of cohesionless soils it is necessary to estimate the effective friction angle..

In the case of cohesive soils the ultimate lateral resistance is based on undrained shear strength (gained from laboratory testing) to be calculated as [4]:

$$p_u = \begin{cases} (3 \cdot c_u + \gamma' \cdot X) \cdot D + J \cdot c_u \cdot X & for \quad 0 < X < X_R \\ 9 \cdot c_u \cdot D & for \quad X > X_R \end{cases} \tag{3}$$

where:

- X depth below soil surface,
- X_R transition depth, below which the value $((3c_u+\gamma' X)D+Jc_uX)$ exceeds $9c_u$ [m],

$$X_R = \frac{6 \cdot D}{\gamma' \cdot \frac{D}{c_u} + J} \tag{4}$$

where:

- D pile diameter,
- c_u undrained shear strength for undisturbed soil samples,
- γ' effective unit weight of soil,
- J empirical coefficient with values from 0,25 to 0,50 (the upper limits value is assumed for soft, normally consolidated cohesive soils).

In static loading:

$$p_u = \begin{cases} \frac{p_u}{2} \cdot \left(\frac{y}{y_c}\right)^{\frac{1}{3}} & for\ y \leq 8 \cdot y_c \\ p_u & for\ y > 8 \cdot y_c \end{cases} \tag{5}$$

In cyclic loading and $X>X_R$:

$$p_u = \begin{cases} \frac{p_u}{2} \cdot \left(\frac{y}{y_c}\right)^{\frac{1}{3}} & for\ y \leq 3 \cdot y_c \\ 0{,}72 \cdot p_u & for\ y > 3 \cdot y_c \end{cases} \tag{6}$$

In cyclic loading and $X \leq X_R$:

$$p = \begin{cases} \frac{p_u}{2} \cdot \left(\frac{y}{y_c}\right)^{\frac{1}{3}} & for\ y \leq 3 \cdot y_c \\ 0{,}72 \cdot p_u \cdot \left(1 - \left(1 - \frac{X}{X_R}\right)\right) \cdot \frac{y - 3 \cdot y_c}{12 \cdot y_c} & for\ 3 \cdot y_c < y \leq 15 \cdot y_c \\ 0{,}72 \cdot p_u \cdot \frac{X}{X_R} & for\ y > 15 \cdot y_c \end{cases} \tag{7}$$

with $y_c = 2{,}5 \cdot \varepsilon_c \cdot D$

where:

- ε_c strain which occurs at one-half the maximum stress in laboratory undrained compression tests.

In the case of cohesionless soils, p-y curves are also non-linear but can be approximated [4]:

$$p = A_i \cdot p_u \cdot \tanh\left(\frac{k_{init} \cdot X}{A_i \cdot p_u} \cdot y\right) \tag{8}$$

where:

A_i factor to account for static or cyclic loading: $A_i=(3-0,8(\cdot z/D))>0,9$ for static and $A_i=0,9$ for cyclic loading,

p_u static ultimate resistance at depth z,

Y lateral deflection,

k_{init} initial modulus of subgrade reaction dependent on the friction angle (see [3]).

The value of ultimate lateral resistance is soil depth dependent. It has been found that in the case of shallow waters it can be calculated from eq. (9) in the case of deeper waters from eq. (10). At a particular depth, the equation leading to lower value should be considered decisive [4].

$$p_{us} = (C_1 \cdot X + C_2 \cdot D) \cdot \gamma' \cdot X \tag{9}$$

$$p_{ud} = C_3 \cdot D \cdot \gamma' \cdot X \tag{10}$$

where:

C_1, C_2, C_3 coefficients dependent on the friction angle ϕ' (see [3]).

The friction angle can be obtained from laboratory tests or in-situ data. If the laboratory data is not available, the values of effective friction angle in sands can be evaluated from CPTU using eq. (11) [8] or eq. (12) [9]:

$$\phi'_{CPT} = \arctan\left[0{,}1 + 0{,}38 \cdot \log\left(\frac{q_c}{\sigma'_{v0}}\right)\right] \tag{11}$$

$$\phi'_{CPT} = 17{,}6° + 11{,}0 \cdot \log(q_{c1}) \tag{12}$$

where:

$$q_{c1} = \left[\frac{q_c}{Pa}\right] \cdot \left[\frac{Pa}{\sigma'_{v0}}\right]^{0,5} \tag{13}$$

with $Pa = 100\ kPa$, σ'_{v0} - vertical effective stress [kPa].

Safe estimation of effective angle of internal friction in sands can be also determined from DMT [10]:

$$\phi'_{DMT} = 28° + 14{,}6 \cdot \log K_D - 2{,}1 \cdot \log^2 K_D \tag{14}$$

In the case of undrained shear strength, the following equations can be used for clays:

$$c_{u,CPT} = \frac{q_t - \sigma_{v0}}{N_{kt}} \tag{15}$$

$$c_{u,DMT} = 0{,}22 \cdot \sigma_{v0} \cdot (0{,}5 \cdot K_D)^{1,25} \tag{16}$$

where:

N_{kt} cone factor for clays (default to 15); $N_{kt}=10{,}5+7\cdot\log(F_r)$

F_r normalized friction ratio,

$$F_r = \frac{f_s}{q_t - \sigma'_{v0}} \cdot 100\% \tag{17}$$

K_D	horizontal stress index; $K_D=(p_0-u_0)/\sigma_{v0}$,
p_0	pressure applied to the soil at the start of the expansion,
u_0	hydrostatic pore water pressure.

2.2 Direct approach

This approach uses the results of advanced in-situ soil investigations to predict the p-y curve and pile response to lateral loading. It reduces the errors due to inaccurate estimation of undrained shear strength and friction angle with the use of eq. (11),(12),(15), it also eliminates the need for laboratory testing on undisturbed soil samples [11]. Based on API and DNV methods Lehane and Truong [12] determined the construction of p-y curves directly from CPT data. They suggested the following equations for soft clay [12]:

$$\frac{p}{p_u} = \tanh \cdot \left[\left(026 \cdot I_r + 3{,}98\right) \left(\frac{y}{D}\right)^{0{,}85} \left(\frac{X}{D}\right)^{-0{,}5} \right] for\, 0 < \frac{X}{D} < 3 \tag{18}$$

$$\frac{p}{p_u} = \tanh \cdot \left[\left(0{,}151 \cdot I_r + 2{,}3\right) \left(\frac{y}{D}\right)^{0{,}85} \right] for\, \frac{X}{D} > 3 \tag{19}$$

where:

$$I_r = \frac{1}{3}\left(\frac{E}{q_n}\right) \cdot N_{kt} \tag{20}$$

and

I_r	rigidity index G/c_u,
G	shear modulus (from eq. (22))
q_n	net cone resistance; $q_n=q_t-\sigma_{v0}$,
q_n	cone tip resistance; $q_t=q_c+(1-a)\cdot u_2$,
a	cone coefficient,
E	Young's modulus (from eq. (21)).

$$E = (q_t - \sigma_{v0}) \cdot 0{,}015 \cdot \left[10^{0{,}55 \cdot I_c + 1{,}68}\right] \tag{21}$$

$$G = \frac{\gamma}{9{,}8} \cdot (51{,}6 \ln f_s + 18{,}5)^2 \tag{22}$$

Regarding sands, Lehane and Suryasentana (2014) [13] proposed:

$$\frac{p}{y \cdot X \cdot D} = 2{,}4 \cdot \left(\frac{q_c}{y \cdot X}\right)^{0{,}67} \cdot \left(\frac{x}{D}\right)^{0{,}75} \cdot \left(1 - e^{-0{,}62 \cdot \left(\frac{X}{D}\right)^{-1{,}2} \cdot \left(\frac{y}{D}\right)^{0{,}89}}\right) \tag{23}$$

Another method (Robertson, 1989) [14] to obtain the p-y curves for laterally loaded piles is based on DMT data. For cohesive soils, the equation of p-y curves is the same as in the API Recommendations (eq. (3),(4)), however the equation for the pile deflection changes [15,16] and y_c can be evaluated from:

$$y_c = \frac{23{,}67 \cdot c_u \cdot D^{0,5}}{F_c \cdot E_D} \tag{24}$$

where:

F_c empirical factor $F_c=10$,
E_D dilatometer modulus.

$$E_D = 34{,}7 \cdot (p_1 - p_0) \tag{25}$$

and p_0 and p_1 – corrected readings from DMT.

For cohesionless soils, also the eq. (3) is used, however the ultimate lateral resistance p_u is the lower value from eq. (26) (Reese, 1974) [17] and eq. (27) [18]:

$$p_u = \sigma'_{v0} \cdot \left[D \cdot (K_p - K_a) + x \cdot K_p \cdot \tan\phi' \cdot \tan\beta \right] \tag{26}$$

$$p_u = \sigma'_{v0} \cdot D \cdot \left[K_p^3 + 2 \cdot K_a \cdot K_p^2 \cdot \tan\phi' - \tan\phi' - K_a \right] \tag{27}$$

where:

$$\beta = 45° + \frac{\phi'}{2} \tag{28}$$

$$y_c = \frac{4{,}17 \cdot \sin\phi' \cdot \sigma_{v0}'}{E_D \cdot F_\phi \cdot (1 - \sin\phi')} \cdot D \tag{29}$$

and

$F_\phi = 1$,
K_a active earth pressure coefficient,
K_p passive earth pressure coefficient.

Parameters such as friction angle and undrained shear strength are calculated with the help of equations (14) and (16), applying the data from the DMT test [19].

3 Monopile lateral stiffness

Application of the p-y method, according to the API and DNV guidelines overestimates the subgrade modulus thus produces enormous results if applied to monopiles [1,20]. As a result, many solutions based on both analytical and three-dimensional analysis have been developed [21]. For the purpose of the article, only some of the developed methods are briefly presented. One of the methods, formulated for cohesive soils, was proposed by Dunnavant & O'Neil [22,23]. The main idea of API Recommendation formulas remains the same, however the expression for the static ultimate lateral resistance p_u and critical deflection y_c changes [24]:

$$p_u = c_u \cdot D \cdot \min \cdot \left\{ 2 + \frac{\sigma'_v(X)}{c_{ua}} + \frac{J}{D} \cdot X, 9 \right\} \tag{30}$$

$$y_c = 0{,}0063 \cdot \left(\frac{E_s \cdot L_r^4}{E \cdot I} \right)^{0,875} \tag{31}$$

where:

c_{ua} average shear strength between the layer surface and depth X,
E_s secant modulus,
L_r representative pile length.

$$L_r = \min\left\{ \left(\frac{E \cdot I}{E_s \cdot D^4} \right)^{0,286}, 3D, L \right\} \tag{32}$$

Similar simplified approach was proposed by Mayne et al. (1992) [25].

These alternative formulations improve the monopile design mainly in overconsolidated clays.

For cohesionless soils a new approach was developed by Thieken et al. (2015) [20]. In this method, a deflection-dependent correction factor, is applied to increase the p_u value at shallow depths. Calculations are made by means of an iterative procedure. In the first step, deformations are calculated with the use of basic p-y curves. Next, the correction factor is applied to the curve and the calculation of deflection is repeated. These steps are reiterated until no significant change is noticed [20].

Apart from analytical approaches, numerical methods are also applied to verify their results [26]. Moreover, model tests in a reduced scale are carried out for a better understanding of the pile – soil interaction. Geotechnical centrifuge is used to investigate the response of model monopiles subjected to different loads [27]. Next, model observations can be extrapolate to analyse the prototype behaviour [28].

Despite many attempts, there is still no specific method for estimating lateral stiffness for large diameter piles. This problem remains unsolved and requires more analysis and systematic research.

4 Conclusions

The advantage of direct method to establish p-y curve is the reduction of extent of laboratory tests on high quality samples. The formulations should be however well calibrated and adjusted to given soil conditions. Following direct approach it is impossible to completely reject laboratory testing and the use of correlation between soil strength characteristics and parameters measured by in-situ soil investigation. The well-developed direct and indirect approaches exists for flexible piles up to 2,5 m in diameter. The monopiles used in offshore structures are generally rigid piles, so they cannot be analysed using any standard p-y approach for flexible piles. The case of rigid or intermediate piles needs further efforts considering analytical and numerical solutions combined with advanced physical modelling of centrifuge tests.

References

1. M. Achmus, K. Abdel-Rahman, Y. S. Kuo, 11th Baltic Sea Geotechnical Conf. Geotechnics In Maritime Engineering, **1**, 463-470 (2008)
2. J. van der Tempel, N.F.B. Diepeveen, D.J. Cerda Salzmann, W.E. De Vries, Wind Power Generation and WInd Turbine Design, 559-594, 768 (2010)

3. Det Norske Veritas, *Offshore Standard DNV-OS-J101design of Offshore Wind Turbine Structures,* 207-214 (2014)
4. American Petroleum Institute, *Recommended Practice for Planning, Designing and Constructing Fixed Offshore Platforms – Working Stress Design*, 63-66 (2002)
5. J.L. Briaud, *Geotechnical Engineering: Unsaturated and Saturated Soils*, 599-603 (2013)
6. F.C. Schroeder, A.S. Merritt, K.W. Sørensen, A. Muir Wood, C.L. Thilsted, D.M. Potts, Frontiers in Offshore Geotechnics III, 735-740 (2015)
7. Det Norske Veritas, *Classification Notes No. 30.4 Foundations*, 8, 16-20 (1992)
8. P. K. Robertson, R. G. Campanella, Canadian geotechnical journal, **20(4)**, 718-733 (1983)
9. F. H. Kulhawy, P. W. Mayne, *Manual on estimating soil properties for foundation design,* **EPRI-EL-6800**, (1990)
10. G. Totani, S. Marchetti, P. Monaco, M. Calabrese, Intl. Conf. On In situ Measurement of Soil (2001)
11. S. Malhotra, *Wind Turbines*, 241-247 (2011)
12. P. Truong, B.M. Lehane, 3rd Intl Symp. on Cone Penetration Testing, 975-982 (2014)
13. S. Suryasentana, B.M. Lehane, Géotechnique 64, **3**, 186-194 (2014)
14. P. K. Robertson, M. P. Davies, R. G. Campanella, Geotechnical Testing Journal, **12(1)**, 30-38 (1989)
15. J.B. Anderson, F.C. Townsend, B. Grajales, Proc. from the Second Intl. Flat Dilatometer Conf., 50-53 (2006)
16. S. Marchetti, G. Totani, M. Calabrese, P. Monaco, Proc. 4th Intl Conf. DFI-Deep Foundation Institute USA on Piling and Deep Foundations, **1**, 263-272 (1991)
17. L. C. Reese, W. R. Cox, F. D. Koop, Offshore Technology in Civil Engineering Hall of Fame Papers from the Early Years, 95-105 (1974)
18. J. M. Murchison, M. W. O'Neill, *Analysis and design of pile foundations*,174-191 (1984)
19. S. Marchetti, Journal of the geotechnical engineering division, 299-320 (1980)
20. K. Thieken, M. Achmus, K. Lemke, Frontiers in Offshore Geotechnics III , 741-746 (2015)
21. J. Wiemann, K. Lesny, W. Richwien, Proc. 7th German Wind Energy Conf. (DEWEK) (2004)
22. T.W. Dunnavant, M.W. O'Neil, JoGE, 115(1), 95-114 (1989)
23. T.W. Dunnavant, M.W. O'Neil, JoGE, 117(5), 827-829 (1991)
24. M. Arroyo, D. Abadias, M. Coelho, Frontiers in Offshore Geotechnics III, 693-698 (2015)
25. P.W. Mayne, F.H. Kulhawy, C.H. Trautmann, F. H. Kulhawy, *Experimental study of undrained lateral and moment behavior of drilled shafts during static and cyclic loading*, **EPRI-TR-100221** (1992)
26. M. Achmus, Indian Geotechnical Conference GEOtrendz - 2010, 92-102 (2010)
27. R.T. Klinkvort, M. Poder, P. Truong, V. Zania, Proc. of the 3rd European Conf.on Physical Modelling in Geotechnics (2015)
28. R.T. Klinkvort, O. Hededal, S.M. Springman, IJPMG, **13(2)**, 38-49 (2013)

An algorithm to generate high dense packing of particles with various shapes

Konrad Miśkiewicz[1,*], *Robert* Banasiak[1], *Maciej* Niedostatkiewicz[2], *Krzysztof* Grudzień[1], and *Laurent* Babout[1]

[1]Institute of Applied Computer Science, Lodz University of Technology, ul. Stefanowskiego 18/22, 90-924 Łódź, Poland
[2]Faculty of Civil and Environmental Engineering, Gdansk University of Technology, ul. Narutowicza 11/12, 80-233 Gdańsk, Poland

Abstract. Discrete Element Method (DEM) is one of available numerical methods to compute movement of particles in large scale simulations. The method has been frequently applied to simulate the cases of grain or bulk material as the major research issue. The paper describes a new method of generating high dense packing with mixed material of two different shape used in DEM simulation. The initial packing is an important parameter to control, because have influence on the first few seconds after start the simulation. Some-times when the material in silo is arranged with loose packing before the start, the particles move downward gravity. These changes between the start and the first few seconds in simulations act strongly on the results at the end of a discharging process in silo. At the initial simulation time it is important to prepare proper packing with mixed material, in order to make sure that the particles will not move due to gravity action. This solution is a necessary step to integrate in the simulation procedure in order to compare later the computer simulation with experimental measurements of material discharge in a silo.

1 Introduction

In many cases in order to validate a number of constructions and check a proper technology or construction simulations are prepared. This work deals with software which used many kind of algorithms, with mathematical equations. The major numerical methods are Finite Element Method (FEM)[1], and Discrete Element Method (DEM)[2, 3]. The first numeric method helps to prepare structural analysis, electromagnetic potential or heat transfer. The second one is used to deal with motion, dynamics of small particles in large scale simulations. This method is addressed to prepare flow granular material or particle mechanics for many of problems especially for discharging process in silos.

There is a large number of software to use numerical methods mentioned above. Both commercial and free software is at hand. The free software such like YADE [4] or WooDEM, was created by scientists who work on granular flow. This frameworks explore methods to generate spherical or ellipsoid packing[5, 6]. It is important to

* E-mail: konrad.miskiewicz@p.lodz.pl

emphasize that this method is not intended generate packing with mixed material. One of the methods was implemented, but the result was a small dense packing[7]. The algorithm was expanded on new methods especially in order to generate packing with high density mixed material. A number of algorithms to generate spherical packing, was developed[8, 9, 10]. One algorithm helped in large scale simulations, applying tetrahedral mesh which can be generated by other software: Ghs3D, Gmsh or Netgen[11, 12]. Another algorithm firstly creates particles at random, next starts the compression[13]. The basic method of sphere packing was called Face Centered Cubic(FCC) and Hexagonal Close Packing (HCP)[14].

2 Mathematical formulations

Generation of a mixed material in one simulation with high dense packing is impossible with methods described earlier. In this case the particles showed two different shapes of grains. One of them was a sphere, not difficult to be describe in simulation. This object detects only two parameters to describe its properties, the radius of sphere and sphere positioning in 3D space. However the spheres are packed together with tracking particles the ellipsoids. Filing with the use of the different material results in a few percentage of the whole. Description of this type of material is more difficult. As in the case of spheres, ellipsoid also shows a center saved in the same 3D space. It is important to emphasize that this shape needs more parameters to describe the object. In order to characterize the object magnitude three dimensions are required in the axes of symmetry. This is the main difference between the sphere and the ellipsoid. It is worth noticing that three axial dimensions coinciding, then the turn the ellipsoid into to sphere. Rotation acts strongly on positioning the particles in simulation. This issue is linked with the ellipsoid case, not active while spheres are taken. It is important to emphasize that the ellipsoid rotation is a new issue in high dense packing, acting on free space around this kind of particles. Rotation maybe considered in three ways: using Euler equation, rotation around axis and quaternion. In many cases the last method is a simple and straightforward way to deal with rotations, and it is more realistic than others.

In the previous proposed algorithm the dense packing is not implemented[6]. A solution was developed to assure a high level of the mixed packing. Following the methods, to generate mixed material packing, the algorithm must work in two and three dimensions. The main shape of silo prepared in simulation to packing, is divided to conduct smaller sub-problems. As a result of this solution the boundaries between sub-problems were created. This method is aimed at faster computing and to improve the process of preparing simulations. On the sub-problem boundaries the method to generate high dense packing must work in two dimensions. The particles are generated in x and y axis the z axis is fixed. Particles are moved from free space to a position calculated by an algorithm.

Equation of an ellipsoid is given (1):

$$\frac{x^2}{a^2}+\frac{y^2}{b^2}+\frac{z^2}{c^2}=1 \tag{1}$$

where:

a, b, c principal semi-axes, and $a > b > c$.

When the a; b; c dimensions are the same the ellipsoid are turns into a sphere. With the aim of ordering particles in right place, the algorithm must calculate the vector of translation. However, the points on the surface are investigated first to detect the shortest distance to point the direction of movement (Fig. 1).

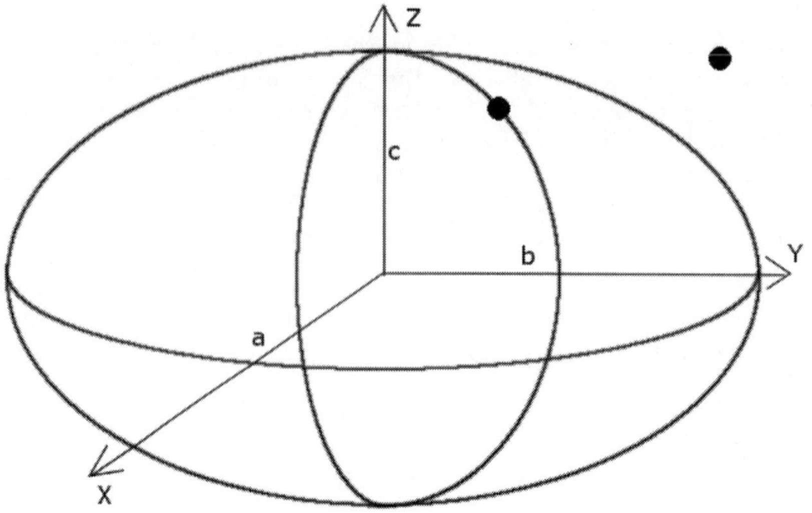

Fig. 1. Schematic picture of ellipsoid and the point on surface.

$$x_p = \frac{a \cdot x_G}{\sqrt{x_G^2 + y_G^2 + z_G^2}} \tag{2}$$

$$y_p = \frac{b \cdot y_G}{\sqrt{x_G^2 + y_G^2 + z_G^2}} \tag{3}$$

$$z_p = \frac{c \cdot z_G}{\sqrt{x_G^2 + y_G^2 + z_G^2}} \tag{4}$$

where:

x_p, y_p, z_p the coordinate of point on the surface on ellipsoid,

x_G, y_G, z_G the point of the particle moved (translated).

The vector of translation was calculated by simple equations:

$$\overrightarrow{PG} = (x_G - x_p, y_G - y_p, z_G - z_p) \tag{5}$$

The equations above are used to change the position of a particle to the nearest boundary of shape, identified by a silo construction. According that ellipsoid could rotation matrix is introduced in the equation. After that the view of equation is presented below (6):

$$\begin{bmatrix} x_p \\ y_p \\ z_p \end{bmatrix}_{True} = \begin{bmatrix} x_0 \\ y_0 \\ z_0 \end{bmatrix} + R \begin{bmatrix} x_p \\ y_p \\ z_p \end{bmatrix}_{Standartd} \tag{6}$$

While x_0, y_0, z_0 marks the centre of an ellipsoid, $\boldsymbol{R}$ is the rotation matrix, and standard matrix means that the point is not rotated.

In addition to positioning particles near the boundary of sub-problems, remember that in the middle of silo or sub-problems the material is intended to detect proper ordering. Each particle must show at least two contact points. In order to calculate the vector of particle translation different equations are applied. The change related to the point is not situated on construction surface of silo, but on the centre of the second particles (Fig.2).

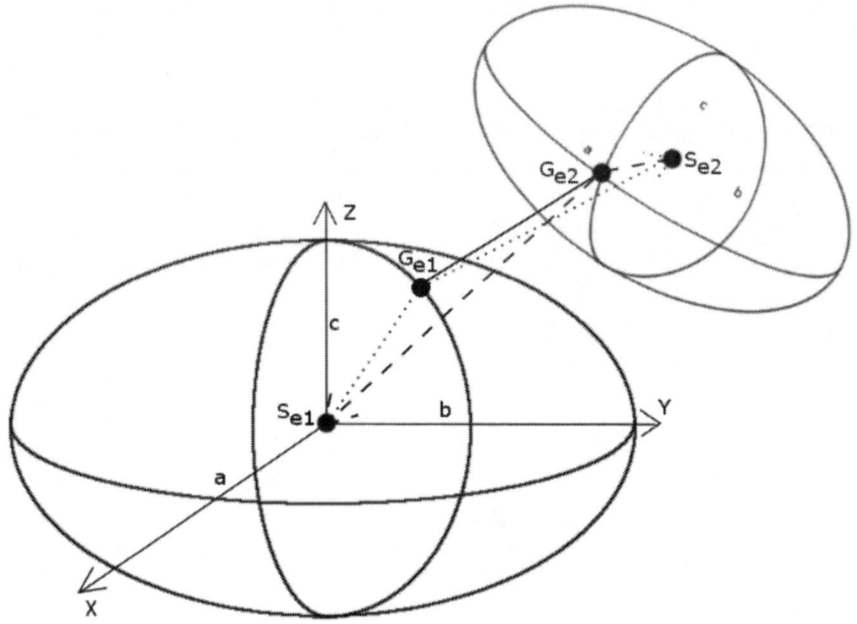

Fig. 2. Schematic picture of two ellipsoid.

3 Algorithm

Some of algorithms generate only one type of material, sometimes the process of filling the model in simulations relies on the positioning shapes in right place. This algorithm is not finally efficient, because while the solution is prepared it is not matched to real result shown a high difference. The work is focused on implementing some features to generating mixed material with high dense packing.

In the first step trust the users describe the layout of sphere or ellipsoid and set the ratio parameter, to describe the proportion between this two material types. Of course add information is required about the object to be filled, for example silo. Next the algorithm calculates the boundaries and execute, the procedure to partition the problem to smaller ones. After that the first particle is generated, and checked that the object is in the silo or shape to be filled with material. If the shape lies in the boundaries, the vector translation is computed to move particle to the boundaries. Before the position changes, an algorithm computes and checks the way to all of boundaries in sub-problems. The nearest way is chosen, so the centre of sphere or ellipsoid is changed.

The next step is to generate random particles whose position is located in the silo. When on entire was entered in the simulation, it started the proper process of ordering particles. The first object that moved to boundaries at the beginning of the process was marked the major one main in this sub-problem or on the boundaries. While changing the position of particles to right places, the method must calculate the vector of translation. It is emphasized that each material shows different directions of movement and different distance to change. In the first step of this complex process is to search nearest particles and then calculate the vector of translation. Of course during this stage, it is checked that the particle does not crossed the boundaries of the silo. When the position of two shapes was changed, algorithm calculate the distance between them to verify the collision. To protect count of spheres or ellipsoids around actually working particle, method calculate the angle ensuing on the main particle in sub-problems or in boundaries. During this process and if all of condition was pass, are saved with map statement, where inform that main particle have this neighbours, and vice versa. Of course the shapes which was moved to the main particle have only one or two neighbours. After this the algorithm change object of working for whom checked next random particles and calculated new vectors of move. The process of computing moves and conditions are repeated for all of material which are in filled object in simulation, to moment when the packing are high dense. At the end if have more space algorithm generate additional particles which was moved immediately to right place.

Above description the method that work on boundary, and on two dimensional area. In three dimension view this process is the same, but all of this particle have big priority to move not in x and y axis but in z axis. It is important because after start the simulation and add gravity, the particles do not moved to bottom in silo. Of course the boundary which generated earlier in this process was mixed and also moved to down, and changed to irregular surface, but this change is very low.

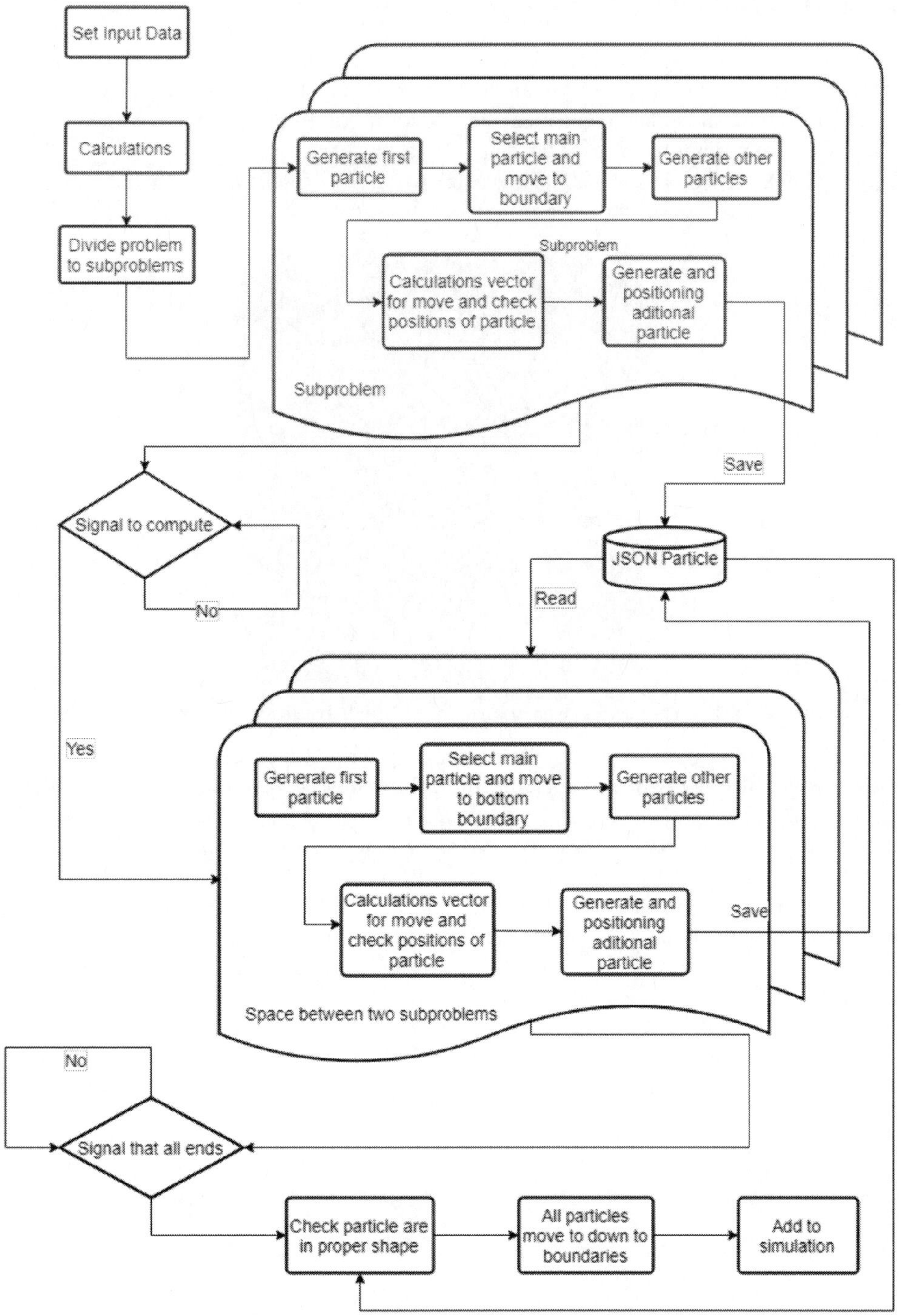

Fig. 3. Schematic picture with algorithm.

4 Results Analysis

Most of algorithm for generate packing with sphere the dense packing is on level average 0:74. It is interesting that some of methods are low level, because one of them generate random packing and other calculates the possible max arrangement of sphere in filled object. It is important to emphasize that generating packing with ellipsoid give different dense, which average is on 0:75. This results is not different as much but the shape of particle are very different, and have influence of dense packing.

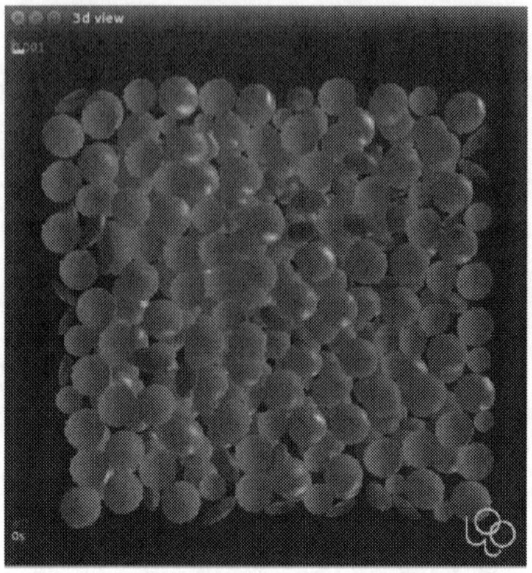

Fig. 4. Schematic picture of packing mixed material of two different shapes.

The proposal algorithm generate dense packing with mixed material of two different shapes, especially with sphere and ellipsoid in one object. It is important to set the ratio between particles that generated in schematic chart about dense packings that the mixed material with two different shapes have influence to the dense packing (Fig. 5). When in the object which filled have only sphere, the ratio is average 0:70. This result is very good for random packing to compare with other algorithms. In case with mixed material this results reaches average about 0:64 for the coefficient of ratio in 25%. It is important to emphasize that this result is for mixed material with two different shape, what means that is very good for the above case.

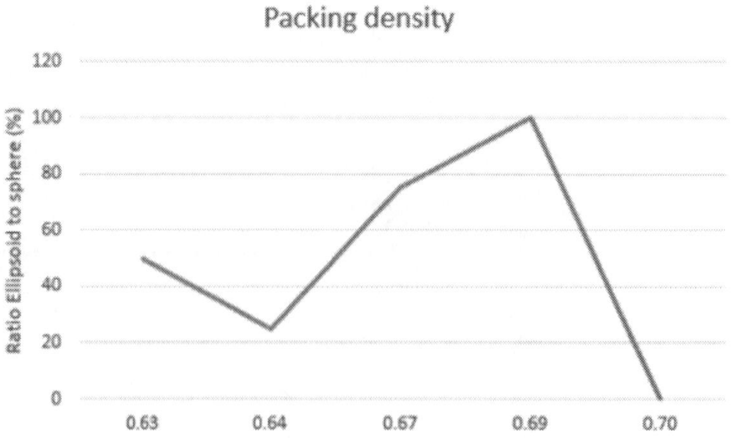

Fig. 5. Schematic chart about dense packing.

5 Conclusion

The algorithms and methods that was proposed for generating particles in high dense packing is not effective with mixed material. Proposal algorithm allow to compute mathematical equations which are helped to positioning mixed material. It is emphasize that this method are used not only in generating particles, but also when the material was saved in database to checking the collision and property of dense packing. Division this problem to smaller one was good efficient, because most of calculation are starting parallel. This solution allow to save time for computing and preparing simulations and are used all power that was in computer. Finally this method helped to compare the real results, which was conducted in X-ray laboratory, with computer simulation.

References

1. P. Wriggers, *Nonlinear Finite Element Methods* (Springer, Berlin Heidelberg, 2008)
2. H. K. Dang, M. Meguid, International Journal of Geomechanics, **10**, 85-91 (2010)
3. M. Jiang, H. Yu, Modern Trends in Geomechanics, 241–269 (2006)
4. J. Kozicki, F. V. Donzé, Computer Methods in Applied Mechanics and Engineering, **197**, 4429–4443 (2008)
5. A. Doney, F. H. Stillinger, P. M. Chaikin, S. Torquato, Physical Review Letters, **251**, 255506-1 (2004)
6. R. K. McGEARY, Journal of the American Ceramic Society, **44** (10), 513–522 (1961)
7. K. Miśkiewicz, R. Banasiak, M. Niedostatkiewicz, K. Grudzień, L. Babout, Proceedings of International Interdisciplinary PHD Workshop, 366-371 (2017)
8. K. Bagi, Granular Matter, **7** (1), 31-43 (2005)
9. B. Harthong, J. F. Jérier, P. Dorémus, D. Imbault, F. V. Donzé, International Journal of Solids and Structures, **46**, 3357–3364 (2009)
10. R. R. Valera, I. P. Morales, S. Vanmaercke, C. R Morfa, L. A. Cortés, H. D. G. Casañas, Computational Particle Mechanics, **2** (2), 161–172 (2015)
11. J. F. Jerier, V. Richefeu, D. Imbault, F. V. Donzé, Computer Methods in Applied Mechanics and Engineering, **199**, 1668–1676 (2010)
12. J. F. Jerier, D. Imbault, F. V. Donze, P. Doremus, Granular Matter, **11** (1), 43–52 (2009)
13. B. D. Lubachevsky, F. H. Stillinger, Journal of Statistical Physics, **60**, 561–583 (1990)
14. J. Conway, J. A. Sloane Neil, *Sphere packings, lattices and groups* (Springer Science+Buisnes Media LLC, New York, 1991)

AUTHOR INDEX